AF790762

Nuclear and Particle Physics

Nuclear and Particle Physics

Claude Amsler

Albert Einstein Center for Fundamental Physics,
Laboratory for High Energy Physics, University of Bern, Switzerland

Emeritus Professor, University of Zurich, Switzerland

IOP Publishing, Bristol, UK

© IOP Publishing Ltd 2015

All rights reserved. No part of this publication may be reproduced, stored in a retrieval system or transmitted in any form or by any means, electronic, mechanical, photocopying, recording or otherwise, without the prior permission of the publisher, or as expressly permitted by law or under terms agreed with the appropriate rights organization. Multiple copying is permitted in accordance with the terms of licences issued by the Copyright Licensing Agency, the Copyright Clearance Centre and other reproduction rights organisations.

Permission to make use of IOP Publishing content other than as set out above may be sought at permissions@iop.org. This textbook was originally published in 2007 by vdf Hochschulverlag AG at ETH Zurich. The current version is a translation into English, which has been substantially expanded.

Claude Amsler has asserted his right to be identified as author of this work in accordance with sections 77 and 78 of the Copyright, Designs and Patents Act 1988.

ISBN 978-0-7503-1140-3 (ebook)
ISBN 978-0-7503-1141-0 (print)
ISBN 978-0-7503-1142-7 (mobi)

DOI 10.1088/978-0-7503-1140-3

Version: 20150501

IOP Expanding Physics
ISSN 2053-2563 (online)
ISSN 2054-7315 (print)

British Library Cataloguing-in-Publication Data: A catalogue record for this book is available from the British Library.

Published by IOP Publishing, wholly owned by The Institute of Physics, London

IOP Publishing, Temple Circus, Temple Way, Bristol, BS1 6HG, UK

US Office: IOP Publishing, Inc., 190 North Independence Mall West, Suite 601, Philadelphia, PA 19106, USA

Contents

Preface

This textbook originated from lectures given at the Physics Institute of the University of Zurich in Switzerland between 1997 and 2012. It is intended as an introduction to nuclear and particle physics for undergraduate students at the bachelor's and master's levels. Students who intend to pursue a career in experimental particle physics will also find valuable information in some of the more advanced sections.

This introduction into the world of nuclei and particles is presented from a historical viewpoint and is phenomenologically oriented. Many of the concepts and phenomena are derived with inductive rather than deductive thinking, following the paths that often led to their final formulation. A balanced blend of experiments and theory is presented. An introduction to the basic concepts of nuclear physics is given before moving on to topics in particle physics which are developed in more detail. Many formulas are derived explicitly from first principles. The reader is expected to master the basic concepts of quantum mechanics.

This textbook was originally published in 2007 in German by vdf Hochschulverlag AG at ETH Zurich. The current version is a translation into English which has been substantially expanded to include very recent developments in neutrino, heavy quark and high energy physics.

The author wishes to acknowledge the contributions from his younger physics colleagues, who carefully read parts of the manuscript: Lucian Ancu, Vincenzo Chiochia, Konrad Nesteruk, Paola Scampoli, Basil Schneider, Sarah Seif El Nasr and Mauro Verzetti.

Author biography

Claude Amsler

Claude Amsler studied experimental physics at the ETH in Zurich, Switzerland. He worked as a research associate at Queen Mary College in London, TRIUMF in Vancouver, the University of New Mexico and Brookhaven National Laboratory. After a CERN fellowship and a brief stay at the University of Munich he joined the Physics Institute of the University of Zurich, where he became full professor of physics in 1999. He has supervised 40 doctoral and master's theses. He is now professor emeritus and a member of the senior staff at the Albert Einstein Center for Fundamental Physics, University of Bern. His main fields of interest are strong interaction physics and meson spectroscopy. He is a member of the CMS Collaboration and of the Particle Data Group. He is currently involved in antihydrogen experiments at CERN. More can be found on his home page http://cern.ch/amsler/.

IOP Publishing

Nuclear and Particle Physics

Claude Amsler

Chapter 1

The discovery of the atomic nucleus

Nuclear physics began on 1 March 1896 with the discovery of radioactivity by Henri Becquerel. Becquerel discovered that uranium salts (such as $K_2UO_2\ (SO_4)_2{\cdot}2H_2O$) generated black tracks on photographic plates. The radiation responsible was first interpreted as penetrating UV-radiation since Wilhelm Röntgen had discovered x-rays the year before. Becquerel called his radiation 'rayons uraniques'. The word 'radioactivity' was introduced later by Marie Curie. Ernest Rutherford and others discovered that radioactivity consisted of at least three components: (i) α-radiation, which is easily absorbed and deflected by magnetic fields, (ii) penetrating β-radiation (which Becquerel had initially observed), which is deflected in the opposite direction, and (iii) γ-radiation, which is not deflected by magnetic fields. α- and β-rays therefore had opposite electric charges, while γ-radiation was electrically neutral.

Rutherford described the energy released as 'atomic' energy[1]. Since atoms were perceived as elementary building blocks[2] the origin of the release of energy was not clear: was the theorem of energy conservation violated or could the atom after all be decomposed into smaller fragments?

More radioactive atoms were soon discovered, including polonium and radium (Pierre and Marie Curie, 1898, see figure 1.1). In 1909 Rutherford and Geiger observed the characteristic spectral lines of helium in a glow-lamp irradiated by a radium source (helium had been discovered in 1895 by Norman Ramsey in uranium ore). They concluded that helium atoms were α particles which had lost their positive charges!

[1] This word was used until the 1950s, after the development of the 'atomic' bomb. The more appropriate word 'nuclear' would have suggested to the media a relationship with the cell nucleus, i.e. with biology (after L Groves, head of the Manhattan Project).

[2] The word 'atomos' was introduced by Democritus. For the Greeks the fundamental building blocks (atoms) were the four elements earth, air, fire and water. The belief that water was elementary lasted until Lavoisier (1743–94). The atom as the fundamental unit of chemical substances was introduced by Dalton (1808). He discovered that the mass of a chemical substance was given by the sum of the masses of its components. Prout (1815) also noticed that the ratio of the mass of an atom to that of the most elementary one—hydrogen—was approximately an integer.

doi:10.1088/978-0-7503-1140-3ch1

© IOP Publishing Ltd 2015

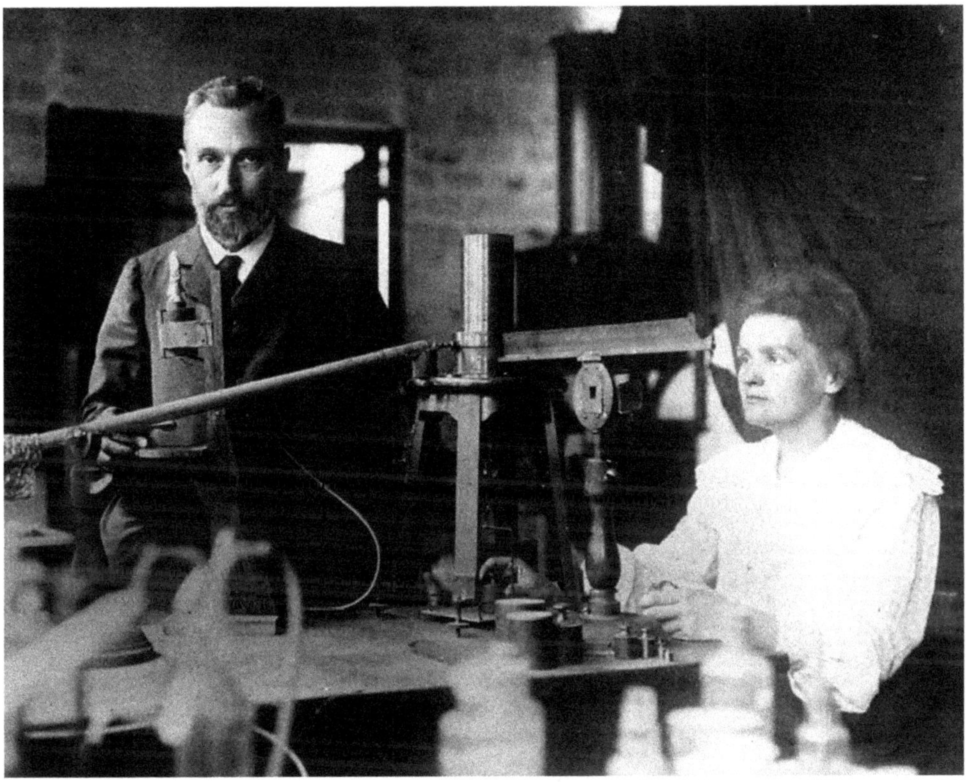

Figure 1.1. Pierre and Marie Curie in their laboratory. Image from Wikimedia Commons.

J J Thomson was convinced, following his discovery of the electron in 1897, that the atom consisted of a 'pudding' of electrons and positive charges, where (strangely enough) only electrons contributed to the mass of the atom. Hydrogen thus contained approximately 2000 electrons. This view was revised in 1906 once the rate for the scattering of x-rays by shell electrons had been been measured and compared with the so-called Thomson cross section. Hydrogen was found to have 1–2 electrons.

After the discovery of the atomic nucleus in 1911 by Rutherford, Geiger and Marsden, which we shall describe in detail, the atom was modelled as a hard core with A positive charges and $(A - Z)$ electrons (total positive charge Ze) surrounded by a cloud of electrons with charge $-Ze$. The electrons in the core were considered to be responsible for the β-radiation. (The neutron was discovered in 1932 and the theory of β-decay as the transformation of a neutron into a proton $(n \rightarrow p)$ was formulated in 1934 by Enrico Fermi.) The simplest atom, hydrogen, consisted of one unit of positive charge, which Rutherford called the 'proton', and one shell electron. The planetary model of the atom as we know it today was introduced by Niels Bohr in 1913. The first artificially produced nuclear reaction $(\alpha^{14}\mathrm{N} \rightarrow p^{17}\mathrm{O})$ was observed in 1914 by Rutherford and Marsden. A comprehensive account of the early days of nuclear and particle physics can be found in [1].

1.1 Rutherford scattering

Hans Geiger and Ernest Marsden, two employees of Rutherford, were investigating α scattering off thin films at the University of Manchester. The apparatus (figure 1.2) consisted of an evacuated chamber, a collimated radium source, R, and a ZnS scintillator, S, which could be swung around a thin gold or silver foil, F, mounted in the centre of the chamber. The kinetic energy of the α particle was $T = 4.78$ MeV.

The scattered α particles produced flashes of light in the ZnS scintillator which could be observed using the microscope, M. The number of flashes for a given measurement time is shown in table 1.1 as a function of scattering angle θ. At small angles the collimation of the source was increased since the human eye cannot resolve counting rates exceeding ~90/min. Geiger and Marsden determined that the rate decreased with increasing scattering angle, following the angular distribution $1/\sin^4 \frac{\theta}{2}$. Surprising was the observation of backward scattered α particles ($\theta > 90°$). This was only possible if the atom was made up of a hard heavy core in which the electric charge was concentrated, the nucleus[3].

Let us now describe the scattering process quantitatively (figure 1.3) by making the following (modern) assumptions:

1. The nucleus is much heavier than the projectile (mass m, velocity v) so that the recoil energy of the nucleus is negligible (that is the velocity v' of the projectile after the collision is $v' = v$). This does not apply to free atoms but to atoms that are bound in a crystal lattice.
2. The core and the projectile are *structureless* (point-like) with charges $+Ze$ and ze, respectively, and have spin 0.
3. The interaction is fully described by the electromagnetic interaction (Coulomb force).
4. The nucleus is not excited in the scattering process (*elastic scattering*) and the projectile does not lose energy by interacting with shell electrons since m is much larger than the mass of the electron.

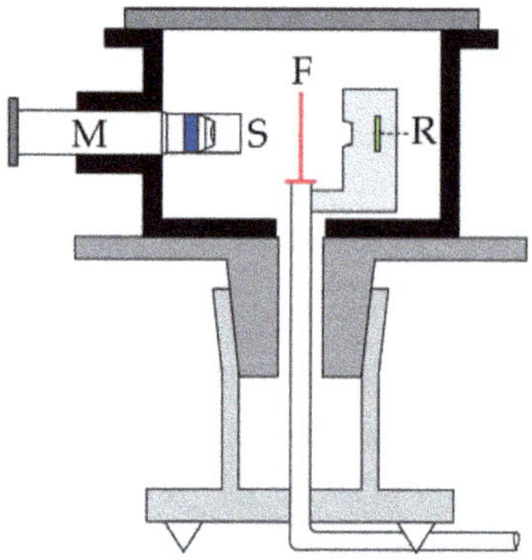

Figure 1.2. The apparatus used by Rutherford, Geiger and Marsden (after [2]).

[3] According to Rutherford the phenomenon was like 'shooting a cannon ball at a piece of tissue paper and have the ball bouncing back at you'.

Table 1.1. Number of counts N as a function of scattering angle θ.

		Silver		Gold	
θ [°]	$\frac{1}{\sin^4 \frac{\theta}{2}}$	N	$N \sin^4 \frac{\theta}{2}$	N	$N \sin^4 \frac{\theta}{2}$
150	1.15	22.2	19.3	33.1	28.8
135	1.38	27.4	19.8	43.0	31.2
120	1.79	33.0	18.4	51.9	29.0
105	2.53	47.3	18.7	69.5	27.5
75	7.25	136	18.8	211	29.1
60	16.0	320	20.0	477	29.8
45	46.6	989	21.2	1435	30.8
37.5	93.7	1760	18.8	3300	35.3
30	223	5260	23.6	7800	35.0
22.5	690	20300	29.4	27300	39.6
15	3445	105400	30.6	132000	38.4
30	223	5.3	0.024	3.1	0.014
22.5	690	16.6	0.024	8.4	0.012
15	3445	93.0	0.027	48.2	0.014
10	17330	508	0.029	200	0.0115
7.5	54650	1710	0.031	607	0.011
5	276300	–	–	3320	0.012

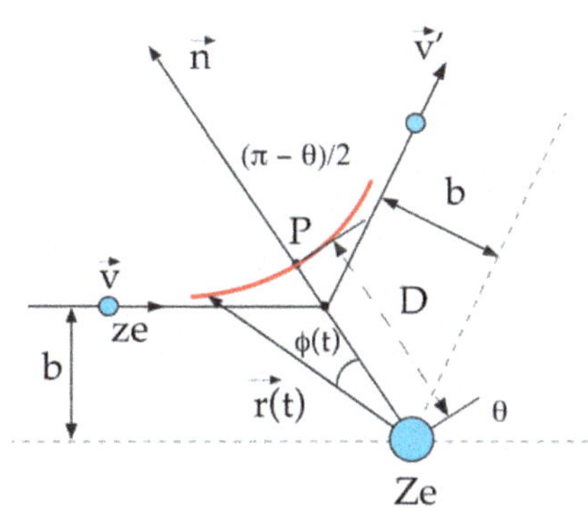

Figure 1.3. Rutherford scattering.

Let $v_\perp = \omega r$ be the component of the velocity perpendicular to $\vec{r}$ (figure 1.3). One obtains for each point on the trajectory from the conservation of angular momentum $\vec{L}$ that

$$\left|\vec{L}\right| = \left|\vec{r} \times m\vec{v}\right| = mvb = mv_\perp r = m\omega r^2 = m\frac{\mathrm{d}\phi}{\mathrm{d}t}r^2 \Rightarrow \frac{\mathrm{d}\phi}{\mathrm{d}t} = \frac{vb}{r^2}. \quad (1.1)$$

The *impact parameter* b is the distance between the incident projectile at asymptotic distances and the beam axis. The scattering angle is denoted by θ, and K_n is the force projected in the direction $\vec{n}$. The projection of the asymptotic momenta on $\vec{n}$ is $-mv\cos(\pi - \theta)/2 = -mv\sin\theta/2$ before, and $mv + \sin\theta/2$ after the collision. Thus along $\vec{n}$ the momentum changes by

$$\Delta p = 2mv\sin\frac{\theta}{2} = \int K_n \mathrm{d}t = \int_{-\infty}^{+\infty} \frac{zZe^2}{4\pi\epsilon_0 r^2}\cos\phi \mathrm{d}t. \tag{1.2}$$

From (1.1) we have that

$$\mathrm{d}t = \frac{r^2}{vb}\mathrm{d}\phi, \tag{1.3}$$

and hence

$$\Delta p = \int_{\frac{-(\pi-\theta)}{2}}^{\frac{(\pi-\theta)}{2}} \frac{zZe^2}{4\pi\epsilon_0 vb}\cos\phi \mathrm{d}\phi = \frac{zZe^2}{4\pi\epsilon_0 vb}2\cos\frac{\theta}{2}. \tag{1.4}$$

Using (1.2) we obtain

$$\tan\frac{\theta}{2} = \underbrace{\left[\frac{zZe^2}{4\pi\epsilon_0 mv^2}\right]}_{K}\frac{1}{b}. \tag{1.5}$$

One finds a one-to-one relation between the impact parameter b and the scattering angle θ. For large (respectively small) distances from the axis,

$$\begin{aligned} b &\rightarrow \infty \Leftrightarrow \theta \rightarrow 0, \\ b &\rightarrow 0 \Leftrightarrow \theta \rightarrow \pi. \end{aligned} \tag{1.6}$$

Consider now a sphere with unit radius (figure 1.4): incoming particles which fly through the ring with area

$$\mathrm{d}\sigma = 2\pi b|\mathrm{d}b| \tag{1.7}$$

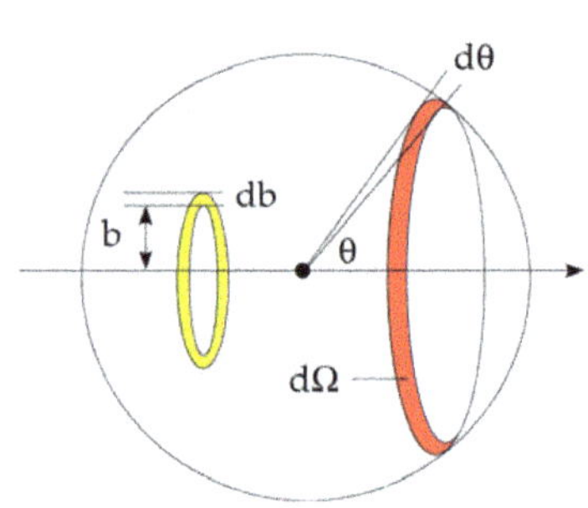

Figure 1.4. The unit sphere with unit radius.

are scattered between θ and $\theta + \mathrm{d}\theta$ into the solid angle

$$\mathrm{d}\Omega = 2\pi \sin\theta \mathrm{d}\theta = 2\pi \mathrm{d}(-\cos\theta). \tag{1.8}$$

The area $\mathrm{d}\sigma$ is referred to as the *differential cross section*. From (1.5) one obtains

$$|\mathrm{d}b| = \left|\mathrm{d}\left(\frac{K}{\tan\frac{\theta}{2}}\right)\right| = \left(\frac{K}{\tan^2\frac{\theta}{2}}\right)\frac{1}{2\cos^2\frac{\theta}{2}}\mathrm{d}\theta, \tag{1.9}$$

hence with (1.7)

$$\mathrm{d}\sigma = 2\pi\left(\frac{K^2}{\tan^3\frac{\theta}{2}}\right)\frac{1}{2\cos^2\frac{\theta}{2}}\mathrm{d}\theta. \tag{1.10}$$

Dividing $\mathrm{d}\sigma$ by $\mathrm{d}\Omega$ from (1.8) one obtains

$$\frac{\mathrm{d}\sigma}{\mathrm{d}\Omega} = \left(\frac{K^2}{\tan^3\frac{\theta}{2}}\right)\frac{1}{2\cos^2\frac{\theta}{2}\sin\theta} = \frac{K^2}{4\sin^4\frac{\theta}{2}} = \frac{1}{16}\left(\frac{zZe^2}{4\pi\epsilon_0 T}\right)^2\frac{1}{\sin^4\frac{\theta}{2}}, \tag{1.11}$$

where $T = mv^2/2$ is the kinetic energy of the projectile. Let us introduce the dimensionless *fine-structure constant*

$$\boxed{\alpha \equiv \frac{e^2}{4\pi\epsilon_0 \hbar c} = \frac{1}{137.036}}, \tag{1.12}$$

as well as the *natural units*

$$\hbar = c = 1. \tag{1.13}$$

Equation (1.11) then takes the simple form known as the Rutherford formula

$$\boxed{\frac{\mathrm{d}\sigma}{\mathrm{d}\Omega} = \frac{z^2Z^2\alpha^2}{16T^2}\frac{1}{\sin^4\frac{\theta}{2}}}, \tag{1.14}$$

which is the differential cross section for scattering into the solid angle $\mathrm{d}\Omega$. As we shall see, formula (1.14) is a measure of the probability for the projectile to be scattered between θ and $\theta + \mathrm{d}\theta$ and therefore describes the *angular distribution* of the scattered particles.

1.1.1 Remarks on units

Energies in nuclear and particle physics are always specified in eV (keV, MeV, GeV, TeV), where 1 eV is the energy increment of an elementary charge in a potential of 1 V: $1\ \mathrm{eV} = 1.6 \times 10^{-19}$ J. How do we then get from (1.14) for $\mathrm{d}\sigma$ the dimension of a surface (e.g. cm^2) when T is given in MeV? With our choice of units (1.13) cm^{-1} is redefined in MeV:

$$\hbar c = 197.3\,\text{MeV fm} = 1, \tag{1.15}$$

where $1\,\text{fm} = 10^{-13}$ cm. Hence

$$\boxed{1\,\text{MeV} \equiv \frac{1}{197.3}\,\text{fm}^{-1}}. \tag{1.16}$$

In a similar manner with $c = 1$ one obtains

$$1\,\text{s} = 2.998 \times 10^{23}\,\text{fm} \tag{1.17}$$

and

$$1\,\text{MeV} = 1.519 \times 10^{21}\,\text{s}^{-1}. \tag{1.18}$$

Hence in all calculations $\hbar$ and c can be set equal to 1. The units obtained in the final result can then be converted into cm, s or MeV using (1.16) or (1.17). Energies are usually given in MeV and masses in MeV c^{-2} since $E^2 = c^2p^2 + m^2c^4$ (or in MeV with $c = 1$). Momenta are normally expressed in MeV c^{-1} to avoid confusion between energy and momentum.

Furthermore, in (1.12) the electric charge e^2 has been replaced by the dimensionless α where

$$e^2 = 4\pi\epsilon_0\alpha, \tag{1.19}$$

since we used SI units for the Coulomb force. Sometimes the Coulomb force is expressed as $e^2/4\pi r^2$ (in the so-called Heaviside–Lorentz units with $\epsilon_0 = 1$). The conversion then reads

$$e^2 = 4\pi\alpha \tag{1.20}$$

for the result to agree with (1.14). Alternatively, in ESU the Coulomb force is expressed as e^2/r^2. The correct substitution is then

$$e^2 = \alpha \tag{1.21}$$

to obtain (1.14). It is therefore mandatory to determine in which units the Coulomb force was written before replacing e^2 by α. It must be emphasized that α always assumes the numeric value $\simeq\frac{1}{137}$ in *any* system of units.

1.1.2 Interpretation of the differential cross section

Consider a beam of cross section F impinging on a target (thickness ℓ) with I_0 incident particles per second (figure 1.5). Let us assume that an incident particle is scattered by (or interacts with) a target nucleus whenever it crosses an effective surface σ. This surface (*total cross section*) depends on the incident energy T, the types of projectile and target, and the nature of the interaction.

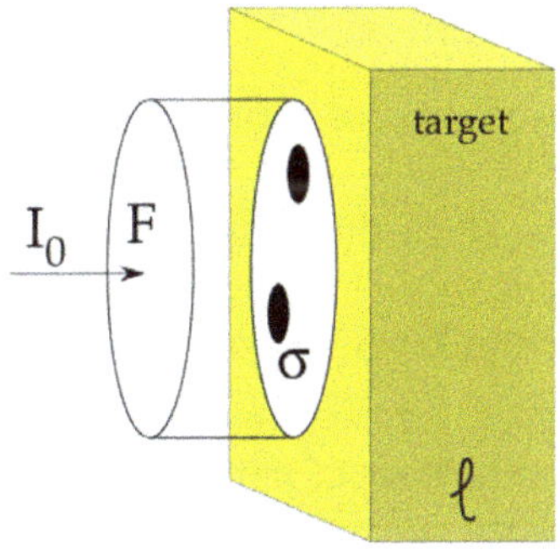

Figure 1.5. Definition of the cross section σ.

The scattering probability for one incident particle and one target nucleus is σ/F. On the other hand, the number of scattering centres crossed by the beam is $n\ell F$, where n is the number of nuclei per unit volume. Assuming that ℓ is small enough so that the surfaces σ do not overlap one obtains the number of scatterings per second:

$$I = \frac{\sigma}{F} n\ell F I_0 = I_0 \sigma n\ell, \tag{1.22}$$

with

$$n = \frac{\rho L}{A}, \tag{1.23}$$

where ρ is the density, L is Avogadro's number and A is the atomic weight. Cross sections σ are usually expressed in units of 'barn':

$$\boxed{1\,\mathrm{b} \equiv 10^{-24}\,\mathrm{cm}^2}. \tag{1.24}$$

(Cross sections for slow neutrons on uranium are very large, they are 'as big as a barn'.) The number of scattered particles per second into the solid angle $\mathrm{d}\Omega$ is then

$$\frac{\mathrm{d}I}{\mathrm{d}\Omega}\mathrm{d}\Omega = I_0 n\ell \frac{\mathrm{d}\sigma}{\mathrm{d}\Omega}\mathrm{d}\Omega \Rightarrow \frac{\mathrm{d}I}{\mathrm{d}\Omega} = I_0 n\ell \frac{\mathrm{d}\sigma}{\mathrm{d}\Omega}. \tag{1.25}$$

The differential cross section $\frac{\mathrm{d}\sigma}{\mathrm{d}\Omega}$ (dimension $\mathrm{b\,sr^{-1}}$) describes the angular distribution of the scattered particles and depends on the physical process. The scattered rate R into a detector of surface S located at a large enough distance d from the target is then with $\Delta\Omega \simeq S/d^2$:

$$R = I_0 n\ell \left(\frac{\mathrm{d}\sigma}{\mathrm{d}\Omega}\right)\frac{S}{d^2}. \tag{1.26}$$

1.1.3 Remarks on Rutherford scattering

The differential cross section (1.14) for Rutherford scattering decreases as $1/\sin^4 \frac{\theta}{2}$ with increasing angle θ, in accordance with the experimental observations of Geiger and Marsden (figure 1.2). The atom thus has a hard core with charge $+Ze$. A series of remarks is in order here:

1. The differential cross section increases with Z^2 for fixed angle θ.
2. $\frac{d\sigma}{d\Omega}$ become infinite when $\theta \to 0$. That should not bother us since for very small scattering angles the impact parameter b exceeds the radius of the outer electronic shell. The charge of the core is then shielded and the factor $Ze \to 0$.
3. Nevertheless, when scattering off a bare (completely ionized) nucleus, the differential cross section diverges for $\theta \to 0$. Furthermore

$$\sigma = \int \frac{d\sigma}{d\Omega} 2\pi \sin \theta d\theta \to \infty, \tag{1.27}$$

   so that the *total* cross section σ for Rutherford scattering is actually not defined. This is a property of the electromagnetic interaction which is related to its infinite range. In a sense the choice of electromagnetic scattering to define the cross section σ in figure 1.5 was inappropriate. However, for interaction potentials decaying faster than $1/r$ (such as nuclear potentials) σ remains finite.
4. The distance of closest approach to the centre of the nucleus is (see figure 1.3)

$$D = \frac{zZ\alpha}{2T}\left(1 + \frac{1}{\sin \frac{\theta}{2}}\right), \tag{1.28}$$

   the derivation of which is left as an exercise for the reader. The minimum is reached for $\theta = 180°$ (backwards scattering):

$$D_{\min} = \frac{zZ\alpha}{T}. \tag{1.29}$$

   The distance of closest approach decreases with decreasing Z and increasing kinetic energy T. When the radius of the nucleus becomes larger than D one expects a deviation from the Rutherford formula (1.14) because the assumption of a point-like nucleus collapses. A good approximation for the nuclear radius is $a \sim 1.2\,\text{fm}\,A^{\frac{1}{3}}$ (see chapter 2). For instance, for gold ($Z = 79$, $A = 197$) $a \sim 7$ fm and for $T = 4.78$ MeV, $z = 2$, the minimum distance is $D_{\min} = 48$ fm (to convert the units use (1.16)). Geiger and Marsden were therefore lucky: with much higher kinetic energies and/or lighter targets their data would not have agreed with (1.14). From the Rutherford formula one can derive an upper limit of 48 fm for the radius of the gold nucleus.

   In 1921 Chadwick and Bieler studied α-proton scattering with a hydrogen target ($Z = 1$). The measured differential cross section departed significantly from the Rutherford formula. (In this case the recoil energy of the proton would have to be taken into account in the derivation of the scattering formula.) Models including finite charge distributions in the α particle and in the proton could not describe the data. The reason for the discrepancy was the onset of the *strong* (nuclear) interaction which only acts at very small distances (~1 fm). A satisfactory explanation for this new force was provided in 1935 by Yukawa who introduced a hypothetical particle, the pion (π) (section 7.3.2). However, the year 1921 can be considered as the actual birth year of particle physics.

The radii of many nuclei (and of the proton) were measured during the 1960s. To avoid complications arising from the finite size of the α particle, point-like projectiles were used, namely electrons. This requires a relativistic derivation of the differential cross section—including the influence of the electron spin—for which the Dirac equation is required (chapter 15). Intimations of this will be given in the next chapter.

5. Could the large angle scattering observed by Geiger and Marsden be due instead to *multiple scattering* in the target? Consider the problem of the one-dimensional random walk with m forward or backward steps of lengths $D_i = \pm 1$, each with 50% probability. After one step the average path length is $\langle D_1 \rangle = 0$, but the mean square distance is

$$\left\langle D_1^2 \right\rangle = 1. \tag{1.30}$$

After $i+1$ steps the mean square distance is with $D_{i+1} = D_i \pm 1$

$$D_{i+1}^2 = D_i^2 + 1 \pm 2D_i \Rightarrow \left\langle D_{i+1}^2 \right\rangle = \left\langle D_i^2 \right\rangle + 1. \tag{1.31}$$

Hence after m steps the mean square distance is

$$\left\langle D_m^2 \right\rangle = m \Rightarrow \sqrt{\left\langle D_m^2 \right\rangle} = \sqrt{m}. \tag{1.32}$$

We can model multiple scattering as a series of increments or decrements θ_i in a two-dimensional plane. From the analogy with the random walk we then expect the mean scattering angle

$$\sqrt{\left\langle \theta_i^2 \right\rangle} = \sqrt{m} \propto \sqrt{\ell}. \tag{1.33}$$

Rutherford and co-workers found, however, that the scattering rate at large angles increased proportionally with target thickness ℓ rather than with $\sqrt{\ell}$, in accordance with formula (1.25).

6. It is easy to show that the Rutherford formula (1.14) also applies to a negative charge ze of the projectile (attractive Coulomb force, $\theta \to -\theta$ in (1.5)). In this case the distance of closest approach becomes

$$D = \frac{zZ\alpha}{2T}\left(-1 + \frac{1}{\sin\frac{\theta}{2}}\right) \to 0 \qquad \text{for } \theta \to 180°. \tag{1.34}$$

References

[1] Pais A 1986 Inward Bound (New York: Oxford)
[2] Geiger H and Marsden E 1913 *Phil. Mag.* **25** 604

IOP Publishing

Nuclear and Particle Physics

Claude Amsler

Chapter 2

The nuclear radius

As discussed in the previous chapter, one does not expect the Rutherford formula to hold at high energies due to the finite size of the nucleus. The projectile penetrates the charge distribution of the nucleus and also feels the strong (nuclear) interaction. Structureless (point-like) projectiles are chosen to measure the *electric charge distribution*, such as the electron or the muon which in addition do not feel the strong interaction. In contrast, the nuclear force can be investigated with electrically neutral strongly interacting particles such as neutrons. In order for the electron to be sensitive to the finite dimension of the nuclear core, its de Broglie wavelength should be comparable to the nuclear radius a, that is

$$\lambda = \frac{2\pi\hbar}{p} = \frac{2\pi}{p} \sim a. \tag{2.1}$$

We shall be more precise about this statement later. For example, with gold ($a = 7$ fm) one needs momenta p of about 1 $\text{fm}^{-1} \simeq 200\,\text{MeV}\,c^{-1}$. Such electrons are highly relativistic ($v \simeq c$ with $m_e = 0.511$ MeV). Accordingly, electron–nucleus scattering must be described relativistically with the additional complication of the electron spin.

2.1 Mott scattering

Deriving the electron–nucleus scattering cross section requires the Dirac equation, hence we shall only show the plausibility of the Mott formula. Let us start from the Rutherford formula (1.14) and define the *momentum transfer*

$$\boxed{\vec{q} = \vec{p} - \vec{p}\,'}, \tag{2.2}$$

doi:10.1088/978-0-7503-1140-3ch2 © IOP Publishing Ltd 2015

where $\vec{p}$ ($\vec{p}'$) are the electron momenta before and after colliding with the nucleus. The modulus of $\vec{q}$ is, neglecting nuclear recoil ($p = p'$):

$$q^2 \equiv \vec{q}^2 = p^2 + p'^2 - 2pp' \cos\theta = 2p^2(1 - \cos\theta) = 4p^2 \sin^2\frac{\theta}{2}$$
$$\Rightarrow q = 2p \sin\frac{\theta}{2}. \tag{2.3}$$

For a non-relativistic electron ($T = p^2/2m_e$) the Rutherford cross section (1.14) becomes

$$\frac{d\sigma}{d\Omega} = \frac{Z^2\alpha^2 16p^4}{16T^2q^4} = \frac{Z^2\alpha^2}{q^4}4m_e^2. \tag{2.4}$$

The relativistic generalization for scattering a spin-$\frac{1}{2}$ electron off a *point-like spin-0* nucleus is obtained by replacing the mass m_e of the electron by its total energy ($E = T + m_e$), and multiplying by the factor $\cos^2\theta/2$. The *Mott* differential cross section reads

$$\left(\frac{d\sigma}{d\Omega}\right)_M = \frac{4Z^2\alpha^2E^2}{q^4}\cos^2\frac{\theta}{2}. \tag{2.5}$$

To interpret the factor $\cos^2\theta/2$ consider backward scattering (figure 2.1) and choose as the z-axis the trajectory of the incident electron. The total angular momentum $\vec{J}$ is composed of the electron spin $\vec{s}$ and the orbital angular momentum $\vec{L}$: $\vec{J} = \vec{s} + \vec{L}$. The orbital angular momentum vanishes on the electron trajectory ($\vec{L} = \vec{r} \times \vec{p} = 0$) so that the projection of the total angular momentum on the z-axis is $M = \frac{1}{2}$ or $-\frac{1}{2}$. Conservation of angular momentum requires that $\Delta M = 0$ during the collision process and therefore the spin projection on the z-axis cannot change, implying that the spin must flip when scattering backwards. However, the Coulomb force does not flip the spin projection on the *flight direction* of the electron (a property that follows from the Dirac equation). Backward scattering (figure 2.1) is therefore forbidden by angular momentum conservation, in agreement with $\cos^2\theta/2 \to 0$ for $\theta \to 180°$ and quite in contrast to α-nucleus scattering.

The cross section (2.5) is applicable to the following scattering processes.

1. The projectile is point-like with spin $\frac{1}{2}$ and is highly relativistic ($v \simeq c$).
2. The nucleus is spinless (otherwise the interaction between the magnetic moment of the electron and that of the nucleus has to be taken into account).

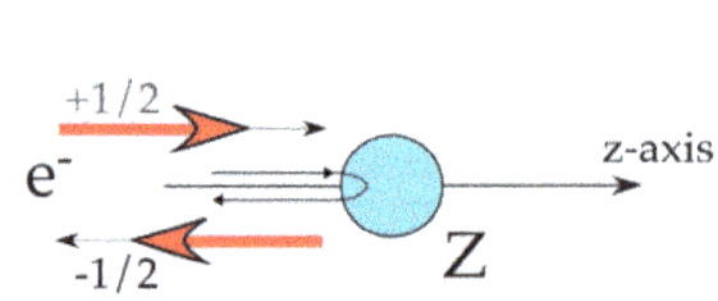

Figure 2.1. Backward electron scattering off a spinless nucleus.

3. The nucleus is point-like (the effect on the finite size will be described in the following section).
4. The atomic number Z of the nucleus is not too large: $(\alpha Z)^2 \ll 1$.
5. The nuclear recoil energy is negligible, therefore the energy of the electron does not change during the collision process.

Nuclear recoil leads to a multiplicative correction factor to the Mott cross section which is usually neglected, except for very light nuclei (see chapter 18):

$$\frac{E'}{E} = \frac{1}{1 + \frac{2E \sin^2 \frac{\theta}{2}}{M}}, \tag{2.6}$$

where E' is the energy of the scattered electron and M the nuclear mass. In addition, the momentum transfer q has to be redefined as (Lorentz-invariant) *4-momentum transfer*. Hence, neglecting the electron mass, q^4 in (2.4) is to be replaced by

$$Q^4 \equiv [(E - E')^2 - (\vec{p} - \vec{p}')^2]^2 = \left[4EE' \sin^2 \frac{\theta}{2}\right]^2, \tag{2.7}$$

which restores the non-relativistic expression for q^4 when the recoil energy is ignored. The Mott cross section in the laboratory now reads

$$\boxed{\left(\frac{\mathrm{d}\sigma}{\mathrm{d}\Omega}\right)_M = \frac{4Z^2\alpha^2 E'^2}{Q^4} \cdot \frac{E'}{E} \cdot \cos^2 \frac{\theta}{2}}. \tag{2.8}$$

2.2 Form factor

Let us now calculate the correction to the Mott cross section (2.5) or (2.8) in the more realistic case of a finite-size nucleus. Consider the diffraction of an incident plane wave on a two-dimensional disc of radius a (figure 2.2).

The intensity distribution at the detector, produced by the beam scattered at the centre of the nucleus (path (a)), is

$$\left(\frac{\mathrm{d}\sigma}{\mathrm{d}\Omega}\right)_M = |A_a(\theta)|^2, \tag{2.9}$$

where $A_a(\theta)$ stands for the (complex) *scattering amplitude*. Scattering along the path (b) is also described by the amplitude $A_a(\theta)$, but with a phase shift δ that depends on the difference in path lengths between (a) and (b):

$$\delta = 2\Delta x k = 2\left(d \sin \frac{\theta}{2}\right)p = qd \tag{2.10}$$

with $\hbar k = p(\hbar = 1$, k is the wave number). The amplitude for path (b) is

$$A_b = A_a \exp(\mathrm{i}\delta). \tag{2.11}$$

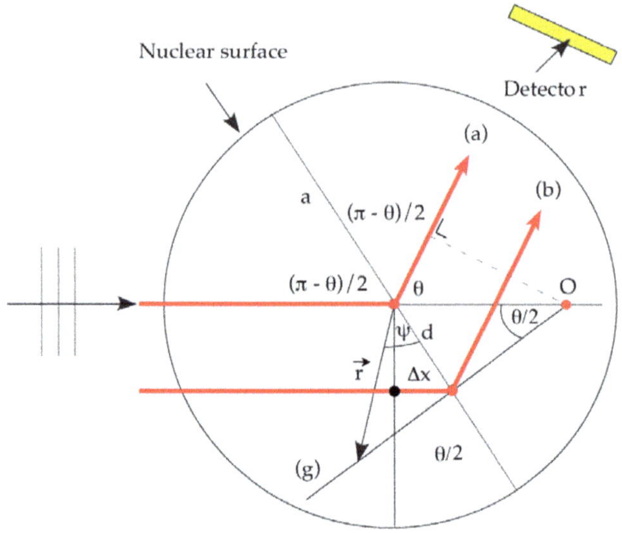

Figure 2.2. Diffraction of a plane wave by the nucleus.

The phase shift δ is the same for all scattering centres on the straight line (g), as is easily seen, for example by considering scattering at point O. The path length difference with respect to the beam (a) is

$$\frac{d}{\sin\frac{\theta}{2}} - \frac{d}{\sin\frac{\theta}{2}}\cos\theta = \frac{d}{\sin\frac{\theta}{2}}(1-\cos\theta) = 2d\sin\frac{\theta}{2} = 2\Delta x, \tag{2.12}$$

as in (2.10). The contribution to the scattering amplitude from any point on (g) is

$$\mathrm{d}A = \rho(\vec{r})\underbrace{r^2\sin\psi\,\mathrm{d}\psi\,\mathrm{d}\beta\,\mathrm{d}r}_{\text{Volume element on (g)}}\exp(\mathrm{i}qd)A_a(\theta). \tag{2.13}$$

The nuclear density $\rho(\vec{r})$ is normalized over the volume of the nucleus:

$$\int \rho(\vec{r})\mathrm{d}^3\vec{r} = 1. \tag{2.14}$$

Integrating (2.13) over the volume of the nucleus gives

$$A = A_a(\theta)\int_0^{2\pi}\int_0^{\pi}\int_0^{\infty}\rho(\vec{r})\underbrace{\exp(\mathrm{i}qd)}_{\exp(\mathrm{i}qr\cos\psi)=\exp(\mathrm{i}\vec{q}\cdot\vec{r})}r^2\sin\psi\,\mathrm{d}\psi\,\mathrm{d}\beta\,\mathrm{d}r$$

$$= A_a(\theta)\int\rho(\vec{r})\exp(\mathrm{i}\vec{q}\cdot\vec{r})\mathrm{d}^3\vec{r}, \tag{2.15}$$

hence

$$\frac{\mathrm{d}\sigma}{\mathrm{d}\Omega} = |A_a(\theta)|^2\left|\int\rho(\vec{r})\exp(\mathrm{i}\vec{q}\cdot\vec{r})\mathrm{d}^3\vec{r}\right|^2, \tag{2.16}$$

or

$$\boxed{\frac{\mathrm{d}\sigma}{\mathrm{d}\Omega}=\left(\frac{\mathrm{d}\sigma}{\mathrm{d}\Omega}\right)_M|F(q)|^2}, \tag{2.17}$$

with the *form factor*

$$\boxed{F(q)=\int\rho(\vec{r})\exp(\mathrm{i}\vec{q}\cdot\vec{r})\mathrm{d}^3\vec{r}}. \tag{2.18}$$

The form factor describing the finite size of the nucleus is the *Fourier transform* of the charge distribution. For a point-like nucleus

$$\rho(\vec{r})=\frac{1}{4\pi}\delta(\vec{r})\Rightarrow F(q)\equiv 1, \tag{2.19}$$

as expected. For small scattering angles, with normalization (2.14) one obtains

$$\theta\to 0\ (q\to 0)\Rightarrow F(0)=1. \tag{2.20}$$

Indeed, for small enough scattering angles the impact parameter is larger than the nuclear radius. The electron flies far away from the core, experiencing only the effect of a point charge. In fact, the form factor (2.18) differs significantly from one when $\vec{q}\cdot\vec{r}\sim 1$, that is when $q\sim 1/r$ everywhere in the integration volume. For example, on the surface

$$q\sim\frac{1}{a}\Rightarrow q\sim 200\ \mathrm{MeV}\ c^{-1} \tag{2.21}$$

for $a = 1$ fm. Momentum transfers in the range of several 100 MeV c^{-1} are needed to resolve the nuclear structure.

The form factor of a *spherical* density distribution $\rho(\vec{r})=\rho(r)$ is, following (2.15),

$$\begin{aligned}F(q)&=\int_0^\infty\rho(r)r^2\mathrm{d}r\underbrace{\int_0^{2\pi}\int_0^\pi\exp(\mathrm{i}qr\cos\psi)\overbrace{\sin\psi\mathrm{d}\psi}^{-\mathrm{d}\cos\psi}\mathrm{d}\beta}_{-\frac{2\pi}{\mathrm{i}qr}[\exp(\mathrm{i}qr\cos\psi)]_1^{-1}}\\&=\int_0^\infty\rho(r)r^2 4\pi\frac{\sin qr}{qr}\mathrm{d}r,\end{aligned} \tag{2.22}$$

hence

$$\boxed{F(q)=\frac{4\pi}{q}\int_0^\infty r\rho(r)\sin(qr)\mathrm{d}r}. \tag{2.23}$$

Consider the special case of a homogeneous density distribution up to the radius a:

$$\rho(r) = \rho_0 = \frac{3}{4\pi a^3}. \tag{2.24}$$

Integrating partially one obtains from (2.23)

$$F(q) = \frac{4\pi}{q}\rho_0 \int_0^a r\,\sin(qr)\mathrm{d}r = \frac{3}{a^3 q}\left(\left[\frac{-r\cos qr}{q}\right]_0^a + \int_0^a \frac{\cos qr}{q}\mathrm{d}r\right)$$
$$= \frac{3[\sin(aq) - aq\cos(aq)]}{(aq)^3}. \tag{2.25}$$

$F(q)$ vanishes whenever

$$\tan(aq) = aq. \tag{2.26}$$

Figure 2.3 shows the expected form factor for $p = 450$ MeV c^{-1} electrons scattered off ^{58}Ni ($a = 4.1$ fm) as a function of momentum transfer q, assuming a homogeneous charge distribution, and as a function of θ (upper scale), using (2.3). The diffraction of the incident wave leads to intensity minima and maxima (circular valleys and troughs) in the plane perpendicular to the beam, akin to Fraunhofer's diffraction pattern of light traversing a hole. However, the zeros are characteristic of diffraction by sharp edges, while the nuclear charge density decreases smoothly towards the surface. This leads to a blurring of the zeros.

Differential cross sections in electron–nucleus scattering are measured using magnetic spectrometers (section 2.4). Figure 2.4 (left) shows the measured cross

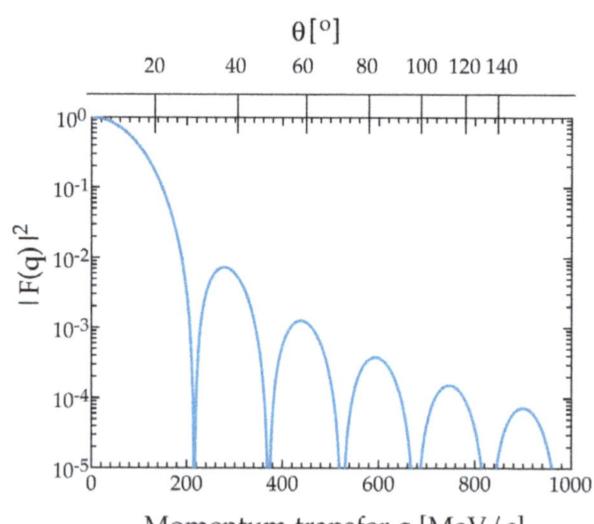

Figure 2.3. Calculated $|F(q)|^2$ for a homogeneous charge distribution of radius $a = 4.1$ fm. The incident electron momentum is $p = 450$ MeV c^{-1}. The top scale shows the corresponding scattering angle.

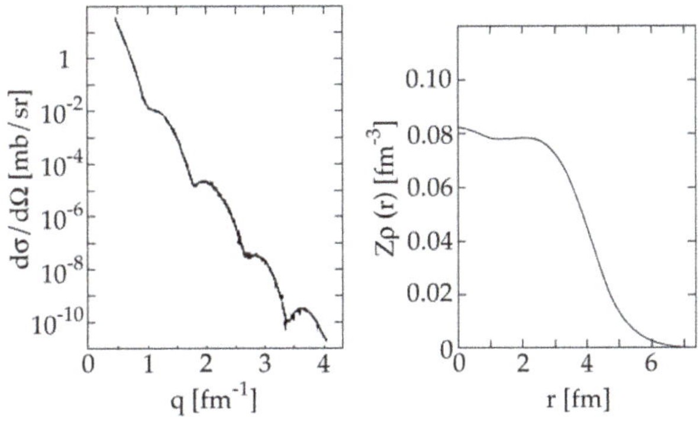

Figure 2.4. Left: differential cross section for electron elastic scattering off ^{58}Ni at 450 MeV c^{-1}. Right: corresponding charge distribution $Z\rho$ with $Z = 28$ (after [1]).

section on ^{58}Ni with 450 MeV c^{-1} electrons. From the intensity minimum around $q = 1.1\ \text{fm}^{-1}$ one finds with the first zero of (2.26) an estimate for the nuclear radius:

$$aq = 4.49 \Rightarrow a = 4.1\ \text{fm}. \tag{2.27}$$

The charge distribution could be determined by the inverse Fourier transform of the measured differential cross section. However, for practical reasons $\mathrm{d}\sigma/\mathrm{d}\Omega$ can only be measured over a limited angular range. One therefore resorts to models of $\rho(r)$ which reproduce the measured differential cross section. A suitable description is given by the *Saxon–Woods* distribution (figure 2.5):

$$\rho(r) = \frac{\rho_0}{1 + \exp\left(\frac{r-a}{d}\right)}. \tag{2.28}$$

A function of this type provides a good approximation for results from more sophisticated analyses, such as those shown for ^{58}Ni in figure 2.4 (right). The *nuclear*

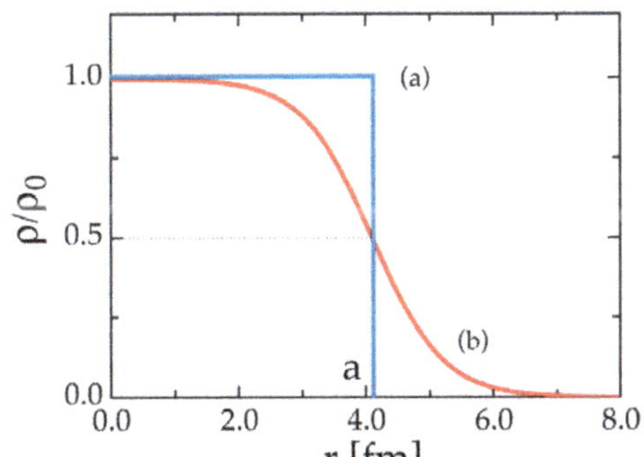

Figure 2.5. Homogeneous charge density distribution (*a*) and Saxon–Woods distribution (*b*) with $a = 4.1$ fm and $d = 0.55$ fm.

radius a and the *skin thickness d* are obtained from numerous fits using the Saxon–Woods distribution:

$$a = 1.18A^{\frac{1}{3}} - 0.48 \text{ fm},$$
$$d = 0.55 \pm 0.07 \text{ fm}, \tag{2.29}$$

for nuclei above $A = 40$.

It is straightforward to show that the Fourier transform (2.23) of an exponential density distribution is the *dipole form factor*:

$$\rho = \frac{1}{8\pi a^3} \exp(-r/a) \Rightarrow F(q) = \frac{1}{\left(1 + a^2q^2\right)^2}, \tag{2.30}$$

where a is a parameter to be extracted from the data. A charge distribution of this type describes the measured proton form factor (section 2.4). In contrast, the density distribution

$$\rho = \frac{1}{4\pi a^2} \frac{1}{r} \exp(-r/a) \tag{2.31}$$

leads to the *monopole form factor*

$$F(q) = \frac{1}{1 + a^2q^2}. \tag{2.32}$$

For light nuclei such as ^{6}Li the charge density distribution is Gaussian, and the corresponding form factor is also Gaussian:

$$\rho = \frac{1}{\left(2\pi a^2\right)^{3/2}} \exp\left(-\frac{r^2}{2a^2}\right) \Rightarrow F(q) = \exp\left(-\frac{a^2q^2}{2}\right). \tag{2.33}$$

Figure 2.6 shows the charge density distribution for various nuclei. For nuclei that are not too light $eZ\rho$ is roughly constant at small radii, approximately equal to $1.1 \times 10^{19}\,\text{C} \cdot \text{cm}^{-3}$ and hence nuclear matter is *incompressible*. With the proton mass $m_p = 1.7 \times 10^{-24}$ g the proton density at the nucleus centre is equal to $eZ\rho_0 (m_p/1.6 \times 10^{-19}\,\text{C}) \simeq 10^{14}\,\text{g cm}^{-3}$!

Finally, expanding the sine-factor in (2.23) in powers of $x = qr$ leads to the approximative expression

$$F(q) \simeq \frac{4\pi}{q} \int_0^\infty r\rho(r)\left(qr - \frac{q^3r^3}{6}\right) \mathrm{d}r = 1 - \frac{4\pi}{6} \int_0^\infty q^2\rho r^4 + \mathrm{d}r$$
$$= 1 - \frac{1}{6}q^2\left\langle r^2\right\rangle, \tag{2.34}$$

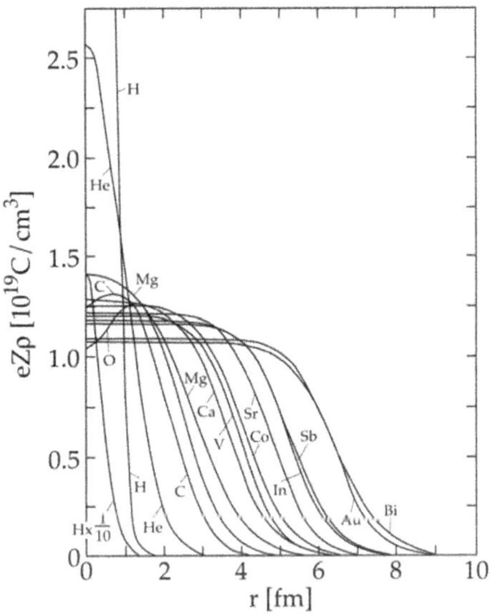

Figure 2.6. Charge density distribution $eZ\rho$ for various nuclei. After [2].

with the mean square radius

$$\left\langle r^2 \right\rangle = 4\pi \int_0^\infty \rho(r) r^4 \mathrm{d}r \simeq -6 \frac{\mathrm{d}F}{\mathrm{d}q^2}\bigg|_{q^2=0}. \tag{2.35}$$

The mean nuclear radius $\sqrt{\langle r^2 \rangle}$ can be obtained directly from the slope of the form factor $F(q^2)$ at forward angles.

2.3 Muonic atoms

The binding energy of the shell electrons depends on the nuclear charge and is quantized (Rydberg formula for the hydrogen atom). The overlap of the electronic and nuclear wave functions causes tiny energy shifts from which the nuclear radius could, in principle, be extracted. The shifts can be measured by observing the photon emission lines between atomic levels. This method is unfortunately not practicable when several electrons shield the nuclear charge, due to uncertainties in the electronic wave function. However, the γ spectra from various isotopes of the same element, which have different numbers of neutrons, can be compared. The observed *isotope shifts* are compatible with Saxon–Woods distributions for the nuclear charge density.

This shielding problem can be avoided with *muonic atoms*. A negatively charged muon (μ^-) stopping in a target is captured by a target atom which ejects a shell electron. The μ^- then cascades down its energy levels by x-ray emission into the innermost Bohr orbit. The radius of the muonic nth orbit is

$$r_n = \frac{n^2}{mZ\alpha} = \frac{n^2}{Z} \times 255 \text{ fm}, \tag{2.36}$$

with

$$m = \frac{m_\mu M}{m_\mu + M} \simeq m_\mu = 106\ \text{MeV}. \tag{2.37}$$

The mass of the muon is m_μ, the mass of the nucleus M and m is the *reduced mass*. Since muons are about 200 times more massive than electrons, the radius of the innermost electronic level corresponds to the *muonic* level $n(\mu^-) = 14$:

$$r_1(e^-) = \frac{1}{m_e Z\alpha} = r_n(\mu^-) \Rightarrow n(\mu^-) = \sqrt{\frac{m_\mu}{m_e}} \sim 14. \tag{2.38}$$

Below $n_\mu = 14$ the muon feels only the bare nuclear charge. For instance, in ^{12}C the lowest orbital radii are, according to (2.36),

$$\begin{aligned} r_3(\mu) &= 383\ \text{fm}, \\ r_2(\mu) &= 170\ \text{fm}, \\ r_1(\mu) &= 43\ \text{fm}, \end{aligned} \tag{2.39}$$

while the nuclear radius is $a \simeq 1.2A^{1/3} = 2.7$ fm. The photon energies to be measured are in the x-ray range, e.g. for the $n = 3$ to $n = 2$ transition,

$$E_\gamma = \frac{m\alpha^2 Z^2}{2}\left(\frac{1}{2^2} - \frac{1}{3^2}\right) = 14\ \text{keV}. \tag{2.40}$$

The γ energy is somewhat lower by ~39 eV due to the finite size of the nucleus, when assuming a homogeneous distribution of the form (2.24). This means that the binding energy of the $n = 2$ level is slightly smaller, the shift of the $n = 3$ level being negligible.

After reaching the lowest orbits the muon is captured by the nucleus through the reaction

$$\mu^- p \to \nu_\mu n. \tag{2.41}$$

In heavy nuclei the $n = 1$ orbit lies even within nuclear matter which is modified by the muon charge. For example, for Pb ($Z = 82$) the radius of the innermost orbit is $r_1 = 3.1$ fm, while the nuclear radius is $a = 6.5$ fm! The muon lives on average $\sim 10^{-8}$ s before being captured through reaction (2.41), during which time it cycles $\sim 10^{12}$ times around the centre of the nucleus. Figure 2.7 shows a photograph of the CERN synchrotron where several studies on muonic atoms were performed.

2.4 Nucleon form factor

As mentioned above, the interaction between magnetic moments must be taken into account when dealing with electron–proton and electron–neutron scattering. The Mott formula (2.8) has to be replaced by the *Rosenbluth formula*. The derivation

Figure 2.7. Several experiments on muonic atoms were performed at CERN's first accelerator, the 600 MeV Synchrocyclotron. The accelerator was built in 1957 and decommissioned in 1990. Image credit: CERN.

which requires the Dirac equation (chapter 15) is given in chapter 18 in the limit of a point-like nucleon. The Rosenbluth formula reads

$$\boxed{\left(\frac{\mathrm{d}\sigma}{\mathrm{d}\Omega}\right)_{eN} = \frac{4\alpha^2 E'^2}{Q^4}\frac{E'}{E}\left[\frac{G_E^2(Q^2) + \tau G_M^2(Q^2)}{1+\tau}\cos^2\frac{\theta}{2} + 2\tau G_M^2(Q^2)\sin^2\frac{\theta}{2}\right]}, \tag{2.42}$$

where

$$\tau = \frac{Q^2}{4M^2} \tag{2.43}$$

and

$$Q^2 = 4EE'\sin^2\frac{\theta}{2}. \tag{2.44}$$

E' is the energy of the scattered electron and M the mass of the nucleon. The functions $G_E(Q^2)$ and $G_M(Q^2)$ are the *electric* and *magnetic* form factors which describe the charge distribution and the distribution of the magnetization in the nucleon.

The magnetic moment of a particle of charge $+e$ is $\mu = g\frac{e}{2m}s$, hence $\mu = \frac{g}{2}\frac{e}{2m}$ for spin $s = \frac{1}{2}$ particles. For point-like spin-$\frac{1}{2}$ particles the g-factor is $g = 2$ as a consequence of the Dirac equation. The nucleon is not point-like but composed of quarks, and hence the *g*-factor differs significantly from the value of 2. The relative difference $\frac{g-2}{2}$ is the *anomalous* magnetic moment. The magnetic moment of the proton is $\mu_p = \frac{g_p}{2}$, that of the neutron $\mu_n = \frac{g_n}{2}$ (both in units of μ_N), where

$$\mu_N = \frac{e}{2m_p} = 3.15 \times 10^{-14}\ \text{MeV T}^{-1} \tag{2.45}$$

is the *nuclear magneton.*

For forward electron–nucleon scattering ($Q^2 \to 0$) one obtains for the proton and the neutron

$$\begin{aligned} G_E(0)_p &= 1, \\ G_M(0)_p &= \frac{g_p}{2} = 2.79, \\ G_E(0)_n &= 0, \\ G_M(0)_n &= \frac{g_n}{2} = -1.91. \end{aligned} \tag{2.46}$$

$G_E(0)$ is the charge in units of e and $G_M(0)$ the magnetic dipole moment in nuclear magnetons μ_N felt by the electron at large distances.

For point-like protons $G_E(Q^2)$ and $G_M(Q^2)$ would be equal to unity, thus recovering the Mott formula (2.8), except for the term in $\sin^2\frac{\theta}{2}$ which describes the interaction between magnetic moments. The Rosenbluth formula can be recast into the simpler form

$$\left(\frac{d\sigma}{d\Omega}\right)_{eN} = \left(\frac{d\sigma}{d\Omega}\right)_M \left[G_E^2(Q^2) + \frac{\tau}{\epsilon} G_M^2(Q^2)\right]\frac{1}{1+\tau}, \tag{2.47}$$

where ϵ is the polarization of the exchanged photon:

$$\epsilon \equiv \frac{1}{1 + 2(1+\tau)\tan^2\frac{\theta}{2}}. \tag{2.48}$$

The form factor is obtained by carrying out measurements of the differential cross section at various angles and incident energies. A spectrometer of the type shown in figure 2.8 is used to measure the form factor of the proton. For *elastic* scattering at a given incident energy, there is a one-to-one relationship between the energy E' of the scattered electron and its scattering angle. Elastic events are selected by a suitable choice of magnetic fields to transmit only scattered electrons (or recoil protons) with momenta that correspond to the angular setting of the spectrometer. This suppresses background from *inelastic* processes such as proton excitation or secondary particle production. The incident intensity is monitored by a Faraday cage. Scattered

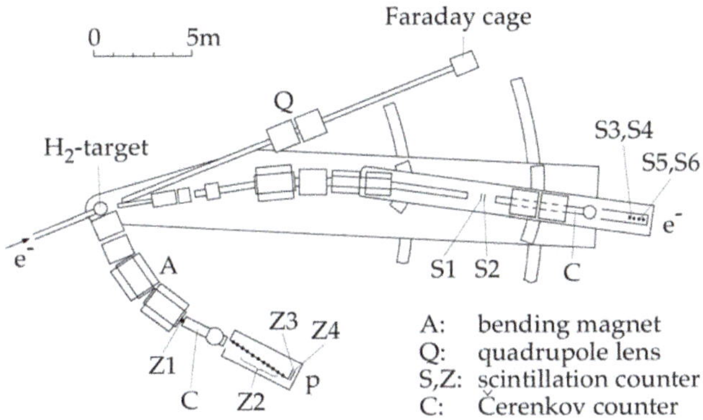

Figure 2.8. A magnetic spectrometer used in electron–nucleus scattering experiments, here in particular for electron–proton scattering. After [3].

electrons are identified and distinguished from other particles by the Čerenkov radiation emitted when traversing a transparent medium (section 9.10).

Figure 2.9 (left) shows the momentum transfer dependence of the proton form factors. A good description of $G_E(Q^2)_p$ and $G_M(Q^2)_p$ is obtained by the *dipole* form factor (2.30):

$$F_D\left(Q^2\right) = \frac{1}{\left(1 + 1.41Q^2\left[\mathrm{GeV}^2\, c^{-2}\right]\right)^2}, \tag{2.49}$$

where $a = 0.234$ fm. The proton is thus an extended structure with distributions of charge and magnetization that decrease exponentially towards the periphery.

The measurement of the neutron form factor is more difficult. Deuterium or ^{3}He targets are used and the contribution from electron–proton scattering is subtracted from the data. At low energies, the scattering of neutrons from nuclear reactors by atomic shell electrons has also been investigated [5, 6]. Figure 2.9 (right) shows data for the neutron magnetic form factor measured at Jefferson Laboratory by

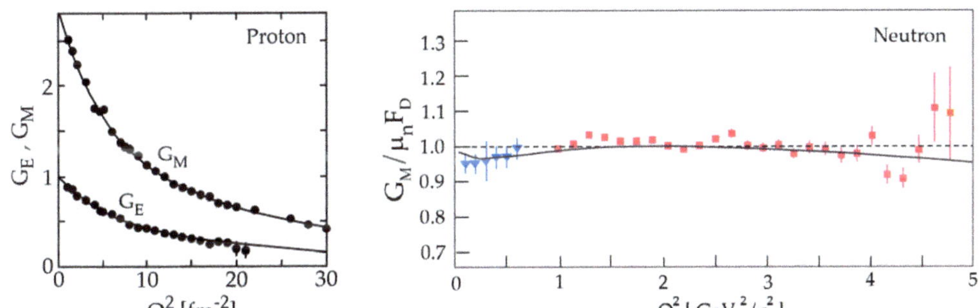

Figure 2.9. Left: form factors of the proton. The curves are the dipole fits (2.30). After [4]. Right: magnetic form factor of the neutron relative to $\mu_n F_D$ ($Q^2 = 1$ GeV^2 c^{-2} corresponds to $\simeq$25 fm^{-2}). The curve is a model prediction. Adapted from [7].

scattering electrons off deuterium, and calculating the ratio of electron–neutron to electron–proton scattering cross sections [7]. Recoil protons were analysed in magnetic spectrometers, neutrons by time-of-flight. The data points in figure 2.9 are divided by $\mu_n F_D(Q^2)$, showing that the magnetization in the neutron is reasonably well described by the dipole form factor (2.49), except for a small deviation below $Q^2 \simeq 1\ \mathrm{GeV}^2\,c^{-2}$.

More precise results on the proton form factors can be obtained by using polarized electrons and measuring the polarization of the recoil proton. This is achieved by measuring the left–right scattering asymmetry (section 11.3) when re-scattering the proton in a hydrogen target. Alternatively, for both protons and neutrons one measures the left–right scattering asymmetry on a polarized hydrogen, deuterium or ^{3}He target. Such experiments are sensitive to the *ratio* $G_E(Q^2)/G_M(Q^2)$ and are particularly advantageous for obtaining data on the neutron electric form factor $G_E(Q^2)_n$, which is notoriously difficult to measure (figure 2.10, left). Such a measurement provides information about the internal (quark) structure of the neutron[1]. On the other hand, a recent double-scattering measurement of this ratio for the proton shows that the dipole form factor does not provide a satisfactory description of the electric and magnetic form factors at high Q^2. The ratio $G_E(Q^2)_p \mu_p / G_M(Q^2)_p$ is not equal to unity at high Q^2 (figure 2.10, right).

Let us estimate the mean square radius of the proton from the dipole form factor (2.49):

$$\left\langle r^2 \right\rangle = 4\pi \int_0^\infty r^2 \rho_0 \exp\left(-\frac{r}{a}\right) r^2 \mathrm{d}r = 96\pi\rho_0 a^5 \quad \text{with} \quad \rho_0 = \frac{1}{8\pi a^3}, \tag{2.50}$$

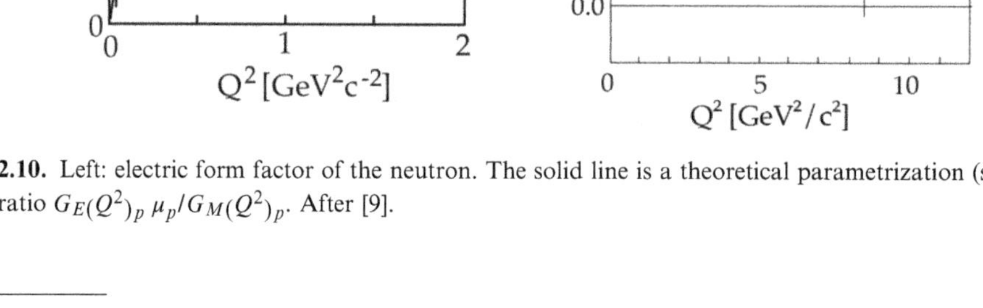

Figure 2.10. Left: electric form factor of the neutron. The solid line is a theoretical parametrization (see [8]). Right: ratio $G_E(Q^2)_p\,\mu_p/G_M(Q^2)_p$. After [9].

[1] The neutron consists of two d quarks (charge $-\frac{1}{3}e$), a u quark (charge $+\frac{2}{3}e$) and a sea of virtual quark–antiquark pairs. Calculations predict that the d quarks are repelled towards the periphery, while the u quarks are attracted towards the centre, thus leading to a (slightly) negatively charged surface and a positively charged core.

hence

$$\left\langle r^2 \right\rangle = 12a^2 \Rightarrow \sqrt{\left\langle r^2 \right\rangle} = 0.81 \text{ fm}. \tag{2.51}$$

More precise measurements of the form factor lead to 0.8775 ± 0.0051 fm [10]. There is currently a discrepancy between electron scattering data and muonic hydrogen data which give a slightly smaller radius of the proton: 0.84087 ± 0.00039 fm [11].

The mean square radius of the neutron is 0.873 ± 0.011 fm [12]. The radii of other composite particles such as the pion (π) and the kaon (K) have also been determined experimentally. This was achieved by scattering high energy pions or kaons off atomic shell electrons [13]. The monopole form factor (2.32) gives the best description of the measured cross section, leading to the mean square radii

$$\begin{aligned} \sqrt{\left\langle r_\pi^2 \right\rangle} &= 0.66 \pm 0.01 \text{ fm}, \\ \sqrt{\left\langle r_K^2 \right\rangle} &= 0.58 \pm 0.04 \text{ fm}. \end{aligned} \tag{2.52}$$

The internal structure of the nucleon is discussed in more detail in chapter 18.

References

[1] Sick I *et al* 1975 *Phys. Rev. Lett.* **35** 910
[2] Hofstadter R 1957 *Ann. Rev. Nucl. Sci* **7** 231
[3] Lohrmann E 1992 *Hochenergiephysik* (Stuttgart: Teubner)
[4] Hughes E B *et al* 1965 *Phys. Rev.* B **139** 458
[5] Köster L *et al* 1976 *Phys. Rev. Lett.* **36** 1021
[6] Krohn V E *et al* 1966 *Phys. Rev.* **148** 1303
[7] Lachniet J *et al* 2009 *Phys. Rev. Lett.* **102** 192001
[8] Bermuth J *et al* 2003 *Phys. Lett.* B **564** 199
[9] Puckett A J R *et al* 2012 *Phys. Rev.* C **85** 045203
[10] Olive K *et al* (Particle Data Group) 2014 *Chin. Phys.* C **38** 090001
[11] Antognini A *et al* 2013 *Science* **339** 417
[12] Kubon G *et al* 2002 *Phys. Lett.* B **524** 26
[13] Amendolia S R *et al* 1984 *Phys. Lett.* B **146** 116
Amendolia S R *et al* 1986 *Phys. Lett.* B **178** 435

Chapter 3

Nuclear masses

The *nuclear mass* is calculated from the proton and neutron masses M_p and M_n, taking into account the binding energy $B(Z, A)$:

$$\boxed{M(Z, A) = ZM_p + \underbrace{(A - Z)}_{=N} M_n - B(Z, A)} \tag{3.1}$$

with

$$\begin{aligned} M_p &= 938.27 \text{ MeV}, \\ M_n &= 939.57 \text{ MeV}, \end{aligned} \tag{3.2}$$

with $c = 1$ and where

$$\begin{array}{ll} Z & \text{is the number of protons (}\textit{atomic number}\text{)}, \\ N & \text{is the number of neutrons}, \\ A & \text{is the number of nucleons} = Z + N \text{ (atomic } \textit{mass number}\text{)}. \end{array} \tag{3.3}$$

Isotopes are nuclei with the same number of protons (Z). *Isotones* have the same number of neutrons (N) and *isobars* the same number of nucleons (A). The *atomic mass* is given by

$$\mathcal{M}(Z, A) = ZM_p + (A - Z)M_n - B(Z, A) + Zm_e - B_e, \tag{3.4}$$

where m_e is the electron mass (0.511 MeV), and where the binding energies B_e of the electrons can often be neglected (e.g. $B_e = 13.6$ eV for the hydrogen atom).

The quantity 'A', which appears in the Mendeleev periodic table of elements (figure 3.1), stands for the *atomic weight* (or relative atomic mass) which is the atomic mass averaged over the isotopic abundance on Earth and relative to $\frac{1}{12}$ of the mass of the ^{12}C atom. The atomic weight 'A' of ^{12}C is defined as 12 g which contains

 © IOP Publishing Ltd 2015

PERIODIC TABLE OF THE ELEMENTS

1 IA	2 IIA	3 IIIB	4 IVB	5 VB	6 VIB	7 VIIB	8	9 VIII	10	11 IB	12 IIB	13 IIIA	14 IVA	15 VA	16 VIA	17 VIIA	18 VIIIA
1 H Hydrogen 1.00794																	2 He Helium 4.002602
3 Li Lithium 6.941	4 Be Beryllium 9.012182											5 B Boron 10.811	6 C Carbon 12.0107	7 N Nitrogen 14.0067	8 O Oxygen 15.9994	9 F Fluorine 18.9984032	10 Ne Neon 20.1797
11 Na Sodium 22.98976928	12 Mg Magnesium 24.3050											13 Al Aluminum 26.9815386	14 Si Silicon 28.0855	15 P Phosph. 30.973762	16 S Sulfur 32.065	17 Cl Chlorine 35.453	18 Ar Argon 39.948
19 K Potassium 39.0983	20 Ca Calcium 40.078	21 Sc Scandium 44.955912	22 Ti Titanium 47.867	23 V Vanadium 50.9415	24 Cr Chromium 51.9961	25 Mn Manganese 54.938045	26 Fe Iron 55.845	27 Co Cobalt 58.933195	28 Ni Nickel 58.6934	29 Cu Copper 63.546	30 Zn Zinc 65.38	31 Ga Gallium 69.723	32 Ge German. 72.64	33 As Arsenic 74.92160	34 Se Selenium 78.96	35 Br Bromine 79.904	36 Kr Krypton 83.798
37 Rb Rubidium 85.4678	38 Sr Strontium 87.62	39 Y Yttrium 88.90585	40 Zr Zirconium 91.224	41 Nb Niobium 92.90638	42 Mo Molybd. 95.96	43 Tc Technet. (97.90722)	44 Ru Ruthen. 101.07	45 Rh Rhodium 102.90550	46 Pd Palladium 106.42	47 Ag Silver 107.8682	48 Cd Cadmium 112.411	49 In Indium 114.818	50 Sn Tin 118.710	51 Sb Antimony 121.760	52 Te Tellurium 127.60	53 I Iodine 126.90447	54 Xe Xenon 131.293
55 Cs Cesium 132.9054519	56 Ba Barium 137.327	57–71 Lantha-nides	72 Hf Hafnium 178.49	73 Ta Tantalum 180.94788	74 W Tungsten 183.84	75 Re Rhenium 186.207	76 Os Osmium 190.23	77 Ir Iridium 192.217	78 Pt Platinum 195.084	79 Au Gold 196.966569	80 Hg Mercury 200.59	81 Tl Thallium 204.3833	82 Pb Lead 207.2	83 Bi Bismuth 208.98040	84 Po Polonium (208.98243)	85 At Astatine (209.98715)	86 Rn Radon (222.01758)
87 Fr Francium (223.01974)	88 Ra Radium (226.02541)	89–103 Actinides	104 Rf Rutherford. (267.122)	105 Db Dubnium (268.125)	106 Sg Seaborg. (271.133)	107 Bh Bohrium (270.134)	108 Hs Hassium (269.134)	109 Mt Meitner. (276.151)	110 Ds Darmstadt. (281.162)	111 Rg Roentgen. (280.164)	112 Cn Copernicium (277)		114 Fl Flerovium (289)		116 Lv Livermorium (288)		

Lanthanide series	57 La Lanthan. 138.90547	58 Ce Cerium 140.116	59 Pr Praseodym. 140.90765	60 Nd Neodym. 144.242	61 Pm Prometh. (144.91275)	62 Sm Samarium 150.36	63 Eu Europium 151.964	64 Gd Gadolin. 157.25	65 Tb Terbium 158.92535	66 Dy Dyspros. 162.500	67 Ho Holmium 164.93032	68 Er Erbium 167.259	69 Tm Thulium 168.93421	70 Yb Ytterbium 173.054	71 Lu Lutetium 174.9668
Actinide series	89 Ac Actinium (227.02775)	90 Th Thorium 232.03806	91 Pa Protactin. 231.03588	92 U Uranium 238.02891	93 Np Neptunium (237.04817)	94 Pu Plutonium (244.06420)	95 Am Americ. (243.06138)	96 Cm Curium (247.07035)	97 Bk Berkelium (247.07031)	98 Cf Californ. (251.07959)	99 Es Einstein. (252.0830)	100 Fm Fermium (257.09510)	101 Md Mendelev. (258.09843)	102 No Nobelium (259.1010)	103 Lr Lawrenc. (262.110)

Figure 3.1. Periodic table of elements (see the text for details). After [1].

L (Avogadro) = 6.022×10^{23} ^{12}C atoms. The atomic weight is therefore the mass of one mole (L atoms) averaged over the abundance of stable isotopes on Earth. The atomic *mass unit u* is defined as

$$1u = \frac{1}{12}\mathcal{M}\left(^{12}\mathrm{C}\right) = 931.494\ \mathrm{MeV} \Rightarrow \mathcal{M}\left(^{12}\mathrm{C}\right) = 12u. \tag{3.5}$$

Since the neutron and proton have almost the same masses and the binding energy $B(Z, A)$ is only about 1% of the nucleon mass, atomic weight and atomic number are approximately equal. The *mass defect* is defined as the difference

$$\begin{aligned}\Delta = \mathcal{M}(Z, A) - A\frac{1}{12}\mathcal{M}\left(^{12}\mathrm{C}\right) &= \mathcal{M}(Z, A) - A \times 931.494\ \mathrm{MeV} \\ &= B\left(^{12}\mathrm{C}\right) - B(Z, A),\end{aligned} \tag{3.6}$$

where the (small) binding energy B_e of the electrons has been neglected. The mass defect is positive (negative) for nuclei that are lighter (heavier) than ^{12}C.

Only stable isotopes are used to obtain the atomic weights shown in figure 3.1, with the exception of Bi, Th, Pa and U, which have long-lived isotopes on geological time scales. Atomic weights in brackets refer to the isotopes with the longest mean lives, where no stable isotope is known. The heaviest element with at least one stable isotope was long believed to be ^{209}Bi (natural abundance of 100%). However, this element decays into ^{209}Tl by α emission with a half-life (chapter 4) of 1.9×10^{19} years [2]. The heaviest stable element is therefore Pb with three stable isotopes (^{206}Pb, ^{207}Pb, ^{208}Pb).

The lighter elements Tc and Pm have no stable isotopes and are produced artificially. The elements 113, 115, 117 and 118 have been reported, but have not been named yet by the International Union of Pure and Applied Chemistry (IUPAC), as the evidence is not considered conclusive enough. The light transuranic elements are obtained by bombarding, for example, ^{238}U with light particles such as neutrons or α, while the heavy ones are produced in heavy ion collisions.

By 1915 the 92 lightest elements were known, except for Pa, Hf, Re, Tc, Fr, At and Pm. Many elements were identified by Henry Moseley who analysed the x-ray spectra from various materials used as cathode targets in x-ray tubes. The energy of the K_α line ($2p$ to $1s$ transition) is proportional to the square of the atomic number (more precisely, proportional to $(Z-1)^2$, due to core shielding by the second electron in the $1s$ shell). In addition, the transition energy is proportional to the reduced mass $\frac{m_e M}{m_e + M}$, so that isotopes of the same element can be resolved.

Nuclear masses (more precisely, ion masses with charge Q) are measured using electrostatic and magnetic mass spectrometers such as the one shown in figure 3.2. The centripetal force on the ion due to the electric field is

$$\frac{Mv^2}{r_E} = QE \Rightarrow \frac{M}{Q} = \frac{Er_E}{v^2}, \tag{3.7}$$

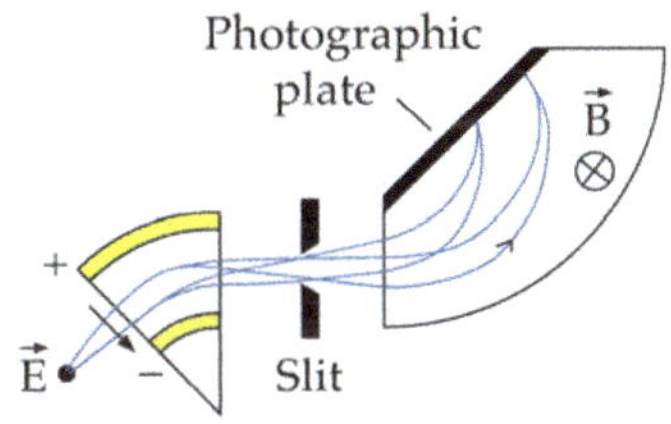

Figure 3.2. The Mattauch spectrometer. The impact coordinate on the photographic plate depends on M/Q.

while in the (homogeneous) magnetic field this force is equal to

$$\frac{Mv^2}{r_B} = QvB \Rightarrow \frac{M}{Q} = \frac{Br_B}{v}. \tag{3.8}$$

Hence in the E-field one measures from the radius of curvature r_E the *kinetic energy* per unit charge, and in the B-field (radius of curvature r_B) the *momentum* per unit charge. From (3.7) and (3.8) one derives the ion mass per unit charge

$$\frac{M}{Q} = \frac{B^2 r_B^2}{E r_E}. \tag{3.9}$$

In the double-focusing spectrometer shown in figure 3.2, ions of a given kinetic energy are first focused on the momentum slit by the appropriately chosen E-field. The ions passing through the slit are then deflected by the B-field towards the photographic plate, where they are sorted according to their momenta, hence according to their ratios M/Q. The latter are obtained by measuring the intensity fringes on the photographic plate, after suitable calibration. Since the ion charge Q is quantized ($Q = Z, Z - 1, Z - 2, \ldots$ in units of e), one actually observes a line spectrum for each atomic species, from which M can be determined.

The resolving power of such a double-focusing spectrometer is very high, typically $\Delta M/M \sim 10^{-8}$. Figure 3.3 shows the measured abundance of elements in the solar system, relative to silicon (see for example [3]). The relative abundance is equal on Earth and in moon rocks, meteorites and cosmic rays, with few exceptions. The elements H, He, C and O predominate. The very light elements (H, He, Li) were produced in the big bang and the elements up to Fe were produced by nuclear fusion in stars (see section 14.3). The elements heavier than $Z \simeq 56$ (Fe) are believed to be produced in stars and supernovae by slow neutron capture (s-process) or rapid multiple neutron capture (r-process), followed by neutron β-decays.

Figure 3.4 shows the binding energy per nucleon B/A for stable isotopes as a function of mass number A. Since all nuclei are unstable above $A = 208$ MeV, the figure shows the isotopes that decay by α emission. Above $A = 20$ (Ne) the ratio B/A becomes nearly constant:

$$\boxed{\frac{B}{A} \simeq 8 \text{ MeV}}. \tag{3.10}$$

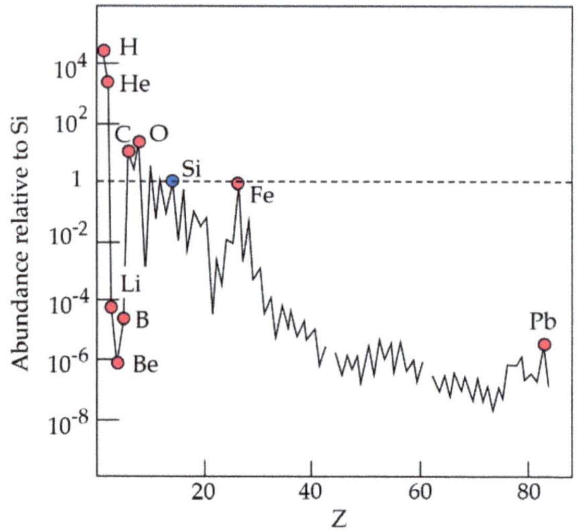

Figure 3.3. The abundance of elements in the solar system, relative to silicon. Adapted from [3].

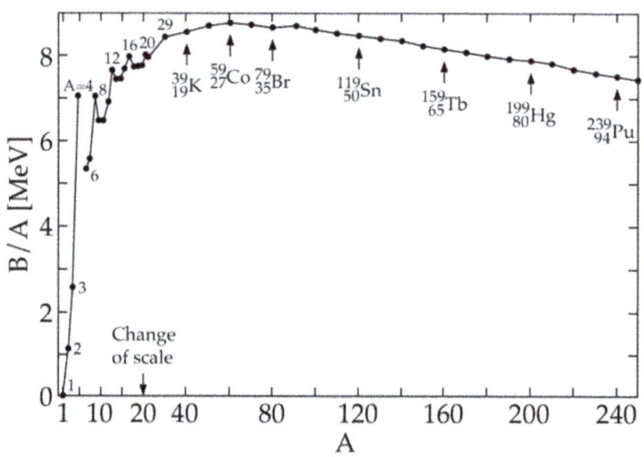

Figure 3.4. Binding energy B/A versus A. After [4, 5].

For a long-range binding force between nucleons one would expect each nucleon to bind with the remaining $A - 1$ ones. The binding energy B should then be proportional to $A(A - 1)$, or in another words $B/A \propto A$. In contrast, (3.10) shows that the interaction acts only on the nearest neighbours, hence that the nuclear force is *short-range*, of the order of the nucleon radius (1 fm).

Figure 3.5 shows the relationship between Z and N for the 288 nuclei known to be stable. The narrow region along the curve corresponds to a *valley of stability* with the most stable nucleus at the bottom (the maximum binding energy or smallest mass). The stability curve can be parametrized as follows:

$$Z = \frac{A}{1.98 + 0.0155\, A^{\frac{2}{3}}}. \tag{3.11}$$

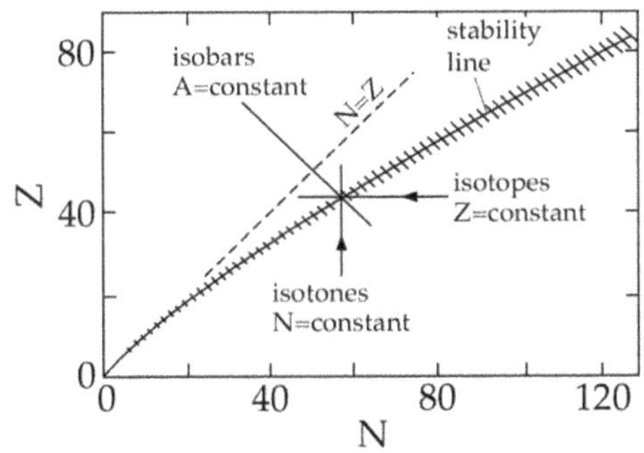

Figure 3.5. The nuclear line of stability (see also figure 6.6 in chapter 6).

As we shall see shortly, the imbalance between neutrons and protons is due to Coulomb repulsion which increases rapidly with Z. Additional neutrons are therefore required to maintain nuclear stability. We now discuss two models to calculate binding energies.

3.1 The Fermi gas model

Let us consider a nucleon confined within the mean nuclear potential generated by the other nucleons, and temporarily ignore all Coulomb interactions between protons. The precise shape of the potential is not known *a priori*. However, we have seen that the binding force is short-range and that the nucleus has relatively sharp edges. As a simple model let us assume that the nucleus is a cube of side length d and consider the square well potential shown in figure 3.6. The nuclear potential is equal to $-V_0$ within the cubic volume and vanishes elsewhere. We consider a potential with infinitely high walls since the nucleons are confined in the well, and thus the nuclear wave function must vanish outside its boundary.

Let us define the energy scale so that the zero point lies at $-V_0$. The solution of the one-dimensional Schrödinger equation (m = nucleon mass, E = kinetic energy)

$$\frac{\Delta\psi}{2m} + E\psi = 0 \tag{3.12}$$

is a standing wave (harmonic oscillation) with vanishing amplitude at the boundaries $x = \pm d/2$, i.e.

$$n_x\frac{\lambda}{2} = d \Rightarrow k_x = \frac{2\pi}{\lambda} = \frac{n_x\pi}{d} \Rightarrow E = \frac{k_x^2}{2m} = \frac{\pi^2 n_x^2}{2md^2}, \tag{3.13}$$

with $n_x = 1, 2, 3$, etc. The energy is quantized and we get, when extending to the three-dimensional case,

$$E = \frac{\pi^2 \overbrace{\left(n_x^2 + n_y^2 + n_z^2\right)}^{\equiv\rho^2}}{2md^2}, \tag{3.14}$$

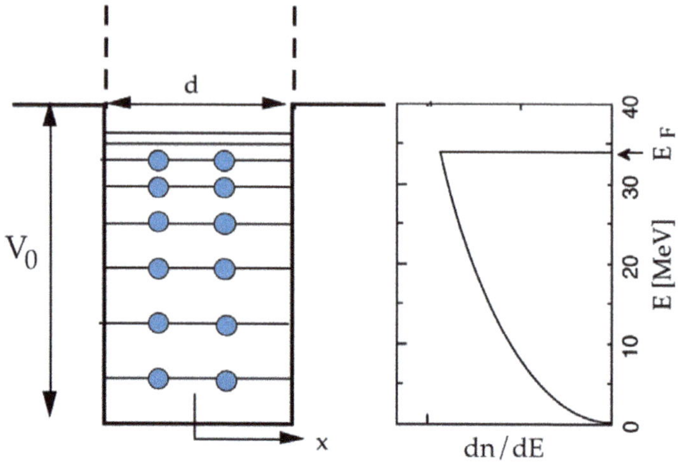

Figure 3.6. The square-well potential.

or

$$\rho^2 = \frac{2md^2E}{\pi^2}. \tag{3.15}$$

Provided that ρ is not too small the number of states between ρ and $\rho + \mathrm{d}\rho$ is

$$\mathrm{d}n = \frac{1}{8}4\pi\rho^2\mathrm{d}\rho. \tag{3.16}$$

The factor $\frac{1}{8}$ arises because n_x, n_y and n_z are positive integers. From (3.15) it follows that

$$2\rho\mathrm{d}\rho = \frac{2md^2}{\pi^2}\mathrm{d}E \Rightarrow \mathrm{d}E = \frac{\pi^2\rho}{md^2}\mathrm{d}\rho, \tag{3.17}$$

and then with (3.16) that

$$\frac{\mathrm{d}n}{\mathrm{d}E} = \frac{4\pi\rho^2}{8\pi^2\rho}md^2 = \frac{1}{2\pi}\frac{\sqrt{2mE}\,d}{\pi}md^2 = \underbrace{\frac{m^{\frac{3}{2}}\overbrace{d^3}^{\tau}}{\sqrt{2}\pi^2}}_{C}\sqrt{E}, \tag{3.18}$$

where τ is the volume of the nucleus. The density of energy levels increases proportionally to $\sqrt{E}$. According to the Pauli principle, two protons (or two neutrons) with opposite spin projections $(\pm\frac{1}{2})$ can occupy each energy level. In the nuclear ground state all levels are filled up to the energy E_F, the *Fermi energy*. The number of occupied energy levels is then

$$n = \int_0^{E_F} 2\frac{\mathrm{d}n}{\mathrm{d}E}\mathrm{d}E = 2\frac{2}{3}CE_F^{\frac{3}{2}}. \tag{3.19}$$

Therefore the Fermi energy is

$$E_F = \left(\frac{3n}{4C}\right)^{\frac{2}{3}} = \left(\frac{n}{\tau}\right)^{\frac{2}{3}} 3^{\frac{2}{3}}\pi^{\frac{4}{3}}\frac{1}{m}\overbrace{\frac{2^{\frac{1}{3}}}{4^{\frac{2}{3}}}}^{\frac{1}{2}}$$

$$= \left(\frac{n}{\tau}\right)^{\frac{2}{3}} 3^{\frac{2}{3}}\pi^{\frac{4}{3}}\frac{1}{2m}. \tag{3.20}$$

To evaluate E_F let us first assume that $n = Z = N = A/2$ and write for the spherical volume

$$\tau = \frac{4\pi r_0^3}{3}A \tag{3.21}$$

with $r_0 = 1.2$ fm. Then

$$E_F = \frac{1}{2m}\left(\frac{1}{8\pi r_0^3}\right)^{\frac{2}{3}}(3\pi)^{\frac{4}{3}} = 33 \text{ MeV}. \tag{3.22}$$

The Fermi energy *does not depend* on A. Each nucleon is thus confined in a potential well of depth

$$\boxed{V_0 = E_F + B/A \simeq 40 \text{ MeV}}. \tag{3.23}$$

Compare V_0 with the Fermi energy of the electron gas in a metal, where $E_F \sim 7$ eV and $B \sim 3$ eV! The order-of-magnitude difference is due to the much larger lattice size for the electron gas, despite the smaller mass m of the electron in (3.22).

The assumption that $N = Z$ was realistic in our simple model without Coulomb repulsion. Indeed, such nuclei would be unstable for $N > Z$ since, following (3.20), $E_F(n) > E_F(p)$. Nuclei would then seek the equilibrium $N = Z$ via neutron β-decay. Hence, without Coulomb repulsion nuclei with $N = Z$ are more stable than those with $N > Z$. Let us calculate the difference in binding energies. The total energy is

$$E_T = \int_0^{E_F} 2E \cdot \overbrace{\frac{dn}{dE}}^{\propto\tau\sqrt{E}\propto A\sqrt{E}} \cdot dE \propto E_F^{\frac{5}{2}}A. \tag{3.24}$$

From (3.20) $E_F \propto \left(\frac{n}{A}\right)^{\frac{2}{3}}$, thus

$$E_T = C' n^{\frac{5}{3}} A^{-\frac{2}{3}}. \tag{3.25}$$

For a symmetrical nucleus one obtains

$$E_{T(N=Z)} = 2C'\left(\frac{A}{2}\right)^{\frac{5}{3}} A^{-\frac{2}{3}}, \tag{3.26}$$

while for $N \neq Z$

$$E_{T(N\neq Z)} = C'\left(Z^{\frac{5}{3}} + N^{\frac{5}{3}}\right)A^{-\frac{2}{3}}. \tag{3.27}$$

The energy difference is

$$\Delta E_T = E_{T(N\neq Z)} - E_{T(N=Z)} = C'A^{-\frac{2}{3}}\left[Z^{\frac{5}{3}} + N^{\frac{5}{3}} - 2\left(\frac{A}{2}\right)^{\frac{5}{3}}\right]. \tag{3.28}$$

Defining the asymmetry parameter

$$\Delta \equiv \frac{1}{2}(Z - N) \tag{3.29}$$

or

$$\begin{aligned} Z &= \frac{A}{2} + \Delta = \frac{A}{2}\left(1 + \frac{2\Delta}{A}\right), \\ N &= \frac{A}{2} - \Delta = \frac{A}{2}\left(1 - \frac{2\Delta}{A}\right), \end{aligned} \tag{3.30}$$

leads to the following expression for ΔE_T:

$$\begin{aligned} \Delta E_T &= C'A^{-\frac{2}{3}}\left(\frac{A}{2}\right)^{\frac{5}{3}}\left[\overbrace{\left(1 + \frac{2\Delta}{A}\right)^{\frac{5}{3}}}^{1+\frac{5.2\Delta}{3A}+\frac{1.5.2}{2.3.3}\left(\frac{2\Delta}{A}\right)^2+\ldots} + \underbrace{\left(1 - \frac{2\Delta}{A}\right)^{\frac{5}{3}}}_{1-\frac{5.2\Delta}{3A}+\frac{1.5.2}{2.3.3}\left(\frac{2\Delta}{A}\right)^2+\ldots} - 2\right] \\ &\propto A^{-\frac{2}{3}}A^{\frac{5}{3}}\Delta^2A^{-2}, \end{aligned} \tag{3.31}$$

hence

$$\Delta E_T \simeq C''\frac{(Z - N)^2}{A} \tag{3.32}$$

with the constant

$$C'' \equiv \frac{5}{9\sqrt{2}}\frac{3}{2^{\frac{2}{3}}10}\left(\frac{9\pi}{4}\right)^{\frac{2}{3}}\frac{1}{mr_0^2}. \tag{3.33}$$

The nuclear mass reaches its minimum (and hence the binding energy its maximum) for $N = Z$, as expected. Now, introducing Coulomb repulsion, one sees that fewer protons than neutrons are accommodated in the square-well potential since at the equilibrium condition $E_F(p) = E_F(n)$ (figure 3.7). The binding energy is correspondingly smaller. We will use (3.32) for the model described in the following section.

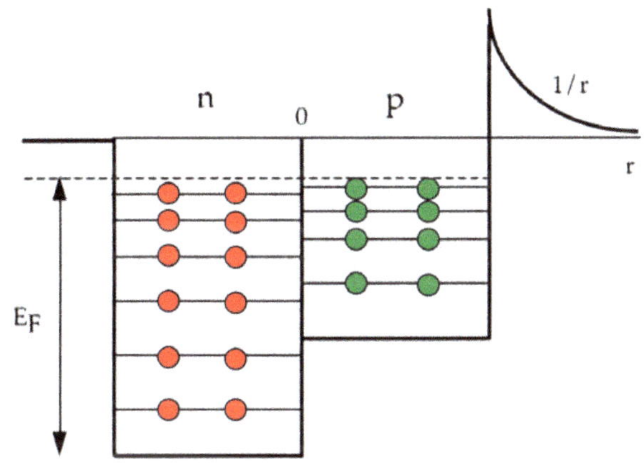

Figure 3.7. The square-well potential for neutrons and protons with a repulsive Coulomb force.

3.2 The liquid drop model

This model was introduced in 1935 by Carl Friedrich von Weizsäcker. We have seen that nuclear matter is incompressible and that the binding energy B/A is approximately constant for not too light nuclei. Similarly, a water droplet is incompressible and the range of the binding (van der Waals) force is short, so that the binding energy b per molecule does not depend on the number of molecules. The total binding energy (the energy required to evaporate the droplet), taking the reduced binding of surface molecules into account, is

$$B = bn - 4\pi R^2\sigma, \tag{3.34}$$

where R is the radius of the droplet containing n molecules, and where σ is proportional to the surface tension. From this analogy the binding energy of the nucleus is determined by its *volume energy* less its *surface energy*:

$$B = a_V A - a_S A^{\frac{2}{3}}, \tag{3.35}$$

where a_V and a_S are positive constants to be determined experimentally.

In addition, the repulsive Coulomb force leads to less binding. Let us calculate the energy stored in a homogeneously charged sphere with radius R and charge density $\rho = \frac{Ze}{4\pi R^3/3}$ (figure 3.8). The required (potential) energy to transport a spherical shell of thickness dr with positive charge $q = 4\pi r^2 \rho \mathrm{d}r$ from infinity to a nuclear core of radius r is

$$\mathrm{d}E = 4\pi r^2 \rho \mathrm{d}r \left(\frac{4\pi r^3 \rho}{3}\right)\left(\frac{1}{4\pi\epsilon_0 r}\right). \tag{3.36}$$

Integrating gives the Coulomb energy

$$E = \int_0^R \frac{4\pi r^4 \rho^2}{3\epsilon_0}\mathrm{d}r = \frac{4\pi R^5 \rho^2}{15\epsilon_0} = \frac{4\pi R^5}{15\epsilon_0}\frac{(Ze)^2}{\frac{(4\pi)^2 R^6}{9}} = \frac{3}{20\epsilon_0}\frac{(Ze)^2}{R\pi}. \tag{3.37}$$

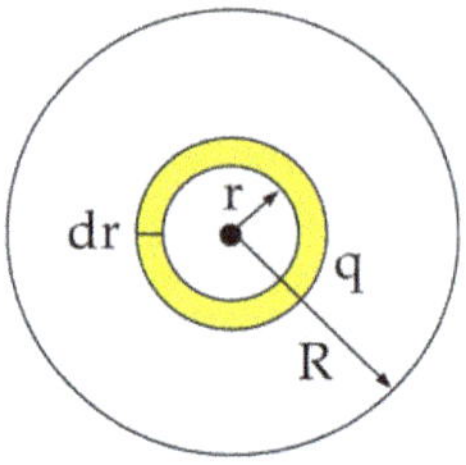

Figure 3.8. Calculation of the Coulomb energy (see the text for details).

We therefore append to the binding energy B, (3.35), the *Coulomb term*

$$-a_C Z^2 A^{-\frac{1}{3}} \ (a_C > 0). \tag{3.38}$$

As explained in section 3.1, the asymmetry $N > Z$ leads to an even smaller binding energy. Hence we add the *asymmetry term* (3.32):

$$-a_A \frac{(Z - N)^2}{A} = -a_A \frac{(A - 2Z)^2}{A} (a_A > 0). \tag{3.39}$$

Figure 3.9 shows the measured energy required to separate a neutron, e.g. from Ba-isotopes (even atomic number $Z = 56$). The binding energies of isotopes with an *even* neutron number are about 2 MeV larger than for those with an *odd* neutron number. In the Fermi gas model the neutrons fill the energy levels pairwise with antiparallel spins, due to the Pauli exclusion principle. Thus, in nuclear matter paired neutrons with antiparallel spins are more strongly bound than those with parallel spins. The same applies to proton pairs. Accordingly, nuclei with an even number of protons and an even number of neutrons (ee-nuclei) are more strongly bound than those with an odd number of protons and an odd number of

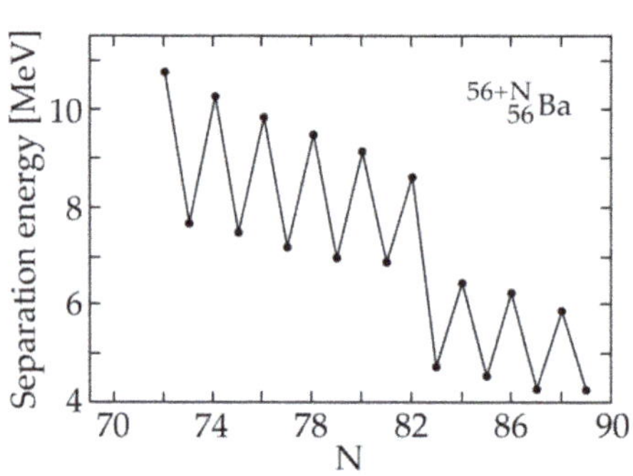

Figure 3.9. Separation energy for various isotopes of Ba. The step at $N = 82$ is due to the closed nuclear shell (see chapter 6). After [4, 5].

neutrons (oo-nuclei)[1]. Therefore the binding energy includes a fifth term, the *pairing term* [6]:

$$\delta(Z, A) = \begin{bmatrix} + \\ 0 \\ - \end{bmatrix} a_P \cdot A^{-\frac{1}{2}} \quad \text{for} \begin{bmatrix} \text{ee} \\ \text{oe} \\ \text{oo} \end{bmatrix} \quad (\text{with } a_P = 12 \text{ MeV}). \tag{3.40}$$

Here oe stands for nuclei with an odd number of neutrons and an even number of protons, or vice versa.

Collecting all terms one obtains for the empirical (Bethe–von Weizsäcker) *mass formula* (3.1) the binding energy

$$\boxed{B(Z, A) = a_V A - a_S A^{\frac{2}{3}} - a_C Z^2 A^{-\frac{1}{3}} - a_A \frac{(A - 2Z)^2}{A} + \delta(Z, A)}. \tag{3.41}$$

Figure 3.10 shows a fit of the data to the calculated binding energies (3.41). The constants a_V–a_P are given in table 3.1. The adjustment is reasonable, except at the so-called *magic* numbers (Z or $N = 20$, 28, 50, 82, 126) which correspond to closed

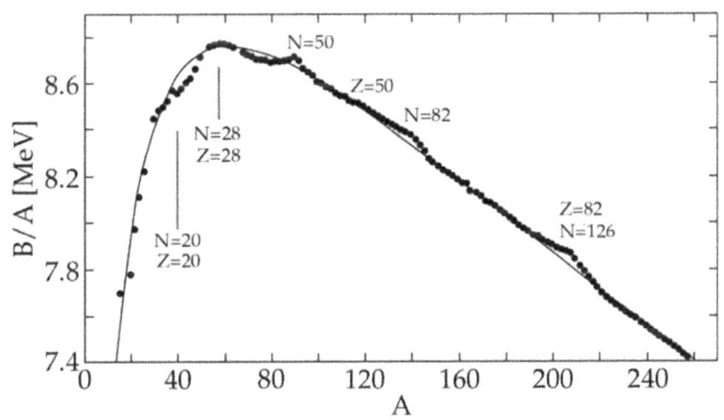

Figure 3.10. Binding energy per nucleon. The measurements (black circles) are compared to a fit with the mass formula (curve). Note the suppressed vertical scale. After [4].

Table 3.1. Adjusted constants in the empirical mass formula.

Term		[MeV]
Volume	a_V	15.56
Surface	a_S	17.23
Coulomb	a_C	0.697
Asymmetry	a_A	23.285
Pairing	a_P	12.0

[1] The deuteron is a stable oo-nucleus, while bound pp or nn-nuclei do not exist. However, binding in the deuteron is provided by the long-range component of the nuclear force.

nuclear 'shells' and for which formula (3.41) underestimates the binding energies (see chapter 6). Obviously, very light nuclei cannot be described by the mass formula. This is also true for excited nuclei and for non-spherical ones.

With decreasing atomic number A the binding energy per nucleon B/A drops, essentially due to the surface term ($B/A \propto A^{-1/3}$). Therefore, *nuclear fusion* of light nuclei releases energy. On the other hand, with increasing mass number A the binding energy B/A decreases for heavy nuclei, mainly because of the Coulomb term ($B/A \propto A^{2/3}$). For heavy nuclei energy is released by *fission*. The binding energy reaches its maximum around $A = 56$ (Fe). This explains why old stars, having exhausted their 'fuel', contain mostly iron.

As already mentioned, due to the short-range force the volume energy is proportional to A and not to $A(A-1)$. The short range of the nuclear interaction may also be seen from the fact that $A = 5$ nuclei (such as ${}^5_2$He) are not bound, in contrast to ${}^4_2$He with $B = 28.3$ MeV. In the case of ${}^5_2$He, and as explained in chapter 6, the unpaired neutron would have to be in a P wave (angular momentum $L = 1$), that is further away from the centre of the nucleus. However, ${}^6_2$He binds with $B = 29.3$ MeV due to two paired neutrons in the P state (although this isotope decays via β emission). Note that formula (3.11) for the stability line in figure 3.5 can be derived with the mass formula (3.1) from the derivative $\partial M/\partial Z = 0$ at fixed A.

References

[1] Olive K *et al* (Particle Data Group) 2014 *Chin. Phys.* C **38** 090001
[2] de Marcillac P *et al* 2003 *Nature* **422** 876
[3] Woosley S E 2007 *Nat. Phys.* **3** 832
[4] Williams W S C 1991 *Nuclear and Particle Physics* (Oxford: Oxford University Press)
[5] Wapstra A H and Audi G 1985 *Nucl. Phys.* A **432** 1
[6] Bohr A and Mottelson B 1969–1975 *Nuclear Structure* (New York: Benjamin)

IOP Publishing

Nuclear and Particle Physics

Claude Amsler

Chapter 4

Radioactive decay

4.1 Decay law

Consider an ensemble of N unstable particles at time t, such as a substance with N radioactive nuclei. The following derivation is valid in general, hence also for unstable elementary particles. The decrease in the number of radioactive nuclei as a function of time during t and $t + \mathrm{d}t$ is given by

$$\mathrm{d}N = -\lambda N \mathrm{d}t. \tag{4.1}$$

The constant λ (dimension s^{-1}) is the *decay constant* which depends on the substance. The number of remaining radioactive nuclei after time t is given by the solution of the differential equation $\frac{\mathrm{d}N}{\mathrm{d}t} = -\lambda N$:

$$\boxed{N(t) = N(0)\exp(-\lambda t)}, \tag{4.2}$$

where $N(0)$ is the number of radioactive nuclei at $t = 0$. Accordingly, the probability for a particular radioactive nucleus to decay in the time interval t to $t + \mathrm{d}t$ is

$$f(t)\mathrm{d}t = \lambda \exp(-\lambda t)\mathrm{d}t \Rightarrow \int_0^\infty f(t)\mathrm{d}t = 1. \tag{4.3}$$

Note that the time $t = 0$, at which the clock is started, is arbitrary and does not necessarily correspond to the time at which the nucleus was produced. Hence the probability for a nucleus to decay at a given time t during the interval $\mathrm{d}t$ does not depend on the time already elapsed after its production. The *mean lifetime* (or mean life) is the reciprocal of the decay constant:

$$\boxed{\tau = \frac{1}{\lambda}}. \tag{4.4}$$

doi:10.1088/978-0-7503-1140-3ch4

© IOP Publishing Ltd 2015

Indeed, the average decay time $\langle t \rangle$ is

$$\langle t \rangle = \lambda \int_0^\infty t \exp(-\lambda t)\, \mathrm{d}t = \frac{1}{\lambda}. \tag{4.5}$$

Examples of extremely long and extremely short mean lives are

$$^{238}\mathrm{U} \to {}^{234}\mathrm{Th} + \alpha, \qquad \tau = 6.5 \times 10^9 \text{ yrs},$$
$$\Delta^{++} \to p\pi^+, \qquad \tau = 6 \times 10^{-24} \text{ s},$$

where the Δ^{++} is an excited state of the proton which will be dealt with in section 12.4.2. The *half-life* $t_{1/2}$ is the average time at which half of the radioactive substance has decayed:

$$N(t_{1/2}) = \frac{N(0)}{2} \Rightarrow \frac{1}{2} = \exp\left(-\lambda t_{1/2}\right) \Rightarrow \boxed{t_{1/2} = \tau \ln 2 = 0.69\tau}. \tag{4.6}$$

Nuclear physicists usually quote half-lives while particle physicists prefer mean lives. The *activity* $I(t)$ of a radioactive source is its *decay rate* (number of decays per second)

$$I(t) = -\frac{\mathrm{d}N}{\mathrm{d}t} = \lambda N(0) \exp(-\lambda t) \Rightarrow I(t) = I(0) \exp(-\lambda t). \tag{4.7}$$

Hence the activity also obeys the decay law (4.2) with

$$I(t) = \lambda N(t). \tag{4.8}$$

The activity is measured in *becquerel* (Bq):

$$1 \text{ Bq} = 1 \text{ decay s}^{-1}.$$

The *curie* (Ci) is an obsolete unit which refers to the activity of 1 g of pure ^{226}Ra:

$$1 \text{ Ci} = 3.7 \times 10^{10} \text{ Bq}.$$

More than one decay mode (or decay 'channel') is often possible, each with probability (*branching ratio*) f_i. For example, ^{212}Bi decays into the following two channels with branching ratios

$$^{212}\mathrm{Bi} \to {}^{212}\mathrm{Po}\ (\beta\text{-decay}), \qquad f_1 = 64\%, \tag{4.9}$$

$$^{212}\mathrm{Bi} \to {}^{208}\mathrm{Tl}\ (\alpha\text{-decay}), \qquad f_2 = 36\%. \tag{4.10}$$

In this case the number $N(t)$ of ^{212}Bi nuclei decreases according to

$$\frac{\mathrm{d}N}{\mathrm{d}t} = -\lambda_1 N - \lambda_2 N \Rightarrow N = N(0) \exp(-[\lambda_1 + \lambda_2]t), \tag{4.11}$$

with the *partial* decay constants λ_1 and λ_2. Hence

$$\lambda = \lambda_1 + \lambda_2 \Rightarrow \tau = \frac{1}{\lambda_1 + \lambda_2}, \tag{4.12}$$

and

$$f_1 = \frac{\lambda_1}{\lambda}, \qquad f_2 = \frac{\lambda_2}{\lambda}. \tag{4.13}$$

The β and α activities of the source are

$$\begin{aligned} I_\beta(t) &= -f_1 \frac{\mathrm{d}N}{\mathrm{d}t} = f_1 \overbrace{N(0)\lambda}^{I(0)} \exp(-\lambda t) = f_1 N\lambda = N\lambda_1, \\ I_\alpha(t) &= -f_2 \frac{\mathrm{d}N}{\mathrm{d}t} = f_2 N(0)\lambda \exp(-\lambda t) = f_2 N\lambda = N\lambda_2, \end{aligned} \tag{4.14}$$

and *not*, e.g., $I_\beta(t) = f_1 I(0) \exp(-\lambda_1 t)$! The mean lifetime of the source is the same for all decay channels.

Consider now a reaction chain. As a first example let us deal with neutron activation and decay, for example

$$^{23}\mathrm{Na} + n \to {}^{24}\mathrm{Na} \xrightarrow{\beta} {}^{24}\mathrm{Mg}. \tag{4.15}$$

The number of ^{24}Na nuclei is determined by the production rate p minus the decay rate. Neglecting the loss of ^{23}Na nuclei ($p =$ constant) the number of ^{24}Na nuclei changes according to

$$\frac{\mathrm{d}N}{\mathrm{d}t} = p - \lambda N \Rightarrow N(t) = \frac{p}{\lambda}[1 - \exp(-\lambda t)], \tag{4.16}$$

where λ is the decay constant of ^{24}Na. Figure 4.1(*a*) shows how $N(t)$ varies with time. At equilibrium ($t \to \infty$) the number of ^{24}Na nuclei reaches saturation with $N(\infty) = p/\lambda$. For $t \to \infty$ the activity of the ^{24}Na source is $I = \lambda N \to p$.

As a second example, consider the nuclear decay chain

$$K_1 \xrightarrow{\lambda_1} K_2 \xrightarrow{\lambda_2} K_3. \tag{4.17}$$

The populations of K_1 and K_2 nuclei change according to

$$\begin{aligned} \frac{\mathrm{d}N_1}{\mathrm{d}t} &= -\lambda_1 N_1, \\ \frac{\mathrm{d}N_2}{\mathrm{d}t} &= \lambda_1 N_1 - \lambda_2 N_2, \end{aligned} \tag{4.18}$$

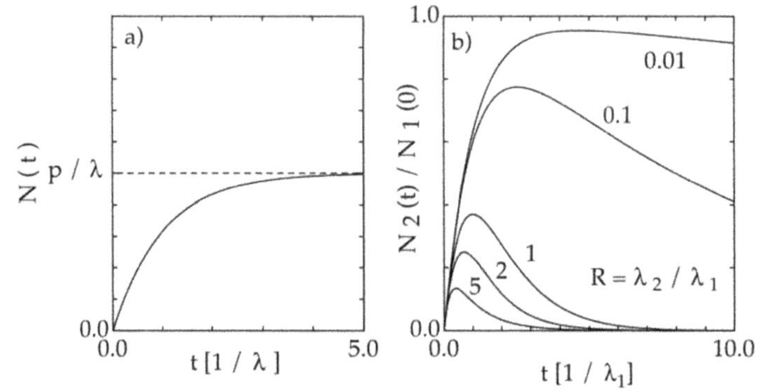

Figure 4.1. (*a*) Number of radioactive nuclei (decay constant λ) produced by a constant activation flux of rate p, as a function of time. (*b*) Relative number of radioactive nuclei K_2 (decay constant λ_2) activated by K_1 decays (decay constant λ_1).

hence

$$
\begin{aligned}
N_1(t) &= N_1(0)\exp(-\lambda_1 t), \\
N_2(t) &= \frac{\lambda_1 N_1(0)}{\lambda_2 - \lambda_1}[\exp(-\lambda_1 t) - \exp(-\lambda_2 t)],
\end{aligned}
\tag{4.19}
$$

as is easily proven by inserting into (4.18).

Figure 4.1(*b*) shows the time evolution $N_2(t)/N_1(0)$ for various ratios $R=\lambda_2/\lambda_1$. The production of K_2 nuclei reaches a maximum when $\mathrm{d}N_2/\mathrm{d}t=0$, hence at time

$$t_m = \frac{1}{\lambda_1 - \lambda_2}\ln\left(\frac{\lambda_1}{\lambda_2}\right). \tag{4.20}$$

For $R \ll 1$, corresponding to very long mean lives of K_2, t_m tends to infinity where $N_2(\infty) = N_1(0)$, and the time dependence is similar to that of figure 4.1(*a*). For the special case $R \to 1$ ($\lambda_1 = \lambda_2 + \epsilon$ with $\epsilon \to 0$) one obtains with $\lambda \equiv \lambda_2$:

$$\frac{N_2(t)}{N_1(0)} = \lambda t\exp(-\lambda t), \qquad t_m = \frac{1}{\lambda}. \tag{4.21}$$

How is a mean lifetime measured? We distinguish four cases:

1. The mean life is *short* with respect to the measurement time and there is only one possible decay channel: one measures the counting rate

$$n(t) = KI = KI(0)\exp(-\lambda t) \tag{4.22}$$

as a function of time. The constant $K \equiv (\Omega/4\pi)\eta$, which depends on the solid angle Ω spanned by the detector and on its efficiency η, does not need to be known. The detector rate $n(t)$ is measured in time bins Δt and the average of the distribution is calculated, or the slope λ is determined by plotting $n(t)$ on a logarithmic scale. The mean lifetime τ is then obtained from (4.4).

2. The mean life is *long* with respect to the measurement time and there is only one possible decay channel. In this case the counting rate stays nearly constant during the experiment. One measures the activity of the source and derives the decay constant λ from (4.8). However, the solid angle Ω and detection efficiency η must be known (factor K), as well as the total number of radioactive nuclei (e.g. from mass spectrometry).
3. The mean life is *short* and there are *several* decay channels: one measures as for case 1 the mean lifetime. The (often channel-dependent) detector efficiency does not matter since λ does not depend on the decay channel. However, since the partial decay constants λ_i are often required for comparison with theoretical predictions, one needs in addition the branching ratios f_i (4.13). The detector then needs to detect all channels with known efficiencies η_i.
4. The mean life is *long* and there are *several* decay channels: one proceeds as for case 2 and determines λ_i from the activity I_i of the source for channel i:

$$I_i = \frac{\lambda_i N}{K_i}. \tag{4.23}$$

This requires the number of radioactive nuclei N to be known, as well as the constants K_i for every channel. The decay constant λ is then determined by summing the partial decay constants λ_i.

4.2 Partial and total widths

The decay constant λ and the partial decay constants λ_i are also called *total* and *partial widths*, respectively. The decay *amplitude* associated with the decay probability $f(t)$ in (4.3) is

$$a(t) = \sqrt{\lambda}\, \exp\left(-\frac{\lambda t}{2}\right), \tag{4.24}$$

since

$$f(t) = a(t)^2. \tag{4.25}$$

The Fourier transform $A(\omega)$ of the amplitude $a(t)$ is

$$A(\omega) = \sqrt{\lambda} \int_0^\infty \exp\left(-\frac{\lambda t}{2}\right) \exp(\mathrm{i}\omega t)\, \mathrm{d}t = \frac{\sqrt{\lambda}}{\frac{\lambda}{2} - \mathrm{i}\omega}. \tag{4.26}$$

The intensity is described by the *Breit–Wigner* (or Cauchy) frequency distribution

$$I(\omega) = |A(\omega)|^2 = \frac{\lambda}{\omega^2 + \frac{\lambda^2}{4}}. \tag{4.27}$$

Let us define ω as the difference between the *observed* energy E and the mean energy E_0 of the decaying state, $\omega \equiv E - E_0$. For $\lambda = 0$ the frequency distribution

(4.27) becomes a Dirac δ-function, $I(\omega) \propto \delta(\omega)$, hence a *stable* state has the *sharp* energy $E = E_0$. On the other hand, for $\lambda \neq 0$ one finds from (4.27) the energy distribution

$$I(E) = \frac{\Gamma}{2\pi}\left(\frac{1}{(E - E_0)^2 + \Gamma^2/4}\right), \tag{4.28}$$

where we have introduced the *full width at half maximum*

$$\Gamma \equiv \lambda, \tag{4.29}$$

and have included a normalization constant $\frac{\Gamma}{2\pi}$, so that

$$\int_0^\infty I(E)\,\mathrm{d}E = 1. \tag{4.30}$$

The Breit–Wigner distribution (4.28) describes the observed energy distribution (line shape) of the decaying state. The decay constant λ, which is equal to the *total* width Γ of the energy distribution (figure 4.2), describes the uncertainty in the energy of the decaying state. The longer the mean life, the smaller the total width, since from (4.4)

$$\boxed{\tau = \frac{1}{\Gamma}}. \tag{4.31}$$

The relation (4.31) between total width and mean life also follows from the uncertainty relation

$$\Delta E \cdot \Delta t \simeq \frac{\hbar}{2} = \frac{1}{2}. \tag{4.32}$$

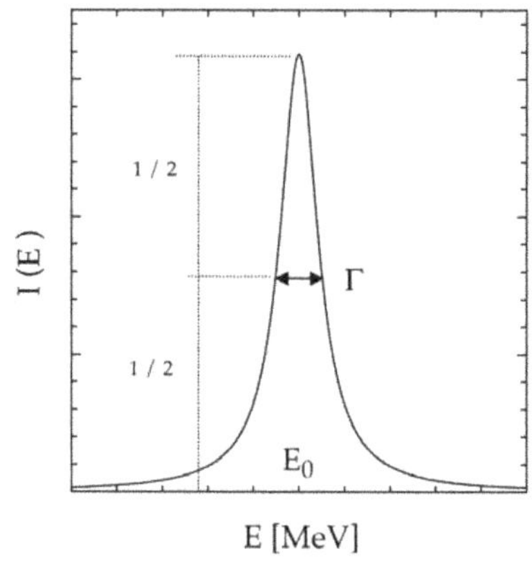

Figure 4.2. Breit–Wigner distribution.

The variance Δt (uncertainty in the decay time) is given by

$$(\Delta t)^2 = \int_0^\infty (t - \tau)^2 \lambda \exp(-\lambda t)\, dt = \tau^2 \Rightarrow \Delta t = \tau. \tag{4.33}$$

Equation (4.31) then follows from (4.32) with $\Delta E = \frac{\Gamma}{2}$.

The partial decay constants λ_i are also called *partial widths*, $\Gamma_i \equiv \lambda_i$. We obtain from (4.12) for n decay channels the total width

$$\Gamma = \sum_{i=1}^{n} \Gamma_i. \tag{4.34}$$

Let us discuss two examples:

1. The mean lifetime of the neutron is $\tau = 880.3 \pm 1.1$ s [1] to which corresponds a total width

$$\Gamma = \frac{1}{\tau} = 1.1 \times 10^{-3}\ \mathrm{s}^{-1} = 3.7 \times 10^{-27}\ \mathrm{fm}^{-1} = 7.3 \times 10^{-25}\ \mathrm{MeV} \tag{4.35}$$

   (with $1\ \mathrm{s} = 3 \times 10^{23}$ fm), which is much smaller than the neutron mass (939 MeV). Hence the mass of the neutron is not better defined than its total width. Performing a more accurate measurement than (4.35) would be senseless (the mass of the neutron is currently known with an experimental uncertainty of ±21 eV [1]).
2. Low energy proton–antiproton annihilation produces mainly pions. Choosing annihilation events with three pions ($\bar{p}p \to \pi^+\pi^-\pi^0$), measuring their momenta $\vec{p}_i$ and plotting the three-pion invariant mass (see section 8.2), one observes intermediate resonances which decay into $\pi^+\pi^-$ (figure 4.3). For example, the peak on the left is the ρ meson with a mass of 770 MeV and a width $\Gamma = 150$ MeV. The experimental resolution on the $\pi^+\pi^-$ mass is much smaller than the resonance width in figure 4.3, hence the

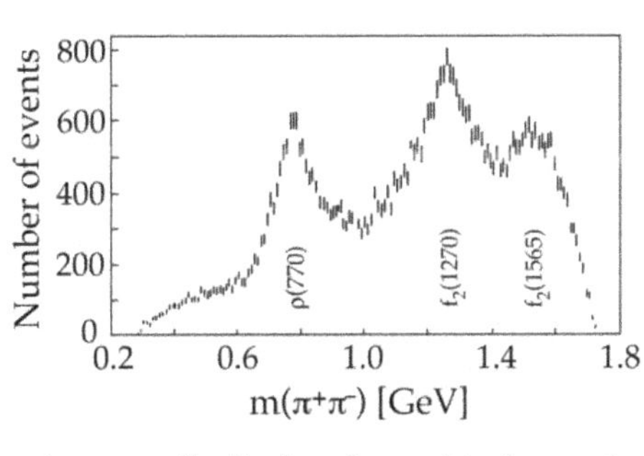

Figure 4.3. Resonances in the $\pi^+\pi^-$ mass distribution observed in the reaction $\bar{p}p \to \pi^+\pi^-\pi^0$ with stopping antiprotons. After [2].

total width can be obtained directly from the data. The mean life of the ρ meson is

$$\tau = \frac{1}{\Gamma} = 6.7 \times 10^{-3}\ \mathrm{MeV}^{-1} = 1.3\ \mathrm{fm} = 4.4 \times 10^{-24}\ \mathrm{s}. \tag{4.36}$$

Hence the mean life of extremely short-lived states is obtained from their total widths.

4.3 Poisson statistics

Consider a radioactive source (decay constant λ) with $N(0)$ radioactive nuclei at time $t = 0$. The average number of counts registered by a detector with detection probability $K = (\Omega/4\pi)\eta$ (solid angle × efficiency) between time t and $t + \Delta t$ is

$$n = N(0)\lambda \exp(-\lambda t) K \Delta t \tag{4.37}$$

(for small enough Δt). The probability $P(n, k)$ that k counts are registered during Δt follows a *Poisson* distribution:

$$\boxed{P(n, k) = \frac{n^k}{k!} \exp(-n)}. \tag{4.38}$$

Indeed, the mean number of counts is

$$\langle k \rangle = \sum_{k=0}^{\infty} k \frac{n^k}{k!} \exp(-n) = n \tag{4.39}$$

with variance

$$\sigma^2 = \sum_{k=0}^{\infty} (k - \langle k \rangle)^2 \frac{n^k}{k!} \exp(-n) = n \Rightarrow \sigma = \sqrt{n}, \tag{4.40}$$

where σ is *one standard deviation* on the number of counts.

Figure 4.4 shows as an example the Poisson distribution for $n = 3.45$. Note the asymmetry of the distribution around its mean value, which is larger than the most

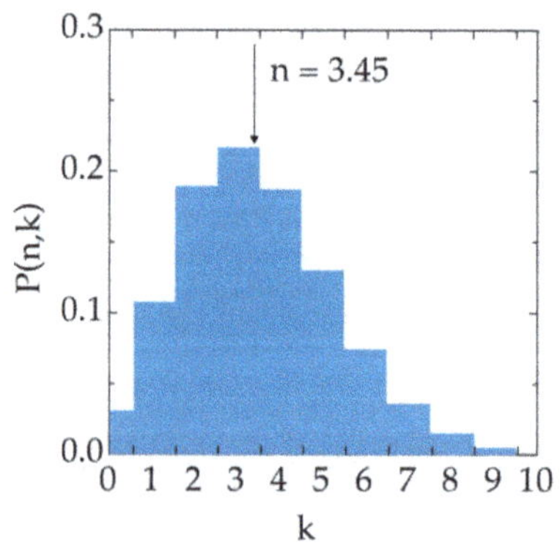

Figure 4.4. Poisson distribution with mean $n = 3.45$.

probable value ($k = 3$). Now, if the experiment is carried out only once and n counts are registered, n is an estimate for the mean and one writes as the *result* of the measurement

$$n \pm \sqrt{n}\,. \tag{4.41}$$

For large n, e.g. $n \geqslant 10$, the Poisson distribution becomes a Gaussian distribution:

$$G(n, k) = \frac{1}{\sqrt{2\pi}\,\sigma} \exp\left[-\frac{(n-k)^2}{2\sigma^2} \right] \quad \text{with} \quad \sigma = \sqrt{n}\,. \tag{4.42}$$

In this case, the *true* number of counts lies within $\pm\sqrt{n}$ of the measured value n, with a probability of 68.3%.

References

[1] Olive K *et al* (Particle Data Group) 2014 *Chin. Phys.* C **38** 090001
[2] May B *et al* 1990 *Z. Phys.* C **46** 191

IOP Publishing

Nuclear and Particle Physics

Claude Amsler

Chapter 5

Nuclear stability

Figure 5.1 shows the *term diagrams* for α-, β- and γ-decays. The atomic number Z is plotted horizontally and the atomic mass difference is plotted vertically. In α-decay (figure 5.1(a)) the radioactive nucleus emits an α particle, whereby Z decreases by two and the mass number A by four. The energy difference between mother and daughter atoms is, with $\mathcal{M}(Z, A) = M(Z, A) + Zm_e$,

$$\begin{aligned} Q &= \mathcal{M}(Z, A) - \mathcal{M}(Z-2, A-4) - \mathcal{M}(^4\text{He}), \\ &= M(Z, A) - M(Z-2, A-4) - M_\alpha, \\ &= B(Z-2, A-4) + B_\alpha(28.3\ \text{MeV}) - B(Z, A). \end{aligned} \tag{5.1}$$

The *Q-value* of an exothermic reaction ($Q > 0$) is the energy released, distributed among all daughter products in the form of kinetic energy. Here Q is equal to the kinetic energy of the α particle plus the recoil energy of the daughter nucleus. Since mother and daughter nuclei have well-defined masses, the α particles are *mono-energetic* with kinetic energies up to a few MeV. This was already known in the 1920s from the constant *range* of α particles observed in cloud chambers and in

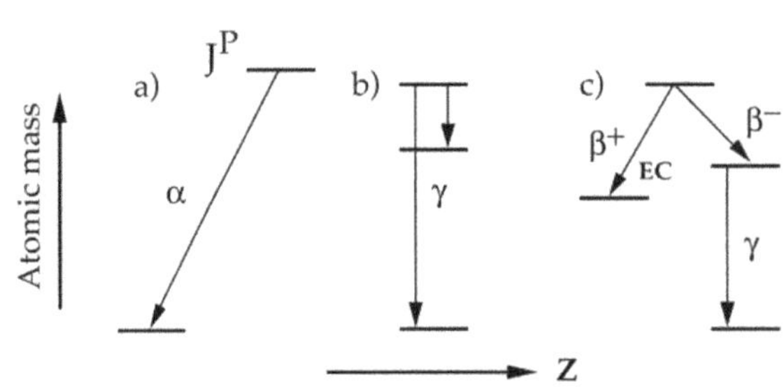

Figure 5.1. Term diagrams for α, γ and β-decay (EC = electron capture). Atomic masses versus atomic number are plotted. The levels have spin and parity J^P.

 © IOP Publishing Ltd 2015

Figure 5.2. A cloud chamber photograph of tracks produced by a polonium α source. Reproduced with kind permission from [1]. Copyright 1926 Springer Science and Business Media.

nuclear emulsions. (Nuclear emulsions were developed by Marietta Blau and Hertha Wambacher at the Institute for Radium Research in Vienna to study cosmic rays.)

Figure 5.2 shows a picture of the typical ionization pattern left by a radioactive α source in a cloud chamber. A cloud chamber contains a gas saturated, for instance, by alcohol vapour. The rapid expansion of the gas leads to condensation along the ionization path. The tracks from the collimated polonium source in figure 5.2 stem from $^{214}\text{Po} \rightarrow {}^{210}\text{Pb} + \alpha$-decays and are all of the same length, showing that the α particles are monoenergetic ($Q = 7.8$ MeV). We shall return to α-decay in section 5.3.

5.1 γ-decay

In γ-decay (figure 5.1(*b*)) a high energy photon is emitted by an excited state of the nucleus. The photon is *monochromatic* with energy E_γ in the range keV to MeV. The energy released (Q, which is also the difference between the nuclear masses) is mainly carried by the photon, while the recoil energy of the daughter nucleus is usually negligible.

The γ radiation can be expanded into multipole moments. The orbital angular momentum $\vec{L}$ carried by the photon with spin $\vec{s}$ is (angular momentum conservation)

$$\vec{J}_i = \vec{J}_f + \vec{\ell}, \tag{5.2}$$

where $\vec{J}_i$ and $\vec{J}_f$ are the nuclear spins in the initial and final states, and $\vec{\ell} = \vec{L} + \vec{s}$. We define the quantization axis z in the direction of the emitted photon. The projection m of the angular momentum $\vec{\ell}$ on the z-axis is given by the projection of $\vec{s}$ ($m = \pm 1$), since $L_z = 0$ and the photon spin is either parallel or antiparallel (but not perpendicular) to its flight direction (see section 11.6 on the helicity of the photon). Since $m = \pm 1$ the quantum number ℓ (*multipole order*) must be at least equal to 1. Then from (5.2) one obtains, following the rules of quantum mechanics,

$$\left|J_i - J_f\right| \leqslant \ell \leqslant J_i + J_f, \quad \ell \geqslant 1, \tag{5.3}$$

where J_i and J_f are the quantum numbers of the mother and daughter nuclei, respectively. Therefore γ transitions between two spin-0 nuclei are impossible.

The nuclear wave functions are either even (+1) or odd (−1) under space inversion (parity). Parity conservation in the electromagnetic interaction (section 11.3) requires the parities of the mother and daughter nuclei to be equal ($\Delta P = P_i P_f = 1$ or $\Delta P =$ 'no') for L odd, or opposite ($\Delta P = -1$ or $\Delta P =$ 'yes') for L even since the internal parity of the photon is −1 (section 11.4). The multipole transitions are labelled $E\ell$ (for electric) and $M\ell$ (for magnetic). The parity changes according to:

$$\begin{aligned} E\ell &: \Delta P = (-1)^{\ell}, \\ M\ell &: \Delta P = (-1)^{\ell+1}. \end{aligned} \tag{5.4}$$

Table 5.1 lists the lowest order multipoles. Since the spin of the photon is $s = 1$ the multipole order ℓ varies from $|L - 1|$ to $L + 1$ with $\ell \geqslant 1$. A few examples are

$$\begin{aligned} 1^+ \to 0^+ &: \ell = 1, \Delta P = \text{no, hence } M1, \\ 1^- \to 0^+ &: \ell = 1, \Delta P = \text{yes, hence } E1, \\ 0 \to 0 &: \gamma \text{ transition forbidden}, \\ \tfrac{3}{2}^+ \to \tfrac{1}{2}^+ &: \ell = 1 \quad \text{or} \quad 2, \Delta P = \text{no, hence } M1 \text{ or } E2. \end{aligned}$$

Highly excited nuclei have short lifetimes (figure 5.3). The mean lifetime increases very rapidly with multipole order and $M\ell$ is suppressed compared to $E\ell$. The mean lifetimes of the electric and magnetic transitions τ_E and τ_M can be approximated by [3, 4]

$$\tau_E(E_\gamma, \ell) \propto \frac{1}{E_\gamma^{2\ell+1}} \frac{1}{A^{\frac{2\ell}{3}}} \quad \text{and} \quad \tau_M(E_\gamma, \ell) \simeq 4.5 A^{\frac{2}{3}} \cdot \tau_E(E_\gamma, \ell). \tag{5.5}$$

High-spin nuclei which decay via high multipole orders (such as $M4$, $E5$, etc), have long lifetimes (typically $\tau \geqslant 10\ \mu s$) and are called *isomers*.

The multipole order of the γ radiation can be established by measuring the *angular distribution* of the photon. This requires the spin of the mother nucleus to be aligned (e.g. by using a magnetic field). The spin of the mother nucleus can then be determined from (5.3), provided that the spin of the daughter nucleus is known.

Table 5.1. Multipole transitions; L is the orbital angular momentum carried by the photon and ℓ is the multipole order.

L	ℓ	ΔP	Multipole
0	1	yes	$E1$
1	1, 2	no	$M1$, $E2$
2	1, 2, 3	yes	$E1$, $M2$, $E3$
3	2, 3, 4	no	$E2$, $M3$, $E4$

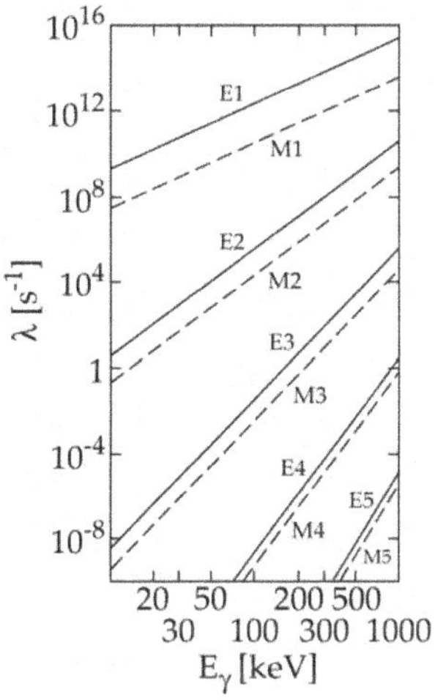

Figure 5.3. Decay constant λ versus E_γ for various multipole transitions and for $A \simeq 100$. The slope increases proportionally to $2\ell + 1$. After [2].

The mean lifetime is determined either by measuring the intensity of the γ line as a function of time, or directly from the line width (Mössbauer resonance absorption). The former method is suitable for excited nuclei produced by a beam impinging on a target. The nuclei emitted from the target are brought to rest in an absorber. One measures the intensity of the Doppler-shifted γ line from decay in flight and compares it with the intensity of the unshifted line from decay at rest in the absorber. The decay constant is then determined by varying the distance between the production target and absorber.

An excited nucleus can also decay by *internal conversion*, in which the energy is transferred directly to a shell electron. The emitted electron is *monoenergetic* with the kinetic energy

$$T_e = Q - B_e, \tag{5.6}$$

where we have neglected the nuclear recoil energy. A discrete spectrum arises with electron emission from the K-shell (1 line), the L-shell (3 lines), the M-shell (5 lines), etc. Internal conversion is particularly important for heavy nuclei (probability proportional to Z^4), at small Q-values, and for high multipole transitions (figure 5.4), for which the contribution from the K-shell dominates. The intensity ratios of the conversion lines from various shells also provide information about the multipole order of the transition.

Following internal conversion, the hole left in the electronic shell is filled with x-ray transitions (monochromatic photons) which in turn may eject electrons from higher shells. The energy of these monoenergetic *Auger* electrons is in the range eV to keV.

Internal pair production is also possible: for $Q > 2m_e$ an e^+e^- pair can be produced in the Coulomb field of the nucleus. For example, the first excited state of ^{16}O (0^+, $Q = 6$ MeV) decays by by e^+e^- emission to the 0^+ ground state since γ transitions between spin-0 nuclei are forbidden.

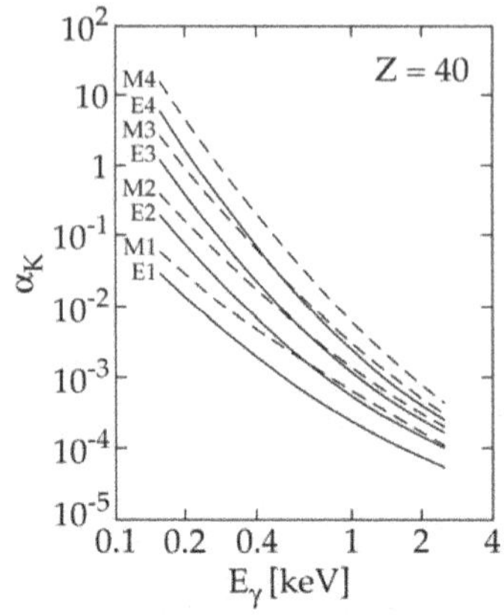

Figure 5.4. Conversion factor for various multipole transitions from the *K*-shell and for $Z = 40$. The ratio of conversion to γ emission probabilities is plotted. After [4].

5.2 β-decay

In β^--decay (figure 5.1(*c*)) the radioactive nucleus emits an electron, thus increasing the atomic number by one unit. Because α and γ spectra had been observed to be discrete, it was assumed at the beginning of the 20th century that electrons from β-decay would also be monoenergetic. Indeed, sharp lines had been observed in spectrographic analyses but, as discovered later, they were due to internal conversion from the excited daughter nucleus left after β-decay (figure 5.5).

Monoenergetic electrons also led to difficulties with angular momentum conservation. For example, the decay

$$^{14}\text{C}\,(1^+) \rightarrow {}^{14}\text{N}\,(0^+) + e^-$$

would produce a monoenergetic electron, but would violate angular momentum conservation due to the fractional electron spin.

James Chadwick proved in 1914 with magnetic analysis and an array of Geiger counters that the β spectrum was *continuous* (figure 5.5). The riddle of angular

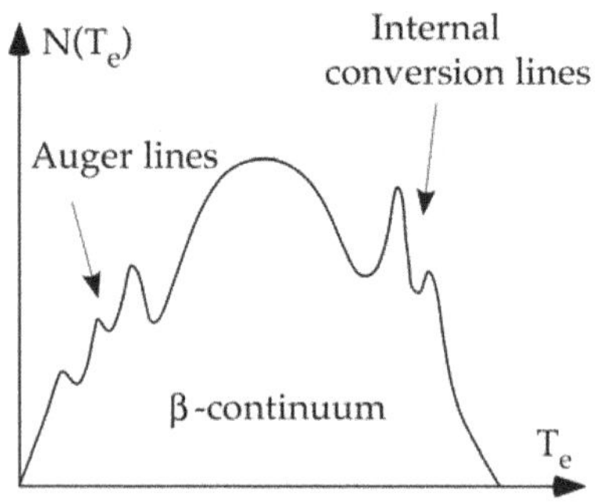

Figure 5.5. Energy distribution of the electron in β-decay with superimposed internal conversion and Auger electrons from the subsequent γ-decay.

momentum conservation was solved in 1930 by Wolfgang Pauli. He postulated that a third invisible object with spin $\frac{1}{2}$ was emitted in the process: the neutrino. The released energy was therefore shared between three particles, which explained why the β spectrum was continuous[1]. In the modern formulation a neutron in the nucleus decays as

$$n \to pe^- \bar{\nu}_e. \tag{5.7}$$

This process is also possible for free neutrons. The mean lifetime of the neutron is 880 s and the Q-value is (assuming that $m_\nu = 0$)

$$Q = M_n - M_p - m_e = 0.78 \text{ MeV}. \tag{5.8}$$

In contrast, the β^+-decay of the proton

$$p \to ne^+ \nu_e, \tag{5.9}$$

with the emission of a positron, is excluded since

$$Q = M_p - M_n - m_e = -1.80 \text{ MeV} < 0. \tag{5.10}$$

The decay (5.9) is nevertheless possible for bound protons inside the nucleus (figure 5.1(c)). The positron was discovered by Carl Anderson in 1932 in cosmic radiation (section 15.2), and β^+-decay was observed by Irène and Frédérique Joliot-Curie in 1934 using artificially produced radioactive elements ($\alpha\,^{27}\text{Al} \to {}^{30}\text{P}\, n$, ${}^{30}\text{P} \to {}^{30}\text{Si}\, e^+ \nu_e$). Positrons from β^+-decay had not been detected at the beginning of the 20th century, because most radioactive nuclei available in nature stem from the α-decays of heavy elements. The neutron excess following α-decay is then compensated by conversion into protons via β^--decay.

The *electron capture* (EC) process

$$pe^- \to n\nu_e \tag{5.11}$$

competes with β^+-decay (figure 5.1(c)). An electron (usually from the K-shell) is captured by a proton in the nucleus. The ensuing gap is filled by x-ray emission (monochromatic γ) or by Auger emission. For electron capture by a free proton the Q-value would be equal to -0.78 MeV, hence the hydrogen atom cannot decay by electron capture. We owe our existence to this tiny negative Q-value!

The process of *internal bremsstrahlung* also occurs during β-decay and electron capture, produced by the acceleration of the e^- (or e^+) in the Coulomb field of the nucleus.

[1] In fact, Pauli postulated that the energy was distributed among an electron in the nucleus, a proton, and a light neutral spin-$\frac{1}{2}$ particle with mass $m \leqslant 10\,m_e$ which he called a 'neutron'. However, as early as 1920, Rutherford had suspected that a neutral constituent in the nucleus (today's neutron) decayed into a proton with the emission of an electron. Reaction (5.7) was established after the discovery of the neutron by Chadwick with Fermi's formulation of a theory for β-decay in 1934. The neutrino was detected directly by Frederick Reines and Clyde Cowan in 1956.

The Q-values for nuclear β-decay are (assuming $m_\nu = 0$)

$$\begin{aligned} Q(\beta^-) &= M(Z, A) - M(Z+1, A) - m_e, \\ Q(\beta^+) &= M(Z, A) - M(Z-1, A) - m_e, \\ Q(\text{EC}) &= M(Z, A) + m_e - M(Z-1, A). \end{aligned} \tag{5.12}$$

In $\beta^\pm$-decay the Q-value is the maximum possible energy of the emitted electron or positron (apart from the usually very small nuclear recoil energy). Computing the atomic masses by adding the contribution from electrons, while neglecting the differences in electronic binding energies, leads to the Q-values

$$\begin{aligned} Q(\beta^-) &= \mathcal{M}(Z, A) - \mathcal{M}(Z+1, A), \\ Q(\beta^+) &= \mathcal{M}(Z, A) - \mathcal{M}(Z-1, A) - 2m_e, \\ Q(\text{EC}) &= \mathcal{M}(Z, A) - \mathcal{M}(Z-1, A). \end{aligned} \tag{5.13}$$

In the term diagram (figure 5.1(c)) the available energy is simply given by the difference of *atomic* masses, except for β^+-decay, for which one has to deduct $2m_e$, $Q(\text{EC}) = Q(\beta^+) + 2m_e$. Electron capture is therefore favoured over β^+-decay.

Let us now move on the $M(Z, N)$ surface for a given isobar (fixed A) around the stability line of figure 3.5, towards larger or smaller values of $N = A - Z$. Consider first nuclei with an odd mass number A, hence odd–even nuclei. The pairing term $\delta(Z, A)$ vanishes according to (3.40) and $M(Z, N)$ describes a parabola as a function of Z, with curvature

$$\frac{\mathrm{d}^2 M}{\mathrm{d}Z^2} = 2a_C A^{-\frac{1}{3}} + 8a_A A^{-1} > 0. \tag{5.14}$$

The stable nucleus is located near the minimum (figure 5.6(a)). Excited nuclei decay by β emission or electron capture and therefore odd–even nuclei have only *one* stable *isobar*.

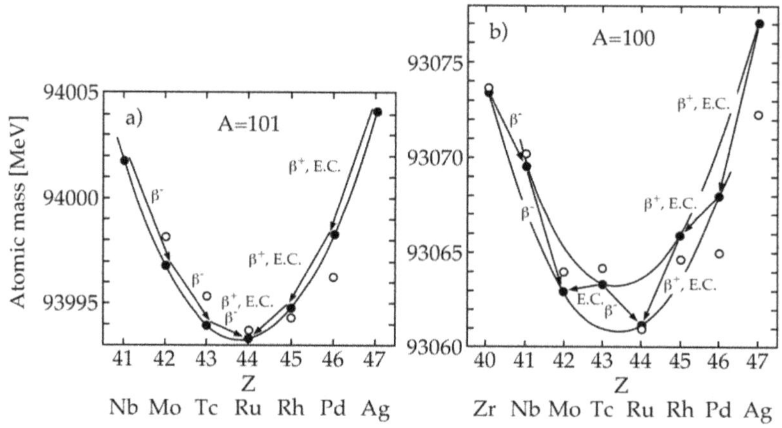

Figure 5.6. Stability parabolas for isobars around $A = 100$: (a) odd–even nuclei; (b) odd–odd nuclei (top), even–even nuclei (bottom). The open circles are measurements and the closed circles are calculations from the mass formula. Note that the vertical scale is suppressed. After [5].

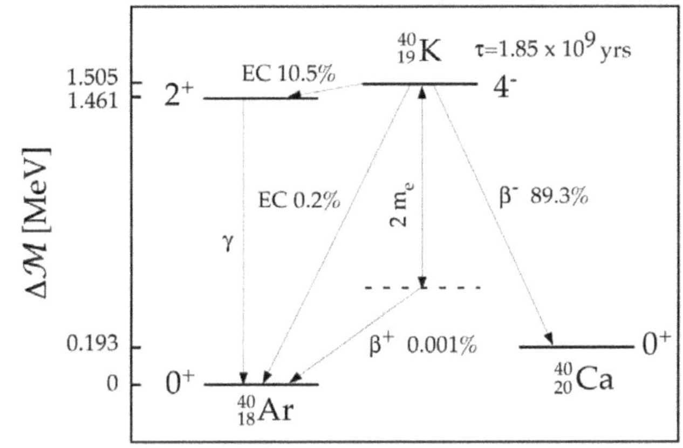

Figure 5.7. Term diagram for $^{40}_{19}$K (difference between atomic masses). The vertical arrow indicates that $2m_e$ must be subtracted to obtain the Q-value. After [5].

For A even the pairing energy is positive for even–even nuclei and negative for odd–odd ones, which leads to two parabolas in the M versus Z graph (figure 5.6(b)). Odd–odd nuclei on the upper parabola do not have have any stable isobar. There are four known exceptions: the light elements $^2_1$H (deuterium), $^6_3$Li, $^{10}_5$B and $^{14}_7$N, for which the liquid drop model is in any case too simplistic. Even–even nuclei on the lower parabola have several stable isobars. These isobars, differing by two units of Z, can nonetheless decay by the simultaneous emission of two electrons (or two positrons) and two neutrinos (*double* β-decay). However, this process is extremely unlikely, so even–even isobars can be considered as stable for practical purposes. Figure 5.6(b) shows the case of

$$^{100}_{42}\text{Mo} \rightarrow {}^{100}_{44}\text{Ru} + 2e^- + 2\bar{\nu}_e$$

with a half-life of about 6×10^{20} years. Another example is

$$^{130}_{52}\text{Te} \rightarrow {}^{130}_{54}\text{Xe} + 2e^- + 2\bar{\nu}_e$$

with a half-life $(7.0 \pm 1.4) \times 10^{20}$ years [6].

Figure 5.7 shows the term diagram for the odd–odd $^{40}_{19}$K nucleus, for which the even–even neighbours to the left and right on the lower parabola are more strongly bound. This nucleus decays by e^- or e^+ emission (or electron capture). As we shall explain later (section 13.2.1) the long lifetime of 1.85×10^9 years is due to the large difference in nuclear spins between $^{40}_{19}$K and $^{40}_{18}$Ar or $^{40}_{20}$Ca. We shall return later to β-decay when discussing weak interaction in more detail (section 13.2).

5.3 α-decay

For large values of A the nucleus decays by α emission. This occurs when the Q-value (5.1) is positive. The condition is fulfilled for $A \geqslant 150$ (Sm). However, the kinetic energy T_α of the α particle is very small for $A = 150$ and hence the decay is strongly suppressed, as we shall show below.

Since $A_\alpha = 4$, there are four decay chains (series) leading to the stable Pb and Bi isotopes (figure 5.8):

$$
\begin{aligned}
&^{238}\text{U} \rightarrow {}^{206}\text{Pb}, \quad && t_{\frac{1}{2}} = 4.5 \times 10^{9}\ \text{years}, \quad && \text{uranium series},\\
&^{235}\text{U} \rightarrow {}^{207}\text{Pb}, \quad && t_{\frac{1}{2}} = 7.1 \times 10^{8}\ \text{years}, \quad && \text{actinium series},\\
&^{232}\text{Th} \rightarrow {}^{208}\text{Pb}, \quad && t_{\frac{1}{2}} = 1.4 \times 10^{10}\ \text{years}, \quad && \text{thorium series},\\
&^{237}\text{Np} \rightarrow {}^{209}\text{Bi}, \quad && t_{\frac{1}{2}} = 2.1 \times 10^{6}\ \text{years}, \quad && \text{neptunium series},
\end{aligned}
\tag{5.15}
$$

where we have started the series from the elements with the longest lifetimes. The isotope ^{209}Bi is quasi-stable with a half-life of 1.9×10^{19} years. The earthly elements ^{238}U, ^{235}U and ^{232}Th have been generated in supernovae. The lifetime of ^{237}Np is short compared to the age of the solar system and therefore the neptunium series is produced artificially.

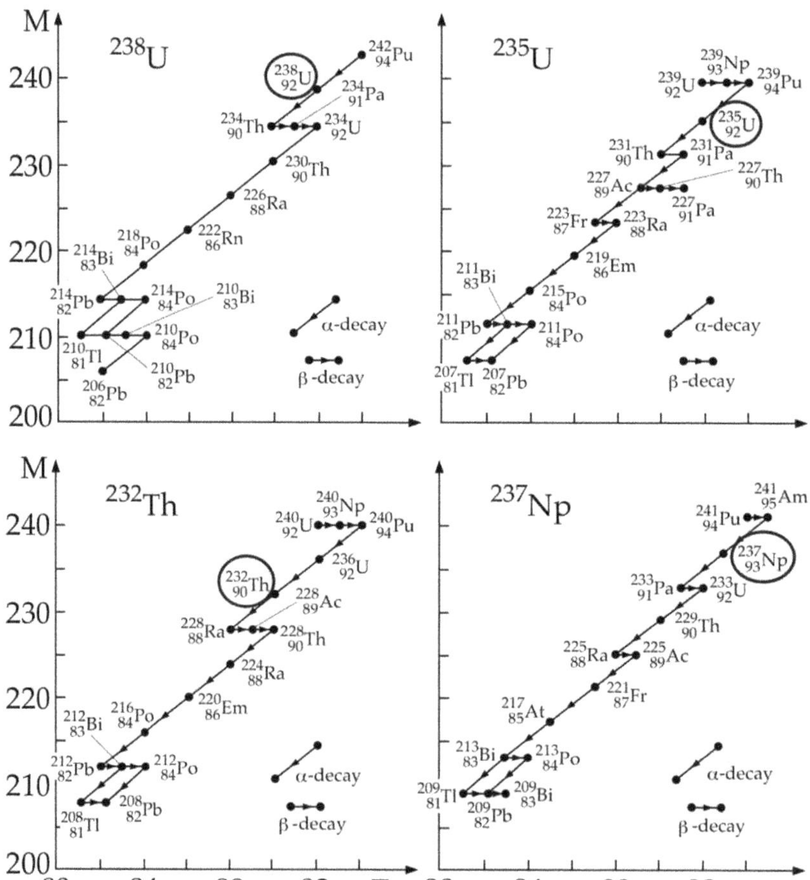

Figure 5.8. Decay charts of the four α series. After [7].

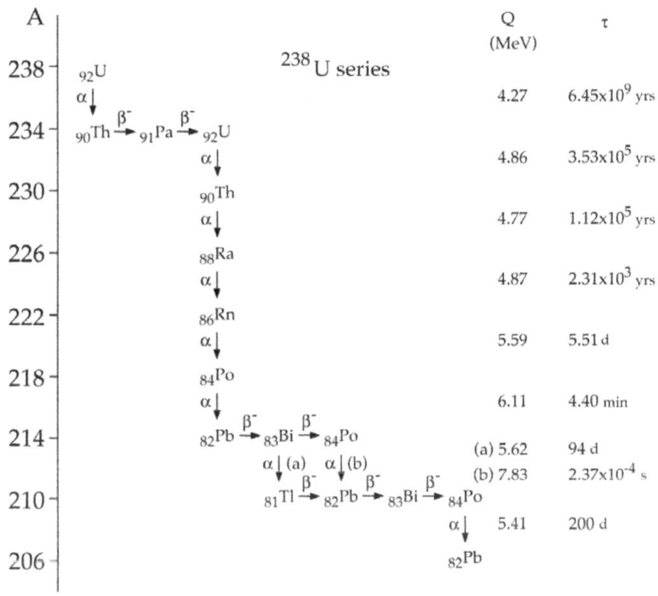

Figure 5.9. Decay chart of the ^{238}U series. After [5].

Figure 5.9 shows the uranium series with the corresponding Q-values and mean lifetimes. Through successive α emissions the nucleus moves away from the stability line. The neutron excess is then reduced by alternating β^--decays. Large differences in lifetimes are observed in the decay chain: a slight increase in Q-value leads to a massive reduction in the mean lifetime. An extreme case can be found in the thorium series:

$$\begin{aligned} ^{232}\text{Th} &\rightarrow {}^{228}\text{Ra} + \alpha : \; t_{\frac{1}{2}} = 1.4 \times 10^{10} \text{ yrs}, \\ ^{212}\text{Po} &\rightarrow {}^{208}\text{Pb} + \alpha : \; t_{\frac{1}{2}} = 3.5 \times 10^{-7} \text{ s}, \end{aligned} \tag{5.16}$$

a difference of 24 orders of magnitude!

The explanation for this astonishing energy dependence was proposed in 1929 by George Gamow. Consider the square-well nuclear potential with Coulomb repulsion shown in figure 5.10. In the nuclear well a pair of protons can fuse with a pair of neutrons to form an α particle with kinetic energy T_α. The nucleus decays immediately when T_α exceeds the potential barrier U_0. When $T_\alpha < U_0$ the α particle can still overcome the potential barrier through the *tunnel effect*. The probability to traverse the barrier increases with T_α and is proportional to the *transmission factor*

$$T = \left| \frac{\alpha_3}{\alpha_1} \right|^2, \tag{5.17}$$

where α_1 and α_3 are the amplitudes of the incoming and outgoing waves, respectively. The mean lifetime is given by

$$\frac{1}{\tau} = \lambda = \lambda_0 \left(\frac{v}{2R} \right) T, \tag{5.18}$$

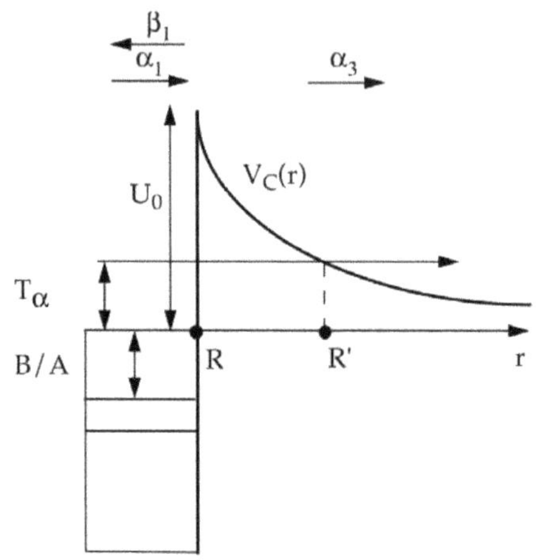

Figure 5.10. Tunnel effect in α-decay.

where λ_0 is the probability of building an α particle in the nuclear well. The factor $\frac{v}{2R}$ describes the number of trials per unit time to overcome the barrier. Let us estimate T by first solving the simpler problem of a square potential of height U_0 (figure 5.11). The solutions of the one-dimensional Schrödinger equation

$$\frac{1}{2m}\frac{d^2\psi}{dx^2} + \left[T_\alpha - V(x)\right]\psi = 0 \tag{5.19}$$

in the intervals I, II, III are

$$\begin{aligned} \psi_I &= \alpha_1 \exp(ik_1x) + \beta_1 \exp(-ik_1x), & k_1 &= \sqrt{2mT_\alpha}, \\ \psi_{II} &= \alpha_2 \exp(-k_2x) + \beta_2 \exp(+k_2x), & k_2 &= \sqrt{2m(U_0 - T_\alpha)}, \\ \psi_{III} &= \alpha_3 \exp(ik_3x), & k_3 &= \sqrt{2mT_\alpha}, \end{aligned} \tag{5.20}$$

for an α particle with mass m.

The amplitudes of the incoming and reflected waves are α_i and β_i, respectively. Continuity at the boundaries, which requires that

$$\begin{aligned} \psi_I &= \psi_{II}, & \psi'_I &= \psi'_{II} \quad \text{at } x = 0, \\ \psi_{II} &= \psi_{III}, & \psi'_{II} &= \psi'_{III} \quad \text{at } x = d, \end{aligned} \tag{5.21}$$

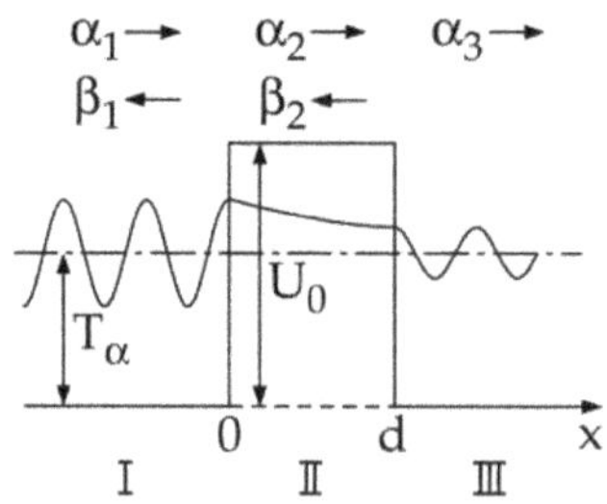

Figure 5.11. Tunnel effect on a square potential barrier.

leads to four equations, hence four ratios between amplitudes. In particular, the transmission factor is

$$T = \left|\frac{\alpha_3}{\alpha_1}\right|^2 = \frac{1}{1 + \frac{U_0^2}{U_0^2 - (2T_\alpha - U_0)^2}\sinh^2(k_2 d)}. \tag{5.22}$$

For a *thick barrier* $y \equiv k_2 d \gg 1$ we have that

$$\sinh y = \frac{\exp(y)[1 - \exp(-2y)]}{2} \simeq \frac{\exp(y)}{2}, \tag{5.23}$$

so that

$$T \simeq 4\left[\frac{U_0^2 - (2T_\alpha - U_0)^2}{U_0^2}\right]\exp(-2k_2 d). \tag{5.24}$$

The momentum-dependent exponential term dominates the lifetime variation and hence in rough approximation

$$T \propto \exp\left[-2\sqrt{2m(U_0 - T_\alpha)}\,d\right]. \tag{5.25}$$

The solution of the three-dimensional problem is derived from the radial equation

$$\frac{1}{2m}\frac{\mathrm{d}^2 u}{\mathrm{d}r^2} + \left[T_\alpha - V(r) - \frac{\ell(\ell+1)}{2mr^2}\right]u = 0 \tag{5.26}$$

with $u(r) = rR(r)$ and $\psi(\vec{r}) = R(r)Y_\ell^m(\theta, \phi)$. This equation has same form as (5.19), apart from the *centrifugal term* which increases the barrier height. The quantum number ℓ is the angular momentum carried by the α particle. Therefore, our derivation can be applied directly as long as $\ell = J_i - J_f = 0$ since the spin of the α particle is zero. This is, for example, the case for transitions between spin-0 nuclei. We now decompose the Coulomb barrier in figure 5.10 into a series of square potentials (heights $V_C(r)$ and widths $\mathrm{d}r$) and multiply the corresponding transmission probabilities. From (5.25) one obtains

$$\boxed{T \propto \exp(-G)} \quad \text{with } G = 2\int_R^{R'} \sqrt{2m\left(\frac{2Ze^2}{4\pi\epsilon_0 r} - T_\alpha\right)}\,\mathrm{d}r, \tag{5.27}$$

where G is the *Gamow factor*. We have used $z = 2$ for the α particle and have approximated the atomic number of the daughter nucleus by $Z - 2 \simeq Z$. Integrating gives

$$G = 2\sqrt{\frac{2m}{T_\alpha}}\frac{2Ze^2}{4\pi\epsilon_0}\left[\arccos\sqrt{\frac{R}{R'}} - \sqrt{\frac{R}{R'}\left(1 - \frac{R}{R'}\right)}\right]. \tag{5.28}$$

For $R' \gg R$ (see figure 5.10) the square bracket is roughly equal to $\pi/2$ and one obtains

$$G = 2\pi Z\alpha\sqrt{\frac{2m}{T_\alpha}} \quad \text{with } T \propto \exp(-G). \tag{5.29}$$

We are now in a position to calculate the decay constant

$$\lambda \simeq C(T_\alpha, R)\exp(-G), \tag{5.30}$$

where the slowly varying function C contains the factors λ_0 and $v/2R$ (5.18), as well as the neglected prefactor in (5.24). Hence

$$\ln(\lambda\tau_0) = C'(T_\alpha, R) - \overbrace{2\pi\alpha\sqrt{2m}}^{a}\frac{Z}{\sqrt{T_\alpha}}, \tag{5.31}$$

where the constant τ_0 (in units of time) is arbitrary. The *Geiger–Nuttal* rule for the mean lifetime reads

$$\boxed{\ln\frac{\tau}{\tau_0} = -C' + a\frac{Z}{\sqrt{T_\alpha}}}. \tag{5.32}$$

The mean lifetime increases *exponentially* with decreasing $\sqrt{T_\alpha}$. The numerical value of the slope a is

$$a = 3.97\ \text{MeV}^{\frac{1}{2}}, \tag{5.33}$$

when T_α is expressed in MeV. Figure 5.12 shows how ln $(\lambda\tau_0)$ varies with $(Z/\sqrt{T_\alpha})$ for the uranium and thorium series. One sees that Gamow's theory reproduces roughly the mean lifetimes of α-decays over 24 orders of magnitude. Local deviations as large as 2–3 orders of magnitude are not surprising in view of the simplifying assumptions made in our derivation (such as the constant term C' in (5.32)). As an illustration, let us calculate the ratio of lifetimes for the decays

$$\begin{aligned} &{}^{238}_{92}\text{U},\ Z-2=90, && T_\alpha \simeq 4.27\ \text{MeV} \Rightarrow \frac{Z}{\sqrt{T_\alpha}} = 43.55\ \text{MeV}^{-\frac{1}{2}},\\ &{}^{226}_{88}\text{Ra},\ Z-2=86, && T_\alpha \simeq 4.87\ \text{MeV} \Rightarrow \frac{Z}{\sqrt{T_\alpha}} = 38.97\ \text{MeV}^{-\frac{1}{2}}. \end{aligned} \tag{5.34}$$

(using the data from figure 5.9). The difference of 4.48 $\text{MeV}^{-\frac{1}{2}}$ leads with (5.32) to the ratio

$$\ln\frac{\tau(\text{Ra})}{\tau(\text{U})} = \ln\tau(\text{Ra}) - \ln\tau(\text{U}) = -3.97\times 4.58 \Rightarrow \frac{\tau(\text{Ra})}{\tau(\text{U})} = 1.3\times 10^{-8}, \tag{5.35}$$

while the measured value is 3.6×10^{-7}.

When several α transitions of the parent nucleus are possible, as in figure 5.13, the α emission with the largest Q-value dominates and λ in (5.31) is the corresponding partial width. For transitions between nuclei with unequal spins the α particle carries

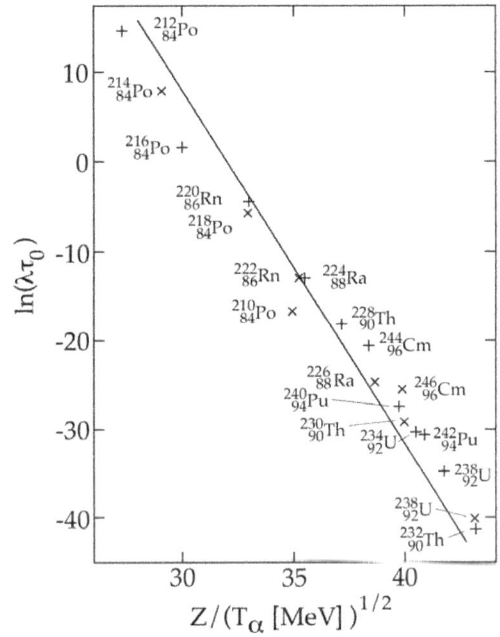

Figure 5.12. Mean lifetime versus $Z/\sqrt{T_\alpha}$ (τ_0 is here equal to 2.7×10^{-8} years). The curve with a slope of $-3.97\,\mathrm{MeV}^{\frac{1}{2}}$ is a fit using (5.31). × = uranium series; + = thorium series. After [5].

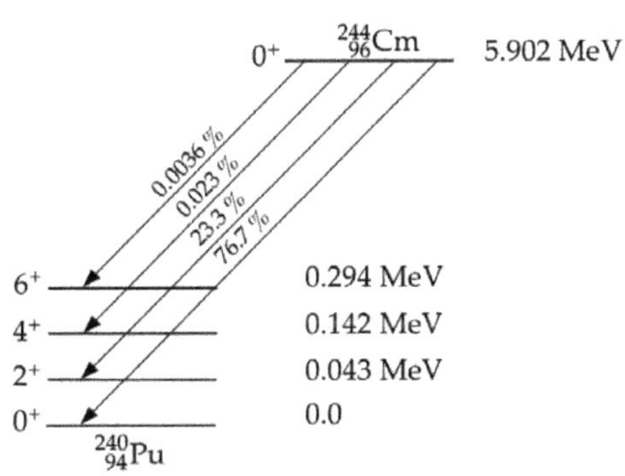

Figure 5.13. Term diagram of ^{244}Cm decay into ^{240}Pu. After [5].

angular momentum ($\ell > 0$) which increases the potential barrier by the centrifugal term in (5.26), so that the partial width and corresponding branching ratio decrease. As an application of α-decay, we describe briefly the production of neutrons from radioactive α sources. Low energy neutrons in the range of 1–10 MeV are conveniently produced in the laboratory with radioactive compounds such as americium–beryllium (Am–Be). Americium is produced from plutonium in nuclear reactors by the reaction

$$^{239}\mathrm{Pu} + 2n \to {}^{240}\mathrm{Pu} + n \to {}^{241}\mathrm{Pu},$$

$$^{241}\mathrm{Pu} \to {}^{241}\mathrm{Am} + e^- + \bar{\nu}_e.$$

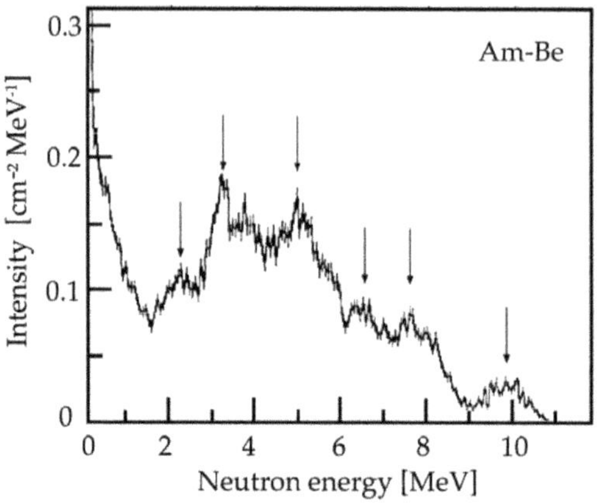

Figure 5.14. Neutron spectrum from an Am–Be source. The arrows indicate established resonances of the α^9Be system. After [8].

The ^{241}Am isotope with a half-life of 432 years emits 5.5 MeV α particles which can be absorbed by ^{9}Be to form a compound nucleus decaying into ^{12}C, plus the emission of a neutron:

$$^{241}\text{Am} \rightarrow {}^{237}\text{Np} + \alpha + \gamma,$$

$$\alpha\,{}^9\text{Be} \rightarrow {}^{12}\text{C} + n + \gamma.$$

Neutrons are often detected by using the reaction $n^3\text{He} \rightarrow p^3\text{H}$ ($Q = 764$ keV). In [8], the neutrons from a 370 GBq Am–Be source are converted into protons in a small proportional chamber operating with ^{3}He-gas, enclosed between two solid-state detectors. The neutron energy can be determined from the kinetic energies of the proton and the tritium nucleus by collecting the ionization charge deposited in the proportional chamber and the solid-state counters. Figure 5.14 shows the continuous energy distribution of the neutron which ranges up to 11 MeV. The arrows indicate well-known resonances of the $\alpha\,{}^9$Be system. The enhancement towards low energies is due to background.

5.4 Fission

5.4.1 Spontaneous fission

Spontaneous fission of heavy nuclei is also possible through the tunnel effect. During the deformation process (figure 5.15) the surface energy increases while the Coulomb energy decreases. The former leads to reduced binding (by $\Delta E_S > 0$), the latter to stronger binding (by $\Delta E_C > 0$). As for α-decay, the potential barrier which is proportional to the difference $\Delta E_S - \Delta E_C$, can be overcome by tunneling. With increasing mass number ΔE_C increases faster than ΔE_S so the barrier decreases [10]. Thus the probability for spontaneous fission increases with mass number A. For ^{238}U the branching ratio for spontaneous fission is very small ($\sim 5 \times 10^{-7}$) compared to α-decay. For super-heavy elements ($Z > 105$) the potential barrier is so small that

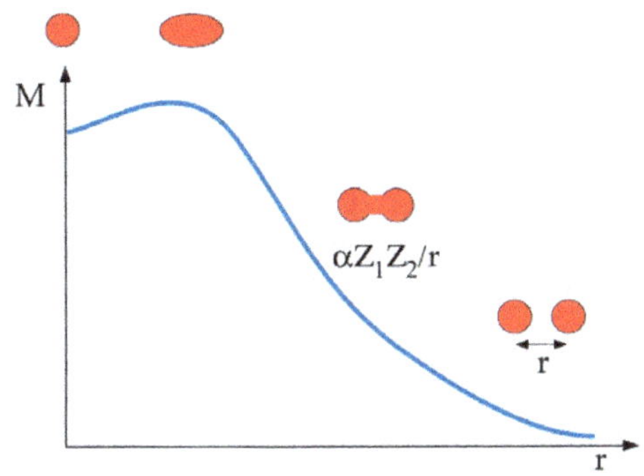

Figure 5.15. Nuclear mass as a function of deformation radius during spontaneous fission.

spontaneous fission dominates, which makes experimental detection difficult. These elements are usually detected by observing the sequence of α-decays to lighter, known nuclei.

5.4.2 Induced fission

We have seen that odd–even nuclei are less strongly bound than neighbouring even–even nuclei due to the pairing energy. Therefore, the absorption of a neutron by an odd–even nucleus with an odd neutron number can lead to an excited even–even nucleus which then falls apart. Examples are $^{235}_{92}$U and the artificially produced isotopes $^{233}_{92}$U and $^{239}_{94}$Pu. The isotope $^{235}_{92}$U fissions with slow neutrons ($T < 1$ eV), e.g. as follows

$$n + {}^{235}_{92}\mathrm{U} \rightarrow {}^{236}_{92}\mathrm{U}^* \rightarrow {}_{36}\mathrm{Kr} + {}_{56}\mathrm{Ba} + 2.43\, n. \tag{5.36}$$

Fission fragments are emitted with mass numbers between 70 and 160. For a given $Z = Z_1 + Z_2$ the Coulomb barrier (proportional to $Z_1 Z_2$) is highest when $Z_1 = Z_2$, hence symmetric fission is disfavoured. The most likely ratio between mass numbers is about 2 : 3 (figure 5.16). In addition 2–3 *prompt* neutrons (on average 2.43) are emitted with energies of, typically, 1 MeV.

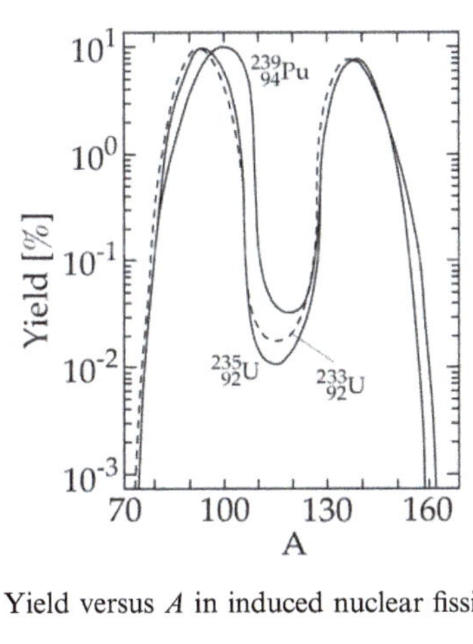

Figure 5.16. Yield versus A in induced nuclear fission. After [7].

Nuclear fission was first observed in 1938 by Enrico Fermi who bombarded $^{238}_{92}U$ with neutrons, but was erroneously interpreted as the production of new trans-uranian elements with $Z = 93$ and 94 (Fermi baptized these elements ausenium and hesperium). According to the later work of Otto Hahn and co-workers, Fermi had detected the fission of the $^{235}_{92}U$ isotope (0.7% content in natural uranium) into lighter elements such as Kr or Ba, see (5.36). The elements with $Z = 93$ (Np) and 94 (Pu) were discovered later by Edwin McMillan and Glenn Seaborg using the reactions (5.42) below (see the excellent essay [9] on the discovery of fission).

The daughter nuclei from reaction (5.36) are excited and decay by cascades (β^--decay, γ-decay and neutron emission). The energy released (~200 MeV) is shared as kinetic energy by the end products. Nuclei obtain approximately 167 MeV, electrons and photons 21 MeV, and antineutrinos 12 MeV (see also figure 14.1 in section 14.1). Neutron emission occurring after β emission is delayed by the relatively long mean life of β-decay. The delayed neutrons (on average $\alpha = 0.0065$ neutrons per fission) play an essential role in the regulation of induced fission in nuclear reactors, as we shall see below.

The emitted neutrons can be re-absorbed by $^{235}_{92}U$-nuclei, inducing a *chain reaction*. Figure 5.17 shows the cross sections for reactions with neutrons on ^{235}U and ^{238}U nuclei. The threshold for ^{238}U fission is around 1 MeV. However most of the emitted neutrons have energies below 1 MeV thus preventing a chain reaction based on ^{238}U. The neutrons are captured and the excited uranium nuclei decay by γ emission.

In contrast, for the ^{235}U-isotope fission dominates at low energy. Thermal neutrons (~30 meV) induce fission with cross sections around 600 b. However, natural uranium contains 99.3% ^{238}U and only 0.7% ^{235}U. Therefore, a chain reaction is possible by either substantially enriching the fraction of ^{235}U, or by moderating the

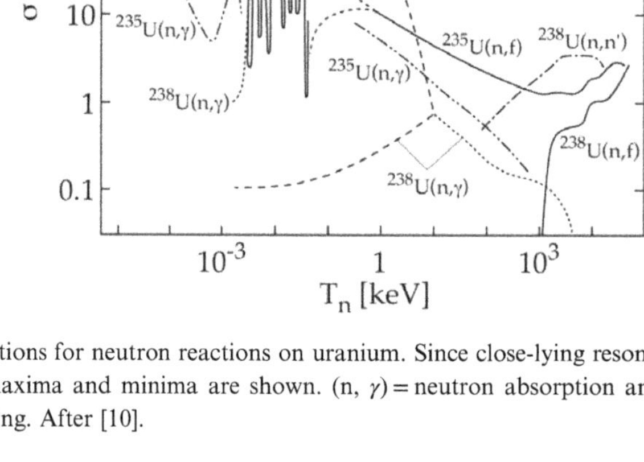

Figure 5.17. Cross sections for neutron reactions on uranium. Since close-lying resonances cannot be drawn, the envelope of the maxima and minima are shown. (n, γ) = neutron absorption and decay by γ emission; f = fission; n′ = scattering. After [10].

neutrons to energies below ~1 eV, where absorption by ^{238}U becomes negligible. The *multiplication factor* k is defined as

$$k = \frac{\text{number of neutrons in one generation}}{\text{number of neutrons in the previous generation}}. \qquad (5.37)$$

The chain reaction stops for $k < 1$ and diverges for $k > 1$. The factor k depends among other things on the ^{235}U content, the type of moderator, and on the precise geometry which affects the outward flow and loss of neutrons. This factor is proportional to the ratio of the volume to surface area of the fissile material. Spherical configurations with large radii are favorable. The *critical mass* ($k = 1$) for pure ^{235}U is about 50 kg, corresponding to a sphere radius of 8.4 cm. In nuclear explosives highly enriched fissile material is used, so the chain reaction develops rapidly without a moderator. The critical mass can be reduced substantially by using a neutron reflector such as ^{9}Be, which at the same time produces further neutrons via the reaction $n^9Be \rightarrow 2n^8Be$. The Hiroshima bomb contained only 1 kg of fissile ^{235}U material.

In nuclear reactors neutrons are slowed down by a moderator surrounding the fuel rods containing the fissile material. Moderators are made of light nuclei such as C or heavy water (D_2O), to maximize the energy loss per collision while at the same time minimizing the loss by absorption. The first chain reaction, observed on 2 December 1942 by Fermi in Chicago, used a graphite moderator. The Canadian CANDU (deuterium-oxide) reactor type uses natural uranium as the fuel and pressurized heavy water as the moderator. Reactors with natural uranium do not work with H_2O as the moderator because neutrons are absorbed by the reaction $np \rightarrow \gamma d$. Modern nuclear power plants use enriched ^{235}U (1–5%) so that normal water can be used which at the same time acts as a coolant (figure 5.18). For security reasons the fuel consists of uranium dioxide since pure uranium reacts with water.

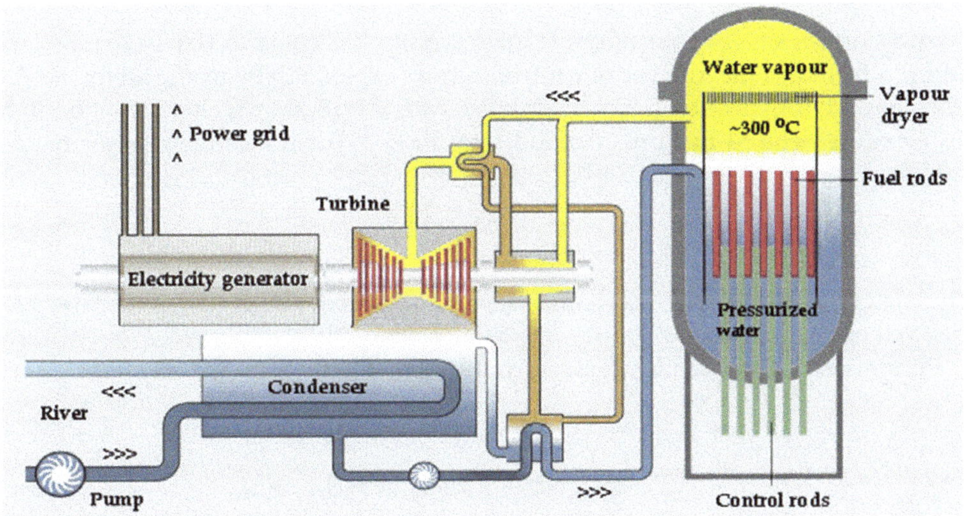

Figure 5.18. Diagram of a pressurized light water reactor.

The critical mass of a natural uranium-powered D_2O reactor is about 2 t, that of a 5% enriched ^{235}U graphite reactor is about 250 kg. The first reactor to be connected to a public electricity network was inaugurated in 1954 in Obninsk near Moscow.

Since the mean life of ^{235}U is shorter than that of ^{238}U, natural uranium previously (e.g. in the Proterozoic Eon) contained much more ^{235}U than today's 0.7%. An uranium mine was found in Oklo (Gabonese Republic) with a fraction of only 0.4% of the ^{235}U isotope. About two 2 billion years ago a natural reactor operated there with 3% ^{235}U, moderated by water, the evaporation of which automatically regulated the fission process.

In nuclear reactors, the stability of the fission process is regulated by control rods made of cadmium for which the neutron absorption cross section is very large (2450 b for thermal neutrons). Let us briefly discuss the time evolution of the fission process. Between t and $t + \mathrm{d}t$ the number of neutrons N changes according to

$$\mathrm{d}N = \left(\frac{k-1}{\tau_p} + \frac{\alpha}{\tau_v}\right)N\mathrm{d}t, \tag{5.38}$$

where $\tau_p \simeq 1$ ms is the mean lifetime of prompt neutrons (between emission and absorption) and $\tau_v \simeq 100$ ms is that of delayed neutrons (fraction $\alpha = 0.65\%$). The integration gives

$$N(t) = N_0 \exp\left(\frac{k-1}{\tau_p} + \frac{\alpha}{\tau_v}\right)t. \tag{5.39}$$

Obviously, the reactor cannot be regulated mechanically without delayed neutrons ($\alpha = 0$) since the *reactor period*

$$T = \frac{\tau_p}{k-1} \tag{5.40}$$

is very short for small values of $k-1$. For example, T is equal to 100 ms for $k = 1.01$, during which time the number of neutrons grows exponentially by the factor $e = 2.7$! Choosing a multiplication factor slightly below $k = 1$ allows the chain reaction to be sustained by delayed neutrons (second term in (5.39)) with a much longer reactor period of

$$T = \frac{\tau_v}{\alpha} = 15\ \mathrm{s}. \tag{5.41}$$

The reactor is regulated mechanically with delayed neutrons in the so-called 'delayed-critical' mode.

Plutonium is produced by neutron capture in ^{238}U through the reactions

$$\begin{aligned} n + {}^{238}_{92}\mathrm{U} &\to {}^{239}_{92}\mathrm{U} + \gamma, \\ {}^{239}_{92}\mathrm{U} &\to {}^{239}_{93}\mathrm{Np} + e^- + \bar{\nu}_e\ (\tau = 34\ \mathrm{min}), \\ {}^{239}_{93}\mathrm{Np} &\to {}^{239}_{94}\mathrm{Pu} + e^- + \bar{\nu}_e\ (\tau = 81\ \mathrm{h}). \end{aligned} \tag{5.42}$$

The mean lifetime of $^{239}_{94}Pu$ is 35 000 years. In a fast *breeder* reactor $^{239}_{94}Pu$ is produced through the capture of fast neutrons by $^{238}_{92}U$. Since $^{239}_{94}Pu$ is fissionable by both thermal and fast neutrons, the breeder reactor simultaneously generates energy and fissile material. Since neutrons would be moderated by light materials one uses instead metals such as liquid sodium as moderator.

5.5 Radiation dosimetry

We have seen in chapter 4 that the activity of a radioactive source is measured in becquerels (Bq) (1 Bq = 1 decay s^{-1}). The activity of the human body (~70 kg) is typically 3700 Bq and stems mostly from ^{40}K (β- and γ-decay) and ^{14}C (β-decay). Soil activities depend on the type of soil but are dominated by the ^{40}K β emitter (typically 400 Bq kg^{-1}) with smaller contributions from ^{226}Ra, ^{232}Th and ^{238}U. Sea water contributes about 10 Bq/ℓ. The activity in a house varies between 100 and 1000 Bq, mainly from the decay of ^{222}Rn (α emitter in the ^{238}U decay chain).

It is important to distinguish the concept of *activity* from the *dose* which is a measure of the radiation damage to a medium. The absorbed dose is the energy deposited per unit mass expressed in *grays* (Gy) (from L H Gray, a British physicist and radiologist, 1905–1965):

$$\boxed{1\ \mathrm{Gy} = 1\ \mathrm{J\,kg^{-1}}}.$$

The 'rad' is obsolete (1 rad = 0.01 Gy). The *fluence* is the number of particles traversing a given medium per unit area. The fluence of charged particles needed to deposit 1 Gy is, with 1 MeV = 1.6×10^{-13} J,

$$\phi\ [\mathrm{cm^{-2}}] = \frac{1[\mathrm{MeV}]}{1.6 \times 10^{-13}} \cdot \frac{1}{10^3[\mathrm{g}]} \cdot \frac{1}{\frac{\mathrm{d}E}{\mathrm{d}x}\left[\mathrm{MeV\ g^{-1}\ cm^2}\right]} = \frac{6.25 \times 10^9}{\frac{\mathrm{d}E}{\mathrm{d}x}}, \tag{5.43}$$

where $\frac{\mathrm{d}E}{\mathrm{d}x}$, expressed in [MeV g^{-1} cm^2], is the energy deposit per unit length times material density (see section 9.2). Fast particles with a charge of $Q = \pm 1$ deposit around 2 MeV g^{-1} cm^2.

The energy deposited does not carry any information about the *biological* damage to a tissue T. The effect of the absorbed dose $D(T, R)$ depends on the type of radiation R and is taken into account by a multiplicative dimensionless factor w_R. The *equivalent* dose H_T is a measure of the overall damage to the tissue (per unit mass) summed over all incurred radiations:

$$H_T = \sum\nolimits_R w_R \cdot D(T, R). \tag{5.44}$$

The equivalent dose (dimension J kg^{-1}) is given in sieverts (Sv) (from R M Sievert, Swedish physicist and radiologist, 1896–1966). The 'rem' is obsolete (1 rem = 0.01 Sv). Figure 5.19 shows the *radiation weighting factor* w_R for various particles and its energy dependence for neutrons. The *effective* dose E is a measure of the exposure of the

Particle type	w_R
γ, e, μ	1
$p, \pi^{\pm}$ (> 2 MeV)	2
α, ions, fission fragments	20
n (see figure right)	5 – 20

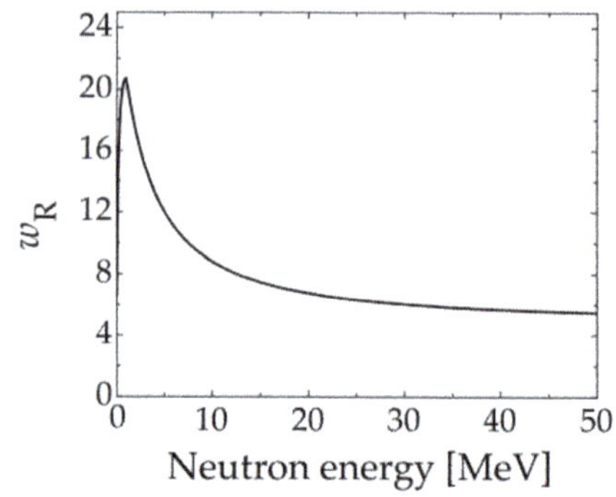

Figure 5.19. Radiation weighting factor for various projectiles (left) and for neutrons versus energy (right).

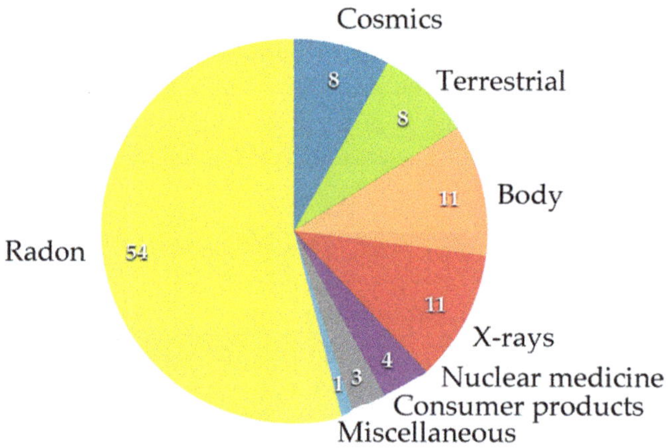

Figure 5.20. Fractional contributions to radiation exposure (in % of the US population).

human body to radiation and is the sum of equivalent doses weighted by the sensitivity of the various organs to radiation damage:

$$E = \sum_T w_T H_T. \tag{5.45}$$

The unit of effective dose is also the sievert. The tissue weighting factor w_T is about 20% for the gonads, 12% each for red bone marrow, the colon, lungs and stomach, 5% each for the breasts, bladder, liver, thyroid and oesophagus, 1% for the skin, and 6% for the remaining tissues and organs.

The annual dose to the human body from natural sources varies between 1 and 13 mSv with a world average of 2.4 mSv. Figure 5.20 shows the fractional contributions which are dominated by ^{222}Rn. At cruising altitudes of 10 km, airline crews and frequent travellers receive an extra dose of about 2 mSv yr^{-1}. The tolerable dose to the general public from artificial sources (such as nuclear power plants and accelerators) is limited to 1 mSv yr^{-1}. The reader of this book, who might soon be working in a particle or nuclear physics laboratory, should not be exposed to more than 20 mSv yr^{-1} (50 mSv yr^{-1} in the US). The whole body lethal dose which induces

Figure 5.21. The 18 MeV proton cyclotron at the University Hospital in Bern is used for radiological diagnostics with positron emission tomography (PET, detection of γ-ray pairs from positron annihilation with a β^+ emitter introduced into the body by means of a biologically active molecule). The ^{18}F isotope is produced with the reaction $p^{18}\text{O} \rightarrow n^{18}\text{F}$ on a water target, 95% enriched in ^{18}O. The ^{18}F isotope decays by β^+ emission into ^{18}O ($t_{1/2} = 110$ min). The cyclotron accelerates H^- ions which are extracted by stripping the electrons in thin carbon foils. The photograph shows the open yoke and the copper 'dees'. Image credit: Saverio Braccini.

a mortality with a 50% probability in 30 days is 2.5–4.5 Gy. Details on dosimetry and radiation protection can be found on p 381 in [11] (see also figure 5.21).

References

[1] Philipp K 1926 *Naturwissenschaften* **14** 1203

[2] Moskowski S A 1955 *β- and γ-Ray Spectroscopy* ed K Siegbahn (Amsterdam: North-Holland)

[3] Blatt J M and Weisskopf V F 1952 *Theoretical Nuclear Physics* (Hoboken, NJ: Wiley)

[4] Marmier P 1960 *Kernphysik* (Zurich: Verlag des Vereins der Mathematiker und Physiker an der ETH Zürich)

[5] Williams W S C 1991 *Nuclear and Particle Physics* (Oxford: Oxford University Press)

[6] Arnold R *et al* 2011 *Phys. Rev. Lett.* **107** 062504

[7] Hänsel H and Neumann W 1995 *Physik: Atome, Atomkerne, Elementarteilchen* (Heidelberg: Spektrum)

[8] Marsh J W, Thomas D J and Burke M 1995 *Nucl. Instrum. Methods Phys. Res.* A **366** 340

[9] Quack M 2013 Error and discovery: why repeating can be new *Angew. Chem. Int. Ed. Engl.* **52** 9362

[10] Mayer-Kuckuk T 1984 *Kernphysik* (Stuttgart: Teubner)

[11] Roesler S and Silari M 2012 *Review of Particle Physics Phys. Rev.* D **86** 010001

IOP Publishing

Nuclear and Particle Physics

Claude Amsler

Chapter 6

The nuclear shell model

We have seen in figure 3.10 that deviations from the empirical mass formula occur for the *magic* numbers

$$\boxed{Z \text{ or } N = 2,\ 8,\ 20,\ 28,\ 50,\ 82,\ 126,\ 184}. \tag{6.1}$$

Nuclei with a magic atomic number Z or neutron number N have many isotopes or isotones, hence must be particularly stable. For example, the enhanced binding of nuclei with a magic neutron number can be seen when plotting the energy of the first excited level as a function of N (figure 6.1).

A similar behaviour is observed for the *electronic* shell when Z takes the values

$$Z = 2,\ 10,\ 18,\ 36,\ 54,\ 80,\ 86. \tag{6.2}$$

Such atoms (noble gases with the exception of ^{80}Hg) have particularly high ionization energies of typically 10–20 eV, compared to only ∼5–10 eV for other elements. These magic numbers occur for *closed* electronic shells in the Coulomb potential. We shall show that magic numbers Z or N correspond to a closed shell in the nuclear potential.

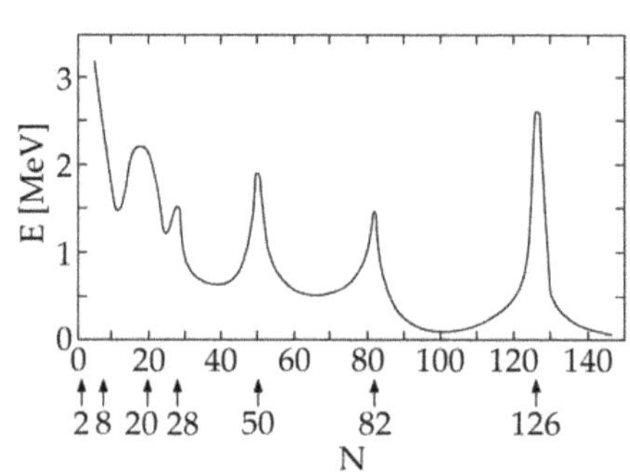

Figure 6.1. Energy needed to reach the first excited level for nuclei with neutron number N. After [1].

 © IOP Publishing Ltd 2015

We first calculate the mass spectrum of the nuclear levels using a realistic model for the nuclear potential $V(r)$, and show that the potential responsible for the short-range nuclear force is proportional to the nuclear charge density $\rho(r)$. The potential between two nucleons takes the form

$$V_{12} = V_0 f(\vec{r}_2 - \vec{r}_1). \tag{6.3}$$

The potential from all nucleons acting on nucleon 1, located at radius $\vec{r}_1$, is

$$V_1(\vec{r}_1) = V_0 \int f(\vec{r}_2 - \vec{r}_1)\rho(\vec{r}_2)\mathrm{d}^3\vec{r}_2. \tag{6.4}$$

The nuclear density can be expanded as a Taylor series around $\vec{r}_1$:

$$\rho(\vec{r}_2) = \rho(\vec{r}_1) + \text{grad}\, \rho \cdot (\vec{r}_2 - \vec{r}_1) + \ldots \tag{6.5}$$

Since the function $f(\vec{r}_2 - \vec{r}_1)$ is short-range we retain only the first term and obtain

$$V_1(\vec{r}_1) \simeq V_0\, \rho(\vec{r}_1) \int f(\vec{r}_2 - \vec{r}_1)\mathrm{d}^3\vec{r}_2 - V_0\, C\rho(\vec{r}_1). \tag{6.6}$$

The nuclear potential is proportional to the nuclear density.

Since the Saxon-Woods density provides a fair description of the nuclear form factors, we try as ansatz a potential (or rather a potential energy) of the form

$$V(r) = -\frac{V_0}{1 + \exp\left(\frac{r-a}{d}\right)}, \tag{6.7}$$

and solve the three-dimensional Schrödinger equation numerically for a nucleon with mass M:

$$\frac{1}{2M}\Delta u + \left[E - V(r) - \frac{\ell(\ell+1)}{2Mr^2}\right]u = 0. \tag{6.8}$$

A typical mass spectrum is shown on the left-hand side of figure 6.2. Each energy level (n, ℓ) can be filled with $2(2\ell+1)$ protons or $2(2\ell+1)$ neutrons. A clear separation can be seen between the levels $1s$ (2 states), $1p$ (6 states) and $1d + 2s$ (12 states), leading to closed shells with magic numbers 2, 8 and 20. The higher magic numbers in (6.1) are, however, not reproduced.

The way out was proposed in 1949 by Maria Goeppert-Mayer, Eugen Wigner and Johannes Jensen. As for the atomic shell, one expects a coupling between the orbital angular momentum $\vec{\ell}$ and the spin $\vec{s}$ of the nucleon (fine structure splitting). The reason for the ℓs coupling in the atomic shell can be understood intuitively as follows: the Lorentz transformation of the electric field into the rest frame of an orbiting electron generates a magnetic field with which the magnetic moment of the electron interacts. This leads to a splitting of the level with total angular momentum $j = \ell \pm \frac{1}{2}$.

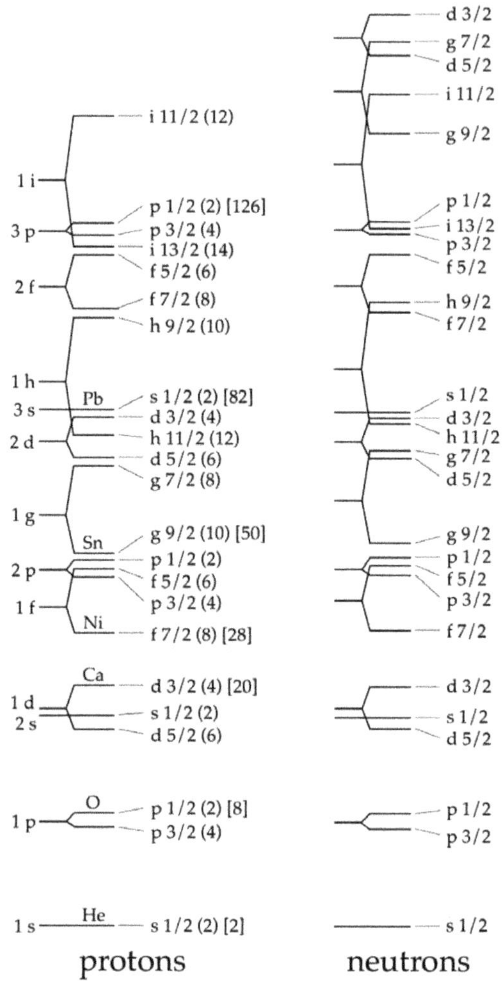

Figure 6.2. Ordering of the nuclear levels in the shell model for protons and neutrons. The unsplit levels have been calculated using the Schrödinger equation with (6.9). The numbers in parentheses are the level multiplicities $2j + 1$. The square bracket is the total number of nucleons up to that level. The spectra are almost identical for neutrons and protons (the latter with Coulomb interaction). Adapted from [2].

The nuclear potential gets an additional term which depends on the strength of the ℓs coupling and now reads

$$V_N(r) = V(r) + V_{\ell s}(r)\vec{\ell} \cdot \vec{s}. \tag{6.9}$$

The spin–orbit coupling cancels in the core for nucleons that are fully surrounded by other nucleons. The main contribution stems mainly from the nuclear periphery where $\rho(r)$ varies quickly:

$$V_{\ell s}(r) \propto \frac{1}{r}\frac{\mathrm{d}\rho}{\mathrm{d}r}. \tag{6.10}$$

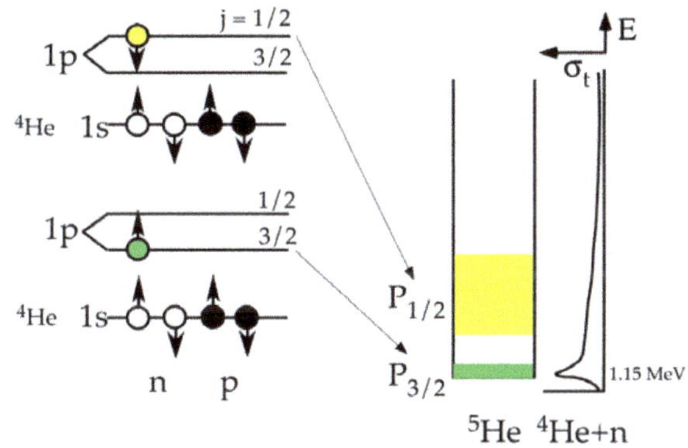

Figure 6.3. Cross section for elastic scattering $n^4\text{He} \rightarrow {}^5\text{He} \rightarrow n^4\text{He}$. After [3].

The fine structure splitting in the atomic shell is negligible when compared to the spacing between undisturbed levels. In contrast the nuclear potential causes a much larger splitting which is comparable to, or even larger than the level spacing. In contrast to the atomic shell the $j = \ell + \frac{1}{2}$ nuclear levels lie *below* the $j = \ell - \frac{1}{2}$ ones. This is due to the repulsive potential between nucleons at short distances which is mediated by ω exchange. This fact can be observed for example in elastic n^4-He scattering: the $j = \frac{3}{2}$ ($\ell = 1$) resonance is excited at lower energies than the $j = \frac{1}{2}$ ($\ell = 1$) resonance (figure 6.3).

Let us compute from (6.9) the level splittings due to spin–orbit coupling. With $\vec{j} = \vec{\ell} + \vec{s}$ and

$$\vec{\ell} \cdot \vec{s} = \frac{1}{2}\left(\vec{j}^2 - \vec{\ell}^2 - \vec{s}^2\right) \tag{6.11}$$

one obtains the expectation value

$$\begin{aligned} \langle \vec{\ell} \cdot \vec{s} \rangle &= \frac{1}{2}\left[j(j+1) - \ell(\ell+1) - \frac{3}{4}\right] \\ &= \frac{1}{2}\left[\left(\ell + \frac{1}{2}\right)\left(\ell + \frac{3}{2}\right) - \ell(\ell+1) - \frac{3}{4}\right] \quad \text{for } j = \ell + \frac{1}{2} \end{aligned} \tag{6.12}$$

and

$$\frac{1}{2}\left[\left(\ell - \frac{1}{2}\right)\left(\ell + \frac{1}{2}\right) - \ell(\ell+1) - \frac{3}{4}\right] \quad \text{for } j = \ell - \frac{1}{2}.$$

Hence for $j = \ell + \frac{1}{2}$,

$$V_N(r) = V(r) + \frac{1}{2}V_{\ell s}(r) \cdot \ell, \tag{6.13}$$

while for $j = \ell - \frac{1}{2}$,

$$V_N(r) = V(r) - \frac{1}{2}V_{\ell s}(r) \cdot (\ell + 1). \tag{6.14}$$

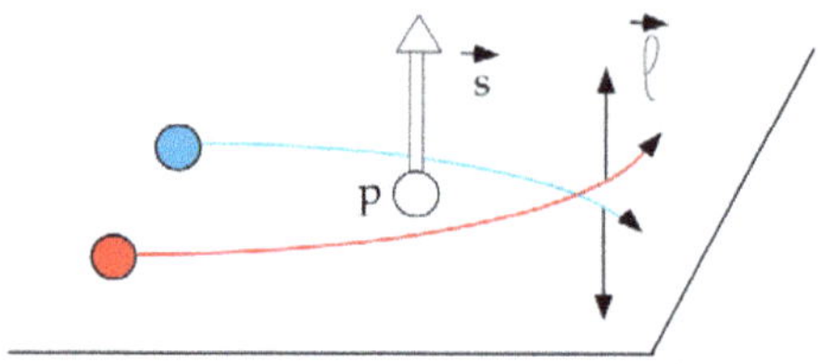

Figure 6.4. $\vec{\ell} \cdot \vec{s}$ dependence in proton scattering off polarized protons: $\vec{\ell} \cdot \vec{s}$ is positive for scattering to the left and negative for scattering to the right.

Table 6.1. Examples of nuclei with a closed shell minus one nucleon or with a single nucleon in the outermost j-shell. Nuclear spins and parities are denoted by I^P.

$^{13}_{6}C$	6p	7n	n:	$p_{\frac{1}{2}}$	$I^P = \frac{1}{2}^-$
$^{13}_{7}N$	7p	6n	p:	$p_{\frac{1}{2}}$	$I^P = \frac{1}{2}^-$
$^{17}_{8}O$	8p	9n	n:	$d_{\frac{5}{2}}$	$I^P = \frac{5}{2}^+$
$^{33}_{16}S$	16p	17n	n:	$d_{\frac{3}{2}}$	$I^P = \frac{3}{2}^+$
$^{87}_{38}Sr$	38p	49n	n:	$g_{\frac{9}{2}}$	$I^P = \frac{9}{2}^+$

The difference is the level splitting which *grows linearly* with ℓ:

$$\boxed{\Delta E \propto 2\ell + 1}. \qquad (6.15)$$

The existence of an $\vec{\ell} \cdot \vec{s}$ dependence of the strong interaction is directly observable, for example by scattering protons on vertically polarized protons: for scattering to the left the term $\vec{\ell} \cdot \vec{s}$ is positive, while for scattering to the right $\vec{\ell} \cdot \vec{s}$ is negative (figure 6.4). This leads to an asymmetry in the scattering rates between left and right.

The resulting term diagram is also shown in figure 6.2. The split (n, ℓ) levels start to overlap for large ℓ (around $^{40}_{20}Ca$) and energetically well-separated shells appear, reproducing the magic numbers (6.1).

Each level in the nuclear shell can accommodate $2j + 1$ protons or $2j + 1$ neutrons with spin projections $-j \leqslant m \leqslant j$. The vector $\vec{j}$ and the associated quantum number j refer to the angular momentum carried by one nucleon. We follow this usage and denote the *nuclear spin* by the vector $\vec{I}$ (quantum number I), while $\vec{J}$ (quantum number J) refers to the total *spin of the atomic shell.*

For even–even nuclei the spins cancel pairwise, leading to a nucleus with spin $I = 0$. From the shell model one can thus predict the nuclear spin I for the ground state nuclei in which a nucleon in the last j-shell is missing, or for which the last j-shell contains only one nucleon: the nuclear spin is $I = j$ and its parity $P = (-1)^{\ell}$, as verified experimentally (table 6.1). As we shall see in section 6.1 nuclear spins can be determined from the hyperfine splitting of the atomic levels.

Many predictions for nuclear properties (such as magnetic dipole and electric quadrupole moments) can be made from the shell model [3, 4]. For deformed nuclei the assumption of a spherically symmetric charge distribution is no longer valid.

This leads to a shift and rearrangement of the levels within the closed shells (the Nilsson model). For extreme deformations even the magic sequence can be disturbed.

The next expected magic number above 126 is for closed shells with $N = 184$ neutrons. This corresponds to $Z \simeq 114$ on the stability line (figure 3.5). Isotopes around this atomic number should be relatively stable, although their half-lives are difficult to predict. One expects them to decay mainly by spontaneous fission and also by α emission. For example, the half-life of $^{287}_{114}$Fl (flerovium) is about 5 s. Super-heavy elements are produced in heavy ion collisions and are detected through their α-decay chains. The heaviest element observed so far carries the atomic number $Z = 118$ and has a half-life of about 1 ms. It was produced by ^{249}Cf ions impinging on a ^{48}Ca target [5]. Figure 6.5 shows as an example the decay chains of two transuranic elements.

Figure 6.6 shows the nuclear 'landscape' Z versus N. The black dots stand for stable elements, the green and yellow zones show the known bound but unstable nuclei. Bound nuclei are predicted to lie within the grey area, delimited by the proton (top) and neutron (bottom) *driplines*. There are four stable doubly magic nuclei, $^4_2$He, $^{16}_8$O, $^{40}_{20}$Ca and $^{208}_{82}$Pb. The doubly magic $^{48}_{20}$Ca and $^{56}_{28}$Ni are unstable but long-lived with half-lives of 4×10^{19} years for the former (double β-decay) and 6 days for the latter.

Heavy elements are thought to be produced in the Sun or in supernovae by r-processes in which several neutrons are captured fast enough, before the nucleus returns to the stability region by multiple β-decays. Such processes occur along the path shown in red (for example starting with the doubly magic ^{132}Sn which has a half-life of 40 s).

As an application of the shell model, we now deal with the interaction of a deformed nucleus with an external electric field.

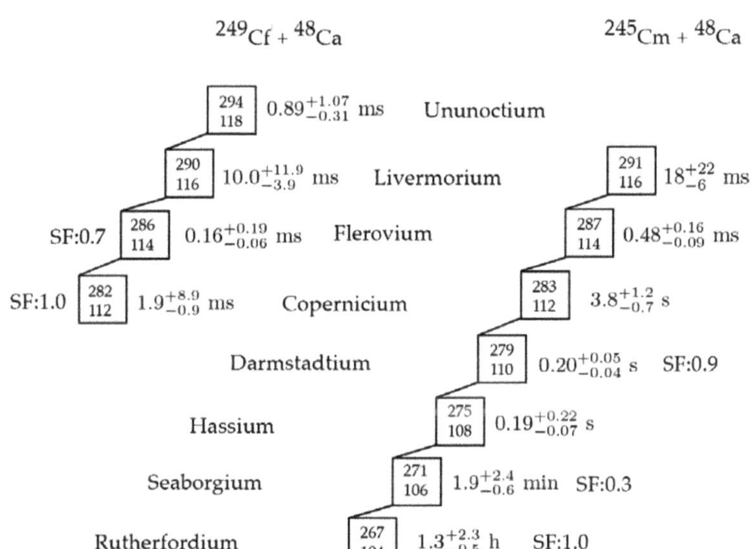

Figure 6.5. α-decay chains of the transuranic elements $Z = 118$ (left) and $Z = 116$ (right) with half-lives. When omitted, the branching ratio for spontaneous fission (SF) is not known. The elements were produced by heavy ion collisions on ^{48}Ca nuclei. After [5].

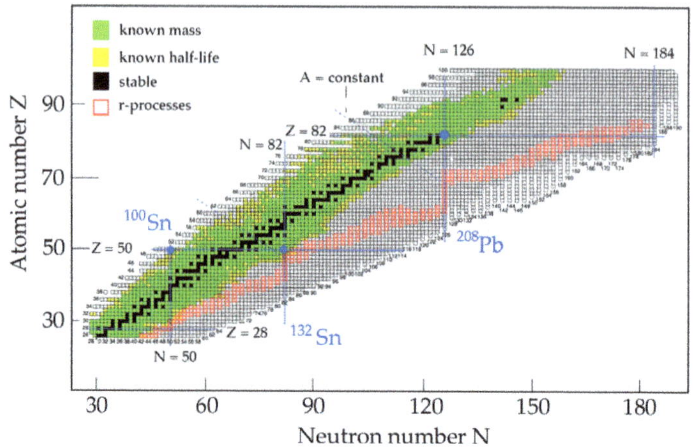

Figure 6.6. The nuclear landscape around the stability line (see the text for details). Adapted from [6] and [7].

6.1 The electric quadrupole moment

Let us calculate the energy of a nucleus with charge distribution $\rho(\vec{r})$ (normalization such that $\int \rho(\vec{r})\mathrm{d}^3\vec{r} = Ze$) in an external electric field $\vec{E}$. The latter could for instance be produced by the electronic shell. We assume that the nucleus is deformed and has the shape of an ellipsoid with the main rotation axis chosen along the z-direction. The potential energy in the electric potential ϕ is

$$W = \int \phi(\vec{r})\rho(\vec{r})\mathrm{d}^3\vec{r}. \tag{6.16}$$

Expanding ϕ as a Taylor series around the centre of the nucleus gives

$$\phi(\vec{r}) = \phi(0) + \sum x_i \frac{\partial \phi}{\partial x_i}\bigg|_0 + \frac{1}{2}\sum x_i x_j \frac{\partial^2 \phi}{\partial x_i \partial x_j}\bigg|_0, \tag{6.17}$$

which we introduce into (6.16). The first two terms are

$$W_{1+2} = Ze\phi(0) - \vec{E}\overbrace{\int \vec{r}\rho(\vec{r})\mathrm{d}^3\vec{r}}^{0}. \tag{6.18}$$

The first term is the potential energy of the nuclear charge in the electric field. The second term is the (static) electric dipole moment which vanishes since $\rho(-\vec{r}) = \rho(\vec{r})$ from parity conservation, and since the integrand is odd under space inversion $\vec{r} \to -\vec{r}$. Nuclei do not possess static electric dipole moments.

We apply the electric field $\vec{E}$ along the z-axis and assume cylindrical symmetry around z. The third term in (6.17) contributes the energy

$$W_3 = -\frac{1}{2}\int \left[\frac{\partial E_x}{\partial x}\bigg|_0 x^2 + \frac{\partial E_y}{\partial y}\bigg|_0 y^2 + \frac{\partial E_z}{\partial z}\bigg|_0 z^2\right]\rho(\vec{r})\mathrm{d}^3\vec{r}. \tag{6.19}$$

Using cylindrical symmetry we obtain with div $\vec{E} = 0$

$$\left.\frac{\partial E_x}{\partial x}\right|_0 = \left.\frac{\partial E_y}{\partial y}\right|_0 = -\frac{1}{2}\left.\frac{\partial E_z}{\partial z}\right|_0, \tag{6.20}$$

hence

$$W_3 = -\frac{1}{2}\left.\frac{\partial E_z}{\partial z}\right|_0 \int \left[-\frac{1}{2}\underbrace{\left(x^2 + y^2\right)}_{r^2 - z^2} + z^2 \right] \rho(\vec{r})\,\mathrm{d}^3\vec{r}, \tag{6.21}$$

or

$$E_Q \equiv W_3 = -\frac{e}{4}\left.\frac{\partial E_z}{\partial z}\right|_0 Q, \tag{6.22}$$

with

$$\boxed{Q \equiv \frac{1}{e}\int \overbrace{\left(3z^2 - r^2\right)}^{r^2\left(3\cos^2\theta - 1\right)} \rho(\vec{r})\,\mathrm{d}^3\vec{r}}\,. \tag{6.23}$$

Q is the *quadrupole moment* of the nucleus with respect to the z-axis. Its dimension is that of a surface and Q is therefore measured in barn (b). For a spherical nucleus (ρ independent of θ) the quadrupole moment vanishes since $\int_{-1}^{1}(3\cos^2\theta - 1)\,\mathrm{d}\cos\theta = 0$. For a *prolate* ellipsoid (cigar-shaped nucleus) $Q > 0$, while for an *oblate* nucleus (lens-shaped nucleus) $Q < 0$ (figure 6.7).

Thus in a *non-uniform* electric field the atomic levels of a deformed nucleus are shifted by the energy E_Q. The potential ϕ may depend on the chemical composition, so the energies may vary due to the monopole term (*isomeric shift*) in (6.18). The shift can be observed using Mössbauer spectroscopy. Since γ transitions are measured, the source and absorber must be made of different chemical compounds. In addition, an asymmetric electronic shell structure generates a field gradient at the nucleus and the levels are further shifted by E_Q when $Q \neq 0$.

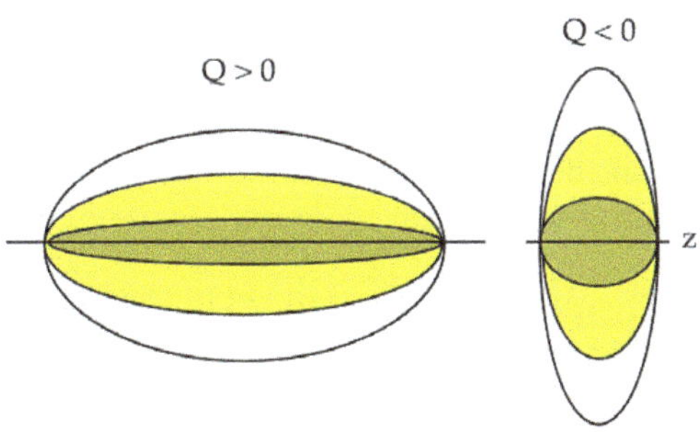

Figure 6.7. Deformed nuclei: prolate ($Q > 0$) and oblate ($Q < 0$) ellipsoids.

The energy shift (6.22) was calculated classically under the assumption that the $\vec{E}$-field was aligned with the z-axis, which is also the direction of the deformation axis. However, the axis of a deformed nucleus is along the nuclear spin $\vec{I}$ while the electric field generated by the electronic shell is parallel to $\vec{J}$, the total angular momentum of the shell electrons. According to the rules of quantum mechanics there are $2I + 1$ projections M of the nuclear spin on the quantization axis. Hence the angle of the ellipsoid main axis is quantized with respect to the z-axis, and depends on the quantum numbers I, J and F, where

$$\vec{F} = \vec{I} + \vec{J} \quad \text{and} \quad |I - J| \leqslant F \leqslant I + J. \tag{6.24}$$

The classical derivation (6.22) is valid for the state with maximum alignment $M = I$. The quantum mechanical expression for the energy shift E_Q has been derived by Casimir [8]:

$$E_Q = \frac{1}{4}\frac{\partial E_z}{\partial z}\bigg|_0 eQ \left[\frac{\frac{3}{2}C(C+1) - 2I(I+1)J(J+1)}{I(2I-1)J(2J-1)} \right], \tag{6.25}$$

where

$$C \equiv F(F+1) - I(I+1) - J(J+1). \tag{6.26}$$

Inserting $I = 0$ or $I = \frac{1}{2}$ one obtains $E_Q = 0$, hence nuclei with spin 0 or $\frac{1}{2}$ do not possess any electric quadrupole moment.

Nuclear spins and electric quadrupole moments can be determined by measuring the hyperfine splitting of the electronic levels, for instance with microwaves. The hyperfine splitting of the electronic level with spin $\vec{J}$ by the nucleus with spin $\vec{I}$ is, for a spherical nucleus,

$$\begin{aligned} E_{hfs}(F) \propto \langle \vec{I} \cdot \vec{J} \rangle &= \frac{1}{2}\langle \vec{F}^2 - \vec{I}^2 - \vec{J}^2 \rangle \\ &= \frac{1}{2}[F(F+1) - I(I+1) - J(J+1)]. \end{aligned} \tag{6.27}$$

The spacing between levels is then proportional to F:

$$\Delta E_{hfs}(F, F-1) = E_{hfs}(F) - E_{hfs}(F-1) \propto F. \tag{6.28}$$

Due to the quadrupole moment the hyperfine levels of deformed nuclei do not obey this *interval rule*, but are slightly shifted by the energy E_Q either upwards or downwards, depending on the sign of the electric field gradient and the sign of Q in (6.25). The measurements are subject to large errors since the gradient is often inaccurately known.

Quadrupole moments can also be determined by using muonic atoms (introduced in section 2.3). One measures the energies of the x-rays emitted while the muon cascades down to the atomic ground state. The transition energies between the lower levels can be computed accurately since the μ^- orbit lies below the $1s$ atomic level for

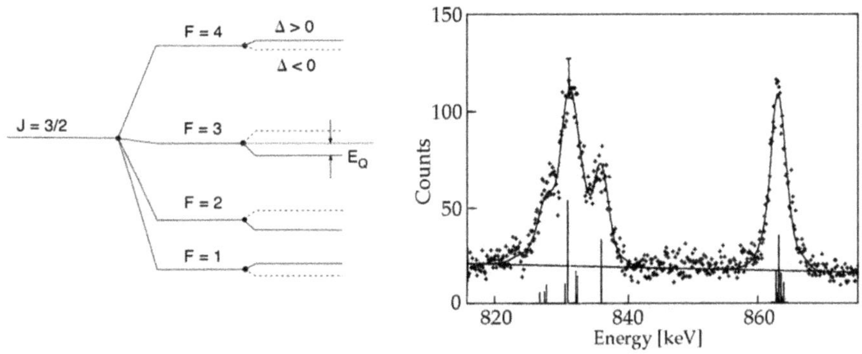

Figure 6.8. Left: term diagram of the $J = \frac{3}{2}$ muonic level in ^{105}Pd, including hyperfine splitting and quadrupole shift (see the text for details, not to scale). Right: $3d$ $(J = \frac{5}{2})$ to $2p$ $(J = \frac{3}{2})$ and $3d$ $(J = \frac{3}{2})$ to $2p$ $(J = \frac{1}{2})$ muonic x-ray transitions. The vertical bars above the horizontal axis show the energies of the ΔF transitions used in the fit (curve). After [9].

not too heavy nuclei, where the nuclear charge is not shielded by the lower electronic shells, and the μ^- is still above the nuclear surface.

Figure 6.8 (left) shows as an example the hyperfine splitting of the electronic $J = \frac{3}{2}$ level in ^{105}Pd $(I = \frac{5}{2})$ and the level shifts due its nuclear quadrupole moment. The levels are shifted upwards or downwards, depending on the sign of $\Delta = \frac{\partial E_z}{\partial z} Q$, see (6.25). Figure 6.8 (right) shows a measurement of the muonic x-ray transitions from the $3d$ $(J = \frac{5}{2})$ to the $2p$ $(J = \frac{3}{2})$ level (left peak) and from the $3d$ $(J = \frac{3}{2})$ to the $2p$ $(J = \frac{1}{2})$ level (right peak). The x-rays are recorded using a Ge(Li) semiconductor detector. The structures in the left peak are due to the hyperfine splitting of the $2p$ $(J = \frac{3}{2})$ level, shifted by the nuclear quadrupole moment (according to (6.25) $E_Q = 0$ for the $2p$ $(J = \frac{1}{2})$ level). The fine splitting between the $2p$ $(J = \frac{3}{2})$ and the $2p$ $(J = \frac{1}{2})$ is 34 keV. The hyperfine splitting of the $2p$ $(J = \frac{3}{2})$ level leads to the quadrupole moment $Q = 0.66$ b [8].

Let us discuss two examples of nuclei with closed shells plus a proton or minus a proton (a hole in the shell). For the nuclear density one expects that

$$\rho \propto |Y_\ell^m(\theta, \phi)|^2, \tag{6.29}$$

where ℓ denotes the angular momentum of the proton or the hole. The value of the magnetic quantum number m is such that the projection M of $\vec{I} = \vec{j}$ on the main ellipsoid axis is maximum. Hence for $j = \ell + \frac{1}{2}$ we have $m = \ell$ so that $M = m + \frac{1}{2} = I$, while for $j = \ell - \frac{1}{2}$ there are two possibilities:

$$m = \ell \Rightarrow M = m - \frac{1}{2} = I \quad \text{or} \quad m = \ell - 1 \Rightarrow M = m + \frac{1}{2} = I. \tag{6.30}$$

- $^{63}_{29}$Cu with $j = \frac{3}{2}$, $\ell = 1$ $(I^P = \frac{3}{2}^-$, $Z = 29$ protons, see figure 6.2), hence $m = 1$: the modulus of the spherical harmonics $|Y_1^1|^2$ in (6.29) is proportional to

$\sin^2\theta$, therefore the additional proton is mostly located at large angles with respect to the z-axis. The nucleus is slightly lens-shaped and the measured quadrupole moment is $Q = -0.15$ b.

- $^{39}_{19}$K with $j = \frac{3}{2}$, $\ell = 2$ ($I^P = \frac{3}{2}^+$, $Z = 19$ protons), hence m takes two possible value, $m = 1$ or 2: the modulus of the spherical harmonics $|Y_2^1|^2$ for the proton hole is proportional to $\sin^2\theta\cos^2\theta$. On the other hand, for the solution $m = 2$ the modulus $|Y_2^2|^2$ is proportional to $\sin^4\theta$. In both cases one expects a deficit of protons at large angles. The nucleus is slightly cigar-shaped, as confirmed by the measurement: $Q = +0.11$ b.

Finally, let us define a mean radius R such that the volume of the ellipsoid is equal to $4\pi R^3/3$ and label the axes a (main axis taken in the z-direction) and b:

$$a = R + \Delta R \qquad \text{and} \qquad b = R - \frac{\overbrace{\Delta R}^{a-b}}{2}. \tag{6.31}$$

For a homogeneously charged ellipsoid with charge Ze one finds from (6.23) that

$$Q = \frac{2}{5}Z(a^2 - b^2) \simeq \frac{6}{5}ZR^2\left(\frac{\Delta R}{R}\right). \tag{6.32}$$

For example, the quadrupole moment of the strongly deformed $^{175}_{71}$Lu nucleus is $Q = 5.68$ b. From (6.32) the deformation is

$$\frac{\Delta R}{R} = 0.15 \Rightarrow a \simeq 7.7\,\text{fm}, \qquad b \simeq 6.2\,\text{fm}, \tag{6.33}$$

with $R \simeq 6.7$ fm. Figure 6.9 shows the measured values Q/ZR^2 for odd–even nuclei. The majority of the deformed nuclei are cigar-shaped ($Q > 0$). The measured values

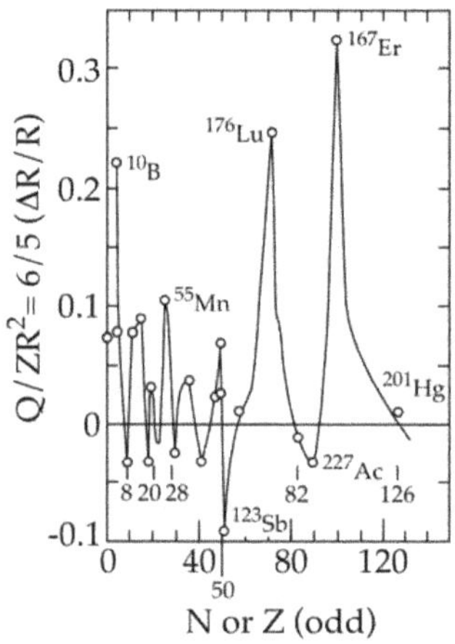

Figure 6.9. Measured electric quadrupole moments for even–odd nuclei. Further measurements along the curves have been omitted. After [3].

for the quadrupole moments vanish at the magic numbers, in accordance with expectations for closed shell nuclei.

References

[1] Marmier P 1960 *Kernphysik* (Zurich: Verlag des Vereins der Mathematiker und Physiker an der ETH Zürich)
[2] Klinkenberg P F A 1952 *Rev. Mod. Phys.* **24** 63
[3] Mayer-Kuckuk T 1984 *Kernphysik* (Stuttgart: Teubner)
[4] Mayer-Kuckuk T 1985 *Atomphysik* (Stuttgart: Teubner)
[5] Oganessian Yu Ts *et al* 2006 *Phys. Rev.* C **74** 044602
[6] Grawe H, Langanke K and Martinez-Pinedo G 2007 *Rep. Prog. Phys.* **70** 1525
[7] Pearson J M, Nayak R C and Goriely S 1996 *Phys. Lett.* B **387** 455
[8] Casimir H B G 1936 *Arch. Musée Teyler* **8** 201
[9] Vuilleumier J L *et al* 1978 *Nucl. Phys.* A **294** 273

Chapter 7

Elementary particles

7.1 Introduction

Particle physics deals with the study of the fundamental constituents of matter—the elementary particles—and their interactions. Up to now all experimental findings have been consistent with the standard model of particle physics, dealt with in the following chapters. However, with the higher energies that will be available in future accelerators, one expects to see deviations from the standard model which should already be observable today at low energies, e.g. in rare decay processes. The absence of anomalies so far is perhaps due to the limited sensitivity of today's experiments.

The standard model distinguishes three types of elementary particles:

1. **The spin-$\frac{1}{2}$ fermions.**
   - There are three families of *leptons*, the electron and its neutrino, the muon and its neutrino, and the tauon and its neutrino:

$$(\nu_e, e^-), \left(\nu_\mu, \mu^-\right), (\nu_\tau, \tau^-), \tag{7.1}$$

   with measured masses

$$\begin{aligned} &m(\nu_e) < 2\,\text{eV}, && m(\nu_\mu) < 190\,\text{keV}, && m(\nu_\tau) < 18\,\text{MeV}, \\ &m(e^-) = 511\,\text{keV}, && m(\mu^-) = 106\,\text{MeV}, && m(\tau^-) = 1.777\,\text{GeV}, \end{aligned}$$

   to be compared with the mass of the proton, $m_p = 938.27$ MeV. The upper limits are derived from direct measurements. The observation of neutrino oscillations (section 14.6) implies that neutrinos have finite masses and, together with $m(\nu_e) < 2$ eV, that also $m(\nu_\mu, \nu_\tau) \lesssim 2$ eV.

   The wave equation for spin-$\frac{1}{2}$ fermions (the Dirac equation) predicts for each fermion the existence of an antifermion with the charge $\overline{Q}e = -Qe$, mass $\bar{m} = m$ and magnetic moment $\bar{\mu} = -\mu$ (the magnetic moment of the

Table 7.1. Lepton numbers L_e, L_μ and L_τ.

	L_e	L_μ	L_τ
ν_e	1	0	0
$\bar{\nu}_e$	–1	0	0
ν_μ	0	1	0
$\bar{\nu}_\mu$	0	–1	0
ν_τ	0	0	1
$\bar{\nu}_\tau$	0	0	–1
e^-	1	0	0
e^+	–1	0	0
μ^-	0	1	0
μ^+	0	–1	0
τ^-	0	0	1
τ^+	0	0	–1

neutrino is tiny in the standard model, $\mu_\nu = 3 \times 10^{-19}$ Bohr magnetons $\times$ m_ν [eV]). The corresponding families of antileptons are:

$$\left(\bar{\nu}_e, e^+\right), \left(\bar{\nu}_\mu, \mu^+\right), \left(\bar{\nu}_\tau, \tau^+\right). \tag{7.2}$$

The quantum numbers associated with leptons are the *lepton numbers* L_e, L_μ and L_τ which take the values +1 for leptons and −1 for antileptons (table 7.1). The lepton numbers of all other particles are equal to zero.

- The *quarks* q also occur in three families (or three *generations*):

$$(u, d), (c, s), (t, b), \tag{7.3}$$

the up, down, charm, strange, top (or true) and bottom (or beauty) *flavours*. The electric charges of the quarks are fractional:

$$\left(\frac{2}{3}, -\frac{1}{3}\right), \left(\frac{2}{3}, -\frac{1}{3}\right), \left(\frac{2}{3}, -\frac{1}{3}\right), \tag{7.4}$$

expressed in units of e, and their masses are

$$m(u) \sim 350\,\text{MeV}, \quad m(d) \sim 350\,\text{MeV}, \quad m(s) \sim 550\,\text{MeV},$$
$$m(c) \sim 1800\,\text{MeV}, \quad m(b) \sim 4670\,\text{MeV}, \quad m(t) \sim 173\,\text{GeV}.$$

Free quarks are not observed. The masses above are therefore effective or constituent masses within bound systems, in contrast to the free or 'bare' masses: $m(u) \sim 2\,\text{MeV}$, $m(d) \sim 5\,\text{MeV}$, $m(s) \sim 90\,\text{MeV}$, $m(c) \sim 1275\,\text{MeV}$, $m(b) \sim 4180\,\text{MeV}$ and $m(t) \sim 173\,\text{GeV}$. The t quark is almost as heavy as a gold nucleus.

The corresponding families of antiquarks $\bar{q}$ are

$$(\bar{u}, \bar{d}), (\bar{c}, \bar{s}), (\bar{t}, \bar{b}), \tag{7.5}$$

with opposite charges,

$$\left(-\frac{2}{3}, \frac{1}{3}\right), \left(-\frac{2}{3}, \frac{1}{3}\right), \left(-\frac{2}{3}, \frac{1}{3}\right). \tag{7.6}$$

The *baryon number* B is defined as

$$\begin{aligned} B &= \frac{1}{3} \text{ for quarks,} \\ B &= -\frac{1}{3} \text{ for antiquarks.} \end{aligned} \tag{7.7}$$

The baryon numbers of all other particles are equal to zero.

As will be shown in section 7.3.1, the lepton and baryon numbers are conserved in reactions between particles and in decays. We and our environment consist only of the first families of leptons and quarks (ν_e, e^-) and (u, d). According to present knowledge leptons and quarks are point-like (structureless). The existence of additional families with yet heavier quarks and leptons and massive neutrinos ($m_\nu > 50$ GeV) is not excluded.

2. **The spin-1 gauge bosons.** In quantum mechanics the interaction between particles is mediated by the exchange of field quanta. The gauge bosons are the field quanta responsible for the interaction between quarks and leptons. There are three interactions (classified in decreasing strength):
   - The *strong interaction* is the force between quarks, mediated by *gluons* (g). This interaction is responsible for binding the quarks, e.g. in the proton. Like quarks, free gluons are not observed. In the standard model the 'bare' mass of the gluon is equal to zero.
   - The *electromagnetic interaction* acts between charged particles and is conveyed by massless *photons* (γ). The measured upper limit for the photon mass is

$$m(\gamma) < 10^{-18}\,\text{eV}. \tag{7.8}$$

   - The *weak interaction* is responsible for the force involving leptons, e.g. for nuclear β-decay. The weak interaction is mediated by the exchange of the heavy $W^\pm$ and Z^0 bosons with masses

$$\begin{aligned} m_{W^\pm} &= 80.385 \pm 0.015\,\text{GeV}, \\ m_{Z^0} &= 91.1876 \pm 0.0021\,\text{GeV}, \end{aligned} \tag{7.9}$$

   and widths (1/mean lifetime)

$$\begin{aligned} \Gamma_{W^\pm} &= 2.085 \pm 0.042\,\text{GeV}, \\ \Gamma_{Z^0} &= 2.4952 \pm 0.0023\,\text{GeV}. \end{aligned} \tag{7.10}$$

Since quarks are electrically charged, they also interact electromagnetically among themselves. However, the coupling to the photon is much smaller than the coupling to the gluon. Similarly, charged leptons and quarks also

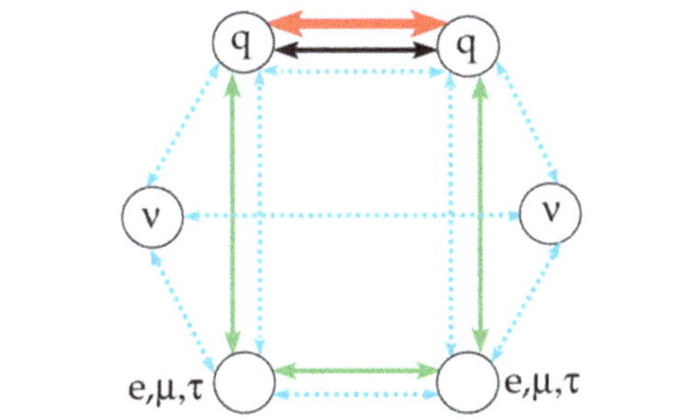

Figure 7.1. The elementary particles and their interactions.

interact through the weak interaction, albeit with considerably reduced strength. Figure 7.1 illustrates schematically the three interactions between elementary particles.

Gravity (exchange of a spin-2 graviton) is hard to unify with the three fundamental forces, and is in any case too weak to be relevant in particle physics.

3. **The spin-0 Higgs boson.** The Higgs boson H endows elementary particles with mass and is responsible for the non-vanishing masses of the W and Z^0 bosons (section 16.7). The Higgs was found at CERN's Large Hadron Collider (LHC) in 2012. Its mass is [1]

$$m_H = 125.7 \pm 0.4 \text{ GeV}. \tag{7.11}$$

Its width, which cannot be measured directly due to insufficient experimental resolution, is estimated to be about 4 MeV. The expected mass range could be estimated earlier from theoretical calculations of 'radiative' corrections to predictable quantities in the standard model (such as the masses of the W and Z^0 bosons), and by comparing with data. Figure 7.2 shows how the χ^2 fit to various measurements varies with Higgs mass. The range below 114 GeV was already excluded by direct searches. The most likely value was around 125 GeV, where the Higgs was also discovered (section 17.3).

Table 7.2 shows the dominant decay channels of elementary particles.

7.2 Mesons and baryons

Quarks and gluons are not observed as free entities. They remain confined in composite structures with finite dimensions called *hadrons*. Hadrons are divided into two classes, *mesons* and *baryons*.

Mesons consist of bound quark–antiquark ($q\bar{q}$) pairs. The baryon number B introduced above is an *additive* quantum number, hence for mesons

$$B(q\bar{q}) = 0. \tag{7.12}$$

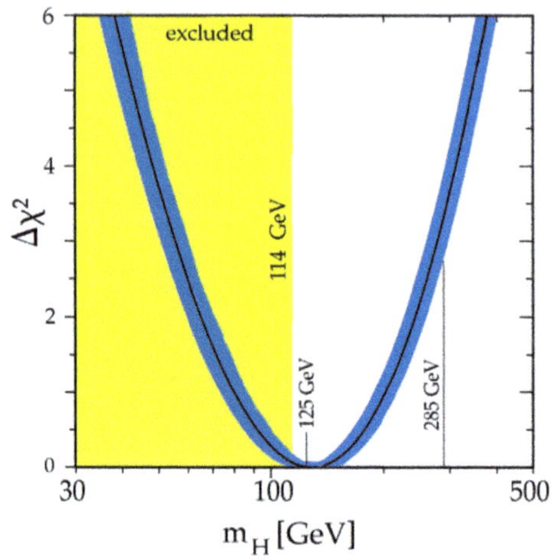

Figure 7.2. Prediction for the Higgs mass calculated from various measurements and employing radiative corrections. The graph shows how the reduced χ^2 changes as a function of Higgs mass. The blue band shows the theoretical uncertainties. The yellow area was excluded at the 95% confidence level by direct searches at CERN's large electron–positron storage ring (LEP). The probability for the mass to lie below 285 GeV was 95%. Adapted from [2].

Table 7.2. Dominant decay modes of elementary particles; *apart from oscillations $\nu_e \leftrightarrow \nu_\mu$, $\nu_\mu \leftrightarrow \nu_\tau$ and $\nu_\tau \leftrightarrow \nu_e$; †via $W^- \to e^-\bar{\nu}_e$; ‡via $W^- \to q\bar{q}'$, $e^-\bar{\nu}_e$, $\mu^-\bar{\nu}_\mu$; $^\$$ordered in decreasing probability.

Mother	Daughter products
ν	stable*
e^-	stable
μ^-	$e^-\bar{\nu}_e\nu_\mu{}^\dagger$
τ^-	$q\bar{q}'\nu_\tau{}^\ddagger$, $e^-\bar{\nu}_e\nu_\tau$, $\mu^-\bar{\nu}_\mu\nu_\tau$
u, d	stable
s	uW^-
c	sW^+
b	cW^-
t	bW^+
γ, g	stable
Z^0	lepton + antilepton, $q\bar{q}$
W^+	$q\bar{q}'$, $e^+\nu_e$, $\mu^+\nu_\mu$, $\tau^+\nu_\tau$
W^-	$q\bar{q}'$, $e^-\bar{\nu}_e$, $\mu^-\bar{\nu}_\mu$, $\tau^-\bar{\nu}_\tau$
H	$b\bar{b}$, W^+W^-, gg, $\tau^+\tau^-$, Z^0Z^0, $c\bar{c}$, $\gamma\gamma^\$$

The spin $\vec{J} = \vec{S} + \vec{L}$ of the meson is constructed from the quark and antiquark spin vectors, $\vec{S} = \vec{S}_q + \vec{S}_{\bar{q}}$, and by adding the relative angular momentum $\vec{L}$. The corresponding quantum numbers are $S = 0$ or 1 and $L = 0, 1, 2$, etc. The meson spin J is a whole number with $|L - S| \leqslant J \leqslant L + S$. Hence mesons are (finite-size) *bosons*. For example, the pion (π^+) is made of the quarks $u\bar{d}$ and has charge $Q = \frac{2}{3} + \frac{1}{3} = +1$ (in units of e) and spin 0.

Baryons are structures consisting of three quarks (qqq), while antibaryons are composed of three antiquarks ($\overline{qqq}$). The baryon numbers are

$$\begin{aligned} B &= 1 \text{ for baryons,} \\ B &= -1 \text{ for antibaryons.} \end{aligned} \tag{7.13}$$

The spin (quantum number) of a baryon is half-integer and equal to $\frac{1}{2}$ or $\frac{3}{2}$ (three spins aligned), or larger when adding angular momentum ($\frac{5}{2}, \frac{7}{2}, \ldots$). Hence baryons are (finite-size) *fermions*. For instance, the proton (p) is a uud system with charge $Q = \frac{2}{3} + \frac{2}{3} - \frac{1}{3} = +1$, the neutron ($n$) a udd system with $Q = \frac{2}{3} - \frac{1}{3} - \frac{1}{3} = 0$. Both are spin-$\frac{1}{2}$ particles. The antiproton ($\bar{p}$) ($\overline{uu}\bar{d}$) has the charge $Q = -1$. A *hyperon* is a baryon containing at least one quark from the second or third generation, such as the (electrically neutral) Λ (uds).

Quarks were proposed in the 1960 s by Murray Gell-Mann, Yuval Ne'eman, André Petermann and George Zweig, initially as a bookkeeping tool to reduce the large number of hadrons (see [1] which also lists the properties of hadrons and other particles). The theory of strong interactions *quantum chromodynamics* (QCD) predicts the existence of composites such as

$$q\bar{q},\ qqq,\ \overline{qqq}, \tag{7.14}$$

while structures such as

$$qq,\ qq\bar{q},\ {}^{235}q \tag{7.15}$$

are excluded. Hadrons composed of

$$\begin{aligned} &qq\overline{qq} \text{ (tetraquark)},\ qqqq\bar{q} \text{ (pentaquark)},\ 6q \text{ (dibaryon)}, \\ &gg \text{ or } ggg \text{ (glueballs)} \end{aligned} \tag{7.16}$$

are also possible in QCD. Several candidates have been observed [1] (for a review see Amsler, p 949 in [3]). We shall return to the spectrum of hadrons in chapter 10.

7.3 Interactions

Particles and their interactions are conveniently depicted by *Feynman* diagrams for which we shall adopt the following simple rule:

> The arrow indicating the direction of time has to be reversed when replacing the particle by its antiparticle.

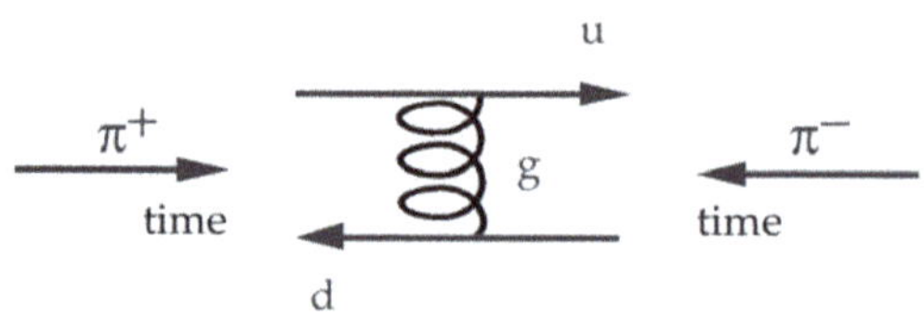

Figure 7.3. Feynman diagram of the pion. The arrows indicate the direction of time for the labelled particles.

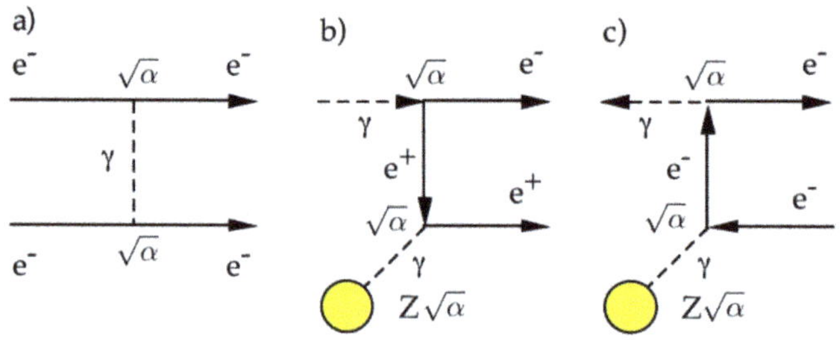

Figure 7.4. Møller scattering $e^-e^- \to e^-e^-$ (*a*), pair production $\gamma Z \to e^-e^+\ Z$ (*b*), and bremsstrahlung $e^-Z \to e^-\gamma Z$ in the Coulomb field of a nucleus with atomic number Z (*c*).

For example, a positron is an electron moving backward in time [4]. Then, its 4-momentum $\mathcal{P} \equiv (E, \vec{p})$ is replaced by $-\mathcal{P}$. An example is shown figure 7.3: if time flows from left to right the figures shows a π^+ (outgoing u and incoming d quark, hence $u\bar{d}$ with $Q=+1$). If time flows from right to left the figure shows a π^- (outgoing d and incoming u quark, hence $\bar{u}d$ with $Q=-1$). The $q\bar{q}$ pair is bound by the exchanged gluon g which is often not shown.

The coupling strength to the photon is determined by the electric charge. Therefore the strength at each vertex in the graph is governed by the amplitude $\sqrt{\alpha}$. Figure 7.4 shows examples of Feynman graphs for the electromagnetic interaction between electrons. The cross section for *Møller* scattering (figure 7.4(*a*)) is proportional to *the square* of the amplitude, hence

$$\frac{d\sigma}{d\Omega} \propto \left(\sqrt{\alpha} \times \sqrt{\alpha}\right)^2 = \alpha^2. \tag{7.17}$$

We have already encountered the α^2-factor in the differential cross section for Rutherford scattering (1.14). Each additional γ exchange contributes another factor α^2 which reduces the cross section. There are further kinematic factors such as the scattering angle and the incident energy.

Pair production (figure 7.4(*b*)) is the creation of an electron–positron pair in the Coulomb field of the nucleus, with differential cross section

$$\frac{d\sigma}{d\Omega} \propto \left(Z\sqrt{\alpha} \times \sqrt{\alpha} \times \sqrt{\alpha}\right)^2 = Z^2\alpha^3. \tag{7.18}$$

This process is not possible with a massless photon in vacuum without violating momentum and energy conservation. The electron can also scatter in the Coulomb field of the nucleus and emit a photon (*bremsstrahlung*, figure 7.4(*c*). Panel (*c*) is

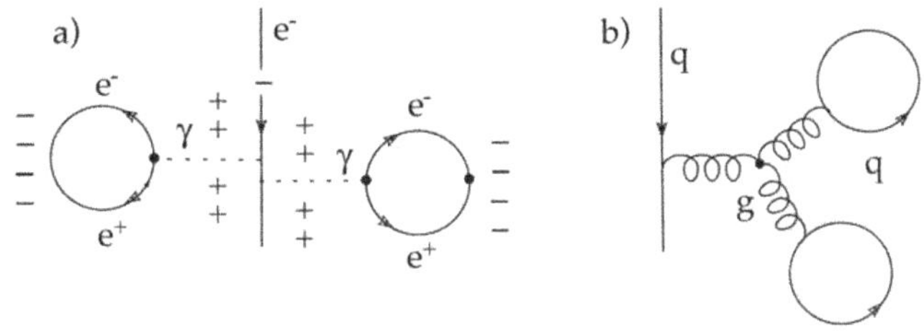

Figure 7.5. Vacuum polarization in the electromagnetic (*a*) and strong (*b*) interactions.

equivalent to panel (*b*) and the cross section is also proportional to $Z^2\alpha^3$. However, they are not equal due to the different kinematic factors.

The coupling constant describing the strength of the interaction is not constant but varies with energy or, more precisely, with 4-momentum transfer Q (see (2.7)). This follows from quantum mechanical corrections: the electromagnetic field of the electron can generate transient (*virtual*) e^+e^- pairs which may exist only during the time Δt, where $2m_e\Delta t \sim \hbar$ (figure 7.5(*a*)). One speaks of higher order corrections to the Feynman graph. The charged cloud of the positrons is attracted by the electron (*vacuum polarization*) and shields its charge. Another particle flying nearby then feels a reduced charge which increases with decreasing distance (increasing momentum transfer Q). The value of the coupling 'constant' at momentum transfer Q can be calculated from its value at momentum transfer μ:

$$\alpha(Q) = \frac{\alpha(\mu)}{1 - \frac{\alpha(\mu)}{\pi}\ln\left(\frac{Q^2}{\mu^2}\right)} \tag{7.19}$$

for vacuum polarization, including the three leptonic loops e, μ and τ. The contributions from quark loops are much smaller. The fine-structure 'constant' α is equal to $\frac{1}{137}$ at $Q = \mu = 1$ MeV c^{-1} and increases with momentum transfer to $\frac{1}{129}$ at $Q = 100$ GeV c^{-1}.

The reverse effect of *antishielding* occurs for the strong interaction. The gluon field radiated by the quark emits further gluons and virtual $q\bar{q}$ pairs that stretch the strong interaction field to larger distances (figure 7.5(*b*)). A neighbouring antiquark feels a growing attractive force with increasing distance, similar to the force acting on a stretched string. The attraction becomes so large that quarks and antiquarks cannot be separated from one another. Quarks and gluons are *confined* in hadrons. In contrast, the binding force weakens with decreasing distance (*asymptotic freedom*). The strength of the strong interaction is determined by the strong coupling 'constant' α_s. The latter can be measured from the ratio R of cross sections for $e^+e^- \to q\bar{q} \to$ hadrons and $e^+e^- \to \mu^+\mu^-$:

$$R = 3\sum_i q_i^2\left[1 + \frac{\alpha_s(Q)}{\pi}\right], \tag{7.20}$$

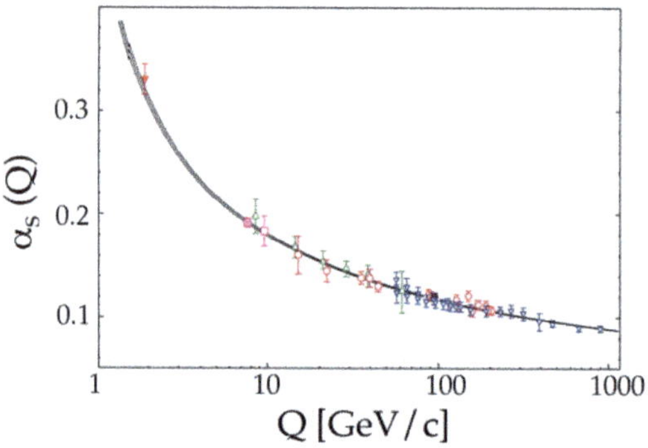

Figure 7.6. Measured dependence of α_s on Q from various processes; $\alpha_s = 0.1185 \pm 0.0006$ at the mass of the Z^0. After [1].

where the sum runs over all kinematically allowed $q\bar{q}$ flavours with charge q_i. The momentum transfer dependence of α_s is given by

$$\alpha_s(Q) = \frac{\alpha_s(\mu)}{1 + \frac{33-2n_f}{12\pi}\alpha_s(\mu)\ln\left(\frac{Q^2}{\mu^2}\right)}, \tag{7.21}$$

with n_f quark flavours (figure 7.6). At very small momentum transfer α_s is equal to ~0.75 [5], roughly 100 times larger than α.

Since α grows with momentum transfer while α_s decreases, one expects the electromagnetic and strong interactions to become equally strong at some very high energy. Indeed, this occurs around 10^{16} GeV, where the weak interaction also reaches the same strength (section 7.4). Such energies were presumably present shortly after the Big Bang. The highest man-made energy that is currently available is 8 TeV at the LHC, to be boosted soon to 13 TeV.

7.3.1 Conservation laws

In the course of this textbook, we shall see which conservation laws apply to which interactions. The relationship between conservation laws and symmetries will be treated in chapter 11. The following three conservation laws hold for all interactions:

1. The electric *charge* is conserved.
2. The (additive) *baryon number* (7.7) is conserved. Quarks and antiquarks are therefore always produced or annihilated pairwise (figure 7.7(*a*)). Inverting the time direction of the antiquark, and replacing $\bar{u}$ by u, shows that a quark line cannot be interrupted without violating baryon number conservation. However, the quark *flavour* may change by absorbing or emitting a $W^{\pm}$, taking conservation of charge into account (figure 7.7(*b*)).

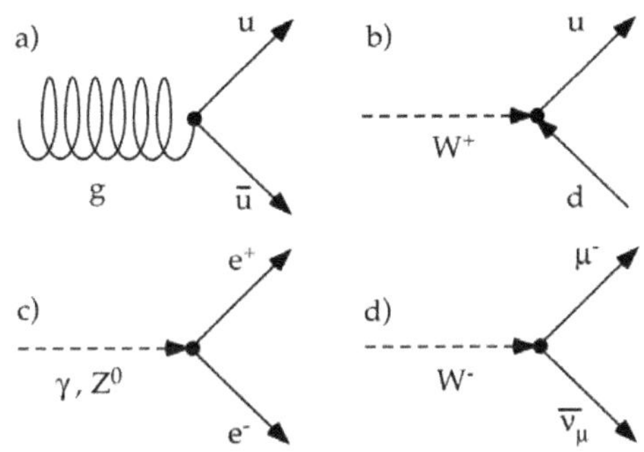

Figure 7.7. Examples of baryon number conservation in the strong interaction (*a*) and in the weak interaction (*b*). Lepton number is conserved in electromagnetic and weak processes (*c, d*).

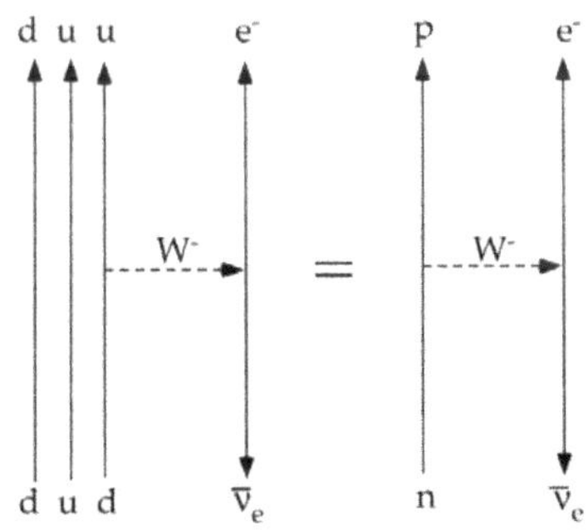

Figure 7.8. Feynman diagram of the neutron β^--decay.

3. The (additive) *lepton number* (table 7.1) is conserved separately in each lepton family:

$$\sum L_e = \text{constant}, \quad \sum L_\mu = \text{constant}, \quad \sum L_\tau = \text{constant}. \tag{7.22}$$

   Leptons and antileptons of the *same* family are always produced or annihilated pairwise, taking charge conservation into account (figures 7.7(*c*) and 7.7(*d*)).

Figure 7.8 shows the Feynman diagram of neutron β^--decay. Electric charge, baryon and lepton numbers are conserved during the decay process

$$n \to pe^-\bar{\nu}_e\left(Q = 0,\, B = 1,\, \sum L_e = 0\right). \tag{7.23}$$

These quantum numbers are also conserved during the underlying elementary interaction between quarks and leptons:

$$d \to u + W^- \to u + e^-\bar{\nu}_e\left(Q = -\frac{1}{3},\, B = \frac{1}{3},\, \sum L_e = 0\right). \tag{7.24}$$

Further examples of allowed and forbidden processes are

$$\mu^- \to e^-\bar{\nu}_e\nu_\mu, \quad Q = -1, \quad B = 0, \quad \sum L_\mu = 1, \quad \sum L_e = 0, \text{ main decay mode of } \mu^-,$$
$$\mu^- \to e^-\nu_e\bar{\nu}_\mu, \quad Q = -1, \text{ forbidden (lepton number conservation)},$$
$$\tau^+ \to \pi^+\bar{\nu}_\tau, \quad Q = 1, \quad B = 0, \quad \sum L_\tau = -1, \text{ allowed},$$
$$\pi^+ \to \mu^+\nu_\mu, \quad Q = 1, \quad B = 0, \quad \sum L_\mu = 0, \text{ main decay mode of the } \pi^+,$$
$$\pi^- \to \mu^-\bar{\nu}_\mu, \quad Q = -1, \quad B = 0, \quad \sum L_\mu = 0, \text{ main decay mode of the } \pi^-,$$
$$\mu^+ \to e^+\gamma, \quad Q = 1, \quad B = 0, \text{ forbidden (lepton number conservation)},$$
$$p \to e^+\pi^0, \quad Q = 1, \text{ forbidden (lepton and baryon number conservation)}. \qquad (7.25)$$

Two important remarks on lepton and baryon numbers should be made.

1. Strictly speaking, only the total lepton number $\sum(L_e + L_\mu + L_\tau)$ is conserved: the first evidence for neutrino oscillations was reported in 1998 (section 14.6). For example, $\bar{\nu}_e$ from nuclear reactors slowly transform into $\bar{\nu}_\mu$ or $\bar{\nu}_\tau$. The wavelength of the oscillation is typically 18 km for reactor antineutrinos. Thus, the lepton number is not conserved in neutrino oscillations. However, this hardly affects the lepton number conservation for charged leptons: the experimental upper limit for the probability of a muon to decay into $e\gamma$ is 5.7×10^{-13} [6], to be compared with the prediction of 10^{-45} from neutrino oscillations.
2. The proton is unstable in grand unified theories (GUTs) which describe the unification of the three forces at very high energy (section 7.4). The decay process $p \to e^+\pi^0$ is possible but violates baryon number conservation. However, the mean lifetime of the proton is extremely long, with a measured lower limit from this channel of $\sim 8 \times 10^{33}$ years [1].

For all practical purposes, baryon and lepton numbers are conserved.

7.3.2 Exchange force

The classic image of a two-body interaction (particle 2 exerts a force $\vec{F} = -\text{grad } V_{12}$ on particle 1) is described in quantum mechanics by the exchange of a third particle with energy E (figure 7.9). To ensure conservation of energy the transit time t of

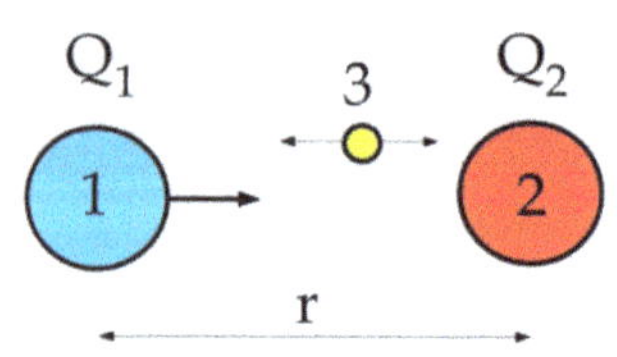

Figure 7.9. The interaction between particles 1 and 2 is mediated by the exchange of the virtual particle 3.

particle 3 needed to cover the distance between 1 and 2 should be short enough to satisfy the uncertainty principle

$$E\ t \simeq \hbar = 1. \tag{7.26}$$

Thus particle 3 exists only during time t and is *virtual*. Consider for instance the electromagnetic interaction between two charges Q_1 and Q_2 separated by the distance r (figure 7.9). From the energy of the exchanged photon and the time t one finds with $c = 1$ that

$$E = q, \qquad t = r \Rightarrow qr \simeq 1, \tag{7.27}$$

using (7.26) and where q is the momentum transferred by particle 1 to particle 2, which is equal to the momentum of the photon. The force is then

$$F = \frac{\mathrm{d}q}{\mathrm{d}t} = \frac{\mathrm{d}}{\mathrm{d}t}\left(\frac{1}{r}\right) = \frac{\mathrm{d}}{\mathrm{d}t}\left(\frac{1}{t}\right) = -\frac{1}{t^2} = -\frac{1}{r^2}, \tag{7.28}$$

in accordance with the expected $1/r^2$ dependence of the Coulomb law. With additional photons the force becomes accordingly larger, with proportionality constant $Q_1 Q_2$.

The strong interaction is *short-range* (not macroscopically perceptible) while the electromagnetic force is *long-range*. How does one then obtain a force that decays faster than $1/r^2$? The solution was proposed by Hideki Yukawa in 1935 who introduced the exchange of a particle 3 with *finite* mass m. The total energy exchanged is

$$E^2 = q^2 + m^2, \tag{7.29}$$

again with $c = 1$. The wave equation of particle 3 is derived by using the correspondence principle

$$E \to \mathrm{i}\frac{\partial}{\partial t}, \qquad \vec{q} \to -\mathrm{i}\,\mathrm{grad} \Rightarrow -\frac{\partial^2}{\partial t^2}\psi = -\Delta\psi + m^2\psi. \tag{7.30}$$

This is the *Klein–Gordon* equation, the relativistic extension of the Schrödinger equation. In the limit $m \to 0$ one recovers the wave equation of the free photon,

$$\Delta\psi = \frac{\partial^2\psi}{\partial t^2}. \tag{7.31}$$

The stationary (time-independent) case $\Delta\psi = 0$ leads to the solution

$$\psi(r) \propto \frac{1}{r} \tag{7.32}$$

which is the Coulomb potential produced by of point-like charge located at $r = 0$. Similarly, let us look for the stationary solution of (7.30) for $m \neq 0$, by expressing the Δ operator in spherical coordinates:

$$\Delta\psi = \frac{1}{r^2}\frac{\mathrm{d}}{\mathrm{d}r}\left(r^2\frac{\mathrm{d}\psi}{\mathrm{d}r}\right) = m^2\psi. \tag{7.33}$$

The solution is

$$\boxed{\psi(r) = \frac{g_s}{4\pi r}\exp\left(-\frac{r}{R}\right)} \quad \text{with} \quad \boxed{R \equiv \frac{1}{m}}, \tag{7.34}$$

as can easily be proven by inserting into (7.33). The potential of the exchanged massive particle decreases faster than $1/r$. The mean *range* R of the force is the *Compton wavelength* of the particle. For vanishing m we find

$$m \to 0 \Rightarrow R \to \infty, \qquad \psi(r) \propto \frac{1}{r}, \tag{7.35}$$

as expected. The coupling constant g_s describes the strength of the strong interaction and plays the role of the electric charge in the electromagnetic interaction. We have seen that at low energy,

$$\alpha_s \equiv \frac{g_s^2}{4\pi} \sim 1 \gg \alpha = \frac{e^2}{4\pi\epsilon_0} = \frac{1}{137}. \tag{7.36}$$

We can now estimate the mass m of the exchanged particle, responsible for the strong interaction force:

$$R \sim 1 \text{ fm} \Rightarrow m \sim 1 \text{ fm}^{-1} \simeq 200 \text{ MeV}. \tag{7.37}$$

In 1936 Carl Anderson and Seth Neddermeyer discovered a particle (the muon) in the cosmic radiation with a mass of 106 MeV, which was initially thought to be Yukawa's particle. However, the interaction of the muon with nuclear matter appeared to be too weak for a strongly interacting particle. Yukawa's particle, the charged pion with a mass $m = 140$ MeV, was discovered in 1947 by César Lattes, Giuseppe Occhialini and Cecil Powell [7, 8], also in cosmic rays (figure 7.10).

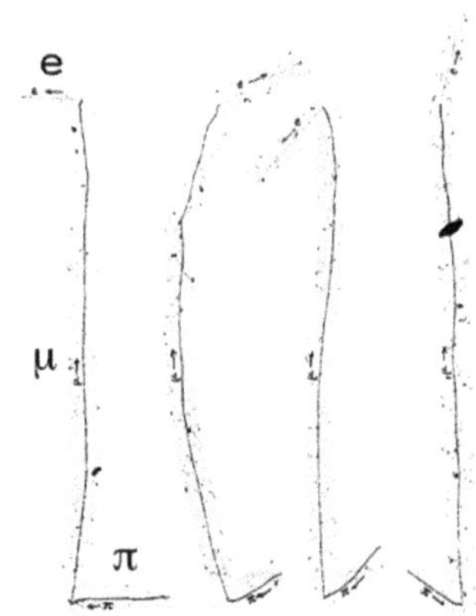

Figure 7.10. The discovery of the π^+ with photographic emulsions at the Pic du Midi (see the text for details). After [8].

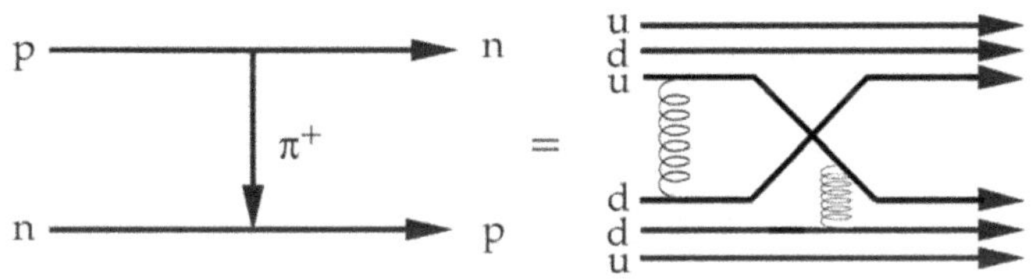

Figure 7.11. *pn* interaction mediated by π exchange (left). The diagram on the right shows the interpretation at the quark and gluon levels.

The short tracks at the bottom are due to stopping pions which decay into muons. The track lengths of the muons are similar ($\simeq$600 μm) in all four events, indicating that the muon is always emitted with the same energy (two-body decay $\pi^+ \to \mu^+ \nu_\mu$). We deal here with positively charged pions since stopping negative pions would be captured and absorbed by nuclei (section 2.3), while our muons decay into lighter particles (positrons, the two neutrinos being invisible). The ionizing power of the stopping π or μ is much larger than that of the faster and lighter positron.

Figure 7.11 shows an example of Feynman diagram depicting the interaction between a proton and a neutron, mediated by the exchanged pion. The 'macroscopic' image (left) shows the strong interaction as a residue (superposition) of the underlying quark and gluon forces (right). This is similar to the van der Waals force in molecules, which is a residue of the underlying Coulomb forces.

The range of the weak interaction is extremely short, nearly point-like:

$$R = \frac{1}{M_W} = \frac{1}{8 \times 10^4}\ \text{MeV}^{-1} = 2.5 \times 10^{-3}\ \text{fm}. \tag{7.38}$$

Finally, let us compare the strength of gravitation with that of electromagnetism. Gravitational and Coulomb forces between two masses M with charges e are equal when

$$K\frac{M^2}{r^2} = \frac{\alpha}{r^2}. \tag{7.39}$$

Using the universal gravitational constant $K = 6.7 \times 10^{-39}$ GeV^{-2} gives $M \sim 10^{18}$ GeV. For $M \sim 10^{19}$ GeV (the Planck mass) gravitational and strong interactions become comparable. Given the huge masses involved here we can safely neglect gravitation in particle physics.

7.3.3 Charged and neutral currents

The weak interaction is mediated by the exchange of a W or a Z^0 boson. The interaction between *charged currents*, such as in neutron β-decay (figure 7.8), involves a W boson. Another example of interaction between charged currents, μ^+ decay, is depicted in figure 7.12(*a*). Because of lepton number conservation, only leptons of the *same* family couple to W bosons.

In contrast, transitions between quark generations are allowed but unlikely (figure 7.12(*b*)–(*d*)). The interaction may change the quark flavour while conserving

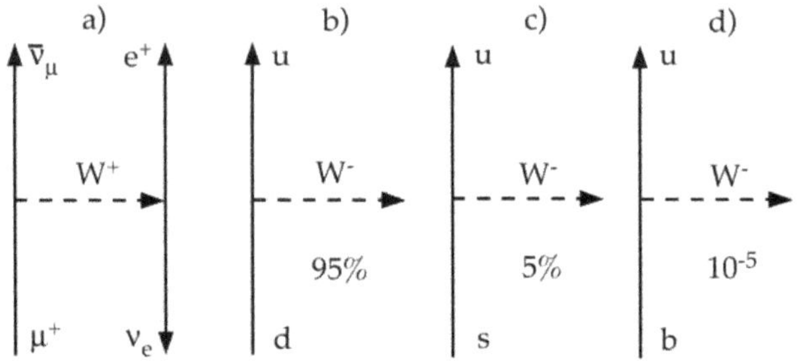

Figure 7.12. Charged current interactions: (*a*) *W* exchange in μ decay; (*b*)–(*d*) *W* emission by *d*, *s* and *b* quarks with relative strengths. The *u* quark couples preferably to the *d*, the *c* quark to the *s*, and the *t* quark to the *b*.

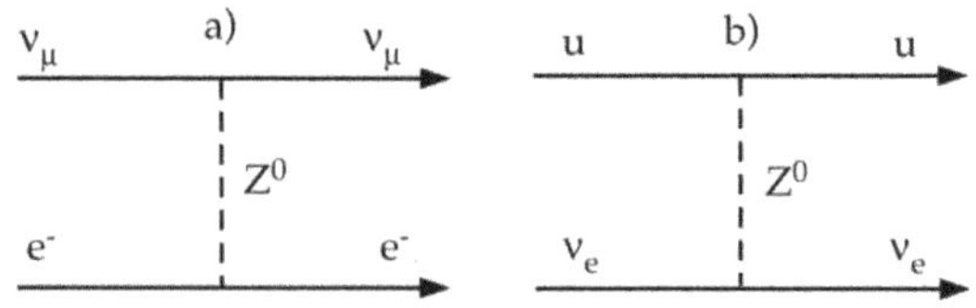

Figure 7.13. Neutral current interactions with Z^0 exchange.

the electric charge. Hence the transition $u \to \bar{d}$ with *W* emission is forbidden by both charge and baryon number conservation. The *Cabibbo–Kobayashi–Maskawa* (CKM) suppression of transitions between quark generations (CKM mechanism) is explained in section 16.3.

The interaction between *neutral currents* occurs through the exchange of a Z^0 boson. The lepton number and the electric charge are conserved (figure 7.13(*a*)). The quark flavour is also conserved (figure 7.13(*b*)), e.g. $u \to c$ transitions are forbidden. Conservation of flavour in neutral current interactions is a consequence of the *Glashow–Iliopoulos–Maiani* (GIM) mechanism explained in section 16.3. Flavour changing neutral currents are also forbidden with gluon exchange (strong interaction) and γ exchange (electromagnetic interaction).

7.3.4 Cross section and mean lifetime

Let us compare the typical values of cross sections and lifetimes for the three interactions. The total cross section for hadronic reactions is in the range mb to b. For example, $\sigma_T \sim 50$ mb in *pp* scattering and $\sigma_T \sim 25$ mb in πp scattering above 10 GeV. Given the typical interaction range $R \sim 1$ fm, one would indeed guess that

$$\sigma_T \sim \pi R^2 = 30 \text{ mb}. \tag{7.40}$$

Similarly, for the weak interaction one expects $\sigma_T \sim 100$ nb with the much shorter range of $R \sim 2.5 \times 10^{-3}$ fm. However, the cross sections are even smaller, e.g. $\sigma_T \sim 0.1$ pb for $\nu_e p$ scattering at 10 GeV.

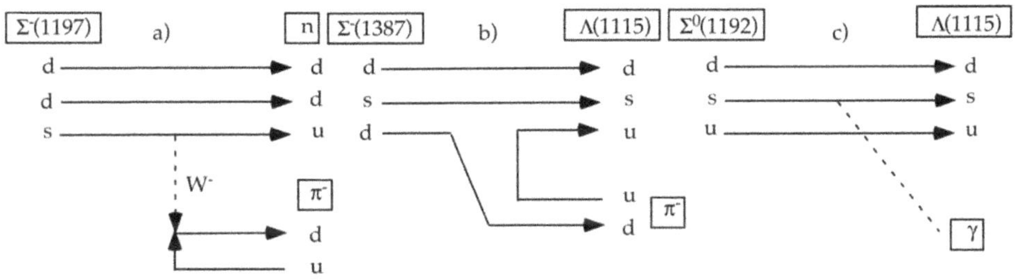

Figure 7.14. (*a*) $\Sigma^-(1197)$ decay; (*b*) $\Sigma^-(1387)$ decay; (*c*) $\Sigma^0(1192)$ decay. The number in parentheses refers to the mass in MeV. The wave functions describing the quark substructures in the Σ^0 and Λ are different.

The typical mean lifetimes of decaying particles are

$$\tau \geqslant 10^{-12}\text{ s for the weak,}$$
$$10^{-20} \leqslant \tau \leqslant 10^{-12}\text{ s for the electromagnetic,}$$
$$\tau \leqslant 10^{-20}\text{ s for the strong}$$

interaction. These values are indicative only. For heavy bosons τ decreases with the third power of the boson mass (section 17.2.1). For example, the W boson decays weakly, but is short-lived ($\tau \sim 3 \times 10^{-25}$ s), due to its very high mass.

Particles decaying only weakly, such as the π, μ, K, Λ, Σ, etc, are often considered as 'stable', their lifetimes being long enough for the production of beams. Figure 7.14 shows a few decay graphs.

1. The spin-$\frac{1}{2}$ $\Sigma^-(1197)$, also known simply as Σ^-, decays as

$$\Sigma^-(1197) \rightarrow n\pi^-. \tag{7.41}$$

   Its mean lifetime is $\tau = 1.5 \times 10^{-10}$ s, corresponding to a weak decay in which the s quark transforms into a u quark (figure 7.14(*a*)).

2. The spin-$\frac{3}{2}$ $\Sigma^-(1387)$ has the same quark content as the Σ^- but decays mainly as

$$\Sigma^-(1387) \rightarrow \Lambda(1115)\pi^- \tag{7.42}$$

   into the neutral $\Lambda(1115)$, also known as Λ (figure 7.14(*b*)). The lifetime $\tau \sim 10^{-23}$ s is typical for a strong interaction. Note that the decay (7.42) is not possible for the $\Sigma^-(1197)$ due to the negative Q-value.

3. The spin-$\frac{1}{2}$ $\Sigma^0(1192)$, also known as Σ^0, decays electromagnetically as

$$\Sigma^0(1192) \rightarrow \Lambda(1115)\gamma, \tag{7.43}$$

   with a mean lifetime of $\tau = 7.4 \times 10^{-20}$ s (figure 7.14(*c*)). This decay mode is not possible for the $\Sigma^-(1197)$ due to charge conservation (a charged Λ does not exist).

Note that the decay mode $\Sigma^+ \rightarrow p\gamma$ ($f = 1.2 \times 10^{-3}$) is a weak process ($s \rightarrow u$ transition).

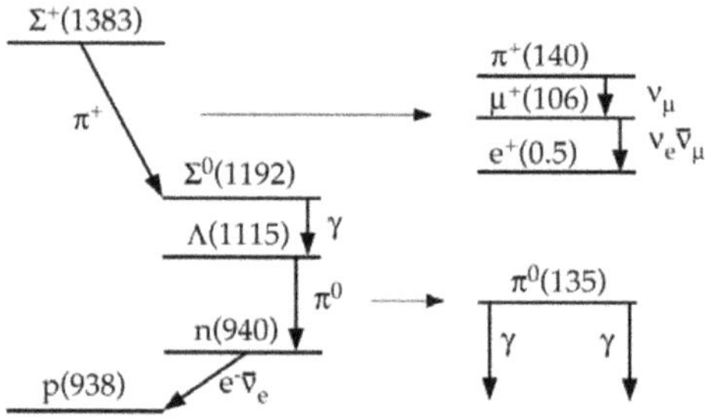

Figure 7.15. $\Sigma^+(1383) \to p\, e^+e^-4\nu\; 3\gamma$. After [9].

Figure 7.15 shows the full decay chain of the $\Sigma^+(1383)$, a spin-$\frac{3}{2}$ *uus* hyperon:

$$\Sigma^+(1383) \to p\; e^+e^-4\nu\; 3\gamma, \tag{7.44}$$

involving all three interactions.

7.4 Beyond the standard model

Although no deviations between experiments and theoretical predictions have been found so far, the standard model is unsatisfactory as it contains many free parameters which need to be fitted to the data. These are for instance the coupling constants of the three interactions, the quark and lepton masses, the weak couplings between quark generations and in neutrino oscillations, as well as parameters describing the asymmetry between matter and antimatter (*CP* violation). Furthermore, we have seen that the strength of the interaction increases with energy for electromagnetic processes, while decreasing for strong ones. Hence the two strengths will become comparable at a certain energy. The question then arises as to whether the weak interaction also reaches the same level. We will see that unification of the three forces is expected at an energy of about 10^{16} GeV.

The first unification model, the GUT was formulated in 1974 by Howard Georgi and Sheldon Glashow [10]. The model is based on the SU(5) symmetry which classifies the leptons and quarks into multiplets. Within a multiplet (quintet or decuplet) leptons transform into quarks by emitting or absorbing heavy ($\sim 10^{13}$ to 10^{17} GeV) bosons with charge $-\frac{1}{3}e$ (*Y* boson) or $-\frac{4}{3}e$ (*X* boson). For example, a *d* quark decays into a positron by emitting an *X* boson, i.e. $d \to e^+X$ (figure 7.16). Therefore, baryon and lepton numbers are not conserved and the proton is unstable.

Experimentally one searches for the decay of the proton in huge (1000 t) water tanks, installed in deep underground mines to reduce the background from cosmic radiation. The signal would be the observation of a positron and a photon from the decay $p \to e^+\pi^0$ ($\pi^0 \to 2\gamma$). The photons then convert into e^+e^- pairs which are identified by their Čerenkov radiation (section 9.10). As mentioned already, the decay of the proton has not been observed so far with a lower limit on the mean lifetime of about 10^{34} years. The theoretical prediction lies in the range of 10^{37} to 10^{38} years, when assuming a unification energy of $\sim 10^{16}$ GeV.

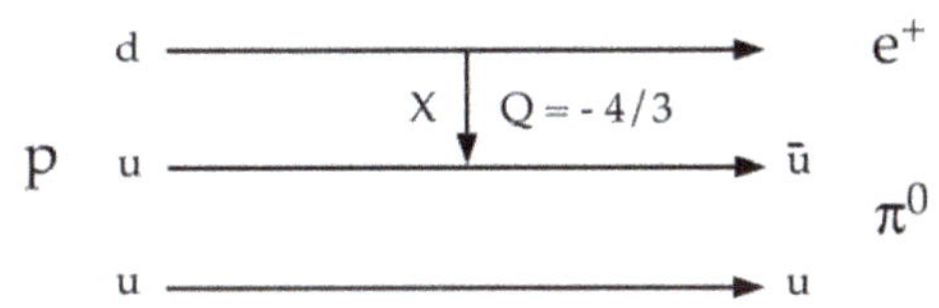

Figure 7.16. Proton decay into $e^+\pi^0$ in GUTs.

The SU(5) model incorporates fractional quark charges and the equality of proton and electron charges, $Q(p) = -Q(e^-)$. Let us define the (energy-dependent) coupling constants.

$$\alpha_1 \equiv \frac{5}{3}\frac{g'^2}{4\pi} = \frac{5}{3}\left(\frac{\alpha}{\cos^2\theta_w}\right) \qquad \text{and} \qquad \alpha_2 \equiv \frac{g^2}{4\pi} = \frac{\alpha}{\sin^2\theta_w}. \tag{7.45}$$

The (energy-dependent) quantity $\sin\theta_w$ is the sine of the Weinberg angle that mixes the U(1)-interaction (coupling g'^2) with the SU(2)-interaction (coupling g^2), see section 16.7. The strengths of the two components α_1 and α_2 of the *electroweak* interaction are then related to the fine-structure constant by

$$\frac{5}{3\alpha_1} + \frac{1}{\alpha_2} = \frac{1}{\alpha}, \tag{7.46}$$

which holds at all energies. Unification $\alpha_1 = \alpha_2 = \alpha_{\text{GUT}}$ is reached when $\alpha_1 = \frac{8\alpha}{3}$, which then leads to the prediction $\sin^2\theta_w = \frac{3}{8} = 0.37$ at the GUT energy (compared to the measured low energy value of 0.23). For the strong interaction one also expects that $\alpha_3 \equiv \alpha_s = \alpha_{\text{GUT}}$ at the unification energy. Taking into account the Q-dependence of α and α_s, see (7.19) and (7.21), as well as the Q-dependence of $\sin^2\theta_w$, one can compute the couplings $\alpha_{1,2,3}$ as a function of momentum transfer Q. Figure 7.17(a) shows that they do not converge to any common value.

The forces can nonetheless be unified in the framework of the minimal supersymmetric standard model (MSSM) [12]. This model eliminates the divergences that arise in the calculation of the Higgs mass. Corrections due to loops of virtual

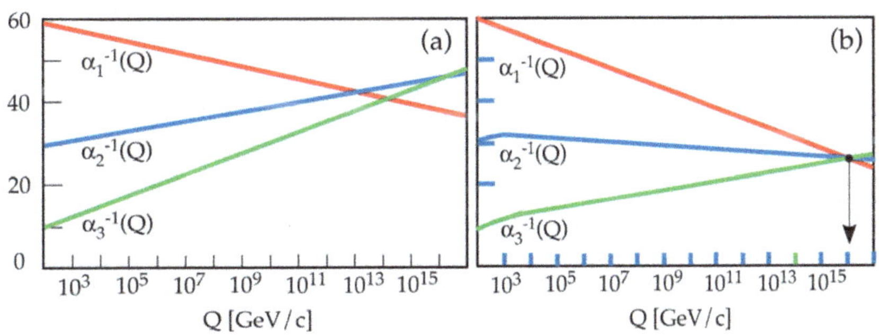

Figure 7.17. Q dependence of the coupling constants for the three interactions in the standard model (a) and in the MSSM (b). After [11].

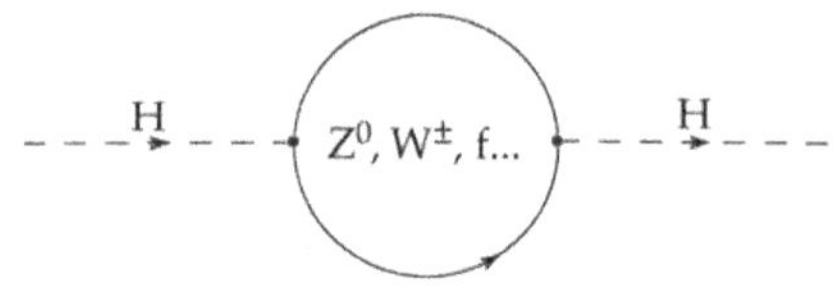

Figure 7.18. Radiative corrections to the Higgs mass.

particle–antiparticle pairs (figure 7.18) move the mass of the Higgs boson towards infinity. This difficulty can be solved by postulating the existence of a spin-0 supersymmetric (SUSY) partner for each spin-$\frac{1}{2}$ fermion. A spin-$\frac{1}{2}$ SUSY partner is also assigned to each spin-1 or spin-0 boson. The contributions of the loops with SUSY particles in figure 7.18 have opposite signs, thus compensating the contributions of standard model loops and stabilizing the Higgs mass.

The SUSY particles are labelled by adding a tilde to the symbol of the standard model particle. The letter 's' is prefixed to the standard model fermions, while the suffix 'ino' is appended to the bosons. The SUSY partners of the leptons and quarks are the spin-0 sleptons and squarks. Examples are the selectron $\tilde{e}^-$ and the stop squark $\tilde{t}$. The SUSY partners of the gauge bosons are the spin-$\frac{1}{2}$ gauginos. These are the gluinos ($\tilde{g}$), photino ($\tilde{\gamma}$), winos ($\tilde{W}^{\pm}$) and zino ($\tilde{Z}^0$), supplemented by the Higgsino ($\tilde{H}$).

The R parity, defined as

$$R \equiv (-1)^{3[B-L]+2S}, \tag{7.47}$$

is conserved, where S is the spin. It is straightforward to verify that for standard model particles $R = +1$, while for SUSY particles $R = -1$. SUSY particles are therefore produced pairwise. Examples of processes are

$$\begin{aligned} q\bar{q} &\to \tilde{q}\bar{\tilde{q}}, \\ \tilde{q} &\to q\tilde{\gamma}. \end{aligned} \tag{7.48}$$

The MSSM contains five Higgs bosons, h^0, H^0, A^0, H^+ and H^-, and correspondingly, the spin-$\frac{1}{2}$ Higgsinos $\tilde{h}^0$, $\tilde{H}_1^0$, $\tilde{H}_2^0$, $\tilde{H}^+$ and $\tilde{H}^-$. The $\tilde{H}_1^0$ and $\tilde{H}_2^0$ mix (in the quantum mechanical sense) with the photino and zino and generate the neutralinos $\tilde{\chi}^0_{1,2,3,4}$. Similarly, the $\tilde{H}^+$ and $\tilde{H}^-$ mix with the two winos, generating the charginos $\tilde{\chi}_1^{\pm}$ and $\tilde{\chi}_2^{\pm}$.

SUSY particles are expected to be heavy, lying roughly in the 100 to 1000 GeV range. Since R parity is conserved they cannot be annihilated by matter but will decay into lighter particles, or scatter. The lightest SUSY particle must be stable, again due to the conservation of R parity. The experimental evidence for SUSY would be the observation of a large missing mass. One of the four neutralinos could be responsible for the missing dark matter in the Universe. Figure 7.17(b) shows how the coupling constants vary, when loops containing SUSY particles are included. The three forces become of equal strengths around 10^{16} GeV when SUSY masses in the 1 TeV range are assumed [11].

The detection of supersymmetry was among the main goals in the construction of the LHC which was commissioned at CERN in 2009. So far, supersymmetry has not been discovered with the LHC running at a 7–8 TeV centre-of-mass energy. Gluinos with masses below 1 TeV are excluded, as well as first and second generation squarks, while squarks from the third generation have been sought without success below 600 MeV. The lower limit for sleptons is around 200 GeV. However, these limits depend on the assumptions made on the SUSY spectrum.

The cross sections for SUSY production and interaction with standard model particles are typical of the weak interaction. The impending operation of the LHC at 13 TeV will boost the production cross section for SUSY particles in the mass range 1.5 to 2 TeV by nearly two orders of magnitude.

References

[1] Olive K *et al* (Particle Data Group) 2014 *Chin. Phys.* C **38** 090001
[2] LEP Electroweak Working Group 2006 *Phys. Rep.* **427** 257
[3] Yao W-M *et al* 2006 *J. Phys. G: Nucl. Part. Phys.* **33** 1
[4] Feynman R 1949 *Phys. Rev.* **76** 749
[5] Barnes T, Close F E and Monaghan S 1982 *Nucl. Phys.* B **198** 380
[6] Adam J *et al* 2013 *Phys. Rev. Lett.* **110** 201801
[7] Lattes C M G, Occhialini G P S and Powell C F 1947 *Nature* **160** 453
[8] Powell C F 1950 *Rep. Prog. Phys.* **13** 350
[9] Williams W S C 1991 *Nucl. Part. Phys* (Oxford: Oxford University Press)
[10] Georgi H and Glashow S 1974 *Phys. Rev. Lett.* **32** 438
[11] Amaldi U, de Boer W and Fürstenau H 1991 *Phys. Lett.* B **260** 447
[12] Dimopoulos S and Georgi H 1981 *Nucl. Phys.* B **193** 150

IOP Publishing

Nuclear and Particle Physics

Claude Amsler

Chapter 8

Relativistic kinematics

The relationship between the *total energy E* of a particle with *rest mass m* and its momentum $\vec{p}$ is $E^2 = c^2\vec{p}^2 + m^2c^4$ or

$$\boxed{E^2 = \vec{p}^2 + m^2} \tag{8.1}$$

in natural units. The squared modulus of a *4-vector* $\mathcal{A} \equiv (a_0, \vec{a})$

$$\mathcal{A}^2 \equiv a_0^2 - \vec{a}^2 \tag{8.2}$$

is invariant under Lorentz transformations. For example, the 4-momentum of a particle

$$\mathcal{P} \equiv \left(E, \vec{p}\right) \tag{8.3}$$

is a 4-vector since its squared modulus

$$\mathcal{P}^2 \equiv E^2 - \vec{p}^2 = m^2 \tag{8.4}$$

does not depend on the reference frame (m being a constant). The kinetic energy expressed in natural units is

$$T = E - m \tag{8.5}$$

and the modulus of the momentum

$$\left|\vec{p}\right| \equiv p = m\beta\gamma, \tag{8.6}$$

where γ is the relativistic factor

$$\boxed{\gamma = \frac{1}{\sqrt{1 - \beta^2}}}, \tag{8.7}$$

doi:10.1088/978-0-7503-1140-3ch8 © IOP Publishing Ltd 2015

and β the speed of the particle (in units of c). Introducing (8.6) into (8.1) leads to

$$E^2 = m^2\beta^2\gamma^2 + m^2 = m^2\left(\frac{\beta^2}{1-\beta^2} + 1\right) = m^2\gamma^2 \tag{8.8}$$

or

$$\boxed{E = m\gamma}. \tag{8.9}$$

In the famous Einstein relation $E = Mc^2$ the variable M refers to the *relativistic* mass $M = m\gamma$ of the particle, while in particle physics one usually works with the rest mass m. One then obtains the two important formulae

$$\boxed{\gamma = \frac{E}{m}} \quad \text{and} \quad \boxed{\beta = \frac{p}{E}}. \tag{8.10}$$

We now consider the special Lorentz transformation from one reference system labelled Σ (e.g. the laboratory system) into another inertial reference frame Σ^* (e.g. the centre-of-mass system of a reaction or the rest frame of an unstable particle). Let Σ^* move in Σ in the $+x$-direction with velocity β. In Σ a particle of mass m would have the 4-momentum $\mathcal{P}$ and in Σ^* the 4-momentum $\mathcal{P}^*$ (figure 8.1). The transformation under the invariance condition $\mathcal{P}^2 = \mathcal{P}^{*2} = m^2$ reads

$$\begin{pmatrix} E \\ p\cos\theta \\ p\sin\theta \\ 0 \end{pmatrix} = \begin{pmatrix} \gamma & \beta\gamma & 0 & 0 \\ \beta\gamma & \gamma & 0 & 0 \\ 0 & 0 & 1 & 0 \\ 0 & 0 & 0 & 1 \end{pmatrix} \begin{pmatrix} E^* \\ p^*\cos\theta^* \\ p^*\sin\theta^* \\ 0 \end{pmatrix}. \tag{8.11}$$

We have chosen the coordinate axes such that $\vec{p}$ lies in the $x - y$ plane. Componentwise one obtains

$$\begin{aligned} E &= \gamma E^* + \beta\gamma p^* \cos\theta^*, \\ p\cos\theta &= \beta\gamma E^* + \gamma p^* \cos\theta^*, \\ p\sin\theta &= p^*\sin\theta^*. \end{aligned} \tag{8.12}$$

The inverse transformation ($\Sigma \to \Sigma^*$) is obtained from (8.11) by matrix inversion, or by replacing β by $-\beta$ since Σ moves in Σ^* with velocity $-\beta$, hence

$$\begin{aligned} E^* &= \gamma E - \beta\gamma p\cos\theta, \\ p^*\cos\theta^* &= -\beta\gamma E + \gamma p\cos\theta, \\ p^*\sin\theta^* &= p\sin\theta. \end{aligned} \tag{8.13}$$

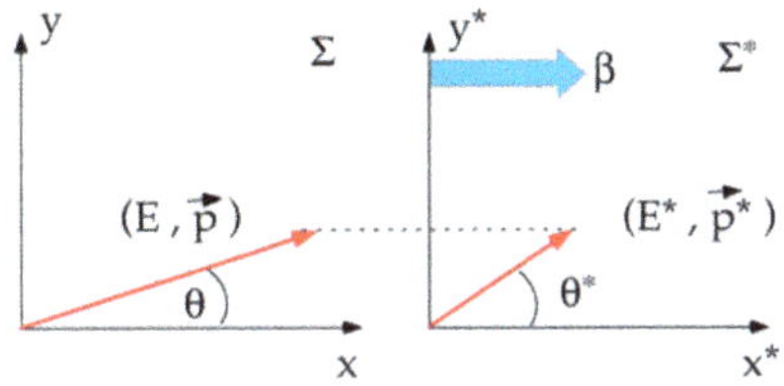

Figure 8.1. Lorentz transformation $\Sigma^* \to \Sigma$.

The correct sign for β follows from the simple consideration that the energy of a forward moving particle in $\Sigma^*(\theta^* = 0)$ is larger in Σ, hence the + sign applies to β in (8.12).

8.1 $\pi^0 \rightarrow 2\gamma$

As a first example, let us discuss π^0 decay into 2γ. The π^0 (system Σ^*) with mass m is moving in the laboratory system Σ with velocity β in the $+x$-direction (figure 8.2(*a*)). In the rest frame of the π^0 the two photons are emitted in opposite directions, each with energy

$$E^* = \frac{m}{2} \tag{8.14}$$

(from energy and momentum conservation) and with the momenta

$$p^* = E^*, \tag{8.15}$$

since photons are massless. Let us denote by θ^* the emission angle of one of the two photons (the second photon is then emitted under the angle $\theta^* + \pi$). According to (8.12) the photon energies in the laboratory system are

$$\begin{aligned} E_1 &= \gamma\frac{m}{2} + \beta\gamma\frac{m}{2}\cos\theta^* = \gamma\frac{m}{2}(1 + \beta\cos\theta^*), \\ E_2 &= \gamma\frac{m}{2} - \beta\gamma\frac{m}{2}\cos\theta^* = \gamma\frac{m}{2}(1 - \beta\cos\theta^*), \end{aligned} \tag{8.16}$$

and therefore we obtain as expected

$$E_1 + E_2 = \gamma m, \tag{8.17}$$

the total energy of the π^0 in the laboratory. Substituting for the energies $h\nu$ and $m/2 = h\nu^*$ in (8.16) one finds the well-known formula for the *Doppler shift* of a moving light source with speed β. The components of the momentum are calculated using (8.12):

$$\begin{aligned} p_1\cos\theta &= \beta\gamma\frac{m}{2} + \gamma\frac{m}{2}\cos\theta^*, \\ p_1\sin\theta &= \frac{m}{2}\sin\theta^*, \end{aligned} \tag{8.18}$$

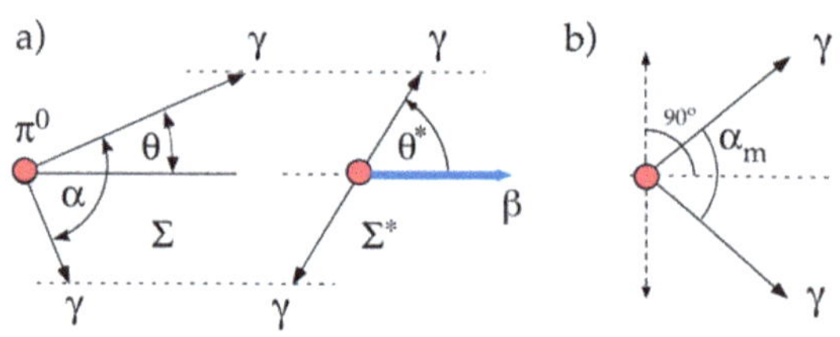

Figure 8.2. (*a*) π^0 decay into 2γ in the laboratory frame (left) and in the π^0 rest frame (right). (*b*) Emission of the 2γ pair under the minimum opening angle in the laboratory.

and similarly for the second photon ($\theta^* \to \theta^* + \pi$). The transformation from the rest frame to the laboratory frame is obtained from the ratio of the last two equations:

$$\tan\theta = \frac{\sin\theta^*}{\gamma(\beta + \cos\theta^*)}. \tag{8.19}$$

In the limit when $\theta^* \to 0$ the lab angle θ also approaches 0. Backward photons in the π^0 rest frame ($\theta^* \to \pi$) appear backwards in the laboratory ($\theta \to \pi$), even with highly relativistic pions.

Consider now the scalar product $\mathcal{P}_1 \cdot \mathcal{P}_2$ expressed in the laboratory:

$$\mathcal{P}_1 \cdot \mathcal{P}_2 = E_1E_2 - p_1p_2\cos\alpha = E_1E_2(1 - \cos\alpha). \tag{8.20}$$

We have denoted by α the *opening angle* that is the angle between the momenta $\vec{p}_1$ and $\vec{p}_2$. This angle takes the value of 180° in the π^0 rest frame, where the scalar product reads

$$\mathcal{P}_1 \cdot \mathcal{P}_2 = 2E_1^*E_2^* = 2\left(\frac{m}{2}\right)^2. \tag{8.21}$$

Since $\mathcal{P}_1 \cdot \mathcal{P}_2$ is Lorentz invariant, terms on the right-hand sides of the last two equations are equal and one obtains with (8.16)

$$1 - \cos\alpha = \frac{m^2}{2E_1E_2} = \frac{2}{\gamma^2\left(1 - \beta^2\cos^2\theta^*\right)}. \tag{8.22}$$

The term (8.22) on the right-hand side is minimum for $\cos\theta^* = 0$ in which case α also reaches its minimum value α_m given by

$$1 - \cos\alpha_m = \frac{2}{\gamma^2}, \tag{8.23}$$

or

$$1 - \cos^2\frac{\alpha_m}{2} + \sin^2\frac{\alpha_m}{2} = 2\sin^2\frac{\alpha_m}{2} = \frac{2}{\gamma^2}, \tag{8.24}$$

hence

$$\boxed{\sin\frac{\alpha_m}{2} = \frac{1}{\gamma}}. \tag{8.25}$$

In the laboratory system the angle between the directions of the two photons is larger than a minimum value which depends only on the energy of the π^0. This minimum angle occurs when the photons are emitted in the rest frame perpendicularly to the flight direction of the laboratory ($\theta^* = 90°$, see figure 8.2(*b*)). For example, for $E_{\pi^0} = 1\,\text{GeV}$ (8.25) gives $\alpha_m = 15.4°$.

We now calculate the *energy distribution* of the photons in the laboratory. Since the π^0 is spinless, the angular distribution in the rest frame is *isotropic*. Isotropy means that the emission probability w into the infinitesimal surface $dF = 2\pi \sin\theta^* d\theta^* = -2\pi d\cos\theta^*$ (the ring on the unit sphere) is constant, hence independent of θ^* (and of the azimuthal angle ϕ^*), therefore

$$\begin{aligned} w(\cos\theta^*)\mathrm{d}\cos\theta^* &= \text{constant} \cdot \mathrm{d}\cos\theta^* \\ &= \frac{1}{2}\mathrm{d}\cos\theta^* \Rightarrow \int_{-1}^{+1} w(\cos\theta^*)\mathrm{d}\cos\theta^* = 1 \end{aligned} \tag{8.26}$$

and not $w(\theta^*)d\theta^* = \text{constant} \cdot d\theta^*$. (An angular distribution should always be plotted as a function of $\cos\theta^*$.) To calculate the energy distribution in the laboratory we perform the variable transformation $\cos\theta^* \to E \equiv E_1$:

$$w(\cos\theta^*)\mathrm{d}\cos\theta^* \equiv n(E)\mathrm{d}E \Rightarrow n(E) = w(\cos\theta^*)\frac{\mathrm{d}\cos\theta^*}{\mathrm{d}E}. \tag{8.27}$$

From the photon energy

$$E = \gamma\frac{m}{2}\left(1 + \beta\cos\theta^*\right) \tag{8.28}$$

follows after performing the derivative that

$$n(E) = w(\cos\theta^*)\frac{2}{\beta\gamma m} = \frac{1}{\beta\gamma m}. \tag{8.29}$$

The energy distribution of the photon is *constant* in the laboratory between the maximum and minimum energies E_+ and E_-, given by

$$E_\pm = \gamma\frac{m}{2}(1 \pm \beta), \tag{8.30}$$

which correspond to $\theta^* = 0$ and $\theta^* = 180°$, respectively. The distribution is a normalized *box* (figure 8.3): $\int_{E_-}^{E_+} n(E)\mathrm{d}E = 1$. We have derived the general rule that monoenergetic particles emitted isotropically in a reference frame Σ^* have a uniform energy distribution (box distribution) in a reference frame Σ moving with velocity β in Σ^*.

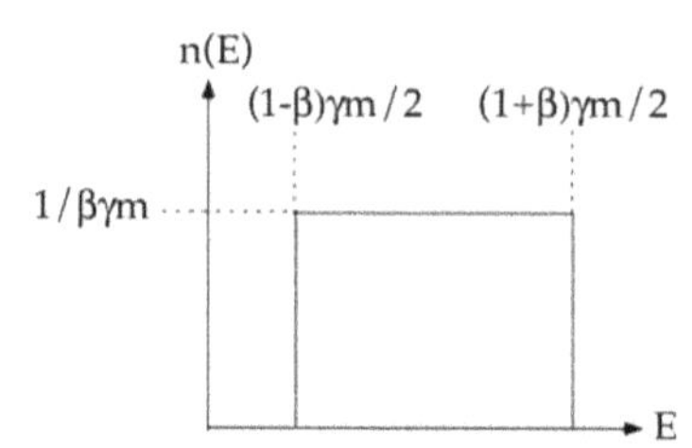

Figure 8.3. Energy distribution of the photons from $\pi^0 \to 2\gamma$ in the laboratory.

The *angular distribution* of one photon in the laboratory is obtained from the variable transformation $\cos\theta^* \to \cos\theta$:

$$N(\cos\theta)\,\mathrm{d}\cos\theta \equiv w(\cos\theta^*)\,\mathrm{d}\cos\theta^* = \frac{1}{2}\mathrm{d}\cos\theta^*, \tag{8.31}$$

hence

$$N(\cos\theta) = \frac{1}{2}\frac{\mathrm{d}\cos\theta^*}{\mathrm{d}\cos\theta} = \frac{1}{2}\frac{\mathrm{d}\Omega^*}{\mathrm{d}\Omega}. \tag{8.32}$$

Expressing in (8.19) $\tan\theta$ and $\sin\theta^*$ through $\cos\theta$ and $\cos\theta^*$, respectively, and calculating the derivative $\frac{\mathrm{d}\cos\theta}{\mathrm{d}\cos\theta^*}$, one obtains after tedious algebraic manipulations the solid angle transformation, or *Jacobian*,

$$\frac{\mathrm{d}\Omega}{\mathrm{d}\Omega^*} = \frac{1+\beta\cos\theta^*}{\gamma^2\left[(\beta+\cos\theta^*)^2 + \frac{\sin^2\theta^*}{\gamma^2}\right]^{\frac{3}{2}}}. \tag{8.33}$$

Figure 8.4 shows the angular distribution (8.32) of one photon in the laboratory system for various π^0 momenta. The photon is emitted preferably in the flight direction of the pion.

We finally calculate the 2γ *opening angle* distribution via the variable transformation

$$\begin{aligned} n(\alpha)\mathrm{d}\alpha = w(\cos\theta^*)\mathrm{d}\cos\theta^* \Rightarrow n(\alpha) &= \frac{1}{2}\frac{\mathrm{d}\cos\theta^*}{\mathrm{d}\alpha} \\ &= \frac{1}{2}\frac{\mathrm{d}\cos\theta^*}{\mathrm{d}\cos\alpha}\cdot\frac{\mathrm{d}\cos\alpha}{\mathrm{d}\alpha} \\ &= -\frac{1}{2}\frac{\mathrm{d}\cos\theta^*}{\mathrm{d}\cos\alpha}\sin\alpha. \end{aligned} \tag{8.34}$$

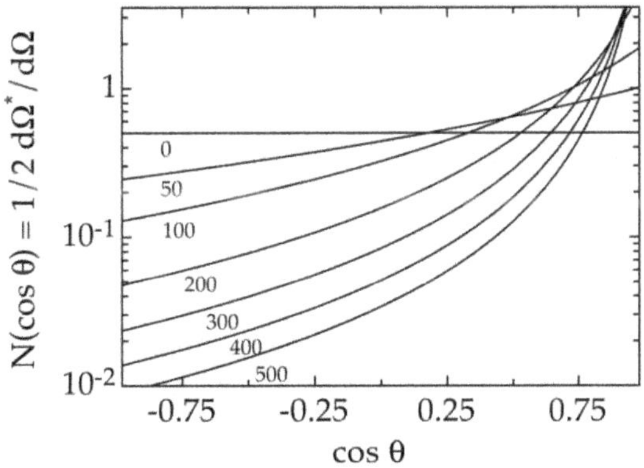

Figure 8.4. $\pi^0 \to 2\gamma$: angular distribution of one γ in the laboratory for various pion momenta in MeV c^{-1}.

The derivative can be computed from (8.22):

$$\frac{\mathrm{d}\cos\alpha}{\mathrm{d}\cos\theta^*} = \frac{2\left[\gamma^2\left(1-\beta^2\cos^2\theta^*\right)\right]'}{\left[\gamma^2\left(1-\beta^2\cos^2\theta^*\right)\right]^2}$$

$$= 2\frac{(1-\cos\alpha)^2}{4}\left(-2\gamma^2\beta^2\cos\theta^*\right)$$

$$= -(1-\cos\alpha)^2\gamma^2\beta^2\cos\theta^*. \tag{8.35}$$

In addition, one obtains from the equality

$$\left(1-\beta^2\cos^2\theta^*\right)(1-\cos\alpha) = \frac{2}{\gamma^2}, \tag{8.36}$$

that

$$\cos^2\theta^* = \frac{(1-\cos\alpha)\frac{\gamma^2}{2}-1}{\beta^2\frac{\gamma^2}{2}(1-\cos\alpha)}. \tag{8.37}$$

Introducing (8.37) into (8.35) leads with $1-\cos\alpha = 2\sin^2\frac{\alpha}{2}$ to

$$\frac{\mathrm{d}\cos\alpha}{\mathrm{d}\cos\theta^*} = -(1-\cos\alpha)^2\gamma^2\beta^2\sqrt{\frac{(1-\cos\alpha)\frac{\gamma^2}{2}-1}{\beta^2\frac{\gamma^2}{2}(1-\cos\alpha)}}$$

$$= -4\beta\gamma^2\sin^4\frac{\alpha}{2}\sqrt{\frac{\gamma^2\sin^2\left(\frac{\alpha}{2}\right)-1}{\gamma^2\sin^2\left(\frac{\alpha}{2}\right)}}. \tag{8.38}$$

Substituting into (8.34) gives the distribution

$$n(\alpha) = \frac{1}{2}\frac{\sin\alpha}{4\beta\gamma^2\sin^3\frac{\alpha}{2}\sqrt{\frac{\gamma^2\sin^2\left(\frac{\alpha}{2}\right)-1}{\gamma^2}}}. \tag{8.39}$$

Finally, we replace $\sin\alpha$ by $2\sin\frac{\alpha}{2}\cos\frac{\alpha}{2}$ and express the square root as

$$\sqrt{\sin^2\left(\frac{\alpha}{2}\right)-\frac{1}{\gamma^2}} = \sqrt{\sin^2\left(\frac{\alpha}{2}\right)-1+\beta^2}, \tag{8.40}$$

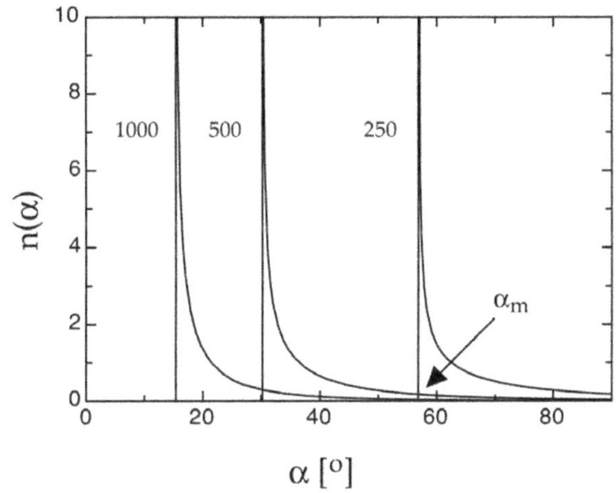

Figure 8.5. Distribution of the 2γ opening angle for various pion momenta in MeV c^{-1}.

to get the final result:

$$n(\alpha) = \left(\frac{1-\beta^2}{4\beta}\right)\frac{\cos\frac{\alpha}{2}}{\sin^2\frac{\alpha}{2}}\frac{1}{\sqrt{\beta^2-\cos^2\frac{\alpha}{2}}}. \tag{8.41}$$

Figure 8.5 shows how the function $n(\alpha)$ varies with pion momentum $p = m\beta\gamma$. The function has a pole at the value of α such that

$$\cos^2\frac{\alpha}{2} = \beta^2 \Rightarrow \boxed{\sin\frac{\alpha}{2} = \frac{1}{\gamma}}. \tag{8.42}$$

The most likely opening angle is just the minimum opening angle α_m, (8.25). The most likely configuration is therefore the one for which the two photons are emitted in the forward hemisphere under the same angle $\theta = \alpha_m/2$, as shown in figure 8.2(*b*) above. This statement refers to the *angular correlation* of the two photons and does not contradict the fact that *one* of the two photons is emitted preferably in the forward direction (figure 8.4).

This neat property can be employed to estimate the energy of neutral pions without actually measuring the energies of the decay photons. It was first used by Jack Steinberger and co-workers [1] to demonstrate the existence of a neutral partner to the charged pion that had been discovered shortly before. The hypothetical π^0 was produced by irradiating a beryllium target with ~300 MeV photons (figure 8.6(*a*)). Neutral pions from the reaction

$$\gamma N \rightarrow \pi^0(\rightarrow \gamma\gamma)N \tag{8.43}$$

are monoenergetic at fixed scattering angles β (neglecting the internal motion of the nucleons in the nucleus). Two scintillation counter telescopes register the two decay photons emitted under the scattering angle β with opening angle α. A lead foil

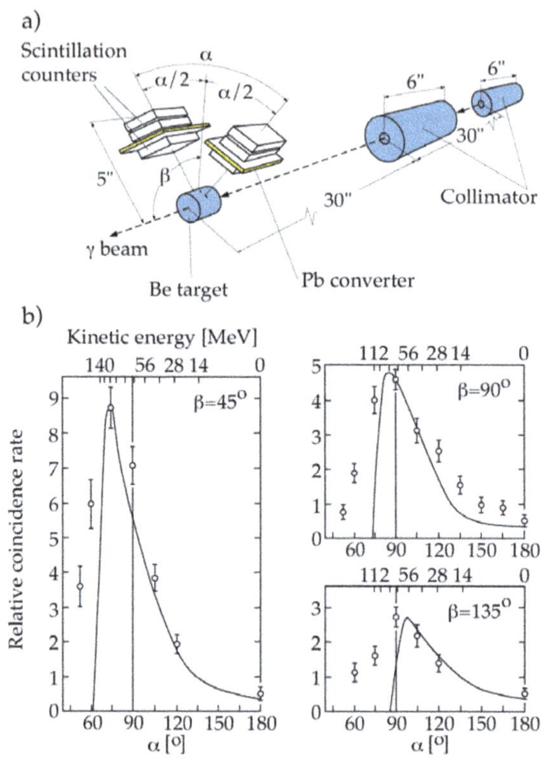

Figure 8.6. (*a*) The experimental set up to demonstrate the existence of the π^0. (*b*) Relative 2γ coincidence rates versus opening angle α for three different inclination angles β of the scintillation counters. The curves show simple predictions omitting scattering and angular resolutions of the counters. After [1].

converts the photons into electron–positron pairs which are detected in the rear counter. The first scintillation counter in each telescope acts as a veto against charged particles emerging from the target.

Figure 8.6(*b*) shows the relative 2γ rates as a function of opening angle α between the telescopes, for three different scattering angles β. A neutral particle decaying into 2γ would lead to an accumulation of events at the minimum opening angle α_m. The curves show the predictions for the decay of a neutral meson with the same mass as the π^+, assuming the same angular distribution as for the analogous reaction $\gamma p \to \pi^+ n$. The minimum opening angle should increase with β since the energy of the π^0 decreases with increasing scattering angle. Figure 8.6(*b*) indeed confirms this behaviour. The distribution is smeared due to the energy spread of the beam and the finite sizes of the detectors.

8.2 Invariant mass

The *invariant mass* squared of a system of *n* particles with 4-momenta $\mathcal{P}_i$ is defined as

$$\boxed{M^2 \equiv \left(\sum_1^n \mathcal{P}_i\right)^2 = \left(\sum_1^n E_i\right)^2 - \left(\sum_1^n \vec{p}_i\right)^2}. \tag{8.44}$$

As the modulus of a 4-vector the invariant mass is Lorentz invariant, it therefore takes the same value in any reference frame. Consider now the decay of a state R with mass M_R into n particles. Its 4-momentum is given by $\mathcal{P} = \Sigma_1^n \mathcal{P}_i$ from energy and momentum conservation. Since $\mathcal{P}^2 = M_R^2$ the mass of the decaying state is equal to the n-body invariant mass. Formula (8.44) provides a quick way to calculate the mass of the decaying state from the energies and momenta of the daughter products, in any reference frame.

For example, let us consider the π^0 decay discussed in the previous section. The square of the 2γ invariant mass in the π^0 rest frame is

$$M^2(2\gamma) = \left(E_1^* + E_2^*\right)^2 - \left(\vec{p}_1^{\,*} + \vec{p}_2^{\,*}\right)^2 = \left(2\frac{m}{2}\right)^2 = m^2, \tag{8.45}$$

the squared mass of the π^0. In the laboratory we also obtain

$$M^2(2\gamma) = (E_1 + E_2)^2 - \left(\vec{p}_1 + \vec{p}_2\right)^2 = 2E_1E_2(1 - \cos\alpha) = m^2 \tag{8.46}$$

according to (8.22), where α is the angle between the momenta of the two photons.

8.3 $\pi \rightarrow \mu\nu$

Consider the two-body decay $a \rightarrow bc$, for example the main decay mode of the charged pion $\pi^\pm \rightarrow \mu^\pm \nu_\mu$. Energy and momentum are conserved, therefore

$$m_a = \sqrt{p_b^{*2} + m_b^2} + \sqrt{p_c^{*2} + m_c^2} \qquad \text{and} \qquad p_b^* = p_c^* \tag{8.47}$$

in the rest frame of particle a. The momenta of the daughters are then given by

$$\boxed{p^* = p_b^* = p_c^* = \frac{\sqrt{\left[m_a^2 - (m_b + m_c)^2\right]\left[m_a^2 - (m_b - m_c)^2\right]}}{2m_a}}. \tag{8.48}$$

For $\pi^\pm \rightarrow \mu^\pm \nu_\mu$ the numerical value is with $m_\mu = 105.7$ MeV, $m_\pi = 139.6$ MeV and assuming a massless neutrino:

$$p^* = p_\mu^* = p_{\nu_\mu}^* = 29.8 \text{ MeV } c^{-1}, \tag{8.49}$$

and therefore the velocities of the muon and neutrino in the rest frame of the pion are

$$\beta_\mu^* = \frac{p^*}{E_\mu^*} = 0.271, \qquad \beta_{\nu_\mu}^* = 1. \tag{8.50}$$

Let the pion move in the laboratory with velocity β. In the rest frame the angular distribution is isotropic since the pion has spin zero. Thus, the energy distributions of the decay muon and neutrino are uniform in the laboratory. For the muon one gets the angle in the laboratory by dividing the last equation in (8.12) by the penultimate equation, and one obtains the generalization of (8.19) for $\beta^* < 1$:

$$\boxed{\tan\theta = \frac{\beta^* \sin\theta^*}{\gamma\left(\beta + \beta^* \cos\theta^*\right)}}, \tag{8.51}$$

with $\beta^* \equiv \beta^*_\mu = 0.271$. Two cases have to be distinguished:

1. $\beta^* < \beta$: the denominator in (8.51) is positive for all values of θ^*, that is, the muon flies into the forward hemisphere in the laboratory below a *maximum* angle, the computation of which is left as an exercise for the reader:

$$\sin\theta_{\max} = \frac{\beta^*\gamma^*}{\beta\gamma}. \tag{8.52}$$

2. $\beta^* > \beta$: the denominator in (8.51) becomes negative for large values of θ^*, and for $\theta^* = \pi$ the angle in the laboratory is also $\theta = \pi$. In this case there is no maximum lab angle.

Figure 8.7 shows how $\cos\theta$ varies with $\cos\theta^*$ for various pion momenta. Two angles in the rest frame correspond to each lab angle below the maximum angle, one for the 'backward' muon and one for the 'forward' muon which are emitted in the laboratory with different momenta. The angular distribution in the laboratory is obtained from (8.51), the generalization of (8.33) for $\beta^* \lessgtr 1$:

$$\boxed{\frac{d\Omega}{d\Omega^*} = \frac{1 + \frac{\beta}{\beta^*}\cos\theta^*}{\gamma^2\left[\left(\frac{\beta}{\beta^*} + \cos\theta^*\right)^2 + \frac{\sin^2\theta^*}{\gamma^2}\right]^{\frac{3}{2}}}}. \tag{8.53}$$

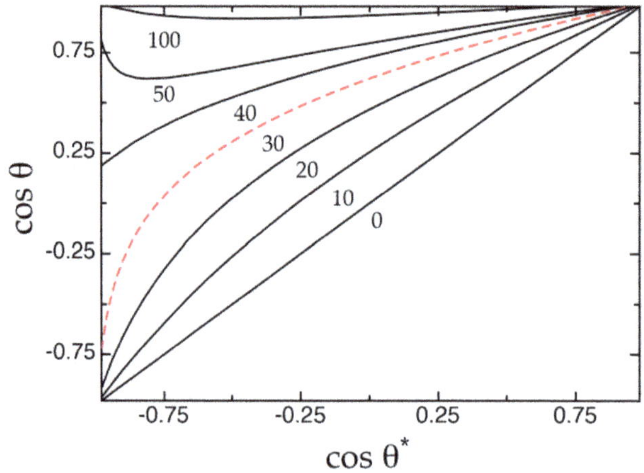

Figure 8.7. $\cos\theta$ versus $\cos\theta^*$ of the muon in pion decay for various pion momenta in MeV c^{-1}. There are two possible values for $\cos\theta$ above $\simeq$30 MeV c^{-1} (dotted curve).

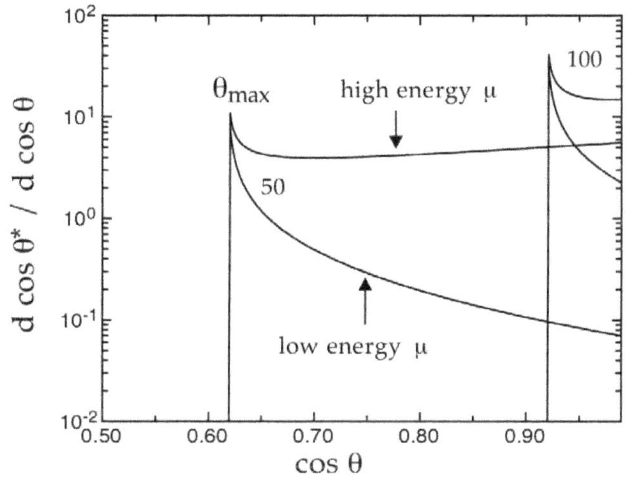

Figure 8.8. Angular distribution $d\Omega^*/d\Omega$ of the muon in the laboratory for 50 and 100 MeV c^{-1} pions. The muon velocity in the pion rest frame is slower than the speed of the latter in the laboratory. There are therefore two branches, one for the forward (high energy) and one for the backward (low energy) muon.

For $\beta^* > \beta$ the angular distribution is similar to that shown in figure 8.4. However, for $\beta^* < \beta$ the behaviour is very different: expressing (8.53) as a function of θ, one finds for each angle θ *two* values of the Jacobian, one for the low energy and one for the high energy muon. The distribution becomes infinite at the maximum angle (figure 8.8).

The laboratory angle of the neutrino and its angular distribution are also given by (8.51) and (8.53) with $\beta^* = 1$ (neglecting the tiny mass of the neutrinos), thus recovering (8.19) and (8.33), respectively. Figure 8.9(*a*) shows the relationship between the neutrino lab angle and its angle in the rest frame of the pion. At high energies neutrinos are emitted mainly in the very forward direction, although there is no maximum laboratory angle.

Finally, we calculate the energy E_ν of the neutrino in the laboratory for ultra-relativistic pions ($\beta \to 1$). From the first equation of (8.12) one obtains

$$\cos\theta^* \simeq \frac{E_\nu - \gamma p^*}{\gamma p^*}. \tag{8.54}$$

Introducing into (8.51) leads after some arithmetic manipulations to

$$E_\nu \simeq \frac{2\gamma p^*}{1 + \gamma^2 \tan^2\theta}. \tag{8.55}$$

Figure 8.9(*b*) shows the relationship between the neutrino lab energy E_ν and the γ-factor of the pion, for various neutrino lab angles. The neutrino energy reaches a maximum when

$$\frac{\partial E_\nu}{\partial \gamma} = 0 \Rightarrow \boxed{\gamma = \frac{1}{\tan\theta}}. \tag{8.56}$$

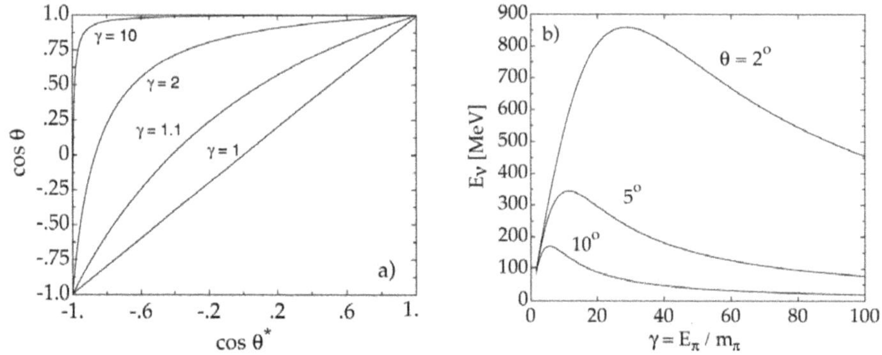

Figure 8.9. (*a*) Relationship between the neutrino lab angle θ and its angle θ^* in the rest frame of the pion for the decay $\pi^\pm \to \mu^\pm \nu$. (*b*) Energy of the neutrino versus γ-factor of the (ultrarelativistic) pion for various angles θ.

The energy of the neutrino is then nearly independent of pion energy: $\Delta E_\nu \simeq \frac{\partial E_\nu}{\partial \gamma}\Delta\gamma = 0$. This property is very useful to produce a nearly monoenergetic neutrino beam (narrow band neutrino beam). Monoenergetic neutrino beams are advantageous for studying neutrino oscillations since the oscillation period depends on E_ν, see section 14.6.1. The price to pay is a lower neutrino intensity, a factor of four lower than for $\theta = 0$. The neutrino detector is installed at a small angle θ with respect to the pion beam, the average energy of which is chosen to fulfil condition (8.56).

8.4 Mandelstam variables *s* and *t*

Let us consider the scattering of a particle with mass m (4-momentum $\mathcal{P}_1$) with a target particle of mass M (4-momentum $\mathcal{P}_2$). After scattering the particles have 4-momenta $\mathcal{P}_3$ and $\mathcal{P}_4$. The *4-momentum transfer*, defined as

$$t \equiv (\mathcal{P}_1 - \mathcal{P}_3)^2 = (\mathcal{P}_4 - \mathcal{P}_2)^2 = (E_1 - E_3)^2 - (\vec{p}_1 - \vec{p}_3)^2 = q^2, \tag{8.57}$$

is also Lorentz invariant, which means that t takes the same value in any reference frame. The quantity

$$s \equiv (\mathcal{P}_1 + \mathcal{P}_2)^2 \tag{8.58}$$

is also Lorentz invariant and is equal to the invariant mass squared of the initial system. In the centre-of-mass frame of reaction $1 + 2 \to 3 + 4$, s is equal to the *square of the total energy W*:

$$\begin{aligned} s &\equiv (E_1^* + E_2^*)^2 - (\vec{p}_1^{\,*} + \vec{p}_2^{\,*})^2 \\ &= (E_1^* + E_2^*)^2 = W^2 \Rightarrow \boxed{W = \sqrt{s}}, \end{aligned} \tag{8.59}$$

since $\vec{p}_1^{\,*} = -\vec{p}_2^{\,*}$. In the laboratory system where target particle 2 is at rest ($\vec{p}_2 = 0$),

$$s = \left(E_1 + M\right)^2 - \vec{p}_1^{\,2} = m^2 + M^2 + 2E_1 M. \tag{8.60}$$

From the last two equations one obtains an important relationship between the total energy E_1 of the projectile in the laboratory and the total energy W in the centre-of-mass system:

$$\boxed{E_1 = \frac{W^2 - M^2 - m^2}{2M}}. \tag{8.61}$$

This formula is useful to calculate the projectile energy needed for a reaction in which different particles are produced in the final state. The *threshold* energy—the minimum energy W required for the reaction to occur—is equal to the sum of the rest masses of the final state particles. Consider for example the reaction

$$\bar{p}p \to \bar{\Lambda}\Lambda. \tag{8.62}$$

Since $m_\Lambda = 1116\,\text{MeV} > m_p$, the threshold energy is equal to $W = 2m_\Lambda$. At this energy the two Λ are produced at rest in the centre-of-mass system. The minimum lab energy of the incident proton is

$$E_1(\text{min}) = \frac{2m_\Lambda^2 - m_p^2}{m_p} = 1717\,\text{MeV}, \tag{8.63}$$

which corresponds to a kinetic energy of 779 MeV. Although the reaction becomes kinematically possible, the cross section close to threshold is reduced by the small phase space available to the final state particles (see section 12.1).

When W becomes large enough that the masses are negligible, (8.61) becomes

$$\boxed{W = \sqrt{s} \simeq \sqrt{2ME_1}}. \tag{8.64}$$

The energy in the centre-of-mass system grows with the *square root* of the incident energy in the laboratory. Performing very high energy experiments with fixed (stationary) targets is therefore uneconomical since a major fraction of the energy is invested in moving the centre of mass. However, with *colliding* beams of equal energies E_1, the total energy available for the reaction is $W = \sqrt{s} = 2E_1$. A comparison between fixed targets and colliders is shown in table 8.1 (see also figure 8.11).

8.5 Proper time

The mean life of an unstable particle moving in system Σ with velocity β is

$$\tau = \gamma\tau^*, \tag{8.65}$$

where τ^* is the particle mean life in its rest frame Σ^*, also called *proper time*. The mean decay length is (in SI units)

$$\Lambda = c\beta\gamma\tau^*. \tag{8.66}$$

Table 8.1. Comparison between stationary targets and colliders for *pp* reactions. The second column gives the total available energy with a proton of total energy E_1 and a stationary proton target, the third column the corresponding energy in a *pp* collider in which both protons collide frontally with the same energy E_1 (energies are given in GeV).

E_1 (projectile)	$\sqrt{s}$ (target)	$\sqrt{s}$ (*pp* collider)
1	1.9	2
10	4.5	20
100	13.8	200
1 000	43.3	2 000
7 000 (LHC)	114.6	14 000

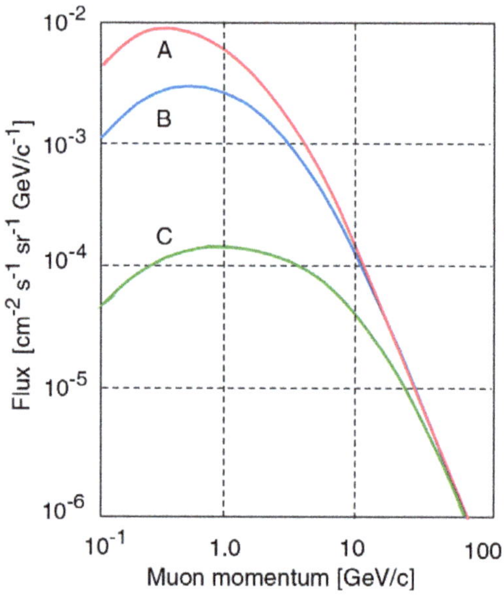

Figure 8.10. Energy distribution of atmospheric muons at the altitude of 3200 m (A) and at sea level, vertically (B) and at 68° (C). After [2].

The mean decay length of unstable ultrarelativistic particles is therefore significantly larger in the laboratory due to the time dilation (8.65). For example, the muon flux produced by cosmic rays impinging in the upper atmosphere reaches a maximum at $p \sim 2\,\mathrm{GeV}\,c^{-1}$. The mean life of such a muon is then $\tau = 41\,\mu\mathrm{s}$ ($\tau^* = 2.197\,\mu\mathrm{s}$) and its mean decay length $\Lambda \simeq 12\,\mathrm{km}$ according to (8.66). Many muons then reach sea level thanks to relativistic time dilation. Figure 8.10 shows the muon spectrum at the altitude of 3200 m and at sea level, vertically and at 68°. At sea level the integrated muon flux is about 2 muons $\mathrm{cm}^{-2}\,\mathrm{min}^{-1}$.

8.6 Rapidity

The interaction between the elementary constituents of matter, the *partons* (quarks and gluons), can be investigated by colliding hadrons at very high energies. Let us

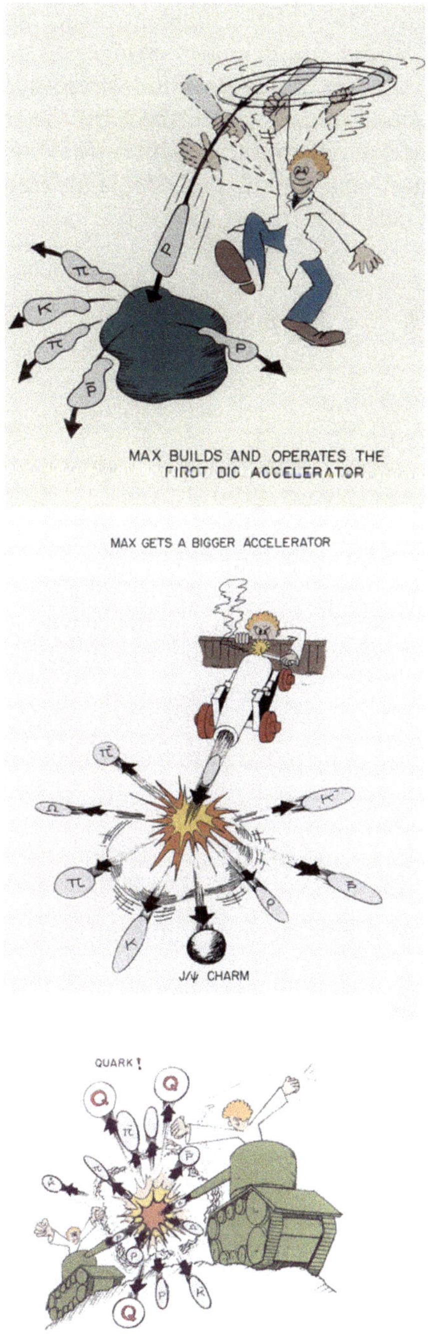

Figure 8.11. Reactions at high $\sqrt{s}$ are achieved with colliding beams. Image courtesy of Brookhaven National Laboratory.

choose as the z-axis the direction of the incident projectiles. Most of the outgoing partons will be emitted under small angles along z. The analysis of the data is facilitated by a suitable Lorentz transformation along this direction, so that most partons fly in (or under small angles with respect to) the transverse plane.

We shall first introduce a kinematic variable—the rapidity—which is additive under Lorentz transformations. Let us consider a reference system (e.g. the rest frame of a decaying particle) which moves with velocity β along z, and define its *rapidity* y as

$$y \equiv \operatorname{atanh}\beta \Rightarrow \beta = \tanh y = \left[\frac{\exp(y) - \exp(-y)}{\exp(y) + \exp(-y)}\right] \tag{8.67}$$

or

$$\gamma = \frac{1}{\sqrt{1 - \tanh^2 y}} = \cosh y. \tag{8.68}$$

At low velocities the rapidity is roughly equal to the velocity, $y \simeq \beta$. Now consider a daughter product in the rest frame of the mother, moving with speed β^* along the z-direction (energy E^* and momentum p^*), hence with rapidity $y^* = \operatorname{atanh}\beta^*$. In the laboratory system the energy and momentum of the daughter are:

$$E = \gamma E^* + \beta\gamma p^*, \qquad p = \beta\gamma E^* + \gamma p^*, \tag{8.69}$$

from which follows its velocity β_L in the laboratory, according to the known rule for the relativistic addition of velocities:

$$\beta_L = \frac{p}{E} = \frac{\beta + \beta^*}{1 + \beta\beta^*}. \tag{8.70}$$

The rapidity of the daughter particle in the laboratory is then given by

$$\tanh y_L = \beta_L = \frac{\tanh y + \tanh y^*}{1 + \tanh y \cdot \tanh y^*} = \tanh(y + y^*), \tag{8.71}$$

or simply

$$y_L = y + y^*. \tag{8.72}$$

This shows that rapidities are *additive* under Lorentz transformations, in contrast to velocities.

In particle physics one defines as rapidity

$$\boxed{y \equiv \operatorname{atanh}\beta_z}. \tag{8.73}$$

The particle moves at an angle θ with velocity $\beta_z = \beta\cos\theta$ along the incident beam direction (taken as the z-axis). The derivation above, and in particular the additive rule (8.72), is also valid for Lorentz transformations along the z-axis. It follows that the difference Δy (or infinitesimally dy) between two rapidities is invariant under Lorentz transformations. Differential cross sections such as $d\sigma/dy$ are then *independent of the reference frame.* Solving

$$\tanh y = \frac{\exp(2y) - 1}{\exp(2y) + 1} = \beta \cos \theta \tag{8.74}$$

for y gives

$$y = \frac{1}{2} \ln \left[\frac{1 + \beta \cos \theta}{1 - \beta \cos \theta} \right] \simeq -\ln \tan \frac{\theta}{2} \tag{8.75}$$

for ultrarelativistic particles. Thus we define as *pseudorapidity* η the rapidity in the limit $\beta \to 1$:

$$\boxed{\eta = -\ln \tan \frac{\theta}{2}}, \tag{8.76}$$

which depends only on the angle θ. Using (8.73) the angular dependence of the pseudorapidity can be rewritten as

$$\tanh \eta = \cos \theta \qquad \text{or} \qquad \frac{1}{\cosh \eta} = \sin \theta. \tag{8.77}$$

The pseudorapidity of a particle flying in the plane transverse to the z-axis is $\eta = 0$, while $\eta = \pm\infty$ for a particle flying along $\pm z$.

At colliders the conservation of momentum cannot be applied to the projections $p_z(i)$ on the beam axis, because particles emitted in the very forward direction are not detected due to the holes in the detector for the beamline. However, their contributions to the total momentum in the transverse plane can be neglected (see also section 17.2). Therefore one uses for the outgoing particles i their *transverse* momenta $p_\perp(i)$, which are the projections of their momenta into the plane perpendicular to the z-axis. From momentum conservation $\Sigma \vec{p}_\perp(i) = 0$ one can then calculate the transverse momentum of an invisible particle (such as a neutrino).

We now express the invariant mass M (section 8.2) of a two-body state with particles masses m_i, momenta $\vec{p}(i)$ and energies $E(i)$ $(i = 1, 2)$ as a function of the transverse momenta. From

$$\begin{aligned} &\vec{p}(i) = \vec{p}_\perp(i) + \vec{p}_z(i) \\ \text{and} \qquad &E(i) = \sqrt{\vec{p}_\perp^{\,2}(i) + m_i^2 + p_z^2(i)} \equiv \sqrt{E_\perp^2(i) + p_z^2(i)} \end{aligned} \tag{8.78}$$

one obtains the invariant mass squared

$$\begin{aligned} M^2 &= [E(1) + E(2)]^2 - [\vec{p}(1) + \vec{p}(2)]^2 = m_1^2 + m_2^2 + 2E(1)E(2) - 2\,\vec{p}(1) \cdot \vec{p}(2) \\ &= m_1^2 + m_2^2 + 2\sqrt{E_\perp^2(1)E_\perp^2(2) + E_\perp^2(1)p_z^2(2) + E_\perp^2(2)p_z^2(1) + p_z^2(1)p_z^2(2)} \\ &\quad - 2\,\vec{p}_\perp(1) \cdot \vec{p}_\perp(2) - 2p_z(1)p_z(2). \end{aligned} \tag{8.79}$$

Let us choose the reference frame in which $p_z(1) = 0$, hence $y(1) = 0$. We find for the (Lorentz invariant) difference $\Delta y = y(2) - y(1)$, together with (8.68):

$$\cosh \Delta y = \cosh y(2) = \frac{1}{\sqrt{1 - \beta_z^2(2)}} = \frac{1}{\sqrt{1 - p_z^2(2)/E^2(2)}} = \frac{E(2)}{E_\perp(2)}, \tag{8.80}$$

or

$$E(2) = E_\perp(2) \cosh \Delta y. \tag{8.81}$$

With

$$p_z^2(2) = E^2(2) - E_\perp^2(2) = E_\perp^2(2)(\cosh^2 \Delta y - 1) \tag{8.82}$$

the invariant mass squared (8.79) reads

$$\boxed{M^2 = m_1^2 + m_2^2 + 2\left[E_\perp(1)E_\perp(2) \cosh \overbrace{\Delta y}^{\simeq \Delta\eta} - \vec{p}_\perp(1) \cdot \vec{p}_\perp(2)\right]}. \tag{8.83}$$

This formula contains only invariant quantities against Lorentz transformations and can hence be computed in any reference frame, for instance in the laboratory for collider experiments. It is particularly useful when dealing with neutrinos (or other invisible particles) for which the longitudinal momenta cannot be determined from total momentum conservation.

References

[1] Steinberger J, Panofsky W K H and Steller J 1950 *Phys. Rev.* **78** 802
[2] Hayakawa S 1969 *Cosmic Ray Physics* (New York: Wiley-Interscience)

IOP Publishing

Claude Amsler

Chapter 9

Accelerators and detectors

The fundamentals of accelerators and detectors are described in this chapter. Details on accelerators can be found for example in [1, 2] and more information on detectors in [3–6].

9.1 Accelerators

The first accelerators were electrostatic devices. In 1921 Heinrich Greinacher developed the high voltage cascade generator. The first accelerator-induced nuclear reaction ($p^7Li \rightarrow 2\alpha$) was produced in 1932 by 400 keV protons from a Cockcroft–Walton cascade generator. In 1930 the first *Van de Graaff* accelerator (Van de Graaff, 1929) reached a voltage of 1.5 MeV. In this type of electrostatic accelerator the high voltage is achieved by transporting electric charges on a textile belt. Some of these accelerators are still in operation today, accelerating, for example, protons to typically 5 MeV. Protons can be accelerated up to about 20 MeV with two or more Van de Graaff connected in series (*tandem* accelerator).

The linear accelerator (LINAC) was proposed in 1924 by Gustav Ising and used for the first time by Rolf Widerøe in 1928 to accelerate ions. The principle is shown in figure 9.1. In an evacuated tube, the ions pass through electrically conducting cylindrical electrodes connected to a high frequency source with alternating phase shift π. The distance d is selected in such a way that the particles passing between the electrodes always feel an accelerating electric field. In Alvarez-LINACs

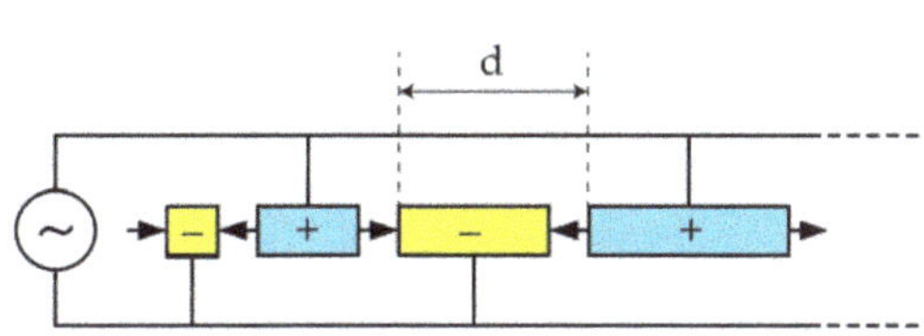

Figure 9.1. The principle of the LINAC. The distance d is equal to $\beta\frac{\lambda}{2}$ where λ is the wavelength of the high frequency source.

doi:10.1088/978-0-7503-1140-3ch9 © IOP Publishing Ltd 2015

the electrodes are in phase and the ions travel from gap to gap during one oscillation period ($d = \beta\lambda$). The electrode lengths increase with increasing particle velocity. The typical energy is around a few MeV m^{-1}. LINACs are now mainly used as injectors for higher energy circular accelerators.

For curved trajectories, for example in ring accelerators, the energy loss by synchrotron radiation grows very rapidly with energy. This effect is devastating at high energies for light particles such as electrons, so LINACs are preferred for accelerating electrons. At the SLAC Linear Collider (SLC) in Stanford, electrons (and positrons) were accelerated to 50 GeV in a 3 km long LINAC, the positrons being produced by bombarding a high-Z target with electrons.

The principle of the *cyclotron* stems from Ernest O Lawrence who built the first cyclotron to accelerate protons up to 1.5 MeV (1932). A gas (e.g. hydrogen) is ionized by a heated filament in the centre of the cyclotron (figure 9.2). The ions (mass m, charge q) are accelerated between two D-shaped 'dees' and follow a spiral in the homogeneous magnetic field B. The centripetal force at radius ρ is equal to the Lorentz force. In SI units,

$$m\frac{v^2}{\rho} = qvB \tag{9.1}$$

or

$$\rho = \frac{p}{qB}. \tag{9.2}$$

With $v = \omega\rho$, the *cyclotron frequency*

$$\boxed{\omega = \frac{qB}{m}} \tag{9.3}$$

does not depend on the radius ρ. The frequency of the alternating high voltage (of several 100 kV) between the dees is adjusted to coincide with the cyclotron frequency ω. Following (9.1) the radius ρ increases linearly with ion momentum and the ions are extracted at the cyclotron periphery by an electrostatic deflector.

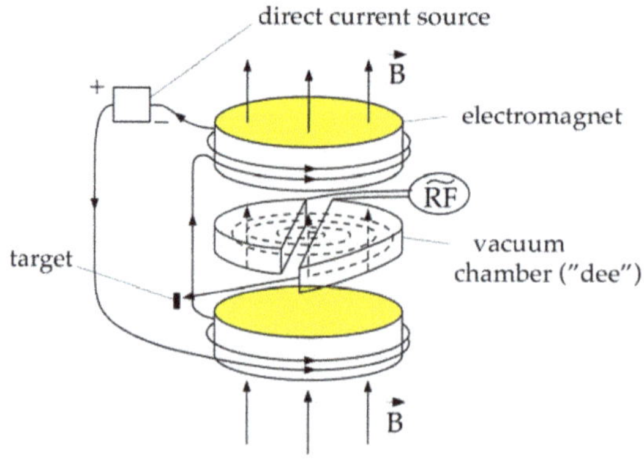

Figure 9.2. The principle of the cyclotron.

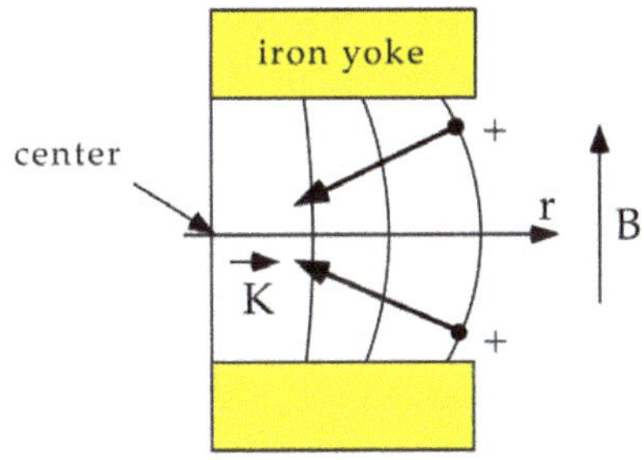

Figure 9.3. In the cyclotron vertical focusing is provided by the slowly decreasing magnetic field (the positive ions fly towards the reader).

A cyclotron with a perfectly homogeneous *B*-field would only provide a vanishingly small intensity since only the ions that leave the source perpendicularly to the magnetic field would be accelerated, while ions emitted up or down would hit the magnet yoke. However, since the *B*-field decreases slightly towards the open periphery, the small vertical component of the Lorentz force $\vec{K}$ (figure 9.3) repels the ions into the central plane, which then perform harmonic oscillations in the vertical direction. The condition for *vertical focusing* is

$$n = -\frac{r}{B}\left(\frac{\partial B_z}{\partial r}\right) > 0, \tag{9.4}$$

where n is the *field index*.

Nonetheless, the field gradient towards the periphery cannot be too large in order to keep the ions around the nominal spiral trajectory in the horizontal plane. The condition for both vertical and horizontal focusing is

$$0 < n < 1. \tag{9.5}$$

This condition has to be fulfilled for all *weak* focusing machines. However, since B decreases with increasing ion energy, the cyclotron frequency (9.3) does as well and hence the ions fall out of phase with the acceleration frequency. One then chooses the cyclotron frequency corresponding to the average *B*-field. An additional phase slip stems from the relativistic mass increase. For relativistic particles (9.3) has to be replaced by

$$\boxed{\omega = \frac{qB}{\gamma m}}. \tag{9.6}$$

Hence, cyclotrons accelerate protons only to about 20 MeV and are useless for electrons. In the *synchrocyclotron* the relativistic mass increase is compensated by slowly reducing the applied high frequency until extraction. Such accelerators (see figure 2.7) deliver macroscopically pulsed bunches of ions. Synchrocyclotrons (McMillan, 1945) have been built to accelerate protons up to about 720 MeV (184-inch Berkeley machine), beyond which the magnet size became prohibitive. In the *isochronous* cyclotron (Thomas, 1938) the time-constant *B*-field grows towards the periphery, proportionally to γ. This is achieved with trim coils located at large radii. Vertical focusing is then restored by using curved magnetic sectors (edge focusing), as shown in figure 9.4 (left). Figure 9.4 (right) shows a photograph of the 600 MeV isochronous proton cyclotron at the Paul-Scherrer Institute (PSI) in Switzerland.

In the *betatron* (figure 9.5, left) acceleration is performed with using rapidly time-varying B-field inducing an electric force along the particle trajectory. Betatrons are mostly used with electrons for medical applications. Figure 9.5 (right) shows a photograph of one of the first betatrons.

Let us derive Widerøe's stability relation between the magnetic field B on the circular electron trajectory (radius ρ) and the average field $\langle B \rangle$ through the loop. From the induction law

$$E\ 2\pi\rho = \frac{\mathrm{d}\phi}{\mathrm{d}t} = \pi\rho^2 \frac{\mathrm{d}}{\mathrm{d}t}\langle B \rangle \tag{9.7}$$

the induced electric field is

$$E = \frac{1}{2}\rho\frac{\mathrm{d}}{\mathrm{d}t}\langle B \rangle = \frac{1}{e}\frac{\mathrm{d}p}{\mathrm{d}t} = \rho\frac{\mathrm{d}B}{\mathrm{d}t}, \tag{9.8}$$

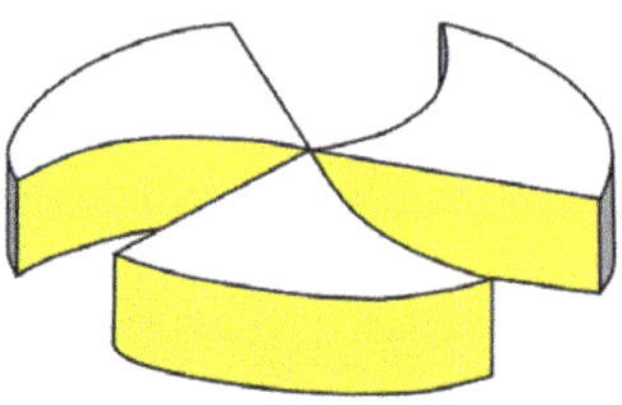

Figure 9.4. Left: in the isochronous cyclotron the average B-field grows with radius. Right: a photograph of the PSI Ring cyclotron with its eight sector magnets and four accelerating cavities. A proton beam of 70 MeV is injected into the machine near its centre. Image credit: Paul Scherrer Institute.

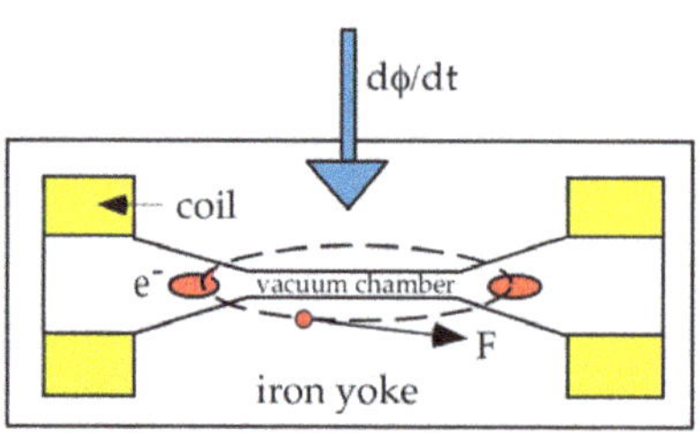

Figure 9.5. Left: the working principle of the betatron. Right: the glass tube of one of the first betatrons. Image credit: Historic Archives ABB.

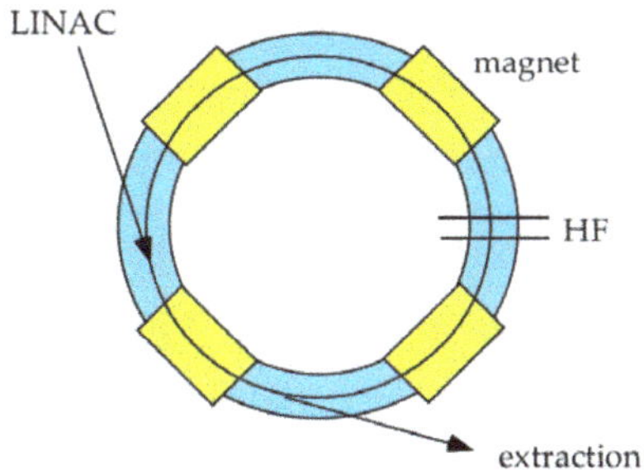

Figure 9.6. The principle of the synchrotron.

using (9.2). Integrating with $B(0) = 0$ and $\langle B\rangle(0) = 0$ gives

$$B(t) = \frac{1}{2}\langle B\rangle(t). \tag{9.9}$$

The field on the electron trajectory must be half the induction field.

According to (9.2) the radius of the accelerator increases linearly with momentum, so the iron yoke becomes uneconomically large at high energies. High energy accelerators are therefore fixed radius ring *synchrotrons* (figure 9.6). In these machines B increases with time until the maximum field is reached. Synchrotrons are therefore macroscopically pulsed, accelerating particle bunches with periods up to several seconds, until the beam is extracted by a fast kicker magnet.

The beam is focused horizontally and vertically, satisfying (9.5). The transverse beam size varies along the circular trajectory and the *betatron amplitudes* determine the dimensions of the vacuum chamber which can be quite large, up to meters. These machines consist of a series of large bending magnets. For example, the Synchrophasotron at the Joint Institute for Nuclear Research in Dubna delivered 10 GeV protons until its decommissioning in 2003. For economic reasons weak-focusing cyclotrons are not manufactured anymore and are no longer in operation.

In a *strongly focusing* synchrotron (Courant, 1952) horizontal and vertical focusing is achieved with strongly inhomogeneous fields from dipole magnets, alternatively with $n \gg 1$ and $n \ll 0$, such as the Alternating Gradient Synchrotron (AGS) in Brookhaven, or the Proton Synchrotron (PS) at CERN. The betatron amplitudes, and thus also the magnet dimensions, are kept small by the strong fields. Modern synchrotrons are *hybrid* machines with dipole magnets for deflection and magnetic lenses for focusing (figure 9.7). Magnetic quadrupole lenses focus the beam

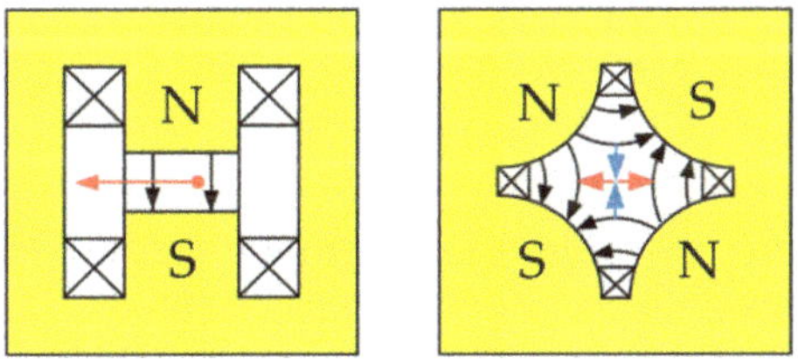

Figure 9.7. Left: dipole magnet. Positive ions moving into the page are deflected by the Lorentz force to the left. Right: quadrupole lens. The positive ions are focussed vertically and defocussed horizontally.

in one direction, while defocusing in the other direction. A pair of lenses with opposite polarities is then able to focus the beam in both directions.

A practical formula may be derived from (9.2) which relates the momentum to the field and curvature radius. To calculate the momentum in eV c^{-1}, divide cp (in J) by the elementary charge e (in C) since 1 eV = 1.6×10^{-19} J. One finds with $q = ez$

$$cp\,[\text{eV}] = czB\rho = 3 \times 10^8 \text{ m s}^{-1}\, zB\,[\text{T}]\rho[\text{m}], \tag{9.10}$$

or

$$\boxed{p[\text{GeV } c^{-1}] = 0.3\, z \cdot B\,[\text{T}]\rho\,[\text{m}]}. \tag{9.11}$$

Hysteresis saturation in iron magnets typically limits the magnetic field to 1.8 T. The radius of curvature for 1 GeV c^{-1} protons ($z = 1$) is then 1.9 m. The superconducting magnets at CERN's Large Hadron Collider (LHC) achieve fields of 9 T. The circumference of the ring is 27 km and the mean field 5.4 T, which lead to a maximum proton momentum of 7 TeV c^{-1}.

Charged particles on *curved* trajectories experience *synchrotron* radiation, leading to energy loss. From the classical theory of electromagnetism one predicts the energy loss per circular orbit [7]

$$\Delta E \propto \frac{1}{\rho}\frac{E^4}{m^4} = \frac{1}{\rho}\gamma^4. \tag{9.12}$$

Due to the γ-factor, synchrotron radiation is particularly important for electrons and positrons, for which

$$\Delta E\,[\text{MeV}] = 8.9 \times 10^{-2}\,\frac{E^4\,[\text{GeV}]}{\rho\,[\text{m}]}. \tag{9.13}$$

Very high energies are produced by using colliders (section 8.4). In CERN's Large Electron–Positron ring (LEP), which was shut down in 2001, 50 GeV electrons and 50 GeV positrons were accelerated and collided in the same vacuum ring. With a circumference of 27 km the energy loss per revolution was 130 MeV. At a 95 GeV per beam the energy loss rose to 1.7 GeV per revolution. In fact, the power loss was slightly higher than given by (9.13) because LEP was only approximately a circular machine. This energy corresponds to a power loss in LEP of 17 MW (for a beam current of 3 mA) which had to be constantly replenished. Superconducting accelerating cavities were used to reduce the power loss through Joule heating. Decisive for the huge diameter of LEP were the synchrotron radiation losses rather than the high beam momenta themselves. Indeed, using (9.11) one finds for the average B-field in LEP only about 0.07 T at 95 GeV.

In e^+e^- and $\bar{p}p$ colliders both beams circulate in the same vacuum tube. For machines in which colliding particles are not charge conjugate, two separate rings are built which intersect at the point of experimentation. The Relativistic Heavy Ion Collider (RHIC) in Brookhaven collides ions of 100 GeV nucleon^{-1}. The LHC is designed to collide two beams of 7 TeV protons. The HERA machine at DESY in Hamburg (decommissioned in 2007) collided 28 GeV electrons (or positrons) with 920 GeV protons. The corresponding centre-of-mass energy was $\sqrt{s} = 320$ GeV (proof left as an exercise for the reader). The electron energy was limited by synchrotron radiation.

Obviously, only collisions between stable particles such as ions, p, $\bar{p}$ and $e^{\pm}$ are possible[1]. The *luminosity* $\mathcal{L}$ is an essential parameter to calculate the reaction rate R at the collision point. For head-on collisions between two beams

$$R = \sigma\mathcal{L}, \quad \text{with } \mathcal{L}\left[\text{cm}^{-2}\text{s}^{-1}\right] = \frac{fnN_1N_2}{4\pi\sigma_x\sigma_y}, \tag{9.14}$$

where σ is the cross section for the reaction of interest. The revolution frequency is f, the number of bunches in each beam around the ring is n, the numbers of particles per bunch are N_1 and N_2, respectively. The size of the beam overlapping surface at the collision point is described by the rms standard deviations σ_x and σ_y.

The first high energy collider was the Intersection Storage Rings (ISR) at CERN, colliding 31 GeV protons (decommissioned in 1984). At the LHC the total *pp* cross section is $\sigma \sim 0.1$ b at $\sqrt{s} = 14$ TeV. With a design luminosity of 10^{34} cm^{-2} s^{-1} one then expects $R = 10^9$ events s^{-1}. This kind of rate, together with the high final state multiplicity (about 100 charged particles at $\sqrt{s} = 14$ TeV), sets drastic demands on the detector design and its readout electronics.

Figure 9.8 shows a sketch of the CERN accelerator complex in 2012, with its flagship, the LHC. Antiprotons ($\bar{p}$) are produced by 26 GeV protons from the PS

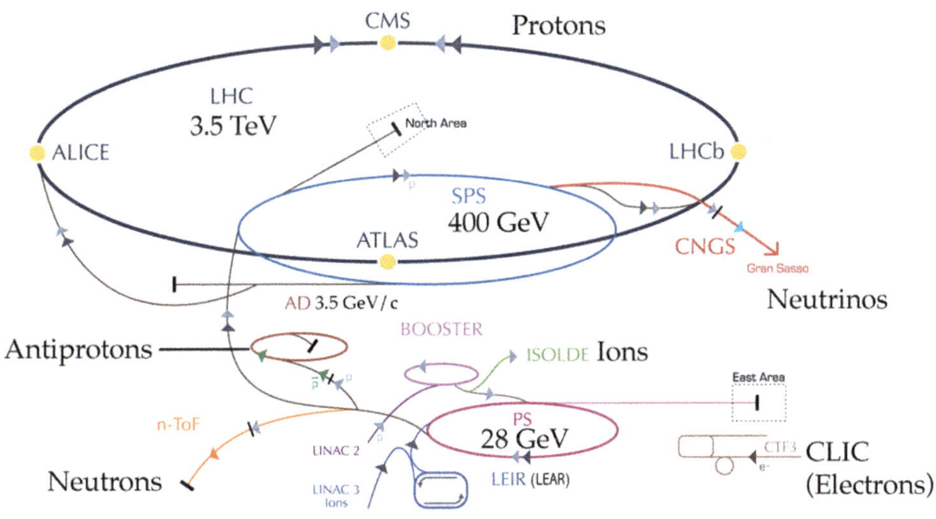

Figure 9.8. CERN's accelerator complex in 2012. AD: Antiproton Decelerator; CNGS: CERN Neutrinos to Gran Sasso; CLIC: Compact Linear Collector; LEIR: Low Energy Ion Ring; LEAR: Low Energy Antiproton Ring, decommissioned in 1996; LHC: Large Hadron Collider; PS: Proton Synchrotron; SPS: Super Proton Synchrotron.

---

[1] The design of a circular $\mu^+\mu^-$ collider ($\sim$2 TeV beam^{-1}) is being discussed. Since muons have a relatively long lifetime at very high energies and are about 200 times heavier than electrons, they emit much less synchrotron light than electrons.

striking a target. The momentum spectrum of the antiprotons reaches a maximum at 3.5 GeV c^{-1}. The antiprotons, which are emitted mainly into the forward hemisphere, are focussed by a magnetic horn (magnetic horns are mostly used to produce high energy neutrino beams, see figure 14.23 in section 14.6.3) and sent to the Antiproton Decelerator (AD) for deceleration to very low momenta (100 MeV c^{-1}), before being extracted for antimatter and atomic physics experiments [8]. Cold antihydrogen atoms were produced for the first time by using antiprotons from the AD [9].

In the 1980s and 90s the Super Proton Synchrotron (SPS) was used as a 270 GeV storage ring, colliding protons with antiprotons. The precursor of the AD was the Antiproton Accumulator (AA). The vacuum chamber of the AA was quite large to collect a maximum of antiprotons. Other unstable secondary particles, such as negative pions and kaons were also collected, but decayed. A pure antiproton beam with small transverse dimensions was generated in the AA by *stochastic cooling* of the antiproton bunches. The principle of stochastic cooling is sketched in figure 9.9. The bunch coordinates and time were sampled at each passage near the pick-up electrodes. A correction signal was then sent across the ring to the kicker. After many orbits the corrections had substantially reduced the transverse dimensions and the momentum byte of the beam. The cooled antiprotons were then stacked in the vacuum chamber. Up to 10^{12} antiprotons per day could be generated and stored. The 3.5 GeV c^{-1} antiprotons were then extracted and injected into the PS, where they circulated in the same beam pipe as the protons, but in the opposite direction. They were then accelerated to 26 GeV and transferred to the SPS for further acceleration to 270 GeV and collisions with 270 GeV protons. The W and Z^0 gauge bosons were discovered at the SPS $\bar{p}p$ collider in 1983 (section 17.2).

The SPS $\bar{p}p$ collider is no longer in operation. On the other hand, the Tevatron at Fermilab, where the top quark was discovered (section 10.5), delivered data in $\bar{p}p$ collisions at $\sqrt{s} \simeq 2$ TeV until 2011.

In the 1980s and 90s the 3.5 GeV c^{-1} antiprotons from the AA could also be decelerated in the PS and stored in the Low Energy Antiproton Ring (LEAR), where the antiprotons were decelerated down to 60 MeV c^{-1} ($T = 2$ MeV) and then slowly extracted with a typical intensity of 10^6 s^{-1}. LEAR hosted several experiments before being decommissioned in 1996, among them the Crystal Barrel studying $\bar{p}p$ annihilation into mesons [10] and CPLEAR studying CP violation in the $K^0 - \overline{K}^0$ system [11].

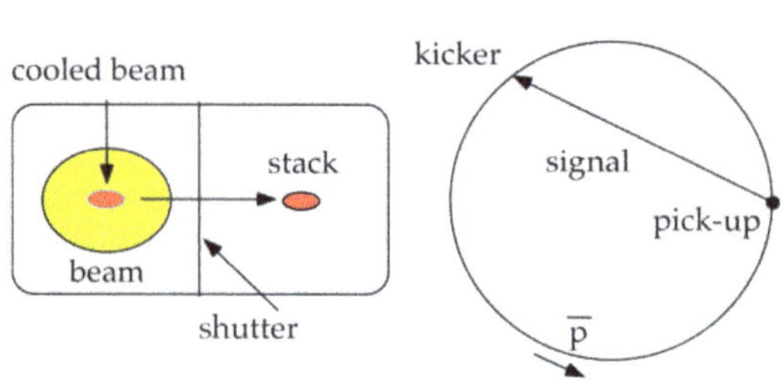

Figure 9.9. The principle of stochastic cooling. The vacuum chamber is shown on the left. The correction signal must arrive at the kicker before the antiproton bunch.

9.2 Charged particle interaction with matter

Let us now discuss the interaction of radiation with matter. Some of the formulae will be given without derivation. A charged particle passing through matter is scattered and loses energy. This results from different processes, such as inelastic scattering off shell electrons (excitation or ionization), nuclear elastic and inelastic scattering, Čerenkov radiation and bremsstrahlung. The various contributions depend on the particle, its energy and the type of absorbing material. Due to their small mass, electrons and positrons have to be dealt with separately from other projectiles such as μ, π, K, p, α and heavy ions. For the latter inelastic scattering off shell electrons is the dominating interaction. Although little energy is transferred by each scattering, the very high cross section ($\sigma \sim 10^{-17} - 10^{-16}\,\mathrm{cm}^2$) leads to a large number of scatterings per unit length. The maximum energy transferred to an electron of mass m_e at rest, by a projectile of mass m and velocity β is

$$T_{\max} = \frac{2m_e\beta^2\gamma^2m^2}{m^2 + m_e^2 + 2\gamma m m_e}. \tag{9.15}$$

$T_{\max}$ is transferred to the electron as kinetic energy (neglecting the small binding energy). For all particles except electrons and positrons, and for not too high energies $E \ll m^2/m_e$, the maximum energy is approximately

$$T_{\max} \simeq 2m_e\beta^2\gamma^2. \tag{9.16}$$

The *stopping power* is defined as the energy loss $\mathrm{d}E$ per unit length $\mathrm{d}x$. Usually $\mathrm{d}x$ is expressed in $\mathrm{g\,cm^{-2}}$:

$$\mathrm{d}x = \rho\,\mathrm{d}s, \tag{9.17}$$

where $\mathrm{d}s$ is the thickness of the absorber in cm and ρ its density in $\mathrm{g\,cm^{-3}}$. Figure 9.10 shows the momentum dependence of the stopping power for positive muons in copper.

The stopping power of a particle with charge ze, traversing a medium with atomic number Z and *atomic weight* A, is given by the *Bethe–Bloch* formula

$$-\frac{\mathrm{d}E}{\mathrm{d}x}\left[\mathrm{MeV\,g^{-1}cm^2}\right] = Kz^2\left(\frac{Z}{A}\right)\frac{1}{\beta^2}\left[\frac{1}{2}\ln\left(\frac{2m_e\beta^2\gamma^2T_{\max}}{I^2}\right) - \beta^2 - \frac{\delta(\beta)}{2}\right] \tag{9.18}$$

or, using (9.16),

$$\boxed{-\frac{\mathrm{d}E}{\mathrm{d}x}\left[\mathrm{MeV\,g^{-1}cm^2}\right] \simeq Kz^2\left(\frac{Z}{A}\right)\frac{1}{\beta^2}\left[\ln\left(\frac{2m_e\beta^2\gamma^2}{I}\right) - \beta^2 - \frac{\delta(\beta)}{2}\right]}, \tag{9.19}$$

valid in the kinematic range $0.05 < \beta\gamma < 500$. The constant K is defined as

$$K = \frac{4\pi L\alpha^2}{m_e} = 7.89 \times 10^{20}\,\mathrm{MeV^{-1}g^{-1}} = 0.307\,\mathrm{MeV\,g^{-1}cm^2}, \tag{9.20}$$

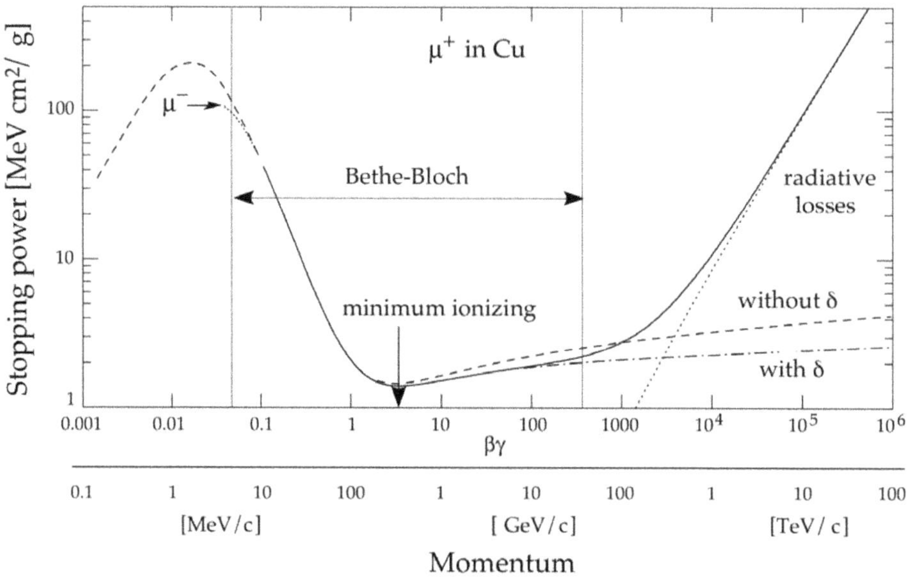

Figure 9.10. Stopping power ($-\mathrm{d}E/\mathrm{d}x$) of positive muons traversing a copper absorber of thickness $\mathrm{d}x$ (expressed in $\mathrm{g\,cm^{-2}}$), as a function of $\beta\gamma$ and as a function of momentum. The full curve shows the sum of all contributions, see formula (9.19). The short dotted line shows the stopping power for negative muons which differs slightly at low energies (Barkas effect). After [3].

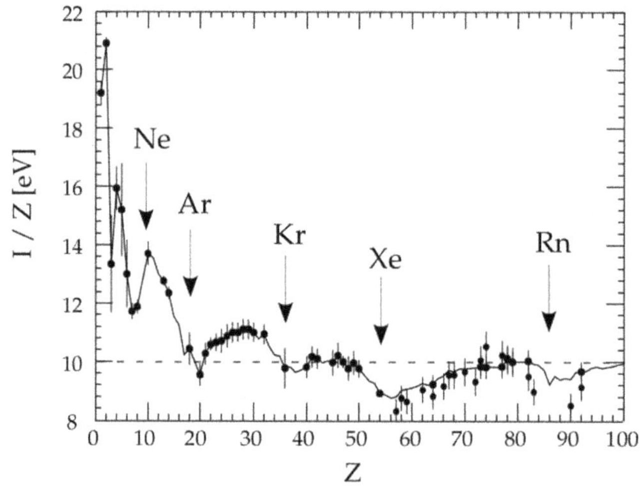

Figure 9.11. Mean excitation energy divided by Z. The line is a guide for the eye. After [3].

where L is the Avogadro number. (We have absorbed the unit of mass from A [g] in K and expressed $1\,\mathrm{MeV}^{-1}$ as $1\,\mathrm{MeV} \times 1\,\mathrm{MeV}^{-2}$ with $1\,\mathrm{MeV}^{-2}$ replaced by $(197.3\,\mathrm{fm})^2$.) I is the mean excitation energy of the atom or molecule in the medium. I/Z is approximately equal to $10\,\mathrm{eV}$ for $Z > 15$ (figure 9.11).

The small correction δ is often neglected in (9.19). It depends on the polarization of the medium which shields the electric field of the projectile. This correction lowers the energy loss and is important at high energies and for large electron densities, i.e. in solids. The Bethe–Bloch formula (9.19) holds as long as (i) the particle speed β is substantially larger than the speed of the shell electrons ($Z\alpha$), otherwise shell corrections are needed in the low energy range of figure 9.10, and (ii) for not too relativistic particles ($\gamma \ll 10^3$), otherwise *bremsstrahlung* contributes to the energy loss in the upper energy range. For muons (figure 9.10) the Bethe–Bloch formula is accurate within 1% between 6 MeV and 6 GeV, while above 100 GeV the contribution from bremsstrahlung becomes significant. Bremsstrahlung has to be taken into account for electrons and positrons, even at low energies (section 9.4).

The stopping power, expressed in $\mathrm{MeV\,g^{-1}\,cm^2}$, is approximately equal for all media at fixed β except for hydrogen since Z/A varies only slowly over the periodic table (figure 9.12). To obtain the energy loss per unit path length $\mathrm{d}s$ (in cm), $\mathrm{d}E/\mathrm{d}x$ has to be multiplied by the density:

$$\frac{\mathrm{d}E}{\mathrm{d}s} \equiv \frac{\mathrm{d}E}{\mathrm{d}x} \cdot \rho. \tag{9.21}$$

The origin of the term $1/\beta^2$ in (9.19), which dominates at low velocities, can be understood intuitively as follows: the momentum transfer to the electron is $q = F_C\,\Delta t$ where F_C is the Coulomb force and Δt the transit time of the projectile. The kinetic energy transferred to the shell electron is

$$T \simeq \frac{q^2}{2m_e} \propto \Delta t^2 \propto \frac{1}{\beta^2}. \tag{9.22}$$

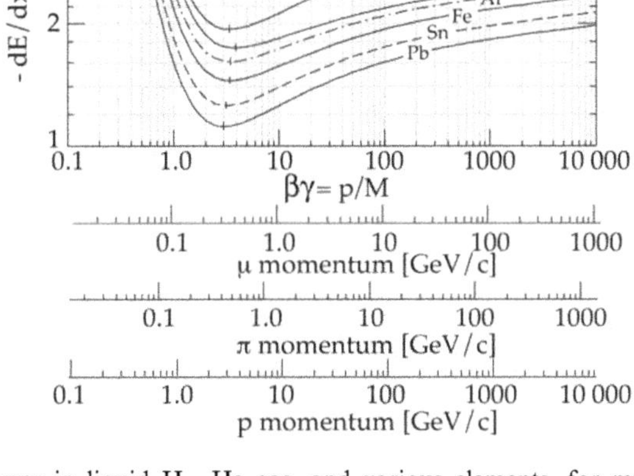

Figure 9.12. Stopping power in liquid H_2, He gas, and various elements, for muons, pions and protons. Radiative effects which become significant above $\beta\gamma = 10^3$ are not included. After [3].

For $\beta\gamma \sim 3$ the stopping power reaches the minimum value

$$\boxed{\frac{\mathrm{d}E}{\mathrm{d}x} \sim 1.5\ \mathrm{MeV\,g^{-1}cm^2}}, \tag{9.23}$$

and then increases slowly with energy. In hydrogen ($Z/A = 1$) the stopping power $\mathrm{d}E/\mathrm{d}x$ is twice as large. Projectiles losing the minimum energy (9.23) are called *minimum ionizing* particles (MIPs). Note that $\mathrm{d}E/\mathrm{d}x$ depends only on the speed of the particle and not on its mass (momentum or energy). This β dependence of the stopping power can be used to identify the type of particle. The particle mass can be determined from a combined measurement of its stopping power and momentum (see section 9.6).

For example, minimum ionizing particles lose in a 1 cm thick lead (Pb) or carbon (C) absorber the energy (see figure 9.12):

$$\begin{aligned}\overline{\Delta E}(\mathrm{Pb}) &\simeq 1.1\ \mathrm{MeV\,g^{-1}cm^2} \times \overbrace{11.3\ \mathrm{g\,cm^{-3}}}^{\rho} \times 1\ \mathrm{cm} = 12.4\ \mathrm{MeV},\\ \overline{\Delta E}(\mathrm{C}) &\simeq 1.7\ \mathrm{MeV\,g^{-1}cm^2} \times 2.3\ \mathrm{g\,cm^{-3}} \times 1\ \mathrm{cm} = 3.9\ \mathrm{MeV}.\end{aligned} \tag{9.24}$$

These are average values of the energy loss ΔE. Fluctuations in the total energy loss are small, because of the high number of atoms along the absorber. Larger fluctuations occur in thin absorbers and at high energies. Due to the smaller number of collisions the distribution of the energy loss becomes asymmetric, with a tail towards high ΔE (Landau–Vavilov distribution, figure 9.13). The rms standard deviation is

$$\sigma[\mathrm{MeV}] \simeq \gamma\sqrt{0.07\,\Delta x\left[\mathrm{g\,cm^{-2}}\right]} \quad \text{for } \frac{40}{3}\frac{\beta^4\gamma^2}{\Delta x} \ll 1, \tag{9.25}$$

and

$$\sigma[\mathrm{MeV}] \simeq \frac{0.15\,\Delta x\left[\mathrm{g\,cm^{-2}}\right]}{\beta^2} \quad \text{for } \frac{40}{3}\frac{\beta^4\gamma^2}{\Delta x} \gg 1, \tag{9.26}$$

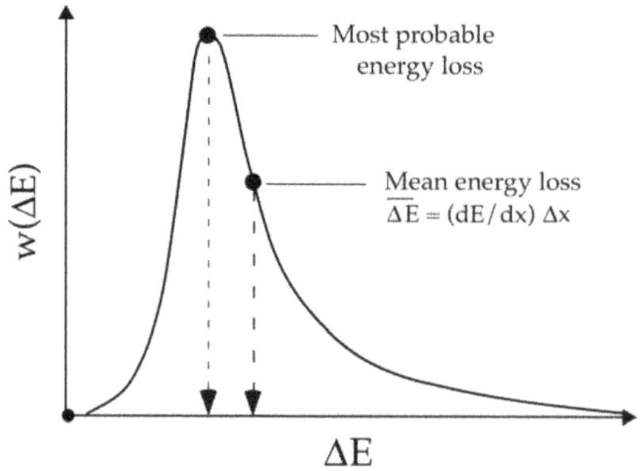

Figure 9.13. Landau–Vavilov energy loss distribution in an absorber of thickness Δx.

where Δx is the absorber thickness. For thick absorbers (large number of collisions) the distribution becomes Gaussian, as expected from the central limit theorem of statistics, with the rms standard deviation $\sigma/\overline{\Delta E} \propto 1/\sqrt{\Delta x}$. However, for minimum ionizing particles and thin absorbers the ratio $\sigma/\overline{\Delta E} \sim 0.1$ does not depend on Δx.

The mean *range* R of the projectile is obtained by integrating the stopping power:

$$R = \int_E^0 \frac{dx}{dE} dE. \tag{9.27}$$

Figure 9.14(*a*) shows the range in lead and in other media (in g cm^{-2}) for various projectiles. For example, the range in cm of a 100 MeV c^{-1} pion in lead is

$$\frac{R}{\rho} = \frac{9 \text{ g cm}^{-2}}{11.3 \text{ g cm}^{-3}} = 0.8 \text{ cm}. \tag{9.28}$$

The range distribution is described by a Gaussian. The relative range fluctuation $\epsilon = \sigma/R$, or *range straggling*, is around 1–3% for protons, depending on the absorber material (figure 9.14(b)). The range straggling ϵ_i of a (single-charged) particle with mass m_i can be calculated from that of a proton of equal velocity β with the help of the relation [14]

$$\epsilon_i(\beta) = \epsilon(\beta)\sqrt{\frac{m_p}{m_i}}. \tag{9.29}$$

For momenta below that of the minimum stopping power, the projectile can be identified by combining a measurement of its ΔE from a thin absorber with its kinetic energy, derived from its range in a thick absorber. According to (9.19)

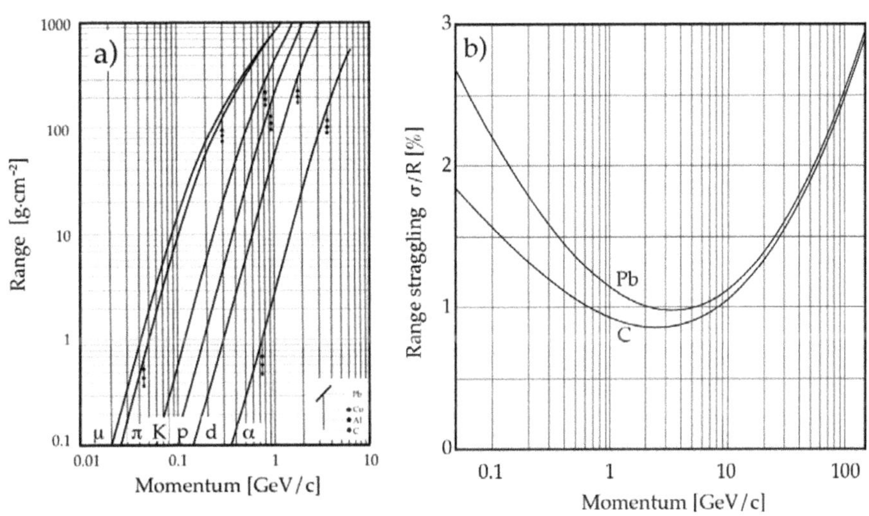

Figure 9.14. (*a*) Range in Pb and other media. After [12]. (Tables of range versus energy are available, see [13].) (*b*) Range straggling of protons in Pb and C. After [14].

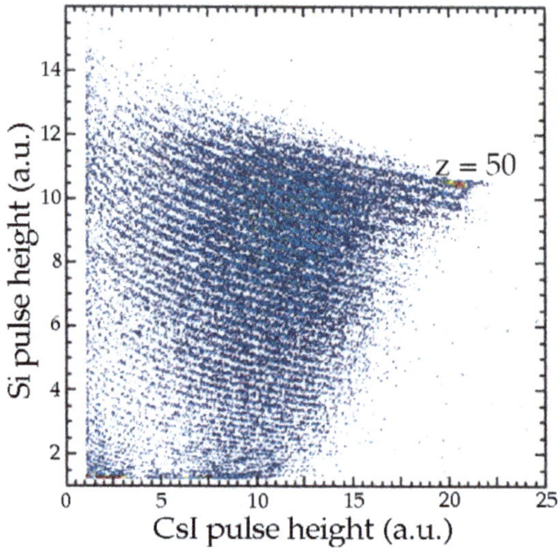

Figure 9.15. ΔE versus E from a heavy ion reaction (in arbitrary units). Note the expected $\Delta E \propto \frac{1}{E}$ dependence (data taken by the CHIMERA detector). After [15].

$\frac{dE}{dx} \propto \frac{z^2}{\beta^2}$ at low energy, while the kinetic energy (here denoted by E) is $\frac{1}{2}m\beta^2$. Therefore

$$\Delta E \propto \frac{z^2}{E} m \sim \frac{z^3}{E}, \tag{9.30}$$

where z and m refer to the atomic number and mass of the projectile. Figure 9.15 shows the stopping power in a thin absorber versus total energy deposit for a ^{124}Sn beam of 35 MeV nucleon^{-1} striking a ^{64}Ni target. A clear separation of the reaction products is observed. The thin absorber is a 300 μm silicon diode, a detector with high energy resolution [4], and the thick absorber is a CsI(Tl) crystal.

9.3 Multiple scattering

A charged particle traversing a medium experiences a series of scatters which modify its direction. Provided that the number of scatters is large (i.e. for thick enough absorbers) the deflection angle at the exit of the absorber is Gaussian distributed. This is a consequence of the central limit theorem. Figure 9.16 shows the distribution of the deflection angle behind the absorber for electrons hitting the entrance surface perpendicularly. The rms standard deviation σ is a measure of the average deflection angle (figure 9.17(a)):

$$\boxed{\sigma[\text{rad}] = \frac{19.2}{\beta p[\text{MeV } c^{-1}]} z \sqrt{\frac{\ell}{X_0}} \left[1 + 0.038 \ln\left(\frac{\ell}{X_0}\right)\right]}, \tag{9.31}$$

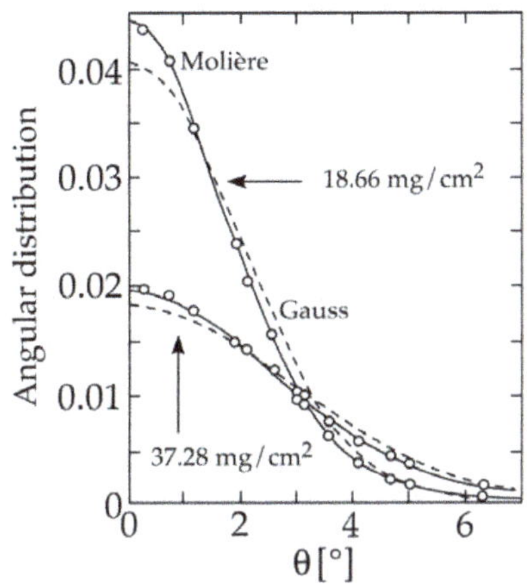

Figure 9.16. Angular distribution of 16 MeV electrons leaving a gold foil. The solid curves interpolate between data points, the dashed curves are Gaussians. After [16].

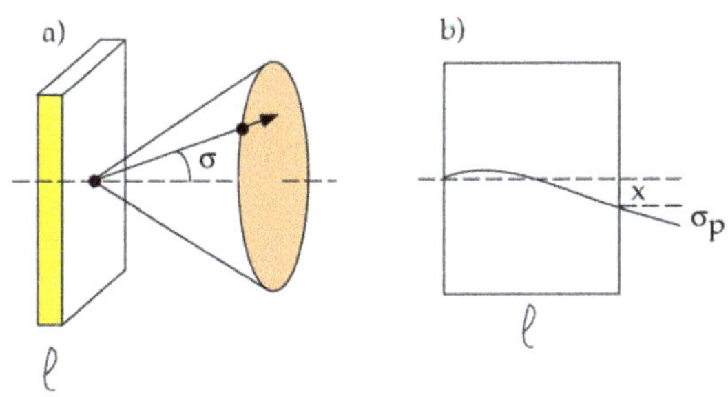

Figure 9.17. Multiple scattering of a charged particle traversing a medium of thickness ℓ.

where ℓ is the absorber thickness, z the charge of the projectile, and X_0 the material-dependent *radiation length*.

The radiation length X_0 has been measured for many elements and can be approximated by

$$X_0\,[\mathrm{cm}] = \frac{1}{\rho} \cdot \frac{716\,\mathrm{g\,cm^{-2}}A}{Z(Z+1)\ln\frac{287}{\sqrt{Z}}} \tag{9.32}$$

(for an analytic expression see (9.54) below). Hence X_0 decreases very roughly as $1/Z^2$ with increasing atomic number. Since the density ρ may depend strongly on temperature and pressure (e.g. for gases) one often quotes as radiation length the quantity

$$X_0'\left[\mathrm{g\,cm^{-2}}\right] = X_0\,[\mathrm{cm}]\,\rho, \tag{9.33}$$

which varies roughly as $\frac{1}{Z}$. Table 9.1 lists values of X_0' and X_0 for various elements.

Table 9.1. Radiation lengths X_0' and X_0 versus Z. For H_2, He and Ar the radiation length X_0 refers to gases at normal temperature and pressure. The critical energy E_c for electrons is defined below (section 9.4).

Element	Z	X_0' [g cm^{-2}]	X_0 [cm]	E_c [MeV]
H_2	1	63.04	7.5×10^5	370
He	2	94.32	5.7×10^5	243
Ar	18	19.55	1.2×10^4	31
C	6	42.70	12.13	84
Al	13	24.01	8.99	50
Fe	26	13.84	1.76	22
Pb	82	6.37	0.56	7

For an estimate the second term in the square brackets in (9.31) can be ignored. In the two-dimensional projection (figure 9.17(*b*)) the mean scattering angle σ is reduced by a factor $\sqrt{2}$:

$$\sigma_p = \frac{\sigma}{\sqrt{2}} \tag{9.34}$$

and the sideways displacement x is given by

$$x = \frac{1}{\sqrt{3}}\ell\sigma_p. \tag{9.35}$$

For example, a 10 GeV c^{-1} pion traversing a 2 cm thick iron absorber ($X_0 = 1.76$ cm) is deflected by the average angle $\sigma_p \simeq 1.5$ mrad.

Multiple scattering limits the accuracy with which the direction or the momentum (curvature) of a charged particle can be measured in a magnetic field. Since the radiation length decreases with increasing Z, detector materials with low Z should be used.

The following rules of thumb are useful to calculate the contribution of multiple scattering to the momentum resolution measured in a homogeneous magnetic field. Let us assume that a particle track of length ℓ is sampled with an rms position resolution ϵ at N equidistant points ($N > 10$). The relative resolution on the momentum p is given by

$$\frac{\sigma_p}{p} \simeq \frac{90\,\epsilon\,[\mathrm{m}]\,p[\mathrm{GeV}\,c^{-1}]}{B[\mathrm{T}]\,\ell\,[\mathrm{m}]^2\,z\sqrt{N+4}} \propto \frac{\epsilon p}{\sqrt{N}B\ell^2}, \tag{9.36}$$

while the contribution from multiple scattering is

$$\frac{\sigma_{p_{ms}}}{p} \simeq \frac{0.053\,z}{\ell\,[\mathrm{m}]\,B[\mathrm{T}]\,\beta}\sqrt{\frac{\ell}{X_0}} \propto \frac{1}{\beta B\sqrt{\ell}}, \tag{9.37}$$

where X_0 is the radiation length of the detector medium and z the charge of the projectile. For example, the momentum resolution on a 1 m long, 1 GeV c^{-1} pion track, sampled every 10 cm with 1 mm precision in a field of 1 T is $\sigma_p/p = 2.3\%$.

If argon is used as the detector gas ($\rho = 1.66 \times 10^{-3}\,\mathrm{g\,cm^{-3}}$ at normal temperature and pressure), multiple scattering spoils the resolution by $\sigma_{p_{ms}}/p = 0.5\%$.

For very thin absorbers the distribution departs from a Gaussian, due to the low number of collisions. The *Molière* multiple scattering distribution is only approximately Gaussian, with a long tail towards large angles (figure 9.16).

9.4 Energy loss of electrons

The Bethe–Bloch formula (9.19) is only approximative for electrons and positrons since the projectile and target (shell) electrons have the same mass. For incident electrons, scattered and recoil electrons are not distinguishable. Replacing $T_{\max}$ in (9.18) by $T/2 \simeq \gamma m_e/2$ one obtains the approximate stopping power for electrons

$$\left(-\frac{\mathrm{d}E}{\mathrm{d}x}\right)_I = K\frac{Z}{A}\frac{1}{\beta^2}\left[\ln\left(\frac{\gamma m_e}{2I}\right) - \beta^2 - \frac{\delta(\beta)^*}{2}\right], \tag{9.38}$$

where the correction δ^* is different than in (9.19). Figure 9.18 shows the energy loss as a function of total energy. At very low energy, Møller ($e^-e^- \to e^-e^-$) and Bhabha ($e^+e^- \to e^+e^-$) scattering contribute as well as positron annihilation. More precise formulae on ionization can be found in [17].

Charged particles are accelerated in the nuclear Coulomb field and may also lose energy as bremsstrahlung. This process is particularly important for light projectiles, electrons and positrons. The energy loss by bremsstrahlung grows linearly with incident energy:

$$\left(-\frac{\mathrm{d}E}{\mathrm{d}x}\right)_B = \frac{E}{X_0'}. \tag{9.39}$$

The energy loss is determined by the radiation length X_0'.

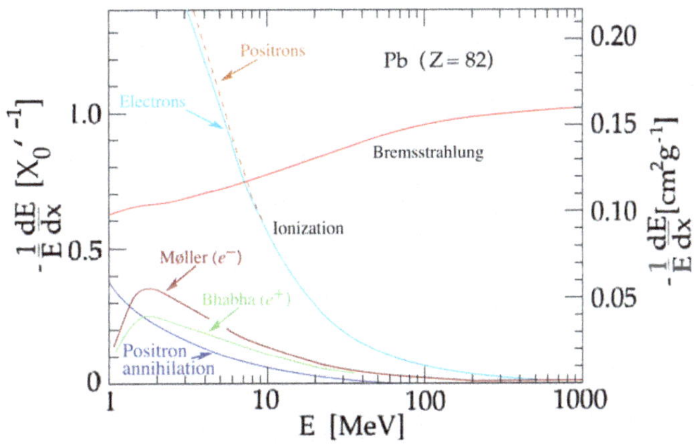

Figure 9.18. Relative energy loss $\frac{\mathrm{d}E}{\mathrm{d}x}/E$ for electrons and positrons in lead, per radiation length on the left and in $\mathrm{cm^2\,g^{-1}}$ on the right. After [3].

Ionization dominates at very low energies ($1/\beta^2$-factor in (9.38)), while bremsstrahlung prevails at high energies (figure 9.18). The *critical* energy E_c is the energy at which the two contributions are equal:

$$\left(\frac{\mathrm{d}E}{\mathrm{d}x}\right)_I = \left(\frac{\mathrm{d}E}{\mathrm{d}x}\right)_B \quad \text{at } E = E_c. \tag{9.40}$$

Above the critical energy (9.38) can no longer be used. In this region the electron loses its energy mainly through the emission of photons. According to (9.39), the energy left after an absorber of thickness $x = s\rho$ is

$$E = E_0 \exp\left(-\frac{x}{X_0'}\right) = E_0 \exp\left(-\frac{s}{X_0}\right). \tag{9.41}$$

The emitted photons generate e^+e^- pairs (pair creation, see section 9.5), which in turn lose energy by bremsstrahlung, creating electromagnetic *showers*. The critical energy is given by the following empirical formulae for solids and gases, respectively:

$$E_c[\mathrm{MeV}] = \frac{610}{Z + 1.24}, \qquad E_c[\mathrm{MeV}] = \frac{710}{Z + 0.92}. \tag{9.42}$$

The critical energies are given in table 9.1 for various elements.

The energy loss by bremsstrahlung is inversely proportional to the square of the projectile mass, so that critical energies can be calculated from (9.42) for heavier particles than electrons. For example, the critical energy for muons is obtained by multiplying the critical energy for electrons by m_μ^2/m_e^2: $E_c(\mu) = 950$ GeV in iron.

9.5 Interaction of photons with matter

Photons interact with matter through three main processes that are relevant to nuclear and particle physics (figure 9.19):

1. **Photoelectric effect.** $\gamma + \text{atom} \to e^- + \text{ion}$. This effect, in which the photon is absorbed by knocking out one of the shell electrons, dominates at photon energies E_γ below 1 MeV. A precise derivation of the cross section for this many-body problem is rather complicated. The following approximation applies at low energies (cross section per atom):

$$\sigma_{p.e.} = \left(\frac{32}{\epsilon^7}\right)^{1/2} \alpha^4 Z^5 \sigma_{\mathrm{Th}}, \tag{9.43}$$

with $\epsilon = E_\gamma/m_e$. The Thomson cross section σ_{Th} which describes γ scattering off free electrons is given by

$$\sigma_{\mathrm{Th}} = \frac{8\pi r_e^2}{3} = 0.665 \text{ b}, \tag{9.44}$$

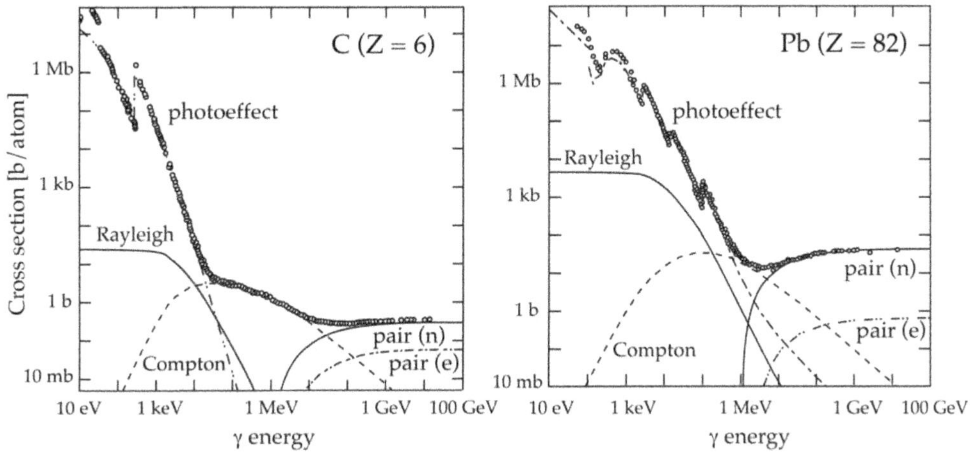

Figure 9.19. Total cross section σ_{tot} for photons in carbon and lead (circles). The contributions of the photoeffect, Rayleigh scattering, Compton scattering, and pair production in the Coulomb field of the nucleus (pair (n)) and in the electron shell (pair (e)) are shown separately. The absorptions edges at low energies (spikes) are due to the ionization of the electronic shell. After [3].

where r_e is the *classical electron radius*

$$r_e = \frac{\alpha}{m_e} = 2.818 \times 10^{-13}\,\text{cm}. \tag{9.45}$$

The cross section grows as Z^5 and decreases as $E_\gamma^{-7/2}$. At higher energies the cross section is given by

$$\sigma_{p.e.} = \frac{4\pi\alpha^6 Z^5}{m_e^2}\frac{1}{\epsilon}. \tag{9.46}$$

2. **Compton scattering.** $\gamma + e^- \to \gamma + e^-$. The Compton effect dominates around 1 MeV. The photon–electron scattering process is quasi-elastic since the binding energies of the shell electrons can be neglected at this energy. The Compton cross section per electron is given by the Klein–Nishina formula:

$$\sigma_C = 2\pi r_e^2 \times \left[\left(\frac{1+\epsilon}{\epsilon^2}\right)\left(\frac{2\{1+\epsilon\}}{1+2\epsilon} - \frac{1}{\epsilon}\ln(1+2\epsilon)\right) + \frac{1}{2\epsilon}\ln(1+2\epsilon) - \frac{1+3\epsilon}{(1+2\epsilon)^2}\right]. \tag{9.47}$$

The total Compton cross section $\sigma_C \cdot Z$ grows linearly with Z. The absorptive part of the cross section decreases below ~10 keV (figure 9.19), while the contribution from coherent Rayleigh scattering increases (at very low

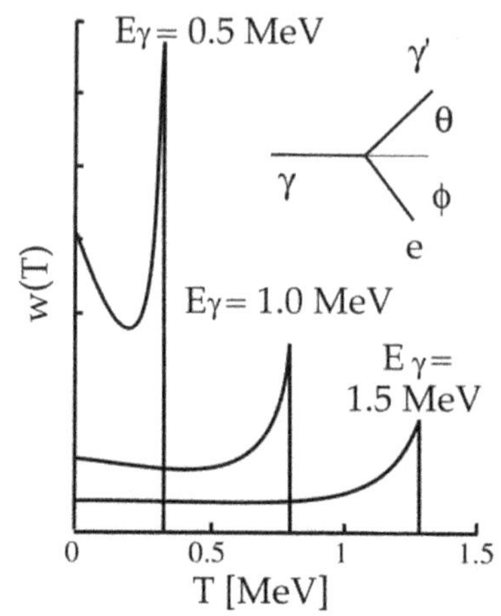

Figure 9.20. Energy distribution of the recoil electron in Compton scattering.

energies the latter is responsible for the blue colour of the sky). The energy of the photon scattered under the angle θ is

$$\boxed{E'_\gamma = \frac{E_\gamma}{1 + \epsilon(1 - \cos\theta)}}. \tag{9.48}$$

The angle of the recoil electron is given by

$$\cot\text{an}\,\phi = (1 + \epsilon)\tan\frac{\theta}{2}. \tag{9.49}$$

Figure 9.20 shows the energy distribution of the electron. The sharp 'Compton edge' corresponds to the highest energy that can be transferred to the electron. The electron energy $T = E_\gamma - E'_\gamma$ reaches this maximum value when the photon is backward-scattered ($\theta = \pi$).

The angular distribution depends on the polarization of the incoming photon. The differential cross section for *unpolarized* photons [18],

$$\frac{d\sigma_C}{d\Omega} = \frac{1}{2}r_e^2\left(\frac{E'_\gamma}{E_\gamma}\right)^2\left(\frac{E_\gamma}{E'_\gamma} + \frac{E'_\gamma}{E_\gamma} - \sin^2\theta\right), \tag{9.50}$$

is shown as polar plot in figure 9.21. The differential cross section for forward-scattered photons (0.078 b sr^{-1} electron^{-1}) does not depend on E_γ. At very low energies Compton scattering goes into Thomson scattering (top green curve in figure 9.21).

3. **Pair production.** $\gamma + Z \rightarrow e^+ + e^- + Z$. This process does not occur in vacuum due to energy and momentum conservation. The lepton pair is produced by the incoming γ in the Coulomb field of the nucleus which absorbs the recoil energy. The energy threshold for pair production is, using (8.61),

$$E_\gamma \simeq 2m_e + 2\frac{m_e^2}{M}. \tag{9.51}$$

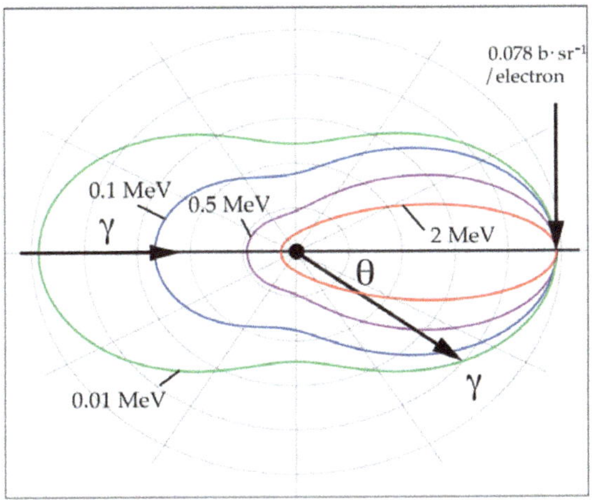

Figure 9.21. Polar plot of the differential Compton cross section for various incident energies E_γ.

Pair production dominates the cross section above ~5 MeV. At low energies ($1 \ll \epsilon < \frac{1}{\alpha Z^{1/3}}$) pair production occurs very close to the nucleus where the photon feels the full charge of the core. The cross section grows with energy according to

$$\sigma_{\text{pair}} = \frac{4\alpha^3 Z^2}{m_e^2}\left(\frac{7}{9}[\ln 2\epsilon - f(Z)] - \frac{109}{54}\right), \tag{9.52}$$

where the correction factor $f(Z)$ is given in [4]. At high energy ($\epsilon \gg \frac{1}{\alpha Z^{1/3}}$), pair production also occurs at larger distances from the core, where the Coulomb field of the nucleus is partially shielded by the electronic shell, so that the cross section saturates:

$$\sigma_{\text{pair}} = \frac{4\alpha^3 Z^2}{m_e^2}\left(\frac{7}{9}\left[\ln\frac{183}{Z^{\frac{1}{3}}} - f(Z)\right] - \frac{1}{54}\right). \tag{9.53}$$

An analytic expression for the radiation length can be found in [4]:

$$\frac{1}{X_0'} \simeq \frac{4\alpha^3 Z^2 L}{m_e^2 A}\left(\ln\frac{183}{Z^{\frac{1}{3}}} - f(Z) + \frac{1}{18}\right). \tag{9.54}$$

Comparing (9.52) and (9.53) and neglecting the small constant terms one obtains at high energies

$$\sigma_{\text{pair}} \simeq \frac{7}{9}\frac{A}{L}\frac{1}{X_0'}. \tag{9.55}$$

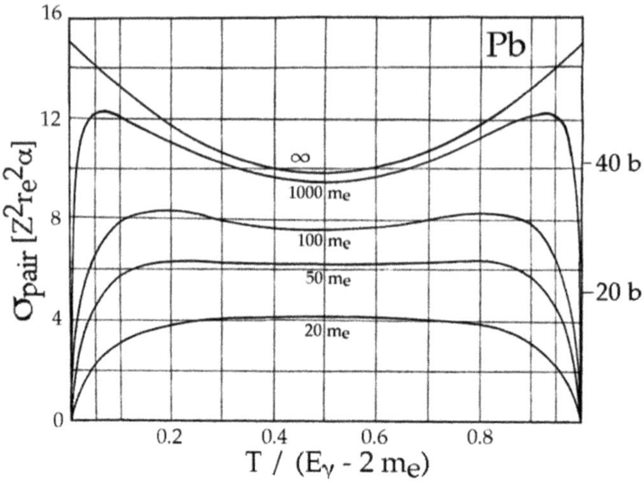

Figure 9.22. Cross section σ_{pair} in lead versus fractional lepton energy for various γ energies. After [19].

Pair production occurs predominantly in heavy materials, as σ_{pair} increases with Z^2, see also (7.18). Figure 9.22 shows the energy sharing between the electron and positron in lead for several γ energies.

Pair production also occurs in the Coulomb field of the shell electrons, albeit with a much smaller cross section (figure 9.19).

The attenuation of the γ intensity I by an absorber of thickness $s = x/\rho$ is with

$$\mathrm{d}I = I\sigma_{\text{tot}}\frac{L}{A}\rho\,\mathrm{d}s, \tag{9.56}$$

$$I = I_0 \exp(-\mu\rho s) = I_0 \exp(-\mu x), \tag{9.57}$$

with the *absorption coefficient*

$$\mu\left[\mathrm{cm}^2\,\mathrm{g}^{-1}\right] \equiv \sigma_{\text{tot}}\frac{L}{A}. \tag{9.58}$$

The energy dependence of μ is shown in figure 9.23 for lead. For example, the intensity I_0 of a 1 MeV γ-ray beam will be attenuated to $I/I_0 \simeq \exp(-0.1 \times 11.3) = 0.3$ in a 1 cm thick lead absorber. Absorption coefficients for other media can be found in reference [3].

The edge around 0.1 MeV in figure 9.23 stems from the K absorption line: below the edge the γ energy is too small to eject an electron from the K-shell (1 s). At lower energies one would observe the three L edges ($2s$, $2p_{1/2}$, $2p_{3/2}$), the five M edges ($3s$, $3p_{1/2}$, $3p_{3/2}$, $3d_{3/2}$, $3d_{5/2}$), etc.

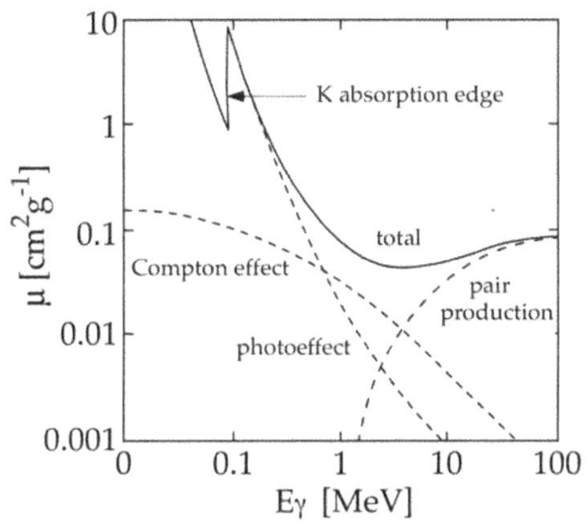

Figure 9.23. Absorption coefficient in lead. The Compton effect curve includes Rayleigh scattering.

At high energies ($\simeq$ 20 MeV in lead) μ saturates when $\sigma_{\text{tot}} \simeq \sigma_{\text{pair}}$. Using (9.55) one obtains a relation between μ and X_0:

$$\mu\rho = \left(\sigma_{\text{tot}}\frac{L}{A}\right)\rho = \frac{7}{9}\frac{A}{L}\frac{1}{X_0'}\frac{L}{A}\rho = \frac{7}{9}\cdot\frac{1}{X_0}. \tag{9.59}$$

From (9.57) one gets

$$\frac{I}{I_0} = \exp\left(-\frac{7}{9}\cdot\frac{s}{X_0}\right) = 0.46 \tag{9.60}$$

for $s = X_0$. Hence, *one radiation length* shields the intensity of a high energy γ source by about 50%.

9.6 Wire chambers

We now describe a few particle detectors. The *multiwire proportional* chamber, invented by Georges Charpak in 1968, measures the spatial coordinates of charged particles (figure 9.24). The gas in the chamber is ionized by the incident particle and the released electrons drift in the electric field towards the signal wire (anode). In the vicinity of the anode wire the field grows as $1/r$. The electrons are accelerated and induce by further ionization a charge avalanche which is collected at the anode wire and amplified. The anode wire is typically 20 μm in diameter, the applied high voltage around 3–4 kV, and the electron multiplication at the wire about 10^5. Anode wires are stretched with a pitch of typically 2 mm. Conductive foils are used for the cathode. The typical distance between the anode and cathode is 5 mm. Instead of using cathode foils a two-dimensional readout is obtained by stretching wires (or laying strips) perpendicularly to the anode wires. The cathode signals are produced by induction from the anode.

Argon at atmospheric pressure is often used as the gas, due to its small electron affinity. The electron avalanche produces a cloud of excited argon ions which decay by γ emission to the ground state. These photons may ionize further atoms

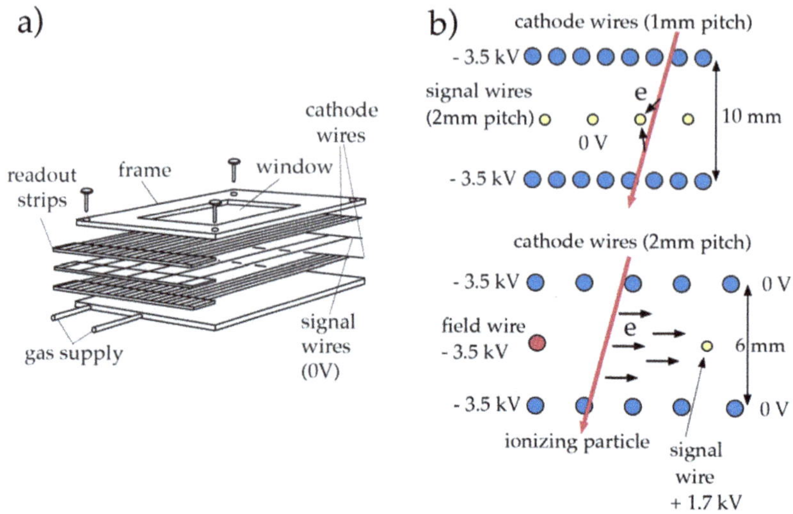

Figure 9.24. The principle of the multiwire proportional chamber and of the drift chamber.

(in particular near the cathode), thereby generating a permanent avalanche. A small admixture of, for example isobutane, is therefore added. Isobutane as the *quench* gas absorbs the photons and dissociates, thus shortening the avalanche time. However, polymer deposits on the wires gradually lead to high voltage discharges and to wire chamber ageing (Malter effect) [20].

Wire chambers provide fast signals with rise times around 0.1 ns and pulse widths of 20 ns. The spatial resolution depends on the wire spacing. With a pitch d of 2 mm, resolutions of $\sigma \simeq 600\ \mu\mathrm{m}$ are achieved (corresponding to the rms standard deviation $\sigma = d/\sqrt{12}$ of a box distribution). The signal height is proportional to the energy deposit ($\mathrm{d}E/\mathrm{d}s \times$ chamber thickness). Hence, multiwire chambers are *proportional* chambers.

A *drift* chamber (figure 9.24, bottom right) measures the drift time of the electrons to the anode. The start time is provided by an external detector such as a scintillation counter (section 9.7) which registers the incoming particle. After suitable calibration of the zero-time and the drift velocity, the distance between the track and the anode can be determined. The drift velocity (1–4 cm $\mu\mathrm{s}^{-1}$) depends on gas mixture, pressure and on the applied electric field. High resolutions are obtained with constant drift velocities, that is in homogeneous electric fields. This can be achieved with an appropriate arrangement of the field wires, as shown in figure 9.24. The typical spatial resolution of a drift chamber is $\sigma = 150\ \mu\mathrm{m}$.

Figure 9.25 shows a drawing of the time projection (drift) chamber (TPC) of the MUNU experiment at the Bugey nuclear reactor, designed to measure the magnetic moment of the antineutrino [21]. This experiment detects the tracks of the recoil electrons in $\bar{\nu}_e e^- \rightarrow \bar{\nu}_e e^-$ scattering. The target electrons are the atomic electrons of the chamber gas (here CF_4). The TPC is about 150 cm long and the applied electric field is 60 kV. The drift electrons travel through a grid to the anode plane. The anode wires are electrically connected to provide only one signal for the entire anode plane.

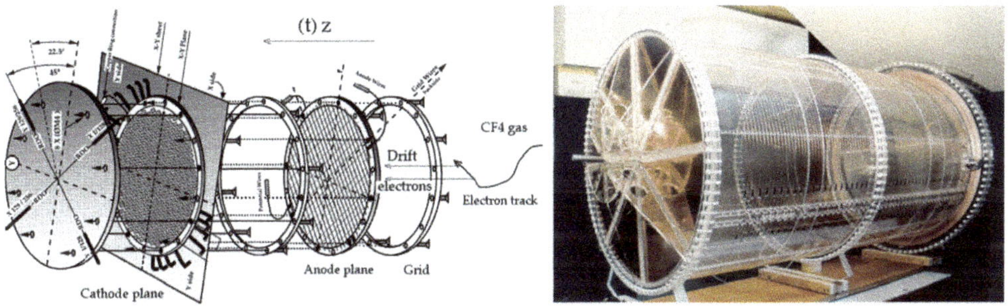

Figure 9.25. The TPC of the MUNU experiment. Left: diagram of the anode plane, grid and cathode strips. Right: a photograph of the acrylic container. Image credit: MUNU Collaboration.

A cathode foil, located behind the anode plane, is coated on both sides with conducting strips (checkerboard-shaped to reduce screening). The strips run on one side of the film in the x-direction and on the other side in the y-direction. The signals on the cathode strips are generated by induction from the anode plane. One measures on the strips the arrival time of the drift electrons (z-coordinate along the TPC). The zero-time is given by those drift electrons that were released from the closest track point to the anode. Thus the entire electron recoil track can be reconstructed in three dimensions (with an overall offset in the z-direction).

The track curvature from a charged particle, hence its *momentum*, can be measured in a magnetic field by using several chamber layers, see (9.36). The *velocity* can also be calculated from the Bethe–Bloch formula by measuring the energy loss in the chambers (signal height). Momentum and velocity together determine the *mass* of the particle. A measurement from the ALICE TPC at the LHC is shown in figure 9.26. This type of particle identification works for relatively slow particles, namely for those below the ionization minimum.

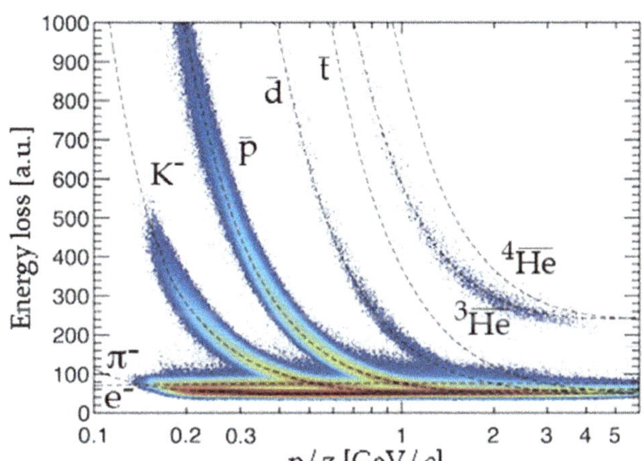

Figure 9.26. Energy loss of negatively charged particles in the ALICE TPC. The data were taken in Pb–Pb collisions at a nucleon–nucleon collision energy of $\sqrt{s} = 2.76$ TeV. Antideuterium, antitritium and anti-^{3}He nuclei are also seen. The dashed curves show the Bethe–Bloch parametrization. Image credit: CERN.

At low energies the time of flight can also be used to determine the particle velocity, but the limited time resolution (typically 100–300 ps) requires a sufficiently long flight path. The velocity can also be measured using Čerenkov counters (section 9.10).

9.7 Scintillation counters

There are two classes of scintillation counters:

1. *Inorganic* scintillators (e.g. NaI, CsI, $Bi_4Ge_3O_{12}$ = BGO, $PbWO_4$) are monocrystals with or without doping (such as Tl-activation, NaI(Tl) or Cs(Tl)). The light is generated by the annihilation of the electron–hole pairs produced by the incident particle. Inorganic scintillation counters have relatively long decay times, typically 20 ns in pure CsI, or 1 μs in the Tl-activated version with much higher light yield. Large crystals are quite expensive. They are mainly used for detecting photons due to their large atomic numbers (e.g. iodine in NaI). The deposited energy is that of the ionizing electron (photoelectric effect at low energy) or that of the electromagnetic shower (pair production and bremsstrahlung at high energy).
2. In *organic* (or plastic) scintillators molecular states are excited by the ionizing particle which disintegrate through fluorescence emission. The admixture of a wavelength shifter converts the UV light into the visible range. Such counters are fast (decay time of ~2 ns), cheap, and can be manufactured as large surfaces. Liquid organic scintillators are also available. Because of the small atomic number ($[CH]_k$ chains) they are not suited for detecting photons.

The light is detected by a photocathode and amplified by a photomultiplier. The photocathode emits 0.2–0.3 photoelectrons per incoming photon. The photoelectrons are accelerated in the vacuum tube and meet a series of electrodes (dynodes), where they produce secondary electron avalanches. The electron multiplication at the output of the photomultiplier is typically 10^8 (figure 9.27).

Figure 9.28 shows a plastic counter with a plexiglas light guide to the photomultiplier. The counter and plexiglas are highly polished, wrapped in aluminium foil and black paper to shield them from external light. Since photomultipliers are

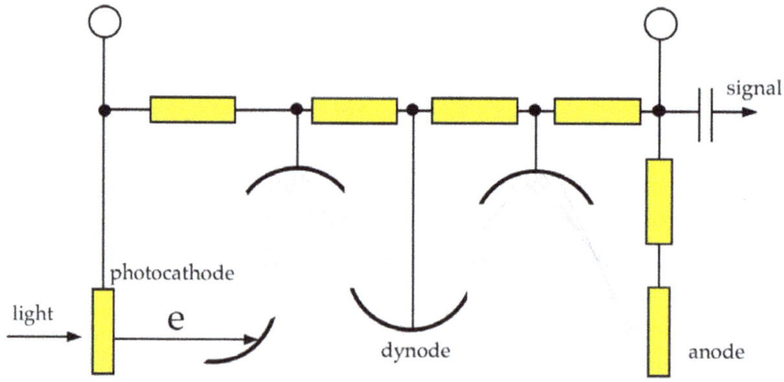

Figure 9.27. Electron multiplication in photomultipliers.

Figure 9.28. A plastic scintillation counter with a photomultiplier, iron shield and high voltage divider. The shining μ-metal cylinder in the middle shields against residual magnetic fields.

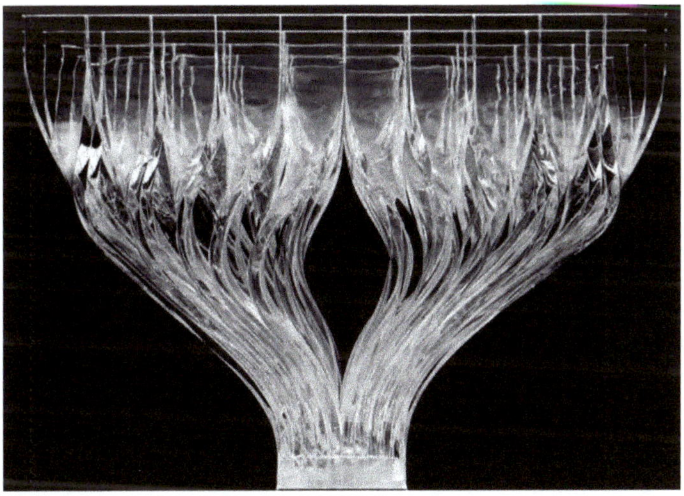

Figure 9.29. Lucite light guides bring the scintillation light from a large scintillator to the photomultiplier tube (bottom) by total internal reflection. Image credit: CERN.

sensitive to magnetic fields, long light guides or optical fibre bundles are used to transfer the light to the photocathode (see also figure 9.29). Alternatively, photomultipliers are replaced by photodiodes for applications in magnetic fields.

Scintillating fibre detectors are also used to track charged particles. Figure 9.30 shows the fibre detector used to measure pion tracks from antiproton annihilation in the AEgIS experiment at CERN [22]. The detector consists of four layers of 1 mm diameter scintillating fibres coupled to clear fibres of the same diameter which direct the optical signals to the photodiodes.

9.8 Bubble chambers

In the bubble chamber, invented by Donald Glaser in 1952, a fluid such as hydrogen, deuterium, propane or freon is expanded and then re-compressed. When a charged particle flies through the chamber during expansion, evaporation bubbles form along

Figure 9.30. Left: the AEgIS scintillating fibre detector. Right: fibres illuminated by UV light.

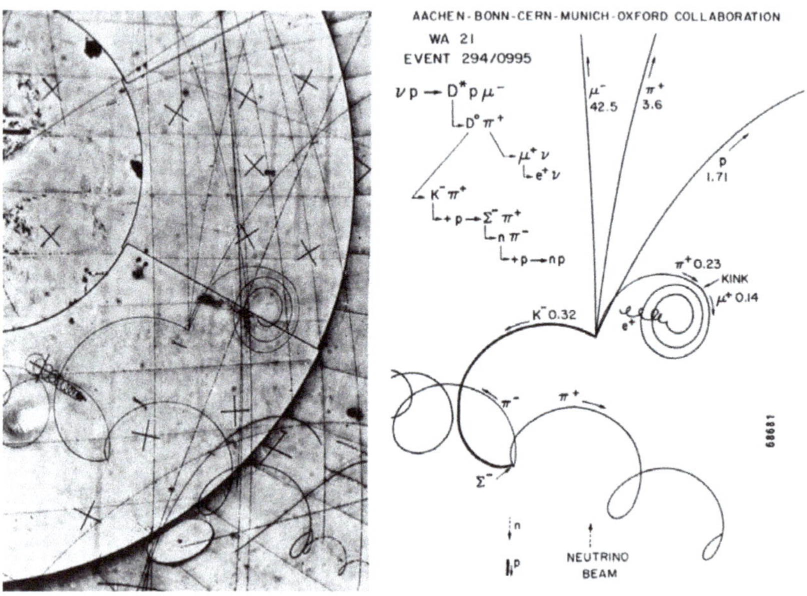

Figure 9.31. Neutrino induced production of charm mesons in the Big European Bubble Chamber (BEBC) at CERN. The crosses are fiducial marks on the inner surface of the container. Image credit: CERN.

its trajectory, which can be photographed from several angles. The pictures are then processed offline. The bubble chamber is usually operated in a magnetic field so that particle momenta can be computed and the kinematics of elementary reactions reconstructed (such as particle decay or interaction with the bubble chamber medium). A fully reconstructed neutrino induced reaction is shown in figure 9.31.

Bubble chambers have been built with diameters up to a few meters. The detection efficiency for photons is modest (small atomic number) and they cannot be triggered externally, so that all expansions are recorded and evaluated. Furthermore, the event rate is very slow, of the order of 1 s^{-1}. For these reasons, and in spite of their excellent tracking capability, bubble chambers are not much in use any more.

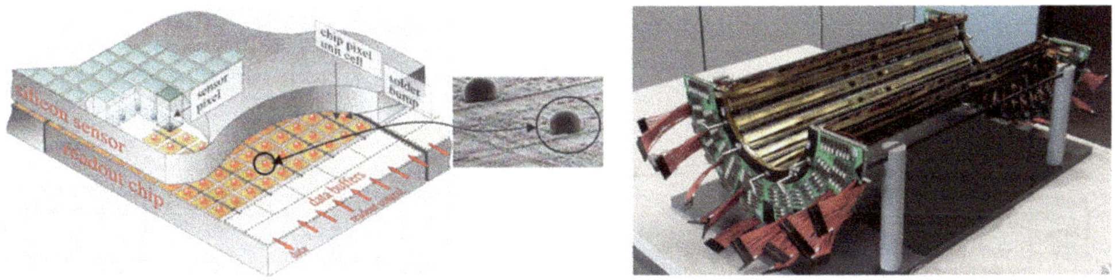

Figure 9.32. Left: the CMS pixel detector layout and electron microscope view of an indium ball for bump bonding. Right: one of the half layers of the CMS pixel detector.

9.9 Segmented silicon detectors

The operating principles of semiconductor radiation detectors are described in many textbooks [4, 23]. Briefly, silicon detectors are based on junctions in which two wafers are brought in contact, one doped with impurities from group 15 (electron donors, *n*-type) such as arsenic or phosphorus, the other with impurities from group 13 (electron acceptors, *p*-type) such as boron or gallium. The electrons from the *n*-type drift into the *p*-type, creating an electric field across the interface. The *depletion* zone is free of mobile charges and is typically 10 μm thick and the potential across the junction is about 500 mV. The depletion zone is increased by applying a voltage across the junction, up to a few hundred volts. The electrons released by the ionization of a charged particle drift towards the *n* region, while holes drift towards the *p* region. The surface of the junction is covered by a highly doped *n* layer on the *n* side and a highly doped *p* layer on the *p* side to provide good conductivity to the metallic readout pad.

A *position sensitive* detector is obtained by segmenting the *n* or *p* layer into strips (one-dimensional readout) or into pixels (two-dimensional readout). The position resolution of a typical strip detector of pitch $d = 50$ μm is $\sigma = \frac{1}{\sqrt{12}}d \simeq 15$ μm. The typical detector thickness is 300 μm. Figure 9.32 (left) shows the layout of the Compact Muon Solenoid (CMS) pixel detector [24]. Each pixel (100×150 μm^2 in its final version) is connected through an indium solder ball to the readout unit, which amplifies the signal. The data are stored on the edge of the chip. Figure 9.32 (right) shows a photograph of one of the half layers surrounding the LHC beam pipe in 2012. The barrel pixel is made of three cylindrical layers containing some 48 million pixels. A position resolution of $\sigma \simeq 10$ μm is achieved in the plane perpendicular to the beam with 100 μm wide pixels by measuring the charge shared between several pixels, as we now describe.

In the presence of a magnetic field *orthogonal* to the electric field, the charge released by ionization does not drift along the electric field, but is deflected by a constant angle θ_L, the *Lorentz* angle which depends on electron mobility and on the magnetic field strength. The force acting on the drift electron with effective mass m is

$$m\frac{\mathrm{d}^2\vec{r}}{\mathrm{d}t^2} = -\frac{e\mu_H}{\mu}\cdot\overbrace{\frac{\mathrm{d}\vec{r}}{\mathrm{d}t}}^{\vec{v}} \times \vec{B} - e\vec{E} - \frac{e}{\mu}\overbrace{\frac{\mathrm{d}\vec{r}}{\mathrm{d}t}}^{\vec{v}}, \tag{9.61}$$

where μ is the electron mobility and μ_H the Hall mobility. The third term on the right-hand side describes the drag force ($\vec{v} = -\mu\vec{E}$ in the absence of a magnetic field). The steady state solution of this equation is [25]

$$\begin{pmatrix} v_x \\ v_y \\ v_z \end{pmatrix} = -\mu_{\text{red}} \begin{pmatrix} 1 + \mu_H^2 B_x^2 & \mu_H^2 B_x B_y - \mu_H B_z & \mu_H^2 B_x B_z + \mu_H B_y \\ \mu_H^2 B_x B_y + \mu_H B_z & 1 + \mu_H^2 B_y^2 & \mu_H^2 B_z B_y - \mu_H B_x \\ \mu_H^2 B_x B_z - \mu_H B_y & \mu_H^2 B_z B_y + \mu_H B_x & 1 + \mu_H^2 B_z^2 \end{pmatrix} \cdot \begin{pmatrix} E_x \\ E_y \\ E_z \end{pmatrix} \tag{9.62}$$

with

$$\mu_{\text{red}} = \frac{\mu}{1 + \mu_H^2 B^2}. \tag{9.63}$$

When $\vec{B}$ and $\vec{E}$ are orthogonal the Lorentz angle is given by

$$\tan\theta_L = \left|\frac{v_x}{v_z}\right| = \mu_H B = r_H \mu B, \tag{9.64}$$

with $\vec{B} = (0, B, 0)$ and $\vec{E} = (0, 0, -E)$ (figure 9.33). The Hall factor $r_H = \frac{\mu_H}{\mu}$ describes the ratio of average times between collisions with the lattice, with and without magnetic field and is of order unity.

The position resolution on the incident track can be improved substantially by measuring the charge distribution among adjacent pixels and computing the coordinate corresponding to the weighted average. On the other hand the Lorentz deflection leads to a displacement of the measured track coordinates, and therefore θ_L needs to be known accurately. The mobility μ depends on the electric field, temperature, doping concentration, and on damage to the lattice caused by irradiation. Therefore θ_L is determined experimentally, for example by searching with the help of an external detector for tracks leading to the smallest cluster sizes for which the incident angle equals θ_L. The mobility is $\mu \sim 10^3\,\text{cm}^2\,\text{V}^{-1}\text{s}^{-1}$ for fields below about $5000\,\text{V}\cdot\text{cm}^{-1}$. Equation (9.64) then gives $\theta_L \simeq 21°$ in the 3.8 T field of CMS, in good agreement with measurements.

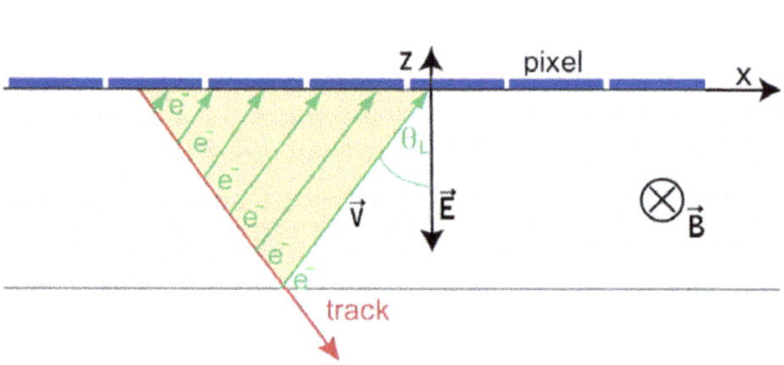

Figure 9.33. For orthogonal $\vec{E}$ and $\vec{B}$ fields the electrons drift under the Lorentz angle θ_L.

9.10 Čerenkov counters

In a medium a transient polarization is produced by the electric field of the traversing charged particle. The temporal charge variation along the particle trajectory leads to dipole radiation when the projectile moves faster than the velocity of light u in the medium (figure 9.34). A shock wave is produced in analogy to the Mach cone from an aircraft flying at supersonic speed.

The condition for coherent radiation is

$$\beta c > u = \frac{c}{n} \Rightarrow \boxed{\beta > \frac{1}{n}}, \tag{9.65}$$

where β is the velocity of the particle and n the refractive index of the medium. The energy of the Čerenkov radiation is a factor of 10^3 to 10^4 smaller than the energy loss by ionization, and is therefore a negligible contribution to the energy deposit in the medium. Čerenkov radiation is emitted mainly in the UV-range and appears as blue light along the particle trajectory. In 'swimming pool' nuclear reactors the blue light surrounding the fuel elements is generated by electrons from β-decays, some of which are highly relativistic, fulfilling condition (9.65) for water ($n = 1.33$).

Using suitable radiators, particles with velocities above a given threshold can be identified with threshold Čerenkov counters. For example, the slower kaons can be separated from the faster pions in a beam of fixed momentum. The threshold for Čerenkov radiation is $\beta = 0.75$ for water and $\beta = 0.67$ for plexiglas ($n = 1.49$). To distinguish highly relativistic particles ($\beta \simeq 1$), media with refractive indices $n \simeq 1$ must be chosen, such as *aerogel* ($n = 1.008$–1.015) or gases. For *gases* n depends on pressure:

$$n - 1 \sim 10^{-5} - 10^{-3}. \tag{9.66}$$

Let us calculate with the help of figure 9.34 the angle θ under which the Čerenkov light is emitted. A shock wave is produced when

$$\beta ct \cos\theta = ut \Rightarrow \cos\theta = \frac{1}{\beta}\frac{u}{c}, \tag{9.67}$$

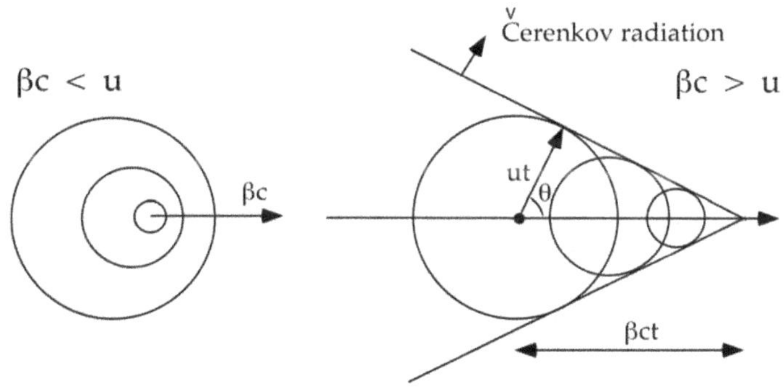

Figure 9.34. Čerenkov radiation: constructive interference of the spherical waves occurs when the speed of the charged particle exceeds that of the light in the medium.

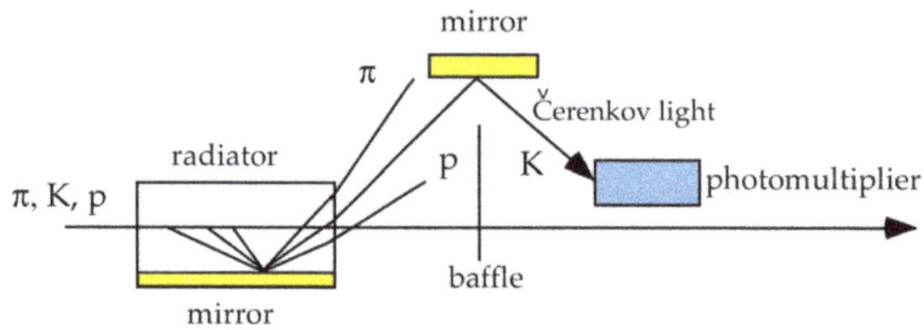

Figure 9.35. Beam particles can be identified with differential Čerenkov counters. The drawing shows how kaons can be selected in a beam also containing pions and protons.

hence

$$\boxed{\cos\theta = \frac{1}{\beta n}}. \tag{9.68}$$

The velocity of the particle can be determined from a measurement of the angle θ and, with known momentum, also the particle type. Figure 9.35 shows the principle of a differential isochronous self-collimating Čerenkov counter (DISC) which selects pions, kaons or protons in a beam of given momentum. The selection occurs by carefully choosing the mirror and baffle geometry.

Figure 9.36 shows the working principle of the Ring Imaging Čerenkov Counter (RICH) to measure the particle velocity. Charged particles emit UV photons in the radiator, which fly along the Čerenkov cone. The photons are converted in the gas chamber (two-dimensional readout) through the photoelectric effect. Ionization circles are observed in the chamber with which the velocity of the particle can be calculated from (9.68). The particle is identified by combining β with a curvature measurement in the magnetic field. The radiator and window materials have to be transparent and chosen to minimize photon absorption. Quartz is transparent in the UV range and is often used. The refractive index of the radiator should not exceed

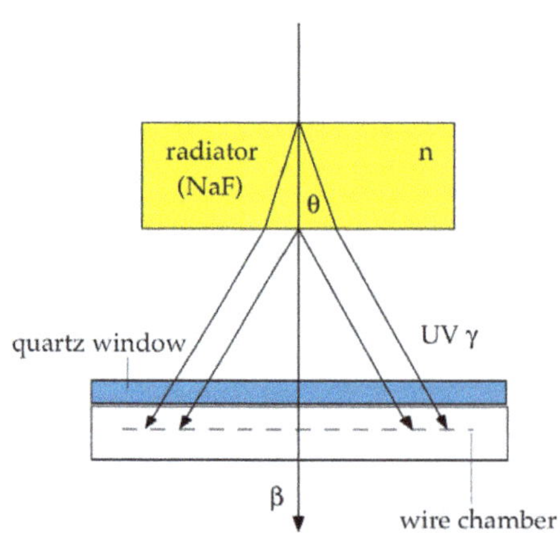

Figure 9.36. The principle of the RICH.

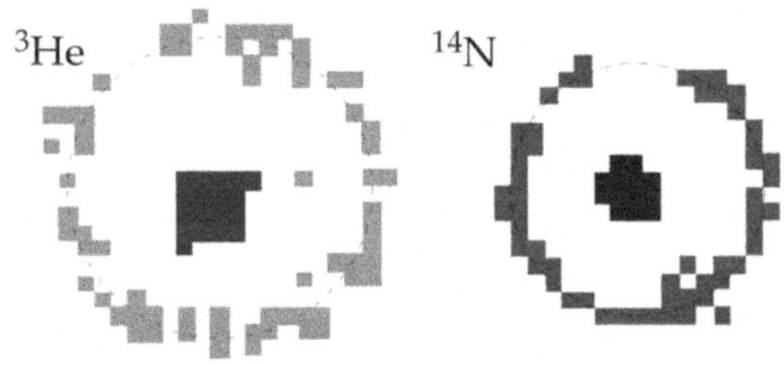

Figure 9.37. Čerenkov rings produced by a ^{3}He ion (left) and a ^{14}N ion (right). The ions fly perpendicularly into the detector surface. The gas is also ionized directly by the traversing ions (signal at the centre).

$n = \sqrt{2}$ to avoid internal reflection for $\beta \simeq 1$ (combine (9.68) with Snell's law). A suitable choice is NaF with $n = 1.336$.

Figure 9.37 displays an example of two electronically reconstructed fully ionized atoms traversing a RICH counter. The detection of a *single* ^{3}He ion with $\beta = 0.92$ is shown on the left and a single ^{14}N ion with $\beta = 0.86$ on the right. The ring radius is larger for the faster ion, as expected from (9.68). Note that more photons are emitted by the ^{14}N ion since the yield of Čerenkov radiation increases quadratically with ion charge. For further examples of Čerenkov rings, see figure 14.18 in section 14.6.

9.11 Transition radiation detectors

Charged particles moving uniformly in vacuum do not radiate. However, charged particles approaching (or leaving) a medium induce free charges (or polarization charges) of the opposite sign on its surface. As the distance between positive and negative charges varies, radiation is generated by the time-dependent electric dipole moment. The *transition* radiation (Ginsburg and Frank, 1945) is emitted mostly in the direction of the projectile, with wavelengths in the far UV to x-ray range, depending on the relativistic γ-factor of the projectile. The average x-ray energy increases with γ and is typically 5 keV for $\gamma = 1000$ [26]. X-rays around 5 keV and below are difficult to detect due to photoelectric absorption in the radiator itself. Therefore materials with low Z, such as lithium ($Z = 3$), mylar or polyethylene are used as the radiators. The radiation is usually detected using drift chambers located behind the foils and running with heavy gases (high Z), such as xenon or krypton, to enhance photoelectric absorption.

The x-ray emission probability is around 1% per interface. Hence many thin foils are stacked along the particle trajectory. The foil thickness is chosen to take advantage of constructive interference (see below). Ultimately, the x-ray intensity detected in the chamber behind the stack is limited by absorption as its thickness increases. Large volume transition radiation detectors (TRD), as operated by ATLAS and ALICE at the CERN LHC, are made of fibres such as polyethylene foam [27, 28].

A γ-factor of 1000 corresponds to 500 MeV electrons or 140 GeV pions. TRDs are therefore mostly used to identify electrons. The radiation is emitted preferably along the surface of a cone around the projectile trajectory, with opening angle $\theta \simeq 1/\gamma$. The overlapping signals from x-rays and from direct ionization therefore

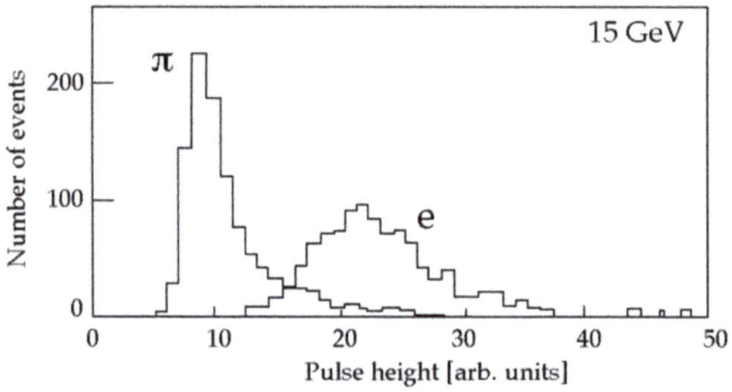

Figure 9.38. Data obtained with a transition radiation detector made of mylar foils. The data were taken alternately with pions, then with electrons. The lower peak is due to ionization from pions traversing the Xe or Kr gas. After [29].

need to be disentangled. This can be performed by analysing the spatial distribution of the charge released: the energy deposited by x-rays in the gas is well localized in space, while ionization by the projectile itself is distributed along its trajectory.

Figure 9.38 shows the performance at 15 GeV of a TRD made of mylar foils, 25 μm thick spaced by 1.5 mm of air. The average x-ray energy deposit is obtained by subtracting the mean dE/dx of the particle traversing the detector gas (Xe or Kr). A fairly good separation between pions and electrons can be achieved.

Let us estimate the optimum foil thickness and consider two x-rays, X_1 and X_2, emitted with the same energy ω under the small angle θ from the front and back surfaces of a radiator of thickness L (figure 9.39). For x-rays emitted at the same time X_1 needs to cover the additional path length $kL\cos\theta$ to be in phase with X_2. However, X_2 is delayed by the transit time $L/\beta c$ of the projectile through the foil.

The phase difference between X_1 and X_2 is

$$\Delta = \omega\frac{L}{\beta c} - \overbrace{\frac{\omega}{u}}^{k} L\cos\theta = \omega\frac{L}{\beta c} - \frac{\omega n(\omega)}{c} L\cos\theta, \tag{9.69}$$

where u is the velocity of light in the medium. To achieve maximum constructive interference between X_1 and X_2 the phase Δ should be equal to 2π. Above absorption

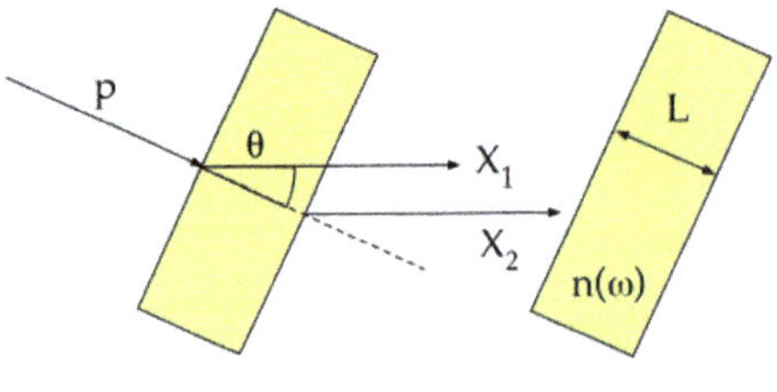

Figure 9.39. Constructive interference in a transition radiation detector (see the text).

frequency the refractive index is smaller than one. By expressing the refractive index as a function of the dielectric constant $\epsilon = 1 - \frac{\omega_p^2}{\omega^2}$, where ω_p is the plasma frequency [28]

$$\omega_p\,[\mathrm{eV}] = 28.8\sqrt{\rho\left[\mathrm{g\ cm^{-3}}\right]\frac{Z}{A}}, \tag{9.70}$$

one obtains the refractive index

$$n(\omega) = \sqrt{1 - \frac{\omega_p^2}{\omega^2}} \simeq \left(1 - \frac{1}{2}\xi^2\right), \quad \text{where } \xi \equiv \frac{\omega_p}{\omega}. \tag{9.71}$$

With the approximations

$$\beta = \sqrt{1 - \frac{1}{\gamma^2}} \simeq 1 - \frac{1}{2\gamma^2} \qquad \text{and} \qquad \cos\theta \simeq 1 - \frac{\theta^2}{2} \tag{9.72}$$

the phase difference (9.69) becomes

$$\Delta \simeq \frac{\omega L}{c}\left(1 + \frac{1}{2\gamma^2}\right) - \frac{\omega L}{c}\left(1 - \frac{\xi^2}{2}\right)\left(1 - \frac{\theta^2}{2}\right) \simeq \frac{\omega L}{c}\left(\frac{1}{2\gamma^2} + \frac{\theta^2}{2} + \frac{\xi^2}{2}\right). \tag{9.73}$$

An appreciable intensity is obtained when $\Delta = 2\pi$. The corresponding *formation length*, defined as $\zeta \equiv \frac{L}{2\pi}$, is

$$\zeta = \frac{2c}{\omega}\left[\frac{1}{\frac{1}{\gamma^2} + \theta^2 + \xi^2}\right]. \tag{9.74}$$

$L = 2\pi\zeta$ is the thickness needed to achieve maximum constructive interference. The formation length as a function of x-ray energy is shown in figure 9.40, calculated

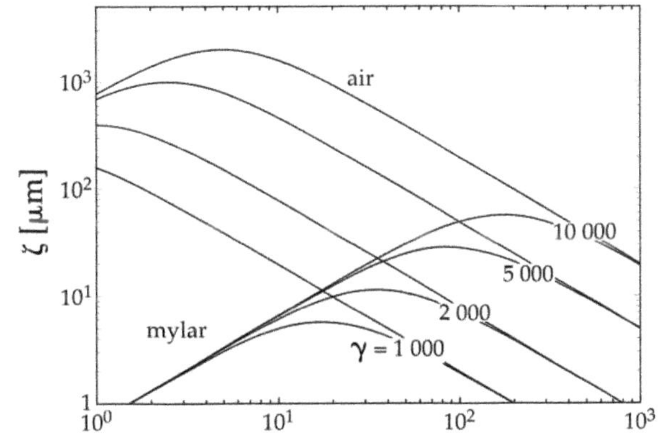

Figure 9.40. Formation length ζ in air ($\omega_p = 0.7$ eV) and in mylar ($\omega_p = 24.4$ eV).

from (9.74) with $\theta = \frac{1}{\gamma}$, for mylar foils and for air gaps. Hard x-rays are suppressed and the formation length of air is one to two orders of magnitude larger than that of mylar. Typical TRDs are made of $\simeq 50$ μm thick mylar foils, separated by 500 μm air gaps.

9.12 Electromagnetic calorimeters

Electromagnetic calorimeters are used to measure the energy of energetic photons, electrons and positrons. As already discussed, these particles generate electromagnetic cascades which are absorbed in the calorimeter. Figure 9.41 shows the simulation of a shower generated by a high-energy electron.

In a *sandwich* calorimeter, conversion plates alternate with scintillation counters (or wire chambers). The electrons and positrons emit bemsstrahlung photons in the conversion plates which turn into e^+e^- pairs. A high-Z material is used such as lead (radiation length $X_0 \propto 1/Z^2$, critical energy $E_c \propto 1/Z$). The calorimeter thickness is chosen to be large enough to contain and absorb the whole shower, so that the energy deposited in the scintillator is proportional to the incident energy, as the simple model below shows.

Energy resolutions of, typically,

$$\frac{\sigma}{E} \simeq \frac{0.1}{\sqrt{E\,[\mathrm{GeV}]}} \tag{9.75}$$

are achieved with lead–scintillator sandwiches. The alternating layers need to be as thin as possible to minimize fluctuations in the energy deposit. Hence *homogeneous* shower counters such as monocrystals (e.g. NaI) offer much higher resolutions:

$$\frac{\sigma}{E} \simeq \frac{0.025}{\sqrt[4]{E\,[\mathrm{GeV}]}}. \tag{9.76}$$

However, large monocrystals are expensive compared to sandwich counters. A lead loaded glass counter records the Čerenkov light emitted by the shower leptons, instead of the ionization energy itself. The typical resolution of *lead glass* counters is

$$\frac{\sigma}{E} \simeq \frac{0.05}{\sqrt{E\,[\mathrm{GeV}]}}. \tag{9.77}$$

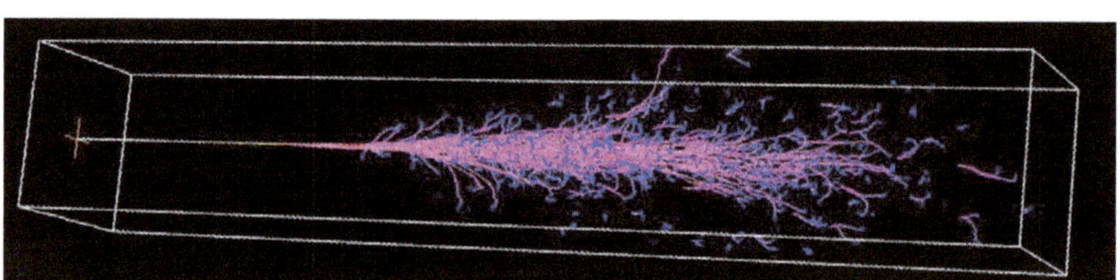

Figure 9.41. The electromagnetic shower induced by a 25 GeV electron in a 30 cm long iron block. The tracks are drawn in blue, red and white for low, medium and high energy electrons or positrons, respectively. Simulation by Sven Menke, www.mppmu.mpg.de/menke.

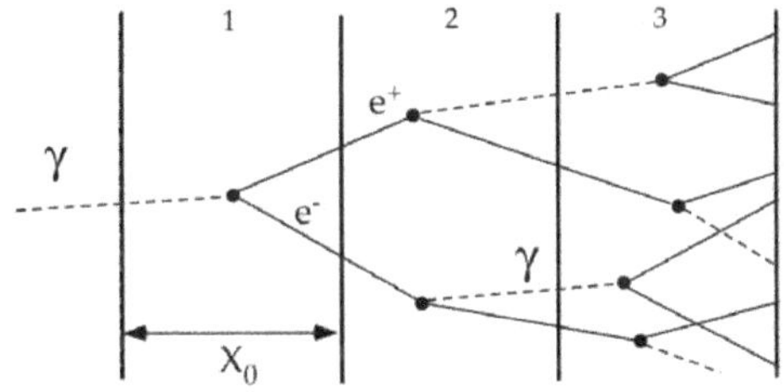

Figure 9.42. In a simplified shower model the number of particles doubles with each additional radiation length.

The main properties of electromagnetic showers can be described using a simplified model. Let us divide the calorimeter into layers of thickness X_0 (figure 9.42). The number of particles (leptons or photons) in the shower doubles after each radiation length. The number of particles $N(t)$ and the mean energy E per particle after t layers are

$$N(t) = 2^t \Rightarrow E \sim \frac{E_0}{2^t}, \tag{9.78}$$

where E_0 is the energy of the incoming projectile. The avalanche reaches a maximum and stops at the thickness t_m, for which E becomes equal to the critical energy E_c:

$$\frac{E_0}{2^{t_m}} = E_c \Rightarrow t_m = \frac{\ln\left(\frac{E_0}{E_c}\right)}{\ln 2}. \tag{9.79}$$

The material thickness $t_m X_0$ at which the shower reaches its maximum development grows logarithmically with incident energy. The total length of the electron and positron tracks is

$$\begin{aligned} L &= \frac{2}{3}\int_0^{t_m} N(t)\mathrm{d}t = \frac{2}{3}\int_0^{t_m} 2^t \mathrm{d}t \\ &= \frac{2}{3}\int_0^{t_m} \mathrm{e}^{t\ln 2}\mathrm{d}t \simeq \frac{2}{3\ln 2}2^{t_m} = \frac{2}{3\ln 2}\left(\frac{E_0}{E_c}\right). \end{aligned} \tag{9.80}$$

Since the signal height is proportional to L (total ionization) the energy deposited in the calorimeter is equal to the incident energy E_0 (after appropriate calibration).

However, the shower does not stop sharply after t_m radiation lengths but decays exponentially (figure 9.43). As the shower must be contained entirely by the calorimeter for optimum energy resolution, the typical thickness is around $t_m = 20$ (see figure 9.43), although $t_m = 5$ radiation lengths already convert 98% of the photons ((9.60) with $s/X_0 = 5$). Electromagnetic calorimeters made of 2 mm thick lead sheets and 5 mm thick scintillators are typically 40 cm long.

The shower develops within a cone with a slowly growing opening angle. The lateral shower distribution can be parametrized by two Gaussians which depend

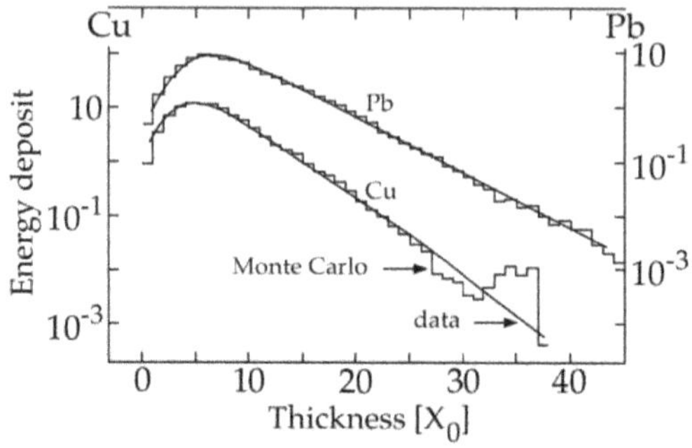

Figure 9.43. Longitudinal particle density for 6 GeV electrons in lead and in copper–scintillator sandwich calorimeters. Plotted is the fractional energy deposit as a function of thickness in radiation lengths. Note that t_m is larger for lead due to its lower critical energy. After [30].

on the energy E_0 and depth t_m. A fraction of 90% of the energy is deposited within a cylinder of radius 3.5 R_M, where R_M is the (energy-independent) Molière radius

$$R_M = \frac{21X_0}{E_c\ [\text{MeV}]}. \tag{9.81}$$

9.13 Hadronic calorimeters

Hadronic calorimeters work on the same principle as electromagnetic calorimeters. Here, the incoming projectile generates secondary particles via inelastic nuclear scattering. The relevant interaction length (replacing X_0) is the mean free path Λ for inelastic scattering,

$$\Lambda = \frac{1}{\sigma n} = \frac{A}{\sigma \rho L}. \tag{9.82}$$

The cross section can be estimated from the disk surface $\sigma \simeq \pi a^2 A^{2/3}$ covered by the nucleus. With $a = 1.2$ fm one finds

$$\rho\Lambda = 35\ \text{g}\,\text{cm}^{-2} A^{\frac{1}{3}}. \tag{9.83}$$

For example, $\Lambda = 18$ cm for lead is much larger than $X_0 = 0.6$ cm, while for iron $\Lambda = 17$ cm. Hadronic calorimeters are correspondingly large. The shower deposits 99% of the initial energy within 10Λ. Such an iron calorimeter has then to be at least 1.80 m thick. Figure 9.44 shows the longitudinal shower development in an iron–scintillator sandwich calorimeter. Our shower model above is too simplistic in this case because high energy photons are also produced by π^0 decays. An electromagnetic shower is then superimposed on the hadronic one, leading to the sharp maximum in figure 9.44.

The typical energy resolution is

$$\frac{\sigma}{E} \simeq \frac{0.5}{\sqrt{E\ [\text{GeV}]}}. \tag{9.84}$$

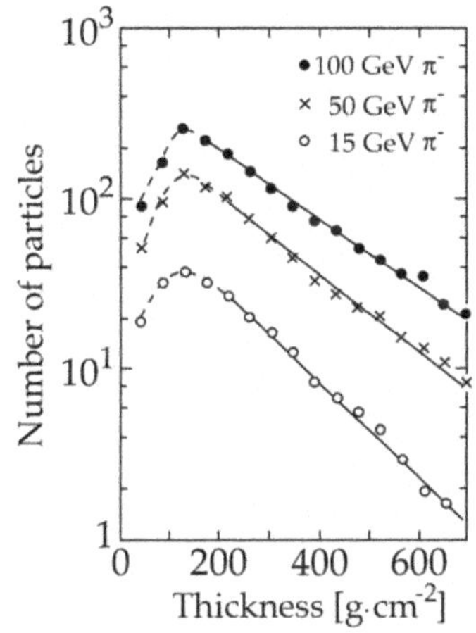

Figure 9.44. Longitudinal development of a hadronic shower in an iron–scintillator sandwich calorimeter. After [31].

This modest resolution makes hadronic calorimeters unsuitable at low energies. The particle momentum is instead measured through its curvature in a magnetic field. However, as the energy increases, the sagitta of the trajectory becomes very small, even with the maximum achievable magnetic fields. Thus, hadronic calorimeters become more advantageous at energies above $\simeq$100 GeV.

References

[1] Wille K 2001 *The Physics of Particle Accelerators: an Introduction* (Oxford: Oxford University Press)

[2] Wiedemann H 2007 *Particle Accelerator Physics* (Berlin: Springer)

[3] Olive K *et al* (Particle Data Group) 2014 *Chin. Phys.* C **38** 090001

[4] Leo W R 2005 *Techniques for Nuclear and Particle Physics Experiments* (Berlin: Springer)

[5] Grupen C and Shwartz B 2008 *Particle Detectors* (Cambridge: Cambridge University Press)

[6] Leroy C and Rancoita P-G 2011 *Principles of Radiation Interaction in Matter and Detection* (Singapore: World Scientific)

[7] Jackson J D 1975 *Classical Electrodynamics* (New York: Wiley)

[8] Amsler C and Scampoli P (ed) 2014 Proceedings of the WAG 2013 Workshop *Int. J. Mod. Phys.: Conf. Series* **30**

[9] Amoretti M *et al* 2002 *Nature* **419** 456

[10] Amsler C 1998 *Rev. Mod. Phys.* **70** 1293

[11] Angelopoulos A *et al* 2003 *Phys. Rep.* **374** 165

[12] Wohl C G *et al* (Particle Data Group) 1984 *Rev. Mod. Phys.* **56** S1

[13] Sternheimer R M 1959 *Phys. Rev.* **115** 137

[14] Sternheimer R M 1960 *Phys. Rev.* **117** 485

[15] Pagano A *et al* 2001 *Nucl. Phys.* A **681** 331c

[16] Hanson A O *et al* 1951 *Phys. Rev.* **84** 634

[17] Seltzer S M and Berger M J 1984 *Int. J. Appl. Radiat. Isot.* **35** 665

[18] Marmier P 1960 *Kernphysik* (Zurich: Verlag des Vereins der Mathematiker und Physiker an der ETH Zürich)
[19] Bethe H and Heitler W 1934 *Proc. R. Soc., London* A **146** 83
[20] Kadyk J A 1991 *Nucl. Instrum. Methods Phys. Res.* A **300** 436
[21] Daraktchieva Z *et al* 2005 *Phys. Lett.* B **615** 153
[22] Storey J *et al* 2013 *Nucl. Instrum. Methods Phys. Res.* A **732** 437
[23] See, for example, Lutz G 2007 *Semiconductor Radiation Detectors* (Berlin: Springer)
[24] Allkofer Y *et al* 2008 *Nucl. Instrum. Methods Phys. Res.* A **584** 25 and references therein
[25] Castoldi A *et al* 1997 *Nucl. Instrum. Methods Phys. Res.* A **399** 227
[26] Cobb J *et al* 1977 *Nucl. Instrum. Methods* **140** 413
[27] Andronic A and Wessels J P 2012 *Nucl. Instrum. Methods Phys. Res.* A **666** 130
[28] Dolgoshein B 1993 *Nucl. Instrum. Methods Phys. Res.* A **326** 434
[29] Cherry M, Müller D and Prince T A 1974 *Nucl. Instrum. Methods* **115** 141
[30] Bathow G *et al* 1970 *Nucl. Phys.* B **20** 592
[31] Holder M *et al* 1978 *Nucl. Instrum. Methods* **151** 69

IOP Publishing

Claude Amsler

Chapter 10

The quark model

10.1 Strangeness

In this chapter we deal with the quark composition of the 200 or more hadrons which have been discovered since 1947. Before the discovery of the π (figure 7.10) only the $e^{\pm}$, the nucleon (p and n), the γ and the $\mu^{\pm}$ were known as elementary particles. In 1947 Rochester and Butler discovered in the cosmic radiation the decay of an electrically neutral particle, the $K^0 \rightarrow \pi^+\pi^-$ (figure 10.1).

The K^0 was subsequently produced at accelerators, for instance with pion beams:

$$\pi^-p \rightarrow K^0\Lambda, \qquad K^0 \rightarrow \pi^+\pi^-, \qquad \Lambda \rightarrow \pi^-p. \tag{10.1}$$

In this reaction the K^0 is produced along with the Λ hyperon (to be precise, the short-lived component $K_S \rightarrow 2\pi$ is observed, see chapter 19). Figure 10.2(a) shows an event of this type. The magnitude of the cross section for producing a $K^0\Lambda$ pair is typical for the *strong* interaction:

$$\sigma(\pi^-p \rightarrow K^0\Lambda) \sim 1 \text{ mb (at 1 GeV } c^{-1}). \tag{10.2}$$

The track lengths of the outgoing particles up to their decay points are in the range of a few centimetres. Accordingly, hyperons and neutral kaons are relatively stable with mean lives

$$\begin{aligned} \tau_{K^0} &= 89.53 \pm 0.05 \text{ ps}, \\ \tau_\Lambda &= 263 \pm 2 \text{ ps}. \end{aligned} \tag{10.3}$$

Therefore they decay through the *weak* interaction. The large difference in interaction strengths between the production and decay of theses particles led physicists of the 1950s to baptize hyperons and kaons '*strange*' particles.

The phenomenon was 'explained' by introducing a new (additive) quantum number, *strangeness S*, that is conserved in production but violated in decay. The violation of strangeness conservation suppresses the decay, which then proceeds

doi:10.1088/978-0-7503-1140-3ch10 © IOP Publishing Ltd 2015

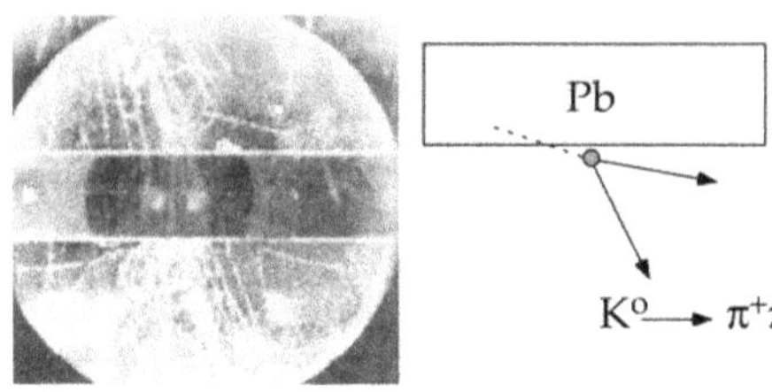

Figure 10.1. The discovery of the $K^0 \to \pi^+\pi^-$ in a cloud chamber (the magnetic field is perpendicular to the picture). The K^0 was produced by cosmic rays impinging on the lead plate. Reproduced with permission from [1]. Copyright 1947 Macmillan Publishers Ltd.

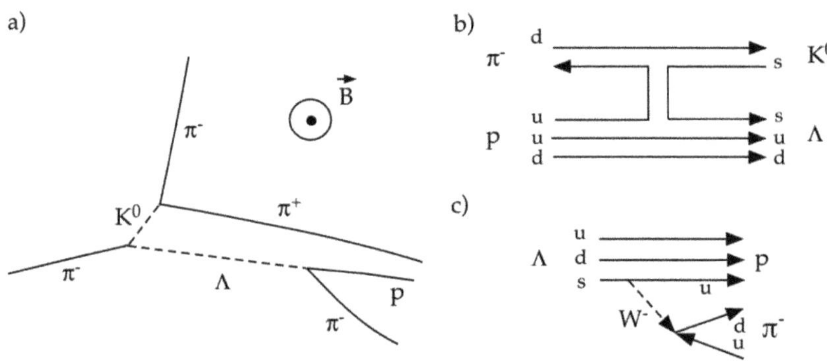

Figure 10.2. (*a*) The reaction $\pi^- p \to K^0\Lambda$ and decays of the invisible neutral K^0 and Λ, as observed in a bubble chamber. The curvatures of the tracks are used to reconstruct the kinematics of the interaction and to measure the momenta of the K^0 and Λ. (*b*) and (*c*) Feynman graphs for $K^0\Lambda$ production and Λ decay, respectively.

through the weak interaction (hence the long lifetime). By convention, the K^0 has strangeness $S = +1$ and the Λ strangeness $S = -1$. The well-known hadrons, such as the π or the p, have strangeness $S = 0$. The total strangeness ($S = 0$) in reaction (10.1) is conserved between the initial and final states. However, in Λ and K^0 decay strangeness is not conserved and changes by one unit ($|\Delta S| = 1$).

We now know that hadrons are made of quarks. We define the strangenesses of the s and the $\bar{s}$ quarks as

$$S(s) \equiv -1, \qquad S(\bar{s}) \equiv +1 \tag{10.4}$$

and attribute to the other quarks the strangeness $S = 0$. Assuming that S is an additive quantum number one can determine the quark content of the Λ and K^0 from their strangenesses:

$$S(\Lambda) = -1 \Rightarrow \text{quark content of the } \Lambda\text{: } uds, \tag{10.5}$$

$$S\left(K^0\right) = 1 \Rightarrow \text{quark content of the } K^0\text{: } d\bar{s}. \tag{10.6}$$

The conservation of strangeness in production is nothing more than the conservation of quark flavour in strong interactions (figure 10.2(*b*)). On the other hand, the violation of strangeness conservation in decays is due to the flavour change of charged currents during the weak interaction, associated with W emission (figure 10.2(c)). We can therefore formulate the $\Delta S = 1$ rule for the interaction between quarks (and between hadrons)[1]:

$$\begin{aligned} \text{strong and electromagnetic interactions} &: \Delta S = 0, \\ \text{weak interaction} &: |\Delta S| \leqslant 1. \end{aligned} \tag{10.7}$$

For example, the neutron β-decay $n \to pe^- \bar{\nu}_e$ is a weak process with $\Delta S = 0$.

10.2 The hadron spectrum

The constituents of hadrons are made of *partons* (quarks, antiquarks and gluons). The *valence* quarks carry the electric charge and the spin of the hadron and determine—at least in first approximation—its structure ($q\bar{q}$, qqq, etc…). Numerous predictions can be made from the quark model, such as the decay rates or the magnetic moments of hadrons.

In addition, *sea* quarks (virtual $q\bar{q}$ pairs) are produced by gluons. Gluons and sea quarks contribute to the hadron mass. In the electromagnetic sector photons also contribute to the mass (for example as binding energy in the hydrogen atom). However, their relative contribution to the mass is much smaller than that of the gluons in hadrons.

We shall now construct the mass spectrum of hadrons by starting with the three lightest quarks u, d and s.

10.2.1 Mesons

The $q\bar{q}$ system in its ground state is either a spin-singlet ($s = 0$) or a spin-triplet ($s = 1$) state. Nine mesons can be constructed with three flavours and three antiflavours. Further *nonets* can be built by switching the angular momentum ℓ between the quark and antiquark (figure 10.3). These are *orbital* excitations of the ground state mesons. *Radial* excitations (vibrations) are described by the radial quantum number n. For $n > 1$ further nonets can be constructed.

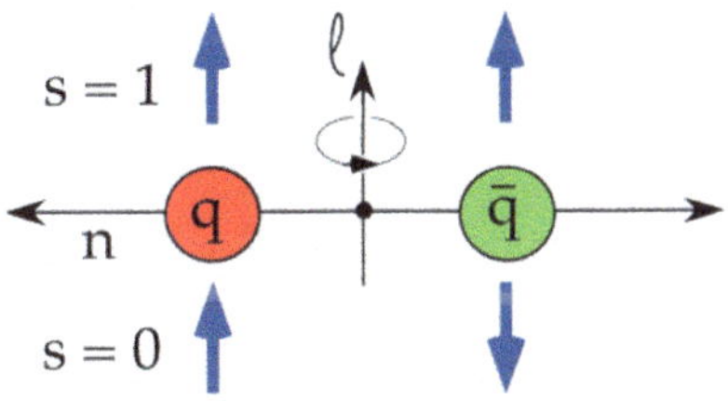

Figure 10.3. A spin-triplet (with parallel spins) and a spin-singlet meson (with antiparallel spins) can be built for each $q\bar{q}$ pair with orbital excitation ℓ (rotation) and radial excitation n (vibration).

[1] With three quark flavours the (negative) strangeness S *increases* by one unit in hyperon decay ($S = -1 \to S = 0$). For hyperons containing heavier quarks S can also *decrease* by one unit. An example is the disintegration of the heavy baryon Ξ_c^+ (section 10.3): $\Xi_c^+(csu)$ $(S = -1) \to \Xi^0(ssu)\pi^+$ $(S = -2)$.

Thus, in analogy to the energy levels of the hydrogen atom, a series of bound $q\bar{q}$ states emerge. The meson spin J is composed of the $q\bar{q}$ spin s and the angular momentum ℓ ($\vec{J} = \vec{s} + \vec{\ell}$). One obtains for the quantum number J

$$|\ell - s| \leqslant J \leqslant \ell + s. \tag{10.8}$$

We denote a nonet by the spectroscopic notation[2]

$$n^{2s+1}\ell_J \quad (\ell = 0 : S, \ell = 1 : P, \ell = 2 : D, \ldots), \tag{10.9}$$

and consider only the ground states with $n = 1$ and $\ell = 0$. For $s = 1$ (hence $J = 1$) we get the 1^3S_1 mesons with SU(3) *flavour* wave functions:

$$\begin{aligned}
\rho^+(770) &= |u\bar{d}\rangle \\
\rho^-(770) &= |\bar{u}d\rangle \\
\rho^0(770) &= \tfrac{1}{\sqrt{2}}(|d\bar{d}\rangle - |u\bar{u}\rangle) \equiv \tfrac{1}{\sqrt{2}}|d\bar{d} - u\bar{u}\rangle \\
K^{*+}(892) &= |u\bar{s}\rangle, \quad \text{strangeness } S = +1 \\
K^{*-}(892) &= |\bar{u}s\rangle, \quad S = -1 \\
K^{*0}(896) &= |d\bar{s}\rangle, \quad S = +1 \\
\overline{K}^{*0}(896) &= |\bar{d}s\rangle, \quad S = -1.
\end{aligned} \tag{10.10}$$

The numbers in parentheses are the masses in MeV. These seven wave functions are normalized and orthogonal: the norms of identical pairs are

$$\langle u\bar{u}|u\bar{u}\rangle = \langle d\bar{d}|d\bar{d}\rangle = \langle s\bar{s}|s\bar{s}\rangle = \langle u\bar{d}|u\bar{d}\rangle = \ldots = 1, \tag{10.11}$$

while the scalar products between different pairs vanish:

$$\langle u\bar{u}|d\bar{d}\rangle = \langle u\bar{d}|\bar{u}d\rangle = \ldots = 0. \tag{10.12}$$

The wave functions need to be symmetrized, see appendix A and [2]. The wave functions of the two additional mesons are linear combinations of the octet and singlet wave functions

$$\begin{aligned}
|8\rangle &= \frac{1}{\sqrt{6}}|u\bar{u} + d\bar{d} - 2s\bar{s}\rangle, \\
|1\rangle &= \frac{1}{\sqrt{3}}|u\bar{u} + d\bar{d} + s\bar{s}\rangle,
\end{aligned} \tag{10.13}$$

[2] In the literature the quantum number $\nu = n + \ell$ is also used instead of n. For example, for the hydrogen atom the lowest states for a given ℓ are denoted as 1s, 2p, 3d, ... instead of the 1S, 1P, 1D, etc, used in the quark model.

and read

$$\begin{aligned}\psi_1 &= |8\rangle\cos\theta - |1\rangle\sin\theta,\\ \psi_2 &= |8\rangle\sin\theta + |1\rangle\cos\theta.\end{aligned} \tag{10.14}$$

We have denoted with the generic names ψ_1 and ψ_2 the two neutral mesons which arise in any $q\bar{q}$ nonet from the mixing of the two wave functions (10.13) with identical quantum numbers (strangeness $S=0$ and isospin $i=0$, see section 10.2.3). The angle θ—the nonet *mixing angle*—is determined experimentally as discussed below. For the 1^3S_1 nonet, as well as for most other well-known nonets with the notorious exception of the 1^1S_0, the mixing angle is very close to the *ideal mixing angle* $\theta = 35.3°$ ($\tan\theta = 1/\sqrt{2}$), for which $s\bar{s}$ decouples from $u\bar{u}$ and $d\bar{d}$ in (10.14). Indeed, with $\cos\theta = \sqrt{2/3}$ and $\sin\theta = \sqrt{1/3}$ one obtains

$$\begin{aligned}\psi_1 &= \frac{1}{3}|u\bar{u} + d\bar{d} - 2s\bar{s}\rangle - \frac{1}{3}|u\bar{u} + d\bar{d} + s\bar{s}\rangle = -|s\bar{s}\rangle,\\ \psi_2 &= \frac{1}{3\sqrt{2}}|u\bar{u} + d\bar{d} - 2s\bar{s}\rangle + \frac{\sqrt{2}}{3}|u\bar{u} + d\bar{d} + s\bar{s}\rangle = \frac{1}{\sqrt{2}}|d\bar{d} + u\bar{u}\rangle.\end{aligned} \tag{10.15}$$

The ψ_1 and ψ_2 states in the 1^3S_1 nonet are the ϕ and ω mesons, respectively, which are nearly ideally mixed:

$$\begin{aligned}\phi(1020) &\simeq -|s\bar{s}\rangle,\\ \omega(782) &\simeq \frac{1}{\sqrt{2}}|d\bar{d} + u\bar{u}\rangle.\end{aligned} \tag{10.16}$$

We now show that the nonet mixing angle can be calculated from the meson masses. In the flavour basis (10.13)

$$|8\rangle = \begin{pmatrix}1\\0\end{pmatrix}, \qquad |1\rangle = \begin{pmatrix}0\\1\end{pmatrix} \tag{10.17}$$

the mass matrix is

$$M = \langle 8,1|\, H\, |8,1\rangle = \begin{pmatrix} m_8 & m_{81}\\ m_{18} & m_1\end{pmatrix}. \tag{10.18}$$

The physical states ψ_1 and ψ_2 and the mass eigenvalues λ_1 and λ_2 are found by diagonalizing M:

$$M\psi = \lambda\psi \Rightarrow \begin{vmatrix} m_8 - \lambda & m_{81}\\ m_{18} & m_1 - \lambda\end{vmatrix} = 0. \tag{10.19}$$

The solution of this secular equation and the relationships between the different meson masses are discussed in detail in appendix B. The eigenfunction ψ_1 with eigenvalue λ_1 and components ($\cos\theta$, $-\sin\theta$) is given by

$$\begin{pmatrix} m_8 & m_{81}\\ m_{18} & m_1\end{pmatrix}\begin{pmatrix}\cos\theta\\ -\sin\theta\end{pmatrix} = \lambda_1\begin{pmatrix}\cos\theta\\ -\sin\theta\end{pmatrix}. \tag{10.20}$$

Multiplying leads for the first row to

$$\tan\theta = \frac{m_8 - \lambda_1}{m_{81}}. \tag{10.21}$$

We now calculate m_8 and m_{18} from the constituent quark masses:

$$m_8 = \frac{1}{6}(1,\, 1,\, -2)\begin{pmatrix} 2m_u & 0 & 0 \\ 0 & 2m_d & 0 \\ 0 & 0 & 2m_s \end{pmatrix}\begin{pmatrix} 1 \\ 1 \\ -2 \end{pmatrix} = \frac{1}{3}(2m_u + 4m_s), \tag{10.22}$$

and

$$m_{81} = \frac{1}{\sqrt{18}}(1,\, 1,\, -2)\begin{pmatrix} 2m_u & 0 & 0 \\ 0 & 2m_d & 0 \\ 0 & 0 & 2m_s \end{pmatrix}\begin{pmatrix} 1 \\ 1 \\ 1 \end{pmatrix} = m_{18} = \frac{2\sqrt{2}}{3}(m_u - m_s), \tag{10.23}$$

where we have assumed that $m_u = m_d$. Using the quark contents (10.10) and replacing the quark masses by the particle masses, one finds that

$$\begin{aligned} m_8 &= \frac{1}{3}\left(m_\rho + 4m_{K^*} - 2m_\rho\right) = \frac{1}{3}\left(4m_{K^*} - m_\rho\right), \\ m_{81} &= \frac{\sqrt{2}}{3}\left(m_\rho - 2m_{K^*} + m_\rho\right) = \frac{2\sqrt{2}}{3}(m_\rho - m_{K^*}). \end{aligned} \tag{10.24}$$

Substituting into (10.21) leads to the *mass formula*

$$\boxed{\tan\theta = \frac{4m_{K^*} - m_\rho - 3m_\phi}{2\sqrt{2}\,(m_\rho - m_{K^*})}} \tag{10.25}$$

with $\lambda_1 \equiv m_\phi$. One obtains $\theta \simeq 36°$ when introducing the numerical values for the meson masses, close to ideal mixing. Since for ideal mixing the ρ and ω mesons would have the same quark content (see (10.10) and (10.16)), one predicts that

$$m_\rho = m_\omega, \tag{10.26}$$

which is almost fulfilled experimentally. Furthermore, for $\tan\theta = \frac{1}{\sqrt{2}}$ (10.25) predicts for an ideally mixed nonet that

$$m_{K^*} = \frac{\overbrace{m_\rho}^{=m_\omega} + m_\phi}{2}, \tag{10.27}$$

which is also in agreement with the measured values.

The mass formula (10.25) can be applied to any nonet by substituting the corresponding mesons. The 1^1S_0 mesons are:

$$\begin{aligned}
\pi^+(140) &= |u\bar{d}\rangle \\
\pi^-(140) &= |\bar{u}d\rangle \\
\pi^0(135) &= \tfrac{1}{\sqrt{2}}|d\bar{d} - u\bar{u}\rangle \\
K^+(494) &= |u\bar{s}\rangle, \quad \text{Strangeness } S = +1 \\
K^-(494) &= |\bar{u}s\rangle, \quad S = -1 \\
K^0(498) &= |d\bar{s}\rangle, \quad S = +1 \\
\overline{K}^0(498) &= |\bar{d}s\rangle, \quad S = -1.
\end{aligned} \tag{10.28}$$

The symmetrized wave functions are given in appendix A. The masses in brackets are usually omitted for these familiar mesons. The ψ_1 and ψ_2 states are the η and η' mesons, respectively:

$$\begin{aligned}
\eta\,(547) &= |8\rangle\cos\theta - |1\rangle\sin\theta, \\
\eta'(958) &= |8\rangle\sin\theta + |1\rangle\cos\theta,
\end{aligned} \tag{10.29}$$

with mixing angle from (10.25)

$$\tan\theta = \frac{4m_K - m_\pi - 3m_\eta}{2\sqrt{2}\,(m_\pi - m_K)} \simeq -11^\circ, \tag{10.30}$$

which is far from ideal mixing. A crude approximation for the quark contents of the η and η' is then

$$\begin{aligned}
\eta(547) &\sim \frac{1}{\sqrt{6}}|u\bar{u} + d\bar{d} - 2s\bar{s}\rangle, \\
\eta'(958) &\sim \frac{1}{\sqrt{3}}|u\bar{u} + d\bar{d} + s\bar{s}\rangle,
\end{aligned} \tag{10.31}$$

which are the flavour octet and flavour singlet states, respectively. More details on the mixing angles can be found in appendix B and in [3], see also Amsler *et al* p 259 in [4].

The spin-triplet mesons (such as the ρ) are heavier than the spin-singlets ones (such as the π). This is due to hyperfine splitting (spin–spin interaction) which is substantially larger for gluon exchange than for photon exchange. Indeed, the 1^1S_0 level of the hydrogen atom lies 5.9×10^{-6} eV below the 1^3S_1 level, a tiny difference compared to the mass of the atom. Figure 10.4 shows the spectrum of light $q\bar{q}$ mesons as it is known today. Details can be found in [4].

10.2.2 Baryons

With three quarks one has to distinguish between spin-$\frac{1}{2}$ and spin-$\frac{3}{2}$ ground state baryons. Angular momentum ℓ can also be switched on. The spin-$\frac{1}{2}$ ground state

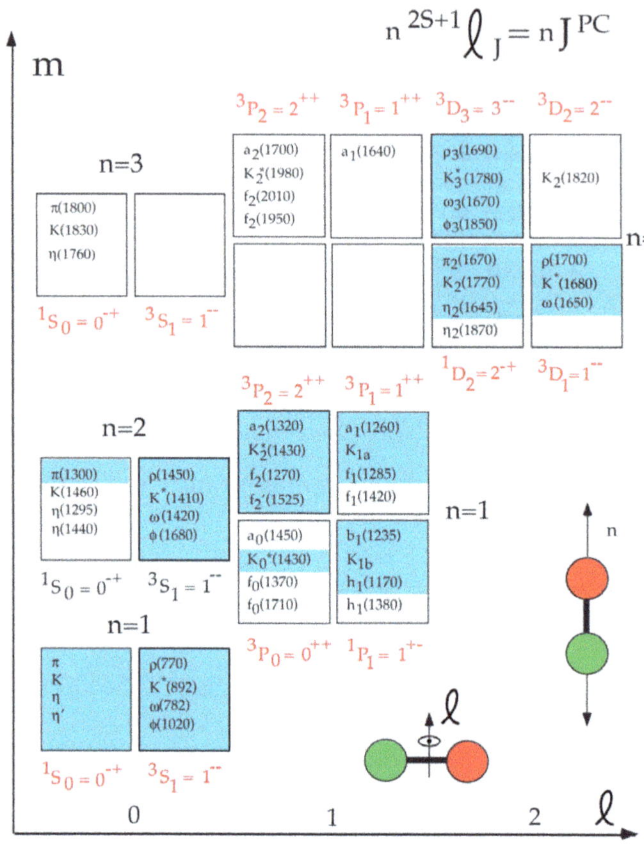

Figure 10.4. The spectrum of the light $q\bar{q}$ mesons (u, d, and s quarks) organized in $n^{2s+1}\ell_J$ nonets (approximate mass scale). The equivalent notation J^{PC} will be explained in sections 11.4 and 11.7. Mesons in the blue areas are well established, while the classification in the white areas is tentative. Candidates for many orbital and radial excitations have not been identified yet. As we shall discuss in section 10.2.3, charged hadrons build subgroups within the nonets (10.10) and (10.28) with identical or comparable masses. The naming convention of hadrons is explained in [4].

($\ell = 0$) baryons are the *eight* p, n, Λ, Σ^+, Σ^0, Σ^-, Ξ^0 and Ξ^- states building an *octet*:

$$
\begin{aligned}
p(938) &= |uud\rangle \\
n(940) &= |udd\rangle \\
\Lambda(1115) &= \tfrac{1}{\sqrt{2}}|(ud - du)s\rangle, \quad S = -1 \\
\Sigma^+(1189) &= |uus\rangle, \quad S = -1 \\
\Sigma^0(1192) &= \tfrac{1}{\sqrt{2}}|(ud + du)s\rangle, \quad S = -1 \\
\Sigma^-(1197) &= |dds\rangle, \quad S = -1 \\
\Xi^0(1315) &= |uss\rangle, \quad S = -2 \\
\Xi^-(1321) &= |dss\rangle, \quad S = -2.
\end{aligned}
\tag{10.32}
$$

For this familiar octet the masses are usually omitted from the baryon names. Note that the antiparticle of the $\Sigma^+ = |uus\rangle$, the negatively charged $\overline{\Sigma^+} = |\overline{uus}\rangle$, is not identical to the $\Sigma^- = |dds\rangle$. The ground state ($\ell = 0$) spin-$\frac{1}{2}$ baryons made of three identical quarks do not exist, nor does the ninth spin-$\frac{1}{2}$ ($\ell = 0$) baryon, due to symmetry properties of the wave functions [2].

The ten *uuu*, *ddd* and *sss* baryons have spin $\frac{3}{2}$ and build a *decuplet*:

$$\begin{aligned}
\Delta^{++}(1230) &= |uuu\rangle \\
\Delta^{+}(1231) &= |uud\rangle \\
\Delta^{0}(1232) &= |udd\rangle \\
\Delta^{-}(1233) &= |ddd\rangle \\
\Sigma^{+}(1383) &= |uus\rangle, \quad S = -1 \\
\Sigma^{0}(1384) &= |uds\rangle, \quad S = -1 \\
\Sigma^{-}(1387) &= |dds\rangle, \quad S = -1 \\
\Xi^{0}(1532) &= |uss\rangle, \quad S = -2 \\
\Xi^{-}(1535) &= |dss\rangle, \quad S = -2 \\
\Omega^{-}(1672) &= |sss\rangle, \quad S = -3.
\end{aligned} \tag{10.33}$$

The flavour wave functions (10.32) and (10.33) show only the quark content and still need to be symmetrized (permutation of three quarks), see appendix C and [2]. The spin-$\frac{1}{2}$ hyperons are the lightest baryons containing *s* quarks. They decay through the weak interaction (with the exception of $\Sigma^0 \rightarrow \Lambda\gamma$) following the $|\Delta S| = 1$ rule (10.7) and are accordingly long-lived. For example,

$$\Sigma^+ \rightarrow p\pi^0,\ n\pi^+, \quad \Xi^0 \rightarrow \Lambda\pi^0. \tag{10.34}$$

However, the spin-$\frac{3}{2}$ baryons (with the exception of the Ω^-) are massive enough to decay through the strong interaction ($\Delta S = 0$) and are correspondingly short-lived. For example,

$$\Delta^{++}(1230) \rightarrow \pi^+ p, \quad \Xi^0(1532) \rightarrow \Xi^0\pi^0,\ \Xi^-\pi^+, \quad \Sigma^+(1383) \rightarrow \Lambda\pi^+. \tag{10.35}$$

More ground state baryons cannot be built with three quarks, and indeed none are found. States such as a Ξ^+ or a Δ^{--} do not exist. The discovery of the (*sss*) Ω^- baryon in 1964 in Brookhaven was a triumph for the quark model introduced a year earlier. Figure 10.5 shows a bubble chamber picture of the reaction

$$K^- p \rightarrow \Omega^- K^+ K^0 \tag{10.36}$$

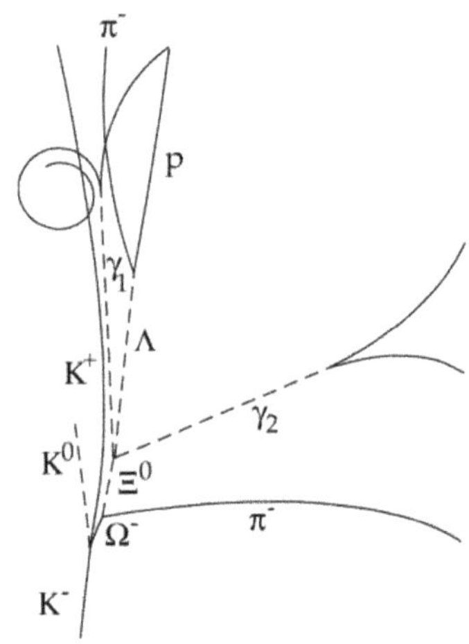

Figure 10.5. The observation of the Ω^- confirmed the quark model. From [5].

with

$$\begin{aligned} \Omega^- &\rightarrow \Xi^0\pi^-, \\ \Xi^0 &\rightarrow \Lambda\pi^0, \\ \Lambda &\rightarrow p\pi^-, \end{aligned} \tag{10.37}$$

and $\pi^0 \rightarrow 2\gamma$, whereby both photons produce e^+e^- pairs. The Ω^- is produced by the strong interaction ($\Delta S = 0$). Note the invisible trajectories (dashed lines) of the weak decaying neutral Ξ^0 and Λ hyperons, each with $|\Delta S| = 1$.

10.2.3 Isospin

We have seen in (10.10), (10.28) and (10.32), (10.33) that mesons and baryons appear with different charges but similar masses in subgroups within SU(3) multiplets. For example

$$\Lambda,\ N = \begin{pmatrix} p \\ n \end{pmatrix}, \qquad \pi = \begin{pmatrix} \pi^+ \\ \pi^0 \\ \pi^- \end{pmatrix}, \qquad \Delta = \begin{pmatrix} \Delta^{++} \\ \Delta^+ \\ \Delta^0 \\ \Delta^- \end{pmatrix}, \tag{10.38}$$

where N stands for the nucleon.

The notion of *isospin* (introduced in 1932 by Werner Heisenberg) follows from an analogy with *spin*: the projection m of the spin vector $\vec{s}$ onto an arbitrary quantization axis takes $2s + 1$ values, $-s \leqslant m \leqslant s$, where s is the spin quantum number. The corresponding $2s + 1$ spin wave functions are degenerate in energy in the absence of external fields. By analogy, the *isospin* quantum number i corresponds to a vector $\vec{i}$ in an abstract space, the isospin space. The projection i_3 of $\vec{i}$ onto a quantization axis in the isospin space takes $2i + 1$ values, $-i \leqslant i_3 \leqslant i$. The $2i + 1$ isospin wave functions correspond to $2i + 1$ hadrons with different charges but equal (degenerate) masses. The slightly different masses in the multiplet stem from the 'bare' quark mass differences. For the multiplets (10.38) one obtains with the following definitions of i and i_3:

$$\begin{aligned}
&\Lambda : i = 0, && i_3 = 0 \\
&N : i = \frac{1}{2}, && i_3 = +\frac{1}{2}\,(p),\ -\frac{1}{2}\,(n) \\
&\pi\ : i = 1, && i_3 = +1\,(\pi^+),\ 0\,(\pi^0),\ -1\,(\pi^-) \\
&\Delta : i = \frac{3}{2}, && i_3 = +\frac{3}{2}\,(\Delta^{++}),\ +\frac{1}{2}\,(\Delta^{+}),\ -\frac{1}{2}\,(\Delta^{0}),\ -\frac{3}{2}\,(\Delta^{-}).
\end{aligned} \tag{10.39}$$

The isospin $\vec{i}$ of the hadron is given by the vector sum of the isospins of its constituents, in analogy to the spin of a multiparticle system. From the quark contents of the multiplets one finds that

$$\boxed{i(u) = i(d) = \frac{1}{2}, \quad i_3(u) = \frac{1}{2}, \quad i_3(d) = -\frac{1}{2}, \quad i(s) = i_3(s) = 0}, \tag{10.40}$$

where i varies between the smallest and the largest possible values, and i_3 is additive. For the antiquarks

$$\boxed{i(\bar{u}) = i(\bar{d}) = \frac{1}{2}, \quad i_3(\bar{u}) = -\frac{1}{2}, \quad i_3(\bar{d}) = +\frac{1}{2}, \quad i(\bar{s}) = i_3(\bar{s}) = 0}. \tag{10.41}$$

The two quarks in the first family (u and d) form a doublet, while the s quark is an isospin singlet. Thus we obtain with an $i = \frac{1}{2}$ quark and an $i = \frac{1}{2}$ antiquark from the first family either $i = 0$ (e.g. the singlet ω) or $i = 1$ (e.g. the triplets π or ρ) with three projections of the quantum number i_3 that correspond to the three electric charges. On the other hand, with three quarks of the first family one obtains either $i = \frac{1}{2}$ (p and n as the nucleon doublet) or $i = \frac{3}{2}$ (the quadruplet Δ with four possible charges). We shall return to isospin in section 12.4.

One represents the SU(3) multiplets by diagrams (weight diagrams of the $SU(3)_f$ flavour symmetry group) in which hadrons are identified by their two coordinates, strangeness S and i_3 (figure 10.6). For $\ell = 0$ one obtains the two meson nonets (10.28), (10.10), 1^1S_0 and 1^3S_1 with $J = 0$ and $J = 1$, respectively, the baryon octet ($J = \frac{1}{2}$) and the baryon decuplet ($J = \frac{3}{2}$), see (10.32), (10.33). Further multiplets can be built with $\ell > 0$ and $n > 1$.

The *Gell–Mann–Nishijima* formula applies to quarks and hadrons with charges Qe:

$$\boxed{Q = i_3 + \frac{B + S}{2}}. \tag{10.42}$$

For instance, one verifies for the u quark and for the Ω^- that

$$u : \frac{2}{3} = \frac{1}{2} + \frac{\frac{1}{3}}{2}, \qquad \Omega^- : -1 = 0 + \frac{1 - 3}{2}. \tag{10.43}$$

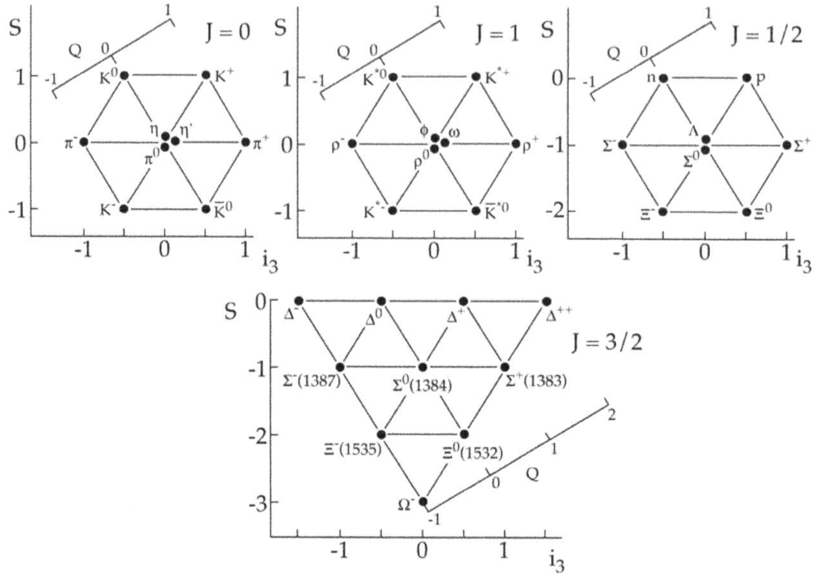

Figure 10.6. SU(3) multiplets of the ground state mesons and baryons ($n = 1$, $\ell = 0$).

Thus a linear relationship holds between i_3, S and Q. Indeed, hadrons of equal charges lie on diagonal lines in figure 10.6.

The Gell–Mann–Nishijima formula can be generalized to six quarks. In analogy to strangeness S one defines for the (isospin $i = 0$) c, b, and t quarks the *flavour* quantum number C, B' and T. The quantum numbers $S(s) = -1$, $B'(b) = -1$, $C(c) = +1$, $T(t) = +1$ and $S(\bar{s}) = 1$, $B'(\bar{b}) = 1$, $C(\bar{c}) = -1$, $T(\bar{t}) = -1$ are defined such that their signs are identical to the signs of the corresponding quark charges. The generalized Gell–Mann–Nishijima formula reads

$$\boxed{Q = i_3 + \frac{B + S + B' + C + T}{2}}. \tag{10.44}$$

The *hypercharge* is defined as [6]:

$$\boxed{Y \equiv B + S - \frac{1}{3}(C - B' + T)}. \tag{10.45}$$

The u and d quarks have the hypercharge $Y = \frac{1}{3}$, the s quark has $Y = -\frac{2}{3}$ and the heavy quarks $Y = 0$. With (10.42) one obtains for hadrons which contain only the lightest u, d or s quarks[3]

$$Q = i_3 + \frac{Y}{2}. \tag{10.46}$$

[3] The definition of hypercharge with heavy quarks is not unique in the literature. The hypercharge is sometimes defined as $Y \equiv B + S + B' + C + T$, in which case (10.46) is valid for all quark flavours.

10.2.4 OZI rule

The following empirical Okubo–Zweig–Iizuka (OZI) rule applies to strong interaction processes:

> Processes described by Feynman graphs, which can be split into two or more *hadrons* without cutting any quark line, are suppressed.

The rule is best explained by a simple example. Figure 10.7(*a*) shows the decay of the ϕ meson into $K\overline{K}$ pairs (K^+K^- or $K^0\overline{K}^0$). The diagram cannot be disassembled into hadrons without cutting a quark line. Figure 10.7(*b*) shows the decay into $\pi^+\pi^-\pi^0$ in which the s quarks annihilate. The vertical dotted line splits the diagram into two hadronic parts without cutting any quark line. This decay is therefore suppressed. Experimentally one finds the branching ratio

$$f(\phi \to K\overline{K}) = 83\%, \qquad f(\phi \to \pi^+\pi^-\pi^0) = 15\%. \tag{10.47}$$

Here the OZI suppression is not dramatic because the mass difference between the initial and final states is much smaller in the former ($Q \simeq 30$ MeV) than in the latter ($\simeq 600$ MeV). The $K\overline{K}$ decay is therefore kinematically disadvantaged by the limited phase space (see section 12.1). In fact, ϕ decay into $\pi^+\pi^-\pi^0$ occurs through the small admixture of $u\bar{u}$ and $d\bar{d}$ in the wave function of the ϕ since the mixing angle is not exactly ideal, as explained above (section 10.2.1). The corresponding Feynman graph is then OZI-allowed.

10.3 Charm and bottom quarks

Due to flavour conservation the c, b and t quarks are always produced as quark–antiquark pairs in strong and electromagnetic interactions. The emitted heavy quark 'dresses' (*hadronizes*) with quarks from the vacuum to form hadrons, with the exception of the t quark which decays instantly into bW^+. Further mesons can be generated with the c quark. For example, the 1^1S_0 nonet of the light quarks is extended to include the D mesons

$$D^0 = |c\bar{u}\rangle, \quad D^+ = |c\bar{d}\rangle, \quad D_s^+ = |c\bar{s}\rangle, \quad \overline{D}^0 = |\bar{c}u\rangle, \quad D^- = |\bar{c}d\rangle, \quad D_s^- = |\bar{c}s\rangle \tag{10.48}$$

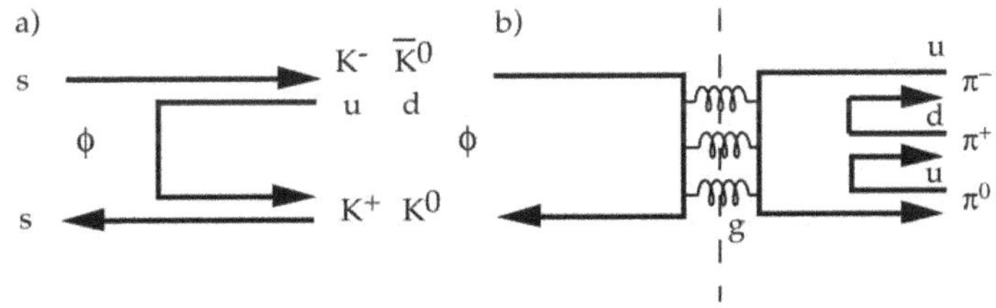

Figure 10.7. OZI-allowed (*a*) and OZI-suppressed decay (*b*) of the ϕ meson.

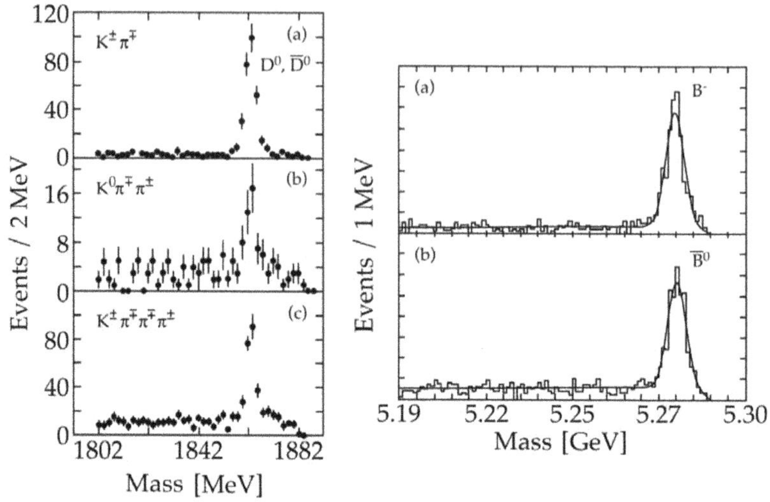

Figure 10.8. Left: D mesons $D^0 \to K^-\pi^+, \overline{K}^0\pi^-\pi^+, K^-\pi^+\pi^+\pi^-$ and $\overline{D}^0 \to K^+\pi^-, K^0\pi^-\pi^+, K^+\pi^-\pi^-\pi^+$. From [7]. Right: B mesons B^- (a) and $\overline{B}^0$ (b) seen in various decay channels such as $D^0\pi^-$, $D^0\rho^-$, and $D^+\pi^-$, $D^+\rho^-$, respectively. From [8].

and the 'hidden' charm state

$$\eta_c = |c\bar{c}\rangle. \tag{10.49}$$

Figure 10.8 (left) shows the D mesons produced in e^+e^- collisions ($e^+e^- \to \psi(3770) \to D\overline{D}$). The c quark in the D meson decays into the s quark in the K meson through the weak interaction.

The corresponding 1^3S_1 mesons are labelled D^*. The 1^3S_1 $c\bar{c}$ ground state is

$$J/\psi(1S) = |c\bar{c}\rangle. \tag{10.50}$$

Baryons may also contain c quarks. Introducing the additive quantum number $C = 1$ for the c quark (and $C = -1$ for the $\bar{c}$ antiquark) one can classify all $\ell = 0$ hadrons in a three-dimensional representation (i_3, S, C) or alternatively (i_3, Y, C). Figure 10.9 shows the weight diagrams of the $SU(4)_f$ flavour symmetry group. All $\ell = 0$ mesons have been detected. All $C = 1$ charm hyperons have also been observed. One of the Ξ_{cc}^+ states has been reported, possibly the spin-$\frac{1}{2}$ dcc baryon with $C = 2$.

Figure 10.10(a) shows a photograph of the reaction

$$\nu_\mu p \to \mu^-\Lambda(\to\pi^-p)\pi^+\pi^+\pi^-\pi^+ \tag{10.51}$$

from a bubble chamber exposed to a neutrino beam in Brookhaven [9]. At first sight this event seemed to violate the $\Delta S = \Delta Q$ rule (chapter 13). Shortly afterwards came the observation of a charmed (anti)hyperon, the ($\bar{u}\bar{d}\bar{c}$) $\overline{\Lambda_c^+}$ in an experiment at Fermilab (figure 10.10(b)). The Λ_c^+ is the lightest charmed baryon with a mass of 2286 MeV and a mean life of 0.2 ps. Since the (uuc) Σ_c^{++} hyperon decays into $\pi^+\Lambda_c^+$, we now know that reaction (10.51) proceeds through the intermediate state $\mu^-\Sigma_c^{++}$ ($\Sigma_c^{++} \to \Lambda_c^+\pi^+ \to \Lambda\pi^+\pi^+\pi^-\pi^+$). Thus figure 10.10($a$) was the first experimental evidence for the existence of charmed baryons.

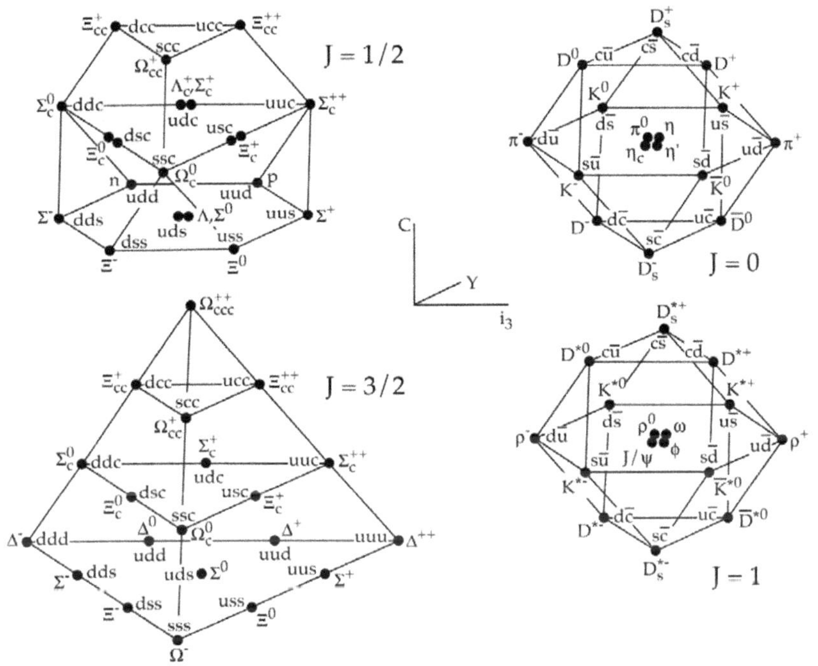

Figure 10.9. SU(4) multiplets of the ground state hadrons including the c quark: 20-plets (left) and 16-plets (right). Plotted is the hypercharge $Y = S + B - \frac{C}{3}$, as defined in (10.45). There are two spin-$\frac{1}{2}$ Ξ_c^0 and two spin-$\frac{1}{2}$ Ξ_c^+ baryons. From [4].

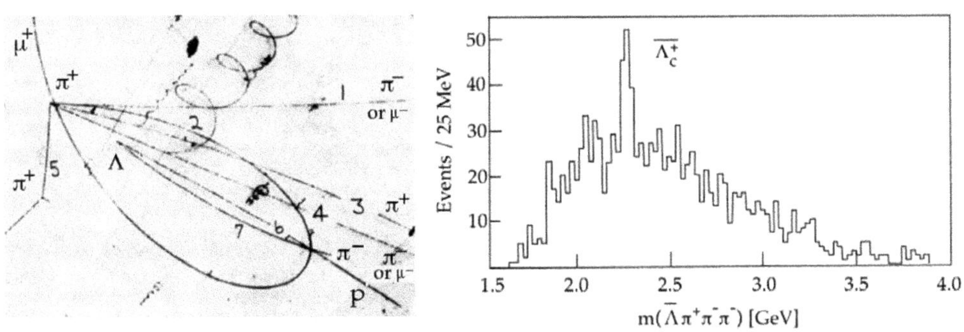

Figure 10.10. Left: a bubble chamber photograph of the reaction $\nu_\mu p \to \mu^- \Lambda(\to \pi^- p)\pi^+\pi^+\pi^-\pi^+$. From [9]. Right: observation of $\overline{\Lambda_c^+}$ decay into $\bar{\Lambda}\pi^+\pi^-\pi^-$ in a broadband photon beam. From [10].

Further mesons can be generated with the b quark, for example the 1^1S_0 B mesons

$$
\begin{aligned}
B^- &= |b\bar{u}\rangle, & \overline{B}_d^0 &= |b\bar{d}\rangle, & \overline{B}_s^0 &= |b\bar{s}\rangle, \\
B^+ &= |\bar{b}u\rangle, & B_d^0 &= |\bar{b}d\rangle, & B_s^0 &= |\bar{b}s\rangle, \\
B_c^- &= |b\bar{c}\rangle, & B_c^+ &= |\bar{b}c\rangle,
\end{aligned}
\tag{10.52}
$$

and the 'hidden' bottom state

$$\eta_b = |b\bar{b}\rangle. \tag{10.53}$$

The corresponding 1^3S_1 $J = 1$ mesons are labelled B^*. The 1^3S_1 $b\bar{b}$ ground state is the 'upsilon'

$$\Upsilon(1S) = |b\bar{b}\rangle. \tag{10.54}$$

The $\ell = 0$ mesons are well known. Figure 10.11 shows the SU(4) weight diagrams of the ground state hadrons made of the u, d, s and b quarks, projected into the $i_3 - Y$ plane. Figure 10.8 (right) shows the signal from B mesons produced in e^+e^- collisions ($e^+e^- \rightarrow \Upsilon(4S) \rightarrow B\overline{B}$). One can see that the b quark in the B meson decays into a c quark in the D meson through the weak interaction. Figure 10.12

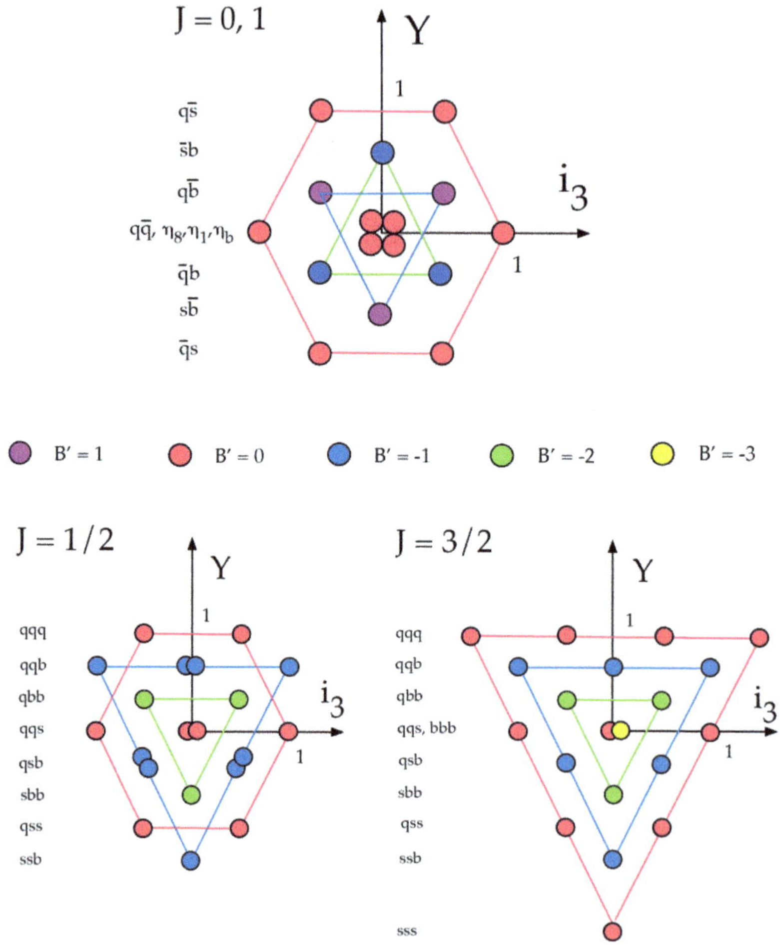

Figure 10.11. Projection into the $i_3 - Y$ plane of the SU(4) multiplets containing the u, d, s and b quarks (q stands for u or d). Top: 16-plet of the ground state $^1S_0$ and $^3S_1$ mesons. Bottom: 20-plets of the ground state baryons.

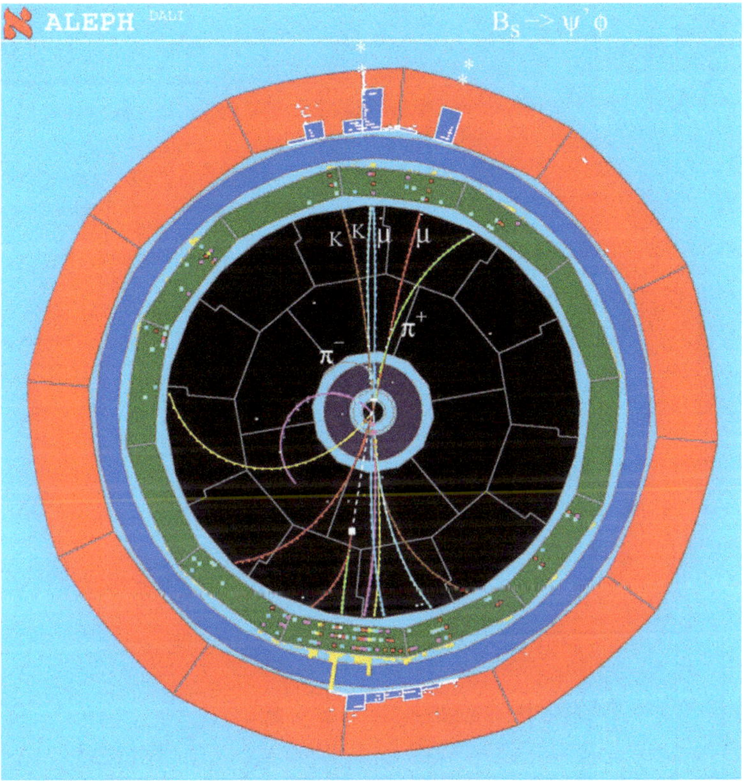

Figure 10.12. An event display from the ALEPH detector at the Large Electron–Positron storage ring (LEP) showing one of the first observed B_s mesons decaying into $\phi(1020)\psi(2S)$. The ϕ decays into K^+K^- and the $\psi(2S)$ into $J/\psi(\rightarrow\mu^+\mu^-)\pi^+\pi^-$. Image credit: CERN.

shows the signal of one of the first observed B_s mesons. So far the $B_c^{\pm}$ is the only known hadron composed of both c and b quarks.

Several spin-$\frac{1}{2}$ hyperons with a b and two light quarks (u, d, s) have been observed [4]. The lightest bottom baryon is the neutral Λ_b (udb) with a mass of 5620 MeV and a mean life of 1.4 ps. A signal was first spotted at the CERN ISR in 1981 [11] but was soon disputed by another experiment [12]. In 1991 the UA1 Collaboration announced the discovery of a baryon decaying into $(J/\psi)\Lambda$ at the CERN $\bar{p}p$ collider [13] (a description of the UA1 experiment is given in section 17.2). The J/ψ was identified by its decay into $\mu^+\mu^-$ pairs and the Λ by its decay into π^-p. Figure 10.13 (left) shows the $(J/\psi)\Lambda$ mass distribution with a clear peak of 16 ± 5 events above background. Although rather modest, this was the first evidence for bottom baryons. Figure 10.13 (right) shows today's state-of-the-art with the Λ_b signal observed at the Large Hadron Collider (LHC) (6870 events).

10.3.1 Charmonium

The c quark was discovered simultaneously at SLAC and Brookhaven in 1974. At SLAC electrons were collided with positrons of equal energy $E_{e^{\pm}}$ in the

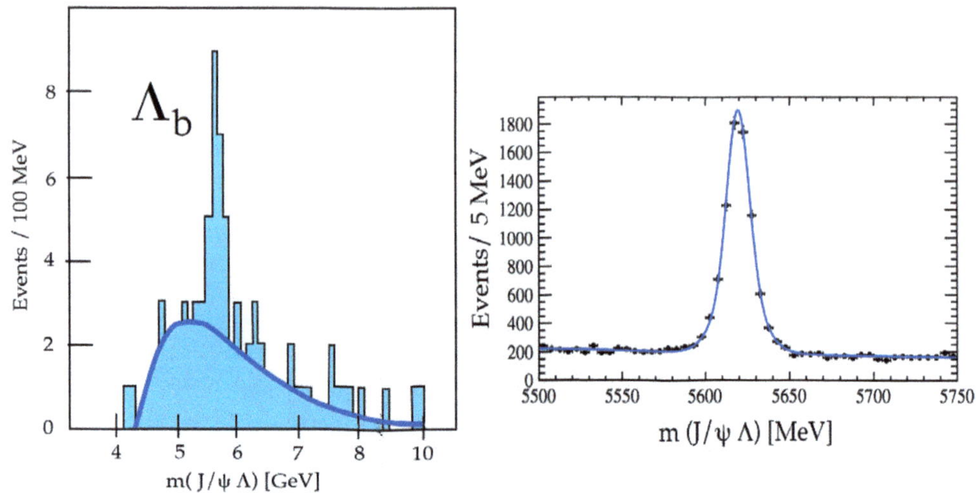

Figure 10.13. Left: $(J/\psi)\Lambda$ mass distribution showing the first observation of the Λ_b above background (blue line). Adapted from [13]. Right: Λ_b signal at the LHC. From [14].

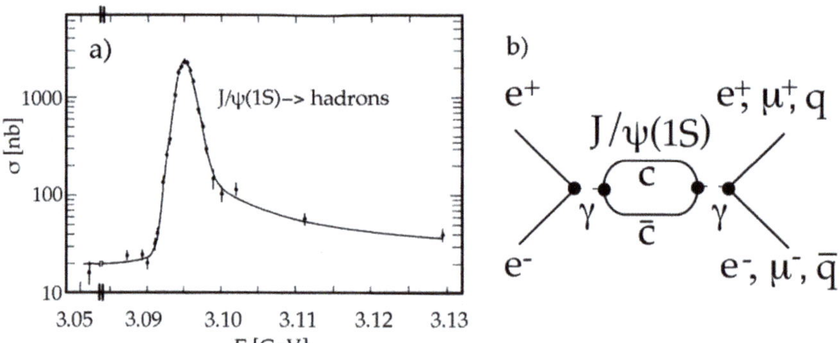

Figure 10.14. (*a*) $J/\psi(1S)$ resonance produced in e^+e^- collisions and decaying into hadrons. The long tail is due to radiative corrections to e^+e^- processes. Adapted from [16]. (*b*) Production and decay of the $c\bar{c}$ resonance.

SPEAR ring. A *sharp* resonance, the $J/\psi(1S)$ decaying into hadrons or $\mu^+\mu^-$ pairs, was observed (figure 10.14(*a*)) at a total energy $E = E_{e^-} + E_{e^+} = 3097$ MeV [15].

Figure 10.14(*b*) shows the production of a $c\bar{c}$ pair through the electromagnetic interaction $e^+e^- \to \gamma \to c\bar{c}$. The $c\bar{c}$ resonance is produced by the virtual photon from e^+e^- annihilation. The spin of the resonance should be equal to 1. This follows from the resonance line shape observed in $\mu^+\mu^-$: the amplitude for $e^+e^- \to J/\psi(1S) \to \mu^+\mu^-$ interferes with the direct amplitude $e^+e^- \to \gamma \to \mu^+\mu^-$ and distorts the line shape of the resonance [16]. Quantum mechanical interference is only possible if the $J/\psi(1S)$ and the γ have the same quantum numbers.

The true width Γ of the resonance is smaller than the observed one which is determined by the 1.3 MeV uncertainty in the beam energies. The true width can be estimated by integrating the observed enhancement in $e^+e^- \to J/\psi(1S) \to$ hadrons (figure 10.14(*a*)), and comparing it with the integral of the Breit–Wigner cross section σ for the $J/\psi(1S)$ resonance:

$$\int_{-\infty}^{\infty} \sigma \mathrm{d}E = \int_{-\infty}^{\infty} \frac{2J+1}{(2s+1)(2s+1)} \cdot \frac{4\pi}{E^2} \cdot \frac{\Gamma_{ee}\Gamma_{\text{hadrons}}}{(E-m)^2+\Gamma^2/4}\mathrm{d}E$$

$$= \frac{3}{4} \cdot \frac{8\pi^2}{m^2}\left(\frac{\Gamma_{ee}\Gamma_{\text{hadrons}}}{\Gamma}\right) = \frac{3}{4} \cdot \frac{8\pi^2}{m^2}\left(\frac{\Gamma_{ee}}{\Gamma}\right)\left(\frac{\Gamma_{\text{hadrons}}}{\Gamma}\right)\Gamma$$

$$= \frac{3}{4} \cdot \frac{8\pi^2}{m^2} f(e^+e^-) f(\text{hadrons})\Gamma$$

$$= \frac{3}{4} \cdot \frac{8\pi^2}{m^2} f(\mu^+\mu^-) f(\text{hadrons})\Gamma. \tag{10.55}$$

The mass of the $J/\psi(1S)$ is m, its spin $J = 1$ and $s = \frac{1}{2}$ is the spin of the leptons. The branching ratios for the decays into e^+e^-, $\mu^+\mu^-$ and into hadrons are $f(e^+e^-) = f(\mu^+\mu^-) \simeq 6\%$ and $f(\text{hadrons}) \simeq 88\%$ respectively. To integrate the cross section we have used the normalized Breit–Wigner (4.28) and have assumed that $4\pi/E^2 \simeq 4\pi/m^2$, which is a good approximation for a narrow resonance. The true width of the resonance ($\Gamma = 93$ keV) turned out to be surprisingly small.

Simultaneously, a resonance was observed in Brookhaven at the same mass in the reaction

$$p\,\mathrm{Be} \to \overbrace{e^+e^-}^{J/\psi(1S)} X \tag{10.56}$$

using 30 GeV protons (figure 10.15, left). In this case the $c\bar{c}$ pair is produced by gluons and $c\bar{c} \to \gamma \to e^+e^-$. The emitted electrons and positrons were analysed by two magnetic spectrometers [17].

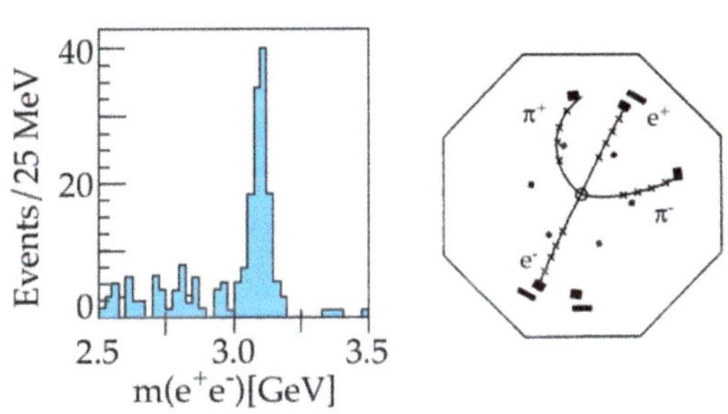

Figure 10.15. Left: the $J/\psi \to e^+e^-$ appears as a narrow resonance in the e^+e^- mass spectrum of reaction (10.56). After [17]. Right: the decay of the $\psi(2S)$ in $J/\psi(\to e^+e^-)\pi^+\pi^-$ seen by the MARK I detector at the SPEAR e^+e^- collider. After [18].

Further higher lying resonances were soon discovered at SPEAR. Figure 10.15 (right) shows the signature of $\psi(2S) \to J/\psi(1S)\,\pi^+\pi^-$. The masses and widths of the spin-1 resonances are:

$$\begin{array}{lll} J/\psi(1S) & M = 3097\ \text{MeV} & \Gamma = \mathbf{93\ keV} \\ \psi(2S) & M = 3686\ \text{MeV} & \Gamma = \mathbf{303\ keV} \\ \psi(3770) & M = 3773\ \text{MeV} & \Gamma = 27\ \text{MeV} \\ \psi(4040) & M = 4039\ \text{MeV} & \Gamma = 80\ \text{MeV} \\ \psi(4160) & M = 4153\ \text{MeV} & \Gamma = 103\ \text{MeV} \\ \psi(4415) & M = 4421\ \text{MeV} & \Gamma = 62\ \text{MeV}. \end{array} \tag{10.57}$$

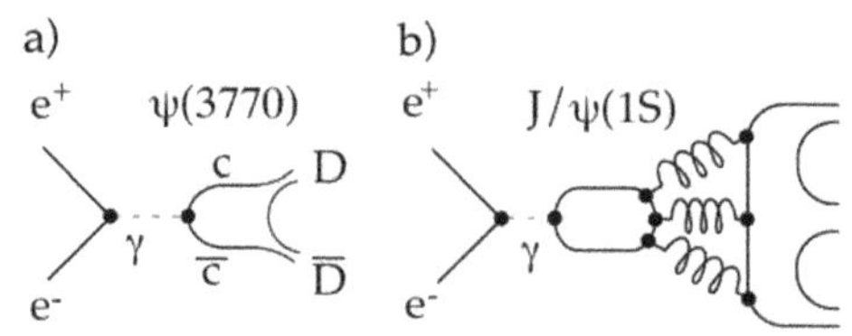

Figure 10.16. Excitation of a ψ resonance in e^+e^- collisions. If the mass of the resonance lies above the $D\overline{D}$ threshold the resonance decays into $D\overline{D}$ (*a*). Below threshold the resonance decays through OZI-suppressed decays (*b*) or electromagnetically (figure 10.14(*b*)).

The astonishing much smaller widths of the two lower states can be explained as follows: on the assumption that the resonance is made of a pair of *new* quarks $c\bar{c}$ the J/ψ should decay through the emission of charmed D mesons (figure 10.16(*a*)). However, the mass of the lightest pair of charmed mesons ($D^0\overline{D}^0$) is 3730 MeV, which is larger than the mass of the two lower states. Therefore the $1S$ and $2S$ states cannot decay into ($D^0\overline{D}^0$) and have to disintegrate through OZI-suppressed processes (figure 10.16(*c*)), which increases their lifetimes substantially and reduces their widths correspondingly.

This OZI suppression of the $J/\psi(1S)$ decay was the smoking gun for the existence of a fourth quark, the charm quark. The $\psi(2S)$, $\psi(4040)$ and $\psi(4415)$ mesons are *radial* excitations of the spin $J=1$ ($\ell=0$, $s=1$) $c\bar{c}$ system. The $\psi(3770)$ and $\psi(4160)$ mesons are $J=1$, $\ell=2$, $s=1$ states (1^3D_1 and 2^3D_1). Since the $\psi(2S)$ also decays as OZI-suppressed its electromagnetic decay occurs as a competing process. Hence the ψ (2S) state should also decay by radiative transitions into the lower $c\bar{c}$ states with quantum numbers $\ell=1$ and $J=0$, 1 or 2. Figure 10.17 shows the *charmonium* spectrum of the $c\bar{c}$ system. Note the analogy to the positronium spectrum (figure 10.18).

Radiative transitions from the $\psi(2S)$ state have been observed at SPEAR by the Crystal Ball detector (figure 10.19, left). The e^+ and e^- beams collide in the centre of the detector and the beam energies (1843 MeV) are tuned such that the $\psi(2S)$ resonance is excited. A ball of 672 NaI (Tl) crystals measures the (monochromatic) photon transitions to the lower $c\bar{c}$ states. The measured spectrum is shown in figure 10.19 (right). Sharp transition lines are indeed observed. The observed line widths are determined by the detector resolution. The background continuum stems from photons from π^0 decays that are emitted by the OZI-suppressed decays of the $\psi(2S)$ resonance.

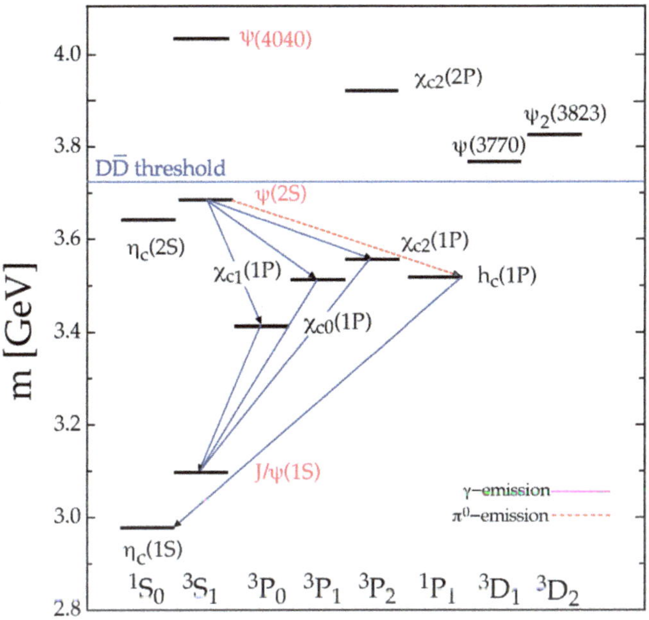

Figure 10.17. The observed states in the charmonium spectrum. The γ transitions $\psi(2S) \to \gamma h_c(1P)$ and $h_c(1P) \to \gamma J/\psi(1S)$ are forbidden by C parity conservation (section 11.7).

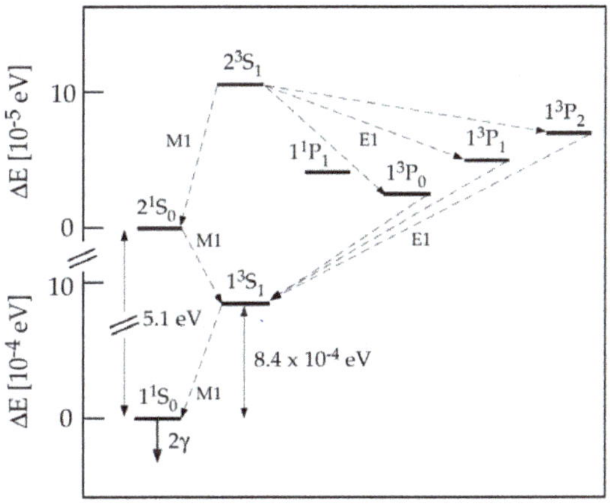

Figure 10.18. The positronium (e^+e^-) spectrum. Note that the energy differences between radial excitations are much larger (~5 eV) than the hyperfine splitting of the $^3S_1$ and $^1S_0$ levels. The levels with equal n are nearly degenerated.

10.3.2 Bottomonium

History repeated itself in 1978 at Fermilab with the discovery of the b quark. Narrow lines were observed at masses around 10 GeV [20] in the $\mu^+\mu^-$ mass spectrum of the reaction

$$pA \to \overbrace{\mu^+\mu^-}^{\Upsilon(1S)} X. \tag{10.58}$$

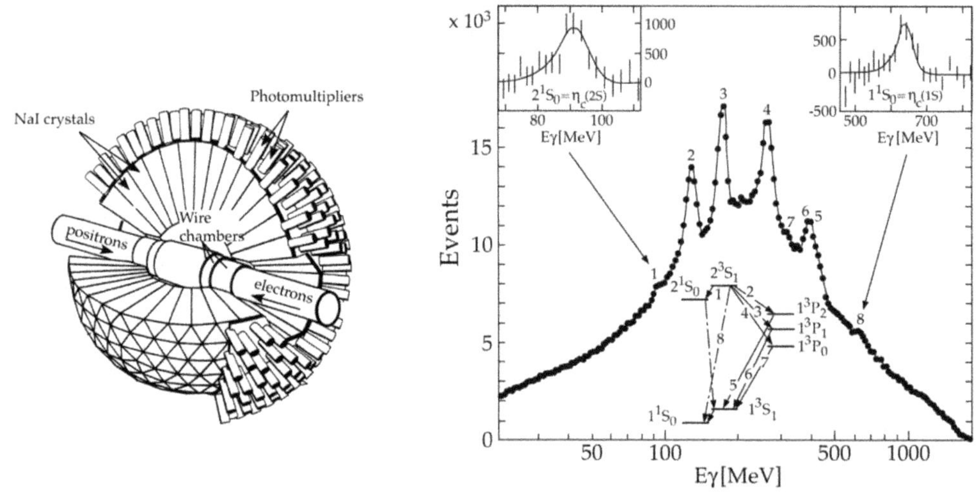

Figure 10.19. Left: the Crystal Ball detector. Right: radiative decay spectrum from the $\psi(2S)$ level. After [19].

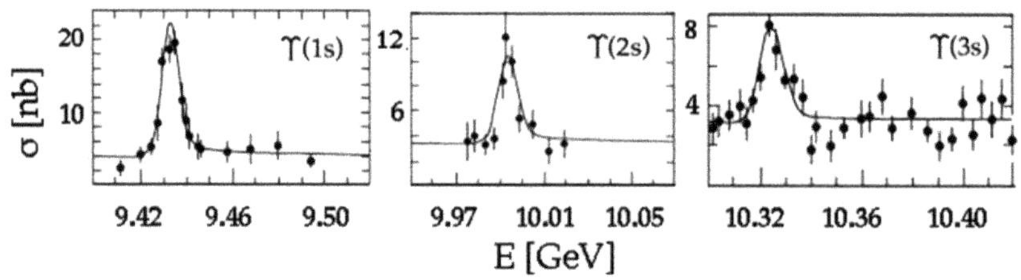

Figure 10.20. ϒ spectrum in $e^+e^- \rightarrow$ hadrons (data collected by the CLEO detector in Cornell [21]).

This was confirmed a few years later by the discovery of sharp resonances in e^+e^- collisions at the CESR and DORIS storage rings at Cornell and DESY, respectively. Figure 10.20 shows the excitation of the ϒ (upsilon) resonances in e^+e^- collisions.

The masses and widths of the (spin-1) ϒ resonances are

$$\begin{array}{lll} \Upsilon(1S) & M = 9\,460\ \text{MeV} & \Gamma = \mathbf{54\ keV} \\ \Upsilon(2S) & M = 10\,023\ \text{MeV} & \Gamma = \mathbf{32\ keV} \\ \Upsilon(3S) & M = 10\,355\ \text{MeV} & \Gamma = \mathbf{20\ keV} \\ \Upsilon(4S) & M = 10\,580\ \text{MeV} & \Gamma = 21\ \text{MeV} \\ \Upsilon(10860) & M = 10\,876\ \text{MeV} & \Gamma = 55\ \text{MeV} \\ \Upsilon(11020) & M = 11\,019\ \text{MeV} & \Gamma = 79\ \text{MeV}. \end{array} \tag{10.59}$$

The slightly smaller masses of the ϒ states in figure 10.20 are due to a calibration inaccuracy of the beam energies. The narrow widths of the three lower states are due to the existence of a fifth quark: the b quark. The kinematic threshold for the decay into a pair of bottom mesons $B\overline{B}$ (10 558 MeV) lies just below the mass of the ϒ(4S). The three lower resonances have substantially lower lifetimes since they have to decay through OZI-suppressed processes. The ϒ(2S) and ϒ(3S) can also decay by

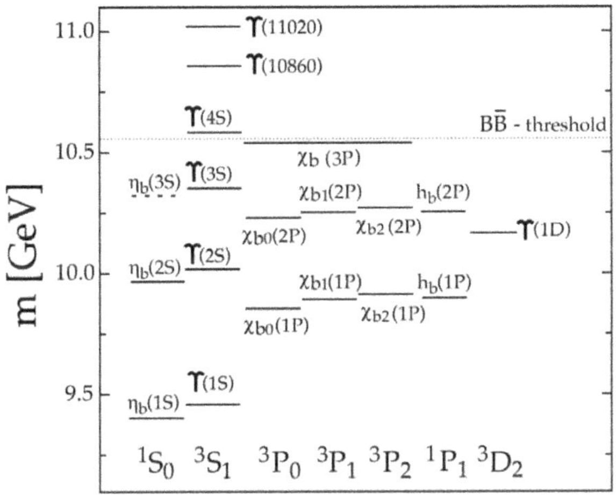

Figure 10.21. The bottomonium spectrum. The $\eta_b(3S)$ has not been identified yet.

radiative transitions into the lower $b\bar{b}$ states with quantum numbers $\ell = 1$ and $J = 0$, 1 or 2. These states have been observed using colliding e^+e^- beams. At Cornell the beams were tuned to the energies of the $\Upsilon(3S)$ and $\Upsilon(2S)$ to observe the radiative transitions to the $\chi_b(2P)$ [22] and $\chi_b(1P)$ states [23], respectively. Figure 10.21 shows the *bottomonium* spectrum of $b\bar{b}$ mesons.

10.3.3 Quarkonium

The observation of γ transitions in the charmonium and bottomonium systems clearly demonstrates that hadrons possess an internal structure made of spin-$\frac{1}{2}$ quarks. Let us qualitatively compare the *quarkonium* spectra with the positronium spectrum of figure 10.18:

1. Relative to their masses, the level spacings and hyperfine splittings are much larger in quarkonium than in positronium. This is due to the binding force between quarks which is mediated by gluon exchange, while positronium is bound by the weaker electromagnetic force.
2. The mass difference between the $2S$ and $1S$ levels is ~600 MeV for both charmonium and bottomonium. For spin-1 'strangeonium' ($s\bar{s}$) the energy difference between the $1S$ state ($\phi(1020)$) and the $2S$ state ($\phi(1680)$) is also ~600 MeV. The $q\bar{q}$ potential generated by the exchanged gluon is therefore flavour-independent, i.e. *universal.*
3. In positronium the binding energies are proportional to $1/n^2$. The level spacing decreases more rapidly with increasing n than for quarkonium.
4. For positronium the centre of gravity of the $1P$ levels almost coincides with the $2S$ level since the $1P$ and $2S$ levels are nearly degenerate. However, for quarkonium the $1P$ levels are located very roughly halfway between the $1S$ and the $2S$ levels.

The levels of a (3D) harmonic potential levels are equidistant and the $1P$ level is located exactly halfway between the $1S$ and $2S$ levels. One therefore guesses that the potential dependence of the $q\bar{q}$ system should lie between $1/r$ and r and makes the following ansatz for the potential energy:

$$\boxed{V(r) = -\frac{4}{3}\frac{\alpha_s}{r} + \kappa r}, \tag{10.60}$$

with κ in GeV fm^{-1}. The first term, which dominates at short distances and is similar to the $1/r$ Coulomb term for photons, describes the exchange of a spin-1 gluon. At large distances, the potential energy grows linearly with r and arises from the exchange of many gluons. Hence the $q\bar{q}$ system remains confined as r grows to infinity ($V(\infty) = \infty$). The solution of the Schrödinger equation describes the quarkonium systems rather satisfactorily with the ansatz (10.60). One finds that

$$\alpha_s \sim 0.3, \qquad \kappa = 14\,\text{t}. \tag{10.61}$$

With increasing r the excitation energy of the $q\bar{q}$ pair is stored as potential energy $V(r)$ until the production of a new $q\bar{q}$ pair from the vacuum becomes more economical (figure 10.22). One speaks of *fragmentation*. The emitted $q\bar{q}$ pairs fly apart (and may also fragment), leading to back-to-back jets of hadrons, as shown in figure 10.23 (left). Sometimes one of the original quarks emits a gluon which turns into a $q\bar{q}$ pair to produce a third jet (figure 10.23, right). From the angular correlations of the jets one can determine the spin of the gluon to be $J = 1$ [24].

For a binding force κ of 14 t and a quark–antiquark distance of 1 fm the potential energy in SI units is

$$V \simeq 10^{-15} \times 14 \times 10^{3} \times 9.81 = 1.4 \times 10^{-10}\,\text{J} \simeq 0.9\,\text{GeV}. \tag{10.62}$$

This corresponds to the typical mass of a meson composed of the lighter quarks. For these mesons the mass is dominated by the gluons, while the contributions from the 'bare' u, d and s quarks are small. In contrast, for heavy mesons the main contribution stems from the bare masses of the c or b quarks.

In a simple model the light meson can be viewed as consisting of a rotating quark–antiquark pair (rotor) connected by a gluon flux tube of length $2R$. The (nearly massless) quarks rotate around the centre of the rotor with the speed of light ($c = 1$). The contribution of the gluons to the meson mass M at a distance r from the centre is $\mathrm{d}M = \kappa \mathrm{d}r$ with $\kappa \simeq 0.9$ GeV fm^{-1}. The mass of the meson is then given by

$$M = 2\int_0^R \frac{\kappa \mathrm{d}r}{\sqrt{1 - v^2}} = \pi R \kappa \tag{10.63}$$

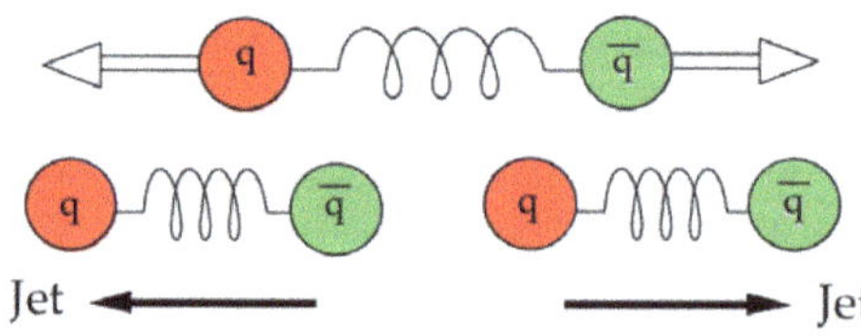

Figure 10.22. Fragmentation of a $q\bar{q}$ pair into a further pair generated from the vacuum. The mesons fly in opposite directions and may fragment into more hadrons, leading to particle jets.

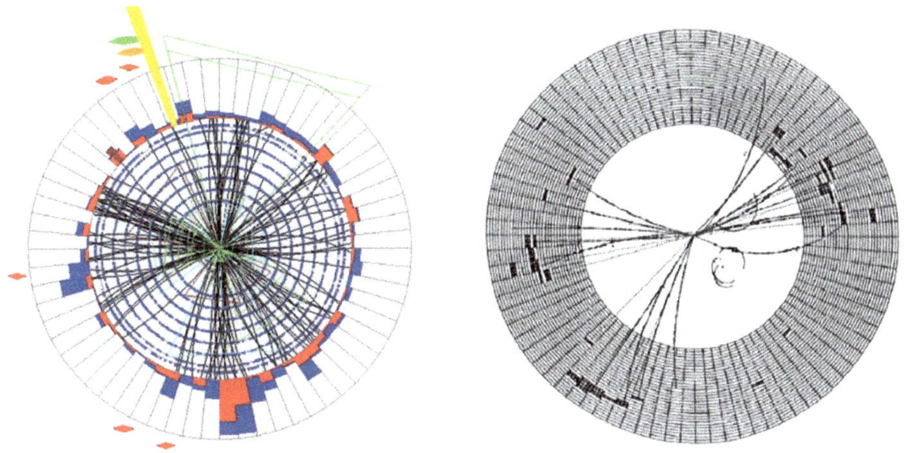

Figure 10.23. Left: a multi-jet event observed at Fermilab by the D0 detector in $\bar{p}p$ collisions at 1.96 TeV. The coloured histogram shows the energy deposits in the calorimeters. Image credit: Fermilab/D0 Collaboration. Right: three-jet event produced in e^+e^- collisions at 31 GeV. The data were taken by the JADE detector at the PETRA storage ring in DESY. Reproduced with permission from [24]. Copyright 2010 Springer Science and Business Media.

with $v = r/R$. The classical angular momentum is

$$L = 2\int_0^R \frac{rv\kappa \mathrm{d}r}{\sqrt{1-v^2}} = \frac{\pi R^2 \kappa}{2}. \tag{10.64}$$

Eliminating R between the last two equations one finds an important relation between L and M,

$$L = \frac{M^2}{2\pi\kappa}, \tag{10.65}$$

or between the spin J and the mass M of the meson:

$$\boxed{J \simeq \frac{M^2}{2\pi\kappa} + \text{const} = 0.9 \times M^2\left[\text{GeV}^2\right] + \text{const}}. \tag{10.66}$$

The constant takes into account the spin of the quark–antiquark pair ($s = 0$ or 1) and the quark masses. Figure 10.24 (Chew–Frautschi plot) shows how J varies with M^2 for $s = 1$ mesons with quantum number ℓ even and $J = \ell + 1$ ($J^{PC} = 1^{--}$, 3^{--}, 5^{--}), and with quantum numbers ℓ odd and $J = \ell + 1$ ($J^{PC} = 2^{++}$, 4^{++}, 6^{++}). The notation J^{PC} is explained in sections 11.4 and 11.7. The relation between J and M^2 (*Regge trajectory*) is indeed linear with a slope of 0.9 GeV^{-2}. This is in good agreement with the rotor model. A similar dependence is found for baryons.

10.4 Colour

The wave function of the (spin-$\frac{3}{2}$) Δ^{++} in the baryon decuplet is *symmetric* under the permutation of two quarks:

$$\Delta^{++} = |uuu\rangle|\uparrow\uparrow\uparrow\rangle|\ell = 0\rangle. \tag{10.67}$$

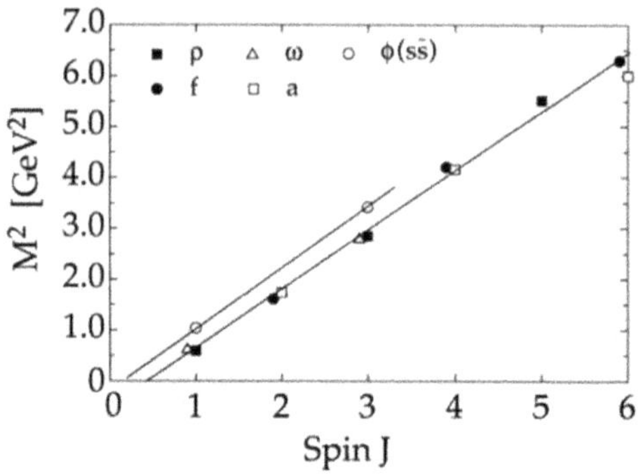

Figure 10.24. Chew–Frautschi plot for the spin triplet mesons with quantum numbers 1^{--}, 3^{--}... (ρ, ω, ϕ) and 2^{++}, 4^{++}... (f, a). The Regge trajectory of the ϕ mesons lies at higher masses due to the heavier s quark.

Hence the Δ^{++} seems to violate the Pauli principle for identical fermions. The same consideration holds for the Ω^- made of three s quarks. The way out introduces a new quantum number and forms the basis of quantum chromodynamics (QCD): quarks and antiquarks exist in three *'colours'*. There are three coloured quarks of the same flavour, red R, green G, and blue B, as well as three antiquarks with anticolours $\overline{R}$, $\overline{G}$, $\overline{B}$. A baryon such as the Λ consists of $u(R)d(G)s(B)$ quarks, or any of the six possible permutations of colour. But since there is only one Λ, the colour wave function is a linear combination of the six permutations and the Λ is colourless. The flavour wave function of the Λ is extended to include the *antisymmetric* colour combinations

$$|uds\rangle \to \beta(\Lambda) = \frac{1}{\sqrt{6}}|u(R)d(G)s(B) + u(B)d(R)s(G) + u(G)d(B)s(R)$$
$$-u(G)d(R)s(B) - u(B)d(G)s(R) - u(R)d(B)s(G)\rangle. \qquad (10.68)$$

Similarly, the colour wave function (10.67) of the Δ^{++} is

$$|uuu\rangle \to \beta(\Delta^{++}) = \frac{1}{\sqrt{6}}|u(R)u(G)u(B) + u(B)u(R)u(G) + u(G)u(B)u(R)$$
$$-u(G)u(R)u(B) - u(B)u(G)u(R) - u(R)u(B)u(G)\rangle, \qquad (10.69)$$

which is antisymmetric under quark permutations.

A meson consists of a linear combination of quarks with colours and antiquarks with the corresponding anticolours, hence the $q\bar{q}$ system is also colourless. For example, the colour wave function of the π^+ is

$$\beta(\pi^+) = \frac{1}{\sqrt{3}}|u(R)\bar{d}(\overline{R}) + u(G)\bar{d}(\overline{G}) + u(B)\bar{d}(\overline{B})\rangle. \qquad (10.70)$$

In nature only colourless configurations of quarks, antiquarks and gluons appear as hadrons. Since quarks carry colour, this also 'explains' why free quarks are not observed.

The interaction between quarks is mediated by gluons through the exchange of colour, which is *conserved* (figure 10.25(*a*)). The underlying symmetry corresponding

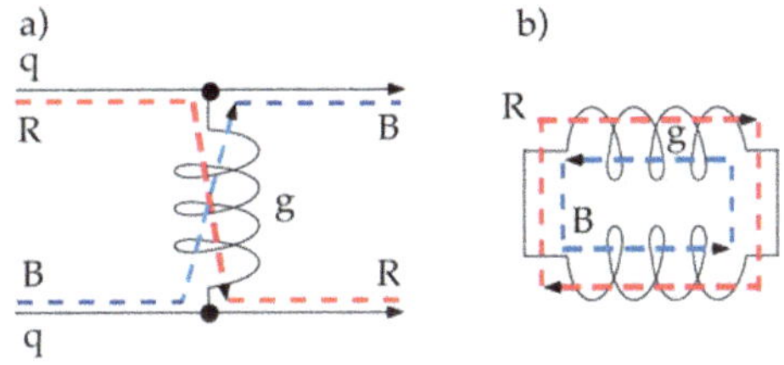

Figure 10.25. (*a*) The interaction between quarks proceeds through the exchange of coloured gluons. The dashed lines show the colour fluxes. (*b*) The coupling between gluons leads to bound states of two or more gluons, the glueballs.

to the conservation of colour is SU(3) colour ($SU(3)_c$). A gluon contains one colour and one anticolour. Nine gluons can be constructed from three colours and three anticolours, with the $SU(3)_c$ wave functions

$$|R\overline{B}\rangle,\ |R\overline{G}\rangle,\ |B\overline{G}\rangle,\ |B\overline{R}\rangle,\ |G\overline{R}\rangle,\ |G\overline{B}\rangle,$$
$$\frac{1}{\sqrt{6}}|R\overline{R} + G\overline{G} - 2B\overline{B}\rangle,\ \frac{1}{\sqrt{2}}|R\overline{R} - G\overline{G}\rangle \tag{10.71}$$

and

$$\frac{1}{\sqrt{3}}|R\overline{R} + G\overline{G} + B\overline{B}\rangle, \tag{10.72}$$

which is colourless. This last state is an $SU(3)_c$ singlet, while the eight gluons in (10.71) build an $SU(3)_c$ octet. Hence the corresponding gluon should be observable as a free entity, but since gluons are confined akin to quarks this gluon cannot exist. The QCD gauge fields are thus the eight coloured gluons with wave functions (10.71).

The qq and $q\bar{q}$ forces in (colourless) mesons and baryons are attractive. In contrast, one can show that these forces are repulsive in coloured systems. Hence nature allows as bound hadronic systems only quark and antiquark configurations that are colourless, such as qqq, $q\bar{q}$, $\overline{qqq}$, $q\bar{q}q\bar{q}$. Configurations such as q, $qq\bar{q}$ or qq do not exist as hadrons.

Mesons made of two quarks and two antiquarks may have been observed. The most prominent candidate is the $X(3872)$ first reported in high energy pN interactions [25] and then firmly established by several experiments at e^+e^-, $\bar{p}p$ and pp colliders [4]. This state is very narrow (1 MeV or less), has quantum numbers 1^{++} and decays into $J/\psi(1S)\ \pi^+\pi^-$. Hence it should contain $c\bar{c}$ pairs. The most likely structure of this state is a bound state of a $D\overline{D}^*$ meson pair ($c\bar{u}u\bar{c}$ or $c\bar{d}d\bar{c}$) weakly bound by one pion exchange [26].

States decaying into charmonium and a charged pion, such as $X(3900)^\pm \rightarrow J/\psi(1S)\pi^\pm$ or $X(4430)^\pm \rightarrow \psi(2S)\pi^\pm$, have also been observed. These mesons are electrically charged and thus contain $c\bar{c}$ pairs *and* charged pairs made of the lighter quarks (for a review see Eidelman *et al*, p 1240 in [4]).

Since the gauge fields themselves carry the charges (colours) of the strong interaction they can bind by sharing their colours. Figure 10.25(*b*) shows a diagram of a colourless bound state of two gluons, a so-called glueball. This hadron does not contain any quarks but only field quanta of the strong interaction. Calculations based on QCD predict that the lightest glueball (with $J^{PC} = 0^{++}$) lies between 1500–1800 MeV [4]. One candidate, the $f_0(1500)$, was discovered in proton–antiproton annihilation at low energies [27, 28].

The following experimental findings clearly demonstrate that three colours must exist for each quark flavour:

1. Provided that the energy is high enough, $q\bar{q}$ pairs ($u\bar{u}$, $d\bar{d}$, $s\bar{s}$, $c\bar{c}$ and $b\bar{b}$) and lepton pairs (e^+e^-, $\mu^+\mu^-$ and $\tau^+\tau^-$) are produced in e^+e^- annihilation by the virtual photon, see figure 10.14(*b*). Resonances such as the ψ or the Υ series are observed in the e^+e^- cross section. The ratio R of the cross sections for hadron production ($q\bar{q}$ pairs) to that for $\mu^+\mu^-$ pairs at the same incident energy is, first assuming colourless quarks and neglecting the small $\frac{\alpha_S}{\pi}$ term in (7.20):

$$R = \frac{\sigma(e^+e^- \to \text{hadrons})}{\sigma(e^+e^- \to \mu^+\mu^-)} = \frac{\alpha^2 \sum q_i^2}{\alpha^2} = \sum q_i^2. \tag{10.73}$$

The numerator sums over all quark charges q_i. At high energies the masses of the outgoing particles are negligible so that the phase space factors cancel in the numerator and denominator. Above the $b\bar{b}$ threshold (~10 GeV) one finds:

$$R = \left(\frac{2}{3}\right)^2 + \left(-\frac{1}{3}\right)^2 + \left(-\frac{1}{3}\right)^2 + \left(\frac{2}{3}\right)^2 + \left(-\frac{1}{3}\right)^2 = \frac{11}{9}. \tag{10.74}$$

Figure 10.26 displays the measurement of R. The measured cross section for hadrons was divided by the calculated cross section for $e^+e^- \to \gamma \to \mu^+\mu^-$ derived from QED. One observes the excitations of the 1^- resonances, in

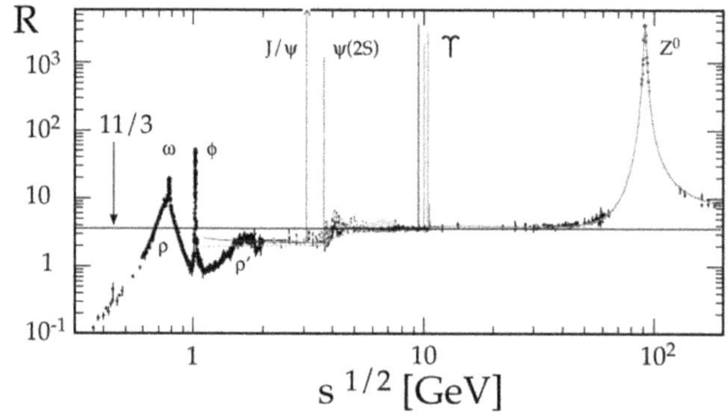

Figure 10.26. Measured cross section for e^+e^- annihilation into hadrons, normalized to $e^+e^- \to \gamma \to \mu^+\mu^-$. From [4].

particular the Υ series around 10 GeV. Above the Υ one finds that $R \simeq 11/3$, in contradiction to the prediction (10.74). However, with three coloured quarks of each flavour one would obtain $R = 11/3$, in agreement with the data.

2. The W^+ boson decays into $e^+\nu_e$, $\mu^+\nu_\mu$ and $\tau^+\nu_\tau$ with measured branching ratios f of $\simeq 11\%$ each, and into hadrons with $f \simeq 66\%$ [4]. Since the W^+ is also expected to decay into $u\bar{d}$ and $c\bar{s}$ with $f = 11\%$ this is only possible if quarks exist in three colours. Note that W^+ decays into $u\bar{s}$, $c\bar{d}$, $u\bar{b}$ and $c\bar{b}$ are 'Cabibbo' suppressed (see section 16.3).

10.5 The top quark

The existence of a sixth quark flavour was predicted from several experimental observations and theoretical considerations. For example, the observed forward–backward asymmetry in $e^+e^- \to b\bar{b}$, measured by the JADE detector at PETRA/DESY (-23% at $\sqrt{s} = 35$ GeV [29]), implied the existence of a partner to the b quark. From the radiative corrections to the W mass (figure 16.25 in section 16.7) the mass of the t quark could be estimated to be around 180 GeV.

The t quark was discovered in 1995 at the Tevatron in $\bar{p}p$ collisions at $\sqrt{s} = 1.8$ TeV [30, 31]. Figure 10.27 shows the main production mechanisms at the Tevatron. The dominant process is the creation of $t\bar{t}$ pairs by one gluon exchange (figure 10.27(*a*)). In contrast, gluon fusion dominates in pp collisions at the LHC (figure 10.27(*b*) and 10.27(*c*)).

The t quark decays almost exclusively into a b quark with the emission of a W^+ boson. Its mean life is extremely short ($\sim 10^{-25}$ s), much shorter than the time scale for hadronization. It can therefore be considered as a nearly free particle. Consider now the production and decay of the pair $t\bar{t} \to (W^+b)(W^-\bar{b})$, leading to two b jets plus leptons or additional jets from W decay (figure 10.28(*a*)). We have seen in the

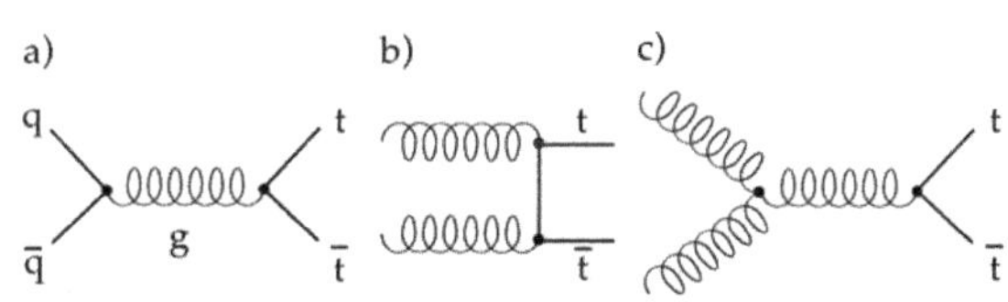

Figure 10.27. t quark production mechanisms in $\bar{p}p$ collisions at high energies. Quark–antiquark annihilation (*a*) dominates with 90% probability at 1.8 TeV, while gluon fusion (*b*) and (*c*) contribute only 10%.

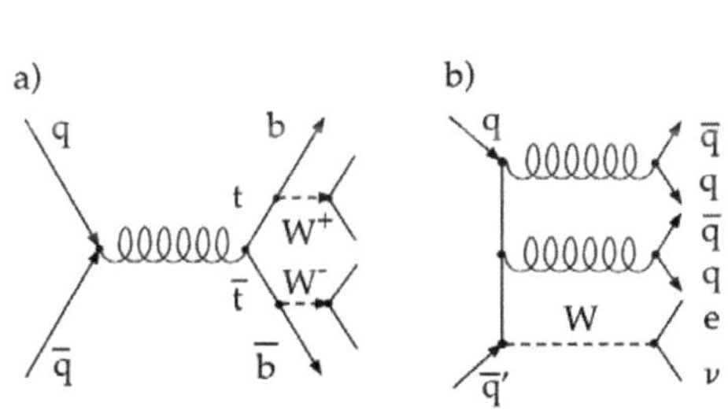

Figure 10.28. (*a*) Feynman graph of $t\bar{t} \to (W^+b)(W^-\bar{b})$ leading to six high energy leptons and jets. (*b*) An example of the QCD background.

Table 10.1. Final state probabilities for the decay of the W^+W^- pair (in units of $\frac{1}{81}$).

	$e^-\bar{\nu}_e$	$\mu^-\bar{\nu}_\mu$	$\bar{u}d$	$\bar{c}s$
$e^+\nu_e$	1	1	3	3
$\mu^+\nu_\mu$	1	1	3	3
$u\bar{d}$	3	3	9	9
$c\bar{s}$	3	3	9	9

previous section that the W^+ decays into $e^+\nu_e$, $\mu^+\nu_\mu$, $\tau^+\nu_\tau$, $c\bar{s}$ (three colours) and $u\bar{d}$ (three colours) with branching ratios of 1/9 each (~11%).

Table 10.1 shows the final state probabilities for W decay. For leptons we limit ourselves to e and μ, which are easier to detect than τ. The probability to generate high energy e^+e^-, $e^\pm\mu^\mp$ or $\mu^+\mu^-$ pairs ('dileptons') from $t\bar{t}$ is $4 \times \frac{1}{81} = 5\%$. Two neutrinos are invisible and hence the kinematics cannot be completely reconstructed. In contrast, the probability to produce a high-energy $e^\pm$ or $\mu^\pm$ and four quark jets from W decays is much larger, $8 \times \frac{3}{81} = 30\%$, and the channel (one missing neutrino) can be fully reconstructed. However, in $\bar{p}p$ collisions the cross section for the production of the t quark ($\simeq$5 pb at 1.8 TeV) is ten orders of magnitude smaller than the total cross section. Hence the four-jet channel is overwhelmed by background from directly produced W bosons and light quark jets (figure 10.28(b)). Therefore the identification of the b quarks from t decays is essential to reduce the background.

Initially, the CDF Collaboration at the Tevatron observed two dilepton events and six events with a lepton (e or μ) and four jets [32]. The b quarks hadronize into b hadrons, such as B mesons which can be detected due to their relatively long lifetimes. One searches for jets containing B mesons with production points well separated from the $p\bar{p}$ collision point. Figure 10.29 (left) shows such an event with a positron and four jets: jets 1 and 4 coming from the decay of b hadrons, jets 2 and 3 from W decay into two quarks. The decay kinematics can be fully reconstructed notwithstanding the missing ν_e. Figure 10.29 (right) shows the distribution of the Wb jet invariant mass. The enhancement is due to the t quark with mass 176 ± 13 GeV [30].

Before concluding this chapter we compare these early data to those obtained recently at the LHC, with a much larger data sample derived from six-jet events in which both W bosons decay into $q\bar{q}$ pairs (see figure 10.28(a)). The probability for this channel is largest: $4 \times \frac{9}{81} = 44\%$ (table 10.1). Figure 10.30 shows the three-jet invariant mass distribution measured with the Compact Muon Solenoid (CMS) detector at 7 TeV (a description of CMS is given in section 17.3). The event selection required two out of six jets to be b jets. The fit led to a t quark mass of 173.49 ± 1.39 GeV [34].

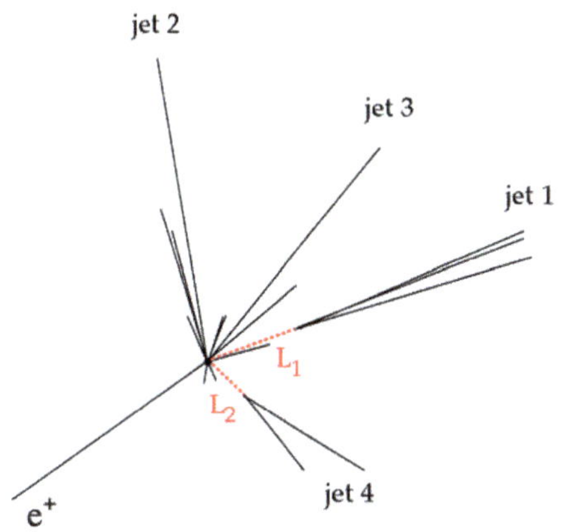

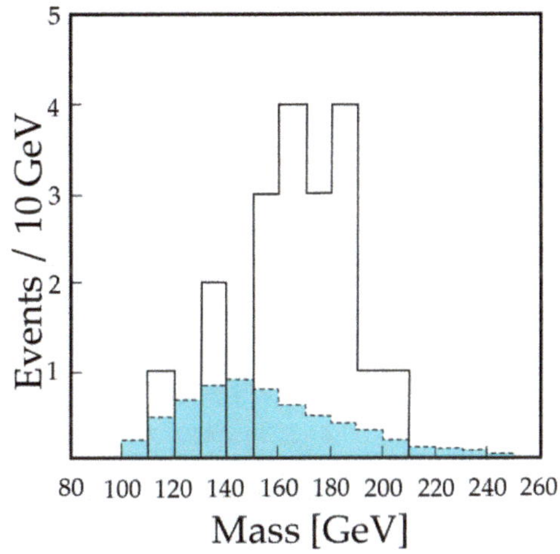

Figure 10.29. Left: production and decay of a $t\bar{t}$ pair seen by CDF at the Tevatron. The $t\bar{t}$ pairs decay into four jets, one positron and one neutrino. Jets 1 and 4 stem from b hadrons which decay at distances $L_1 = 4.5$ mm and $L_2 = 2.2$ mm from the collision point. After [33]. Right: invariant mass distribution showing the t quark signal (19 events). The blue histogram is the background contribution. Adapted from [30].

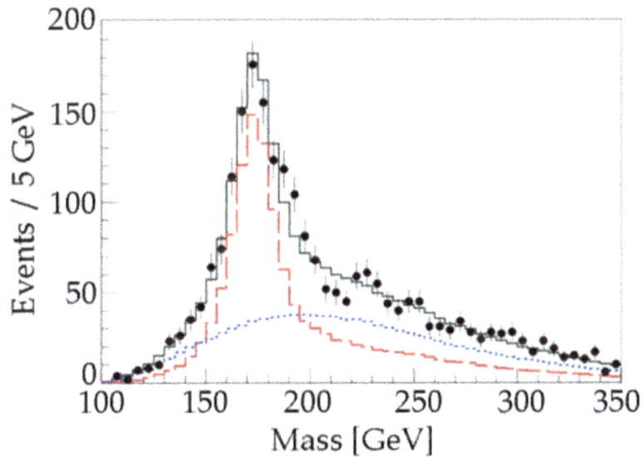

Figure 10.30. Reconstructed t quark mass distribution from the CMS experiment at the LHC (2418 events). The red curve shows the fitted signal contribution and the blue curve the multi-jet background. From [34].

References

[1] Rochester G D and Butler C C 1947 *Nature* **160** 855
[2] Close F E 1979 *An Introduction to Quarks and Partons* (London: Academic)
[3] Amsler C *et al* 1992 *Phys. Lett.* B **294** 451
[4] Olive K *et al* (Particle Data Group) 2014 *Chin. Phys.* C **38** 090001
[5] Barnes V *et al* 1964 *Phys. Rev. Lett.* **12** 204
[6] Flamm D and Schöberl F 1982 *Introduction to the Quark Model of Elementary Particles* (London: Gordon and Breach)
[7] Schindler R H *et al* 1981 *Phys. Rev.* D **24** 78

[8] Alam M S *et al* 1994 *Phys. Rev.* D **50** 43
[9] Cazzoli E G *et al* 1975 *Phys. Rev. Lett.* **34** 1125
[10] Knapp B *et al* 1976 *Phys. Rev. Lett.* **37** 882
[11] Basile M *et al* 1981 *Lett. Nuovo Cimento* **31** 97
[12] Drijard D *et al* 1982 *Phys. Lett.* B **108** 361
[13] Albajar C *et al* 1991 *Phys. Lett.* B **273** 540
[14] Aaij R *et al* (LHCb Collaboration) 2013 *Phys. Rev. Lett.* **110** 182001
[15] Augustin J-E *et al* 1974 *Phys. Rev Lett.* **33** 1406
[16] Boyarski A *et al* 1975 *Phys. Rev. Lett.* **34** 1357
[17] Aubert J J *et al* 1974 *Phys. Rev. Lett.* **33** 1404
[18] Abrams G S *et al* 1974 *Phys. Rev. Lett.* **34** 1181
[19] Partridge R *et al* 1980 *Phys. Rev. Lett.* **45** 1150
[20] Herb S W *et al* 1977 *Phys. Rev. Lett.* **39** 252
[21] Andrews D *et al* 1980 *Phys. Rev. Lett.* **44** 1108
[22] Han K *et al* 1982 *Phys. Rev. Lett.* **49** 1612
[23] Klopfenstein C *et al* 1983 *Phys. Rev. Lett.* **51** 160
[24] Söding P 2010 *Eur. Phys. J.* H **35** 3
[25] Antoniazzi L *et al* 1994 *Phys. Rev.* D **50** 4258
[26] Törnqvist N A 1992 *Phys. Rev. Lett.* **67** 556
[27] Amsler C *et al* 1995 *Phys. Lett.* B **342** 433
[28] Amsler C and Close F E 1996 *Phys. Rev.* D **53** 295
[29] Bartel W *et al* 1984 *Phys. Lett.* B **146** 437
[30] Abe F *et al* 1995 *Phys. Rev. Lett.* **74** 2626
[31] Abachi S *et al* 1995 *Phys. Rev. Lett.* **74** 2632
[32] Abe F *et al* 1994 *Phys. Rev.* D **50** 2966
[33] Bhat P C, Prosper H B and Snyder S S 1998 *Int. J. Mod. Phys.* A **13** 5113
[34] Chatrchyan S *et al* (CMS Collaboration) 2014 *Eur. Phys. J.* C **74** 2758

IOP Publishing

Nuclear and Particle Physics

Claude Amsler

Chapter 11

Conservation laws

Consider the wave function $\psi(\vec{r})$ of a particle, or of a system of particles. The probability density is $|\psi(\vec{r})|^2$ where $\psi(\vec{r})$ is an eigenfunction of the operator H:

$$H\psi = E\psi. \tag{11.1}$$

For closed systems (H independent of t) the time evolution of the system is given by the time-dependent wave function

$$\Psi(\vec{r}, t) = \mathrm{e}^{-\mathrm{i}Ht}\psi(\vec{r}) = \mathrm{e}^{-\mathrm{i}Et}\psi(\vec{r}). \tag{11.2}$$

The expectation value of an operator A (measurement result) is

$$\langle A\rangle(t) = \int \Psi^* A\Psi \mathrm{d}V = \int \psi^* \mathrm{e}^{\mathrm{i}Ht} A \mathrm{e}^{-\mathrm{i}Ht}\psi \mathrm{d}V, \tag{11.3}$$

since H is Hermitian: $H^\dagger = H$. Hermiticity ensures that the probability density is conserved as a function of time, and that the energy spectrum is real. The time evolution of $\langle A\rangle$ is

$$\frac{\mathrm{d}\langle A\rangle}{\mathrm{d}t} = \int \psi^* \mathrm{e}^{\mathrm{i}Ht}\left(\mathrm{i}HA + \frac{\partial A}{\partial t} - \mathrm{i}AH\right)\mathrm{e}^{-\mathrm{i}Ht}\psi \mathrm{d}V. \tag{11.4}$$

The measured value is constant in time (*conserved*) when

$$\boxed{\frac{\mathrm{d}\langle A\rangle}{\mathrm{d}t} = 0 \Leftrightarrow [H, A] = 0}, \tag{11.5}$$

as long as A does not depend explicitly on t. The measured value $\langle A\rangle$ is conserved when the corresponding operator A commutes with H, in which case a set of common eigenfunctions of A and H can be found. If a transformation $f(A)$ also

 © IOP Publishing Ltd 2015

commutes with H, the wave function $\psi' = f(A)\psi$ also qualifies to describe the physical system. This follows from the commutativity $[f(A), H] = 0$:

$$H\psi' = Hf(A)\psi = f(A)H\psi = f(A)E\psi = E\psi'. \quad (11.6)$$

In the next section we shall consider continuous transformations f (such as a spatial translation or a rotation) that are associated with an operator A (momentum and angular momentum, respectively). If H is invariant (*symmetric*) under such transformations, H commutes with A and the measured value $\langle A\rangle$ is conserved. Sometimes A itself is a symmetry ($f \equiv 1$) under which H is invariant (e.g. parity and C parity). Then the wave functions ψ are to be sought among the eigenfunctions of A [1] and the measured value $\langle A\rangle$ is conserved[1]. Let us summarize:

> To each symmetry $f(A)$ (so that both ψ and $\psi' = f(A)\psi$ describe the physical system) corresponds an operator A with conserved expectation value $\langle A\rangle$. The wave functions ψ (or ψ') of the system are to be found among the eigenfunctions of A (or $f(A)$).

11.1 Momentum conservation

As a first example, let us deal with *spatial translations*. The wave function of a system $\psi(x)$ translated along the x-axis by the infinitesimal amount $\mathrm{d}x$ is

$$\psi'(x) \equiv \psi(x + \mathrm{d}x) \simeq \psi(x) + \mathrm{d}x \cdot \frac{\partial\psi}{\partial x} = \left(1 + \mathrm{d}x \cdot \frac{\partial}{\partial x}\right)\psi(x). \quad (11.7)$$

For a finite translation Δx the operation on the right-hand side can be repeated n times with $\Delta x = n\mathrm{d}x$:

$$\psi'(x) = \psi(x + \Delta x) = \lim_{n\to\infty}\left(1 + \frac{\Delta x}{n} \cdot \frac{\partial}{\partial x}\right)^n \psi(x). \quad (11.8)$$

Introducing the momentum operator

$$p_x \equiv \frac{1}{\mathrm{i}}\frac{\partial}{\partial x} \quad (11.9)$$

leads to

$$\boxed{\psi'(x) = \overbrace{\mathrm{e}^{\mathrm{i}p_x\Delta x}}^{f(p_x)}\psi(x)}. \quad (11.10)$$

The unitary operator $f(p_x)$ is the *representation* of a translation. If $\psi'(x)$ also describes the physical system, H commutes with p_x ($[H, p_x] = 0$) and $\langle p_x\rangle$ is conserved. From the definition of the momentum operator (11.9) this occurs when H does not

[1] To be precise, there exists a set of orthogonal wave functions which are eigenfunctions of H and A. Eigenfunctions of A are also eigenfunctions of H in the absence of degeneracies.

depend on x. Thus, conservation of *momentum* is associated with *symmetry* under space translations. The wave function of a free particle is symmetric under translations and therefore its momentum is conserved.

11.2 Angular momentum

As a second example, consider a *rotation* by an infinitesimal angle $\mathrm{d}\phi$ about the z-axis. The wave function of the rotated system is

$$\begin{aligned}\psi'(\vec{r}) &\equiv \psi(x + \mathrm{d}x, y + \mathrm{d}y, z) \simeq \psi + \frac{\partial\psi}{\partial x} \cdot \mathrm{d}x + \frac{\partial\psi}{\partial y} \cdot \mathrm{d}y \\ &= \left[1 + \left(x\frac{\partial}{\partial y} - y\frac{\partial}{\partial x}\right)\mathrm{d}\phi\right]\psi(\vec{r}),\end{aligned} \tag{11.11}$$

where $\mathrm{d}x$ and $\mathrm{d}y$, expressed in polar coordinates, are

$$\begin{aligned}&\mathrm{d}x = -y\mathrm{d}\phi, \qquad \mathrm{d}y = x\mathrm{d}\phi \\ \text{with} \qquad &x = \rho\cos\phi, \qquad y = \rho\sin\phi.\end{aligned} \tag{11.12}$$

Introducing the angular momentum operator

$$J_z \equiv (\vec{r} \times \vec{p})_z = \frac{1}{\mathrm{i}}\left(x\frac{\partial}{\partial y} - y\frac{\partial}{\partial x}\right) \tag{11.13}$$

gives

$$\psi'(\vec{r}) = (1 + \mathrm{i}J_z \cdot \mathrm{d}\phi)\psi(\vec{r}). \tag{11.14}$$

A finite rotation $\Delta\phi$ is obtained by applying the operation (11.14) n times with $\Delta\phi = n\mathrm{d}\phi$:

$$\psi'(\vec{r}) = \lim_{n\to\infty}\left(1 + \mathrm{i}J_z \cdot \frac{\Delta\phi}{n}\right)^n\psi(\vec{r}), \tag{11.15}$$

so that

$$\boxed{\psi'(\vec{r}) = \overbrace{\mathrm{e}^{\mathrm{i}J_z\Delta\phi}}^{f(J_z)}\psi(\vec{r})}\,. \tag{11.16}$$

The unitary operator $f(J_z)$ is the *representation* of a rotation around the z-axis. In the case that $\psi'(\vec{r})$ also describes the physical system, H commutes with J_z ($[H, J_z] = 0$) and $\langle J_z\rangle = M$ is conserved. Since $J_z = -\mathrm{i}\frac{\partial}{\partial\phi}$ this occurs when H does not depend on ϕ. The conservation of the third component of the *angular momentum* is associated with *rotational symmetry* around the z-axis (figure 11.1).

Figure 11.1. A photograph of a snow crystal illustrating the rotational symmetry of order 6. Image credit: http://www.its.caltech.edu/~atomic/snowcrystals/.)

The wave functions $\psi(\vec{r})$ of the system are then to be found among the eigenfunctions of J_z. The eigenfunctions are degenerate (same energy)[2], if in addition

$$\left[H, J_x\right] = 0, \qquad \left[H, J_y\right] = 0, \tag{11.17}$$

i.e. for *spherical symmetry*, that is invariance under rotations around any axis. The transformation then reads:

$$\boxed{\psi'\left(\vec{r}\right) = e^{i\vec{J}\cdot\vec{\chi}}\psi\left(\vec{r}\right)}, \tag{11.18}$$

where $\vec{\chi}$ is the direction of the rotation axis and $|\vec{\chi}|$ the rotation angle.

A *phase* transformation of the form (11.10), (11.16) or (11.18) is called *global*, since the phases Δx, $\Delta\phi$ and $\vec{\chi}$ do not depend on the coordinates $\vec{r}$ and time t. Invariance under a global phase transformation is associated with a *symmetry*. We will show in section 16.4 that in gauge theories a *local* phase transformation $\vec{\chi}\left(\vec{r}, t\right)$ can be compensated by a gauge transformation of the fields, so that the equation of motion of the system (e.g. the Schrödinger or Dirac equation) remains invariant.

11.3 Parity

We now deal with a discrete operation in which the operator A itself describes the symmetry ($f \equiv 1$). Let us define the *space inversion*

$$\psi'\left(\vec{r}\right) = P\psi\left(\vec{r}\right) \equiv \psi\left(-\vec{r}\right). \tag{11.19}$$

[2] The degeneracy is lifted by a magnetic field. For example, the energy of a free electron with angular momentum $\vec{J} = \vec{s}$ is $H = H_0 + \frac{ge}{2m}s_z B$, where the z-axis is chosen along the direction of $\vec{B}$ and $g = 2$. Then $J_x = s_x$ and $J_y = s_y$ do no longer commute with H, spherical symmetry is broken and J_x, J_y are not conserved (precession of the spin about the z-axis). However, $J_z = s_z$ still commutes with H, and J_z is conserved.

If H commutes with P, i.e. $[H, P] = 0$ hence $H(\vec{r}) = H(r)$, the wave functions of the system are the eigenfunctions ψ of P:

$$P\psi(\vec{r}) = \pm\psi(\vec{r}), \tag{11.20}$$

since $P^2\psi(\vec{r}) = \psi(\vec{r})$. The *quantum number* $P = \pm 1$ in (11.20) is the *parity* of the system which is conserved.

As an example, consider the hydrogen atom. The Coulomb potential is spherically symmetric and hence invariant under space inversions. Therefore

$$V(\vec{r}) = V(-\vec{r}) \Rightarrow H(\vec{r}) = H(-\vec{r}) = \frac{\vec{p}^2}{2m_e} + V(r). \tag{11.21}$$

The wave function of the electron (eigenfunction of H)

$$\psi(r, \theta, \phi) = \chi(r)\, Y_m^\ell(\theta, \phi) \tag{11.22}$$

is also an eigenfunction of the parity operator with eigenvalue $(-1)^\ell$. Indeed, for spherical harmonics

$$PY_m^\ell(\theta, \phi) = Y_m^\ell(\pi - \theta, \pi + \phi) = (-1)^\ell Y_m^\ell(\theta, \phi). \tag{11.23}$$

Systems such as free atoms, nuclei and hadrons, for which $H(\vec{r}) = H(-\vec{r})$, have well-defined parities[3].

Under space inversion P the momentum and the spin change as follows:

$$\vec{p} \to -\vec{p}, \qquad \vec{s} \to \vec{s}. \tag{11.24}$$

Space inversion is equivalent to the combined operation of a reflection R from a plane mirror and a rotation of 180° about the z-axis perpendicular to the mirror (figure 11.2(*a*)):

$$P = R\, e^{iJ_z\pi}. \tag{11.25}$$

With J_z conserved the system is rotationally invariant with respect to the z-axis. Conservation of parity is then equivalent to invariance under reflections. The components of the momentum and spin parallel (//) and orthogonal ($\perp$) to the mirror change as follows (figure 11.2(*b*))

$$\boxed{p_{//} \to p_{//}, \qquad p_\perp \to -p_\perp, \qquad s_{//} \to -s_{//}, \qquad s_\perp \to s_\perp}. \tag{11.26}$$

Parity is conserved in strong and electromagnetic interactions since the transition operator from the initial to the final state commutes with P (as we shall see in section 13.3, parity is not conserved in weak interactions). This means that the cross section of a strong or electromagnetic process is equal to that of the space inverted one.

As an example, consider elastic scattering of spin-0 off polarized spin-$\frac{1}{2}$ particles, such as pion scattering off polarized protons ($\pi^\pm p \uparrow \to \pi^\pm p$). Polarized protons have

[3] In contrast, the Hamiltonian of the hydrogen atom in the electric field (Stark effect) comprises the additional term $-eEz$ (z being in the direction of the field) which changes sign under parity, and which therefore does not commute with P. The degeneracy of the first excited levels $(n, \ell, m) = (2, 0, 0)$ and $(2, 1, 0)$ is lifted by the electric field and the corresponding wave functions are mixtures of the $\ell = 1$ and 2 eigenstates of parity [2]. An electric dipole moment is induced and ℓ is no longer a 'good' quantum number.

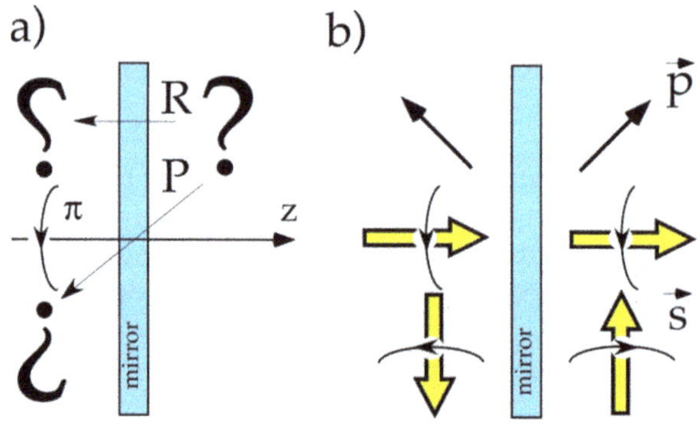

Figure 11.2. (*a*) The parity transformation P can be decomposed into a mirror reflection R and a rotation of 180° about the axis perpendicular to the mirror. (*b*) Mirror reflection of the momentum and spin vectors.

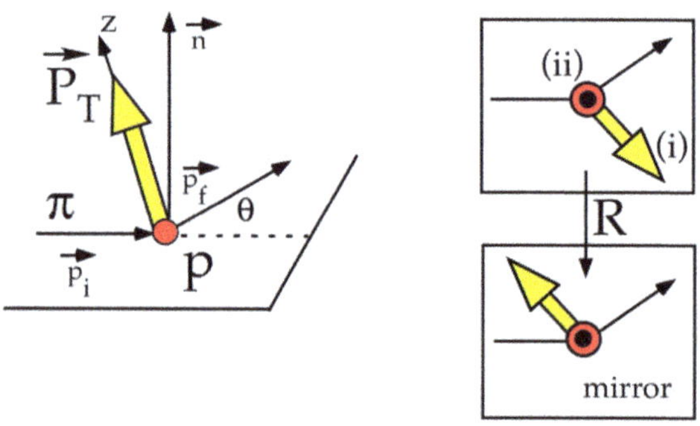

Figure 11.3. Left: $\pi^{\pm}p\uparrow\rightarrow\pi^{\pm}p$ scattering off polarized protons. Right: the reaction and mirror process. The cross section does not depend on the polarization component in the scattering plane (i), while no such statement can be made from R invariance for a component perpendicular to the scattering plane (ii).

their spins $\vec{s}$ aligned along a given direction (figure 11.3, left). With $N_{\uparrow}$ spins lying in the z-direction and $N_{\downarrow}$ spins in the $-z$-direction, the target *polarization* is

$$\vec{P}_T \equiv \frac{N_{\uparrow} - N_{\downarrow}}{N_{\uparrow} + N_{\downarrow}} \cdot \frac{\vec{s}}{|\vec{s}|}. \tag{11.27}$$

If the polarization vector lies in the scattering plane then the differential cross section for scattering under the angle θ does not depend on $\vec{P}_T$, since the mirrored reaction (with reversed polarization) occurs with the same cross section (arrow labelled (i) in figure 11.3, right). This is a consequence of P conservation or R invariance.

However, if the polarization is perpendicular to the scattering plane, then the corresponding mirrored reaction is not the reaction with reversed polarization (arrow labelled (ii) in figure 11.3, right) and the cross sections do not need to be equal. In fact, the differential cross section for $\pi^{\pm}p\uparrow\rightarrow\pi^{\pm}p$ reads

$$\frac{\mathrm{d}\sigma}{\mathrm{d}\Omega} = \left.\frac{\mathrm{d}\sigma}{\mathrm{d}\Omega}\right|_{P_{\mathrm{T}}=0} \left[1 + a_{\pm}(\theta)\vec{P}_T \cdot \vec{n}\right], \quad \text{where } \vec{n} = \frac{\vec{p}_i \times \vec{p}_f}{\left|\vec{p}_i \times \vec{p}_f\right|}. \tag{11.28}$$

The unit vector $\vec{n}$ is perpendicular to the scattering plane. The initial and final momenta of the pion, $\vec{p}_i$ and $\vec{p}_f$, respectively, and the scattering angle θ refer to the centre-of-mass system of the reaction. The functions $a_+(\theta)$ and $a_-(\theta)$ are the *analysing power* for $\pi^+p \to \pi^+p$ and $\pi^-p \to \pi^-p$ scattering, respectively. Note that the rates for scattering to the left and scattering to the right ($\vec{n} \to -\vec{n}$) are not necessarily equal when the proton polarization $\vec{P}_T$ has a component parallel to $\vec{n}$. The numbers N_L and N_R of pions scattered under the angle θ to the left and to the right, are

$$N_L = N_0\big(1 + a_\pm(\theta)P_T\big), \tag{11.29}$$

$$N_R = N_0\big(1 - a_\pm(\theta)P_T\big), \tag{11.30}$$

for protons with polarization $\vec{P}_T$ orthogonal to the scattering plane. The analysing power is then

$$a_\pm(\theta) = \frac{1}{P_T}\left(\frac{N_L - N_R}{N_L + N_R}\right). \tag{11.31}$$

Figure 11.4 shows the angular dependence of $a_+(\theta)$ measured at PSI in Switzerland with 236 MeV pions.

Protons cannot be polarized statically without enormous magnetic fields (or very low temperatures). At $T = 0.5$ K in a $B = 2.5$ T field the polarization of free protons is only $P_T = \tanh(\mu_p B/k_B T) = 0.5\%$. In contrast, the polarization of electrons is 99.7%, thanks to their much larger magnetic moments. Polarized protons are obtained by using a butanol target ($CH_3(CH_2)_2CH_2OH$) operated at 0.5 K in a 2.5 T field [4]. Typically, $P_T = 70\%$ is achieved by means of dynamic nuclear polarization [5] in which the polarization of electrons is transferred to protons by the dipole–dipole interaction.

As a second example of parity conservation, let us deal with an electromagnetic process. The excited state of fluorine $^{19}F^*$ ($J^P = \frac{1}{2}^-$) decays into the ground state ($J^P = \frac{1}{2}^+$) by emitting a 110 keV γ (E1-transition). Let θ be the angle between the

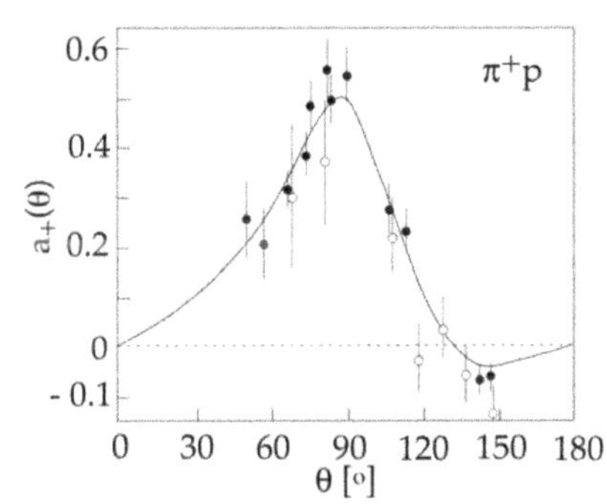

Figure 11.4. Analysing power $a_+(\theta)$ in $\pi^+p\uparrow$ scattering with $T_\pi = 236$ MeV pions in the laboratory. The scattering angle θ is in the centre-of-mass system. After [3].

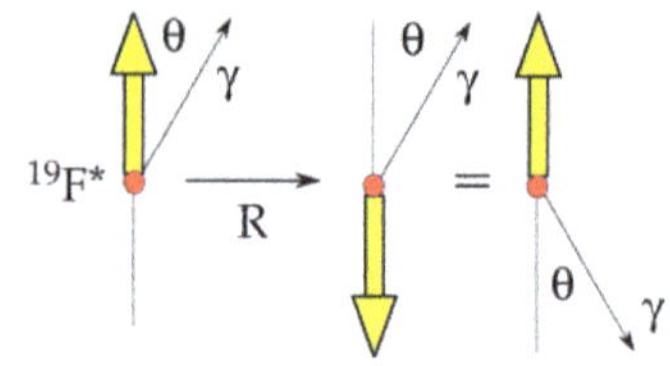

Figure 11.5. γ-decay of polarized $^{19}F^*$ and its mirror image.

spin of $^{19}F^*$ and the γ-direction. The emission probability is not isotropic, but is equal under the angle θ and $\pi - \theta$ due to parity conservation (figure 11.5). Measuring with polarized $^{19}F^*$ nuclei the forward rate R_F ($\theta < 90°$) and comparing with the backward rate R_B ($\theta > 90°$) leads to the (tiny) asymmetry [6]

$$\Delta = \frac{R_F - R_B}{R_F + R_B} = (-18 \pm 9) \times 10^{-5}, \tag{11.32}$$

while parity conservation would require Δ to vanish. The small parity violating effect stems from the weak contribution to the electromagnetic interaction (section 13.3).

It is important to note that parity is a *multiplicative* quantum number: consider for instance a system with wave function ψ, composed of two subsystems a and b with parities P_a and P_b and with relative angular momentum ℓ. The parity of the wave function $\psi = \psi_a \psi_b$ is

$$\boxed{P = (-1)^\ell P_a P_b}. \tag{11.33}$$

11.4 Internal parity

Consider the strong interaction process

$$pp \to pn\pi^+ \tag{11.34}$$

in which a pion is produced. Since parity is conserved between the initial and final states we need to introduce *internal* (or intrinsic) parities for all hadrons, in analogy to the electric charge which is also conserved.

As a consequence of the Dirac equation (section 15.4), the relative parity between a fermion and its antifermion is negative. The experimental proof will be described in section 11.6 below. We choose $P(f) = +1$ for the fermion and $P(\bar{f}) = -1$ for the corresponding antifermion. Since quark flavours do not change in strong or in electromagnetic interactions, the relative parity between quarks q (or antiquarks $\bar{q}$) of different flavours cannot be established. For quarks we therefore *define* the internal parity

$$P(q) \equiv 1 \Rightarrow P(\bar{q}) = -1, \tag{11.35}$$

independently of flavour. Hence one obtains for the ground state baryons, following (11.33):

$$P(p) = +1, \qquad P(n) = +1, \qquad P(\Lambda) = +1, \tag{11.36}$$

and $P = -1$ for antibaryons. The internal parity of a meson with angular momentum ℓ between the quarks is

$$\boxed{P(q\bar{q}) = -(-1)^{\ell}}. \tag{11.37}$$

Let us now determine the internal parity of the pion from the reaction

$$\pi^- d \rightarrow nn, \tag{11.38}$$

with pions stopping in a liquid deuterium (^2H) target. We shall assume that (i) the spin $J(\pi) = 0$ has already been determined experimentally (section 12.2.1) and (ii) that the spin of the deuteron is $J(d) = 1$. We also know that the internal parity of the deuteron must be $P(d) = +1$, this state being a bound system of a proton and a neutron with even angular momentum (dominantly zero).

The stopped pion is captured by an ^2H atom. By observing the x-rays emitted while the pion cascades down to the lower energy levels, one finds that capture by the deuterium nucleus occurs mainly from s levels[4], hence with angular momentum $\ell = 0$. Thus in the initial state $|i\rangle$ the total angular momentum is $J_i = 1$ and the parity

$$P_i = (-1)^{\ell} P(\pi) P(d) - P(\pi). \tag{11.39}$$

We now also determine the parity of the final state $|f\rangle$. The wave function of the neutron pair is composed of the spin wave function $|sm\rangle$ (spin $s = 0$ or 1, projection m) and the orbital wave function $\phi_L(n, n)$:

$$\psi(n, n) = |sm\rangle \phi_L(n, n), \tag{11.40}$$

where L is the angular momentum in the two-neutron pair. The wave function $|jm\rangle$ for a system with angular momentum j can be written as the linear superposition of the wave functions of the two subsystems with angular momenta j_1 and j_2,

$$|jm\rangle = \sum_{m_1, m_2} \langle jm | j_1 j_2 m_1 m_2 \rangle \, | j_1 j_2 m_1 m_2 \rangle, \tag{11.41}$$

or vice versa, using the orthogonality condition,

$$| j_1 j_2 m_1 m_2 \rangle = \sum_j \langle jm | j_1 j_2 m_1 m_2 \rangle \, |jm\rangle. \tag{11.42}$$

The quantum numbers j and m vary between $|j_1 - j_2|$ and $j_1 + j_2$ and between $-j$ and $+j$ respectively. The sum extends over all m_1 and m_2 such that

$$m = m_1 + m_2. \tag{11.43}$$

The Clebsch–Gordan coefficients $\langle jm | j_1 j_2 m_1 m_2 \rangle$ are tabulated (figure 11.6). For our neutron pair $j_1 \equiv s_1 = \frac{1}{2}$ and $j_2 \equiv s_2 = \frac{1}{2}$. To make the equations easier to read,

[4] The pion is captured in the $n = 16$ level of pionic deuterium by ejecting the shell electron. While colliding with ^2H atoms the relatively small pionic atom experiences the neighbouring electric fields which mix its atomic ℓ levels (Stark effect). The finite probability to be in the s states leads to the prompt absorption of the pion by the deuterium nucleus, thus suppressing transitions to the lowest n levels. This effect is particularly pronounced at high densities, that is in liquid deuterium.

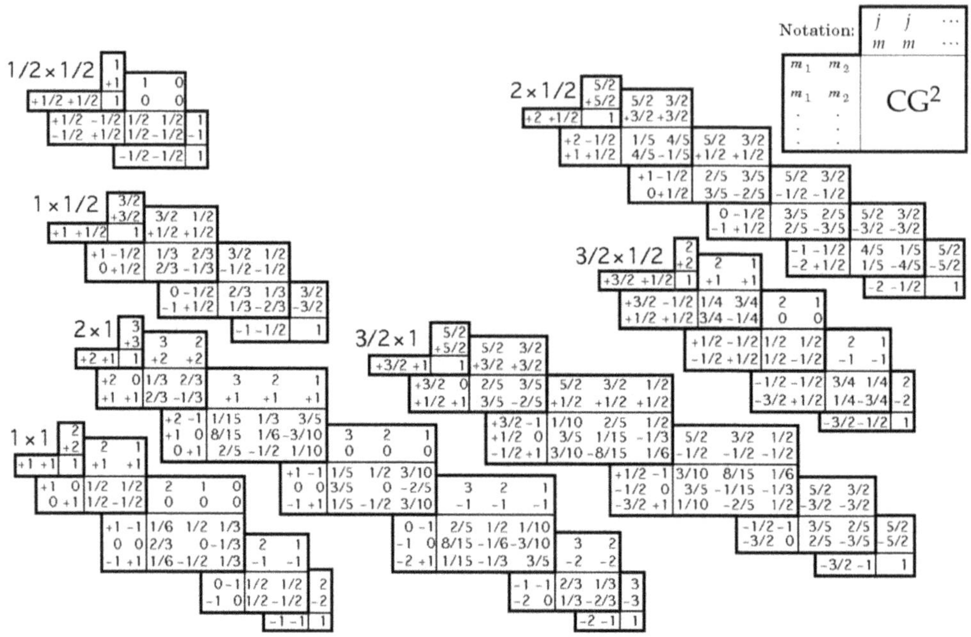

Figure 11.6. Clebsch–Gordan (CG) coefficients (squared) for the couplings $j_1 \times j_2$. The notation '$\mathrm{CG}^2 = -1/2$' means that $\mathrm{CG} = -1/\sqrt{2}$. After [7], where further CG coefficients can be found.

let us simply write $|j_1 j_2 m_1 m_2\rangle \equiv |m_1 m_2\rangle$. We then obtain for $j \equiv s = 1$ the three spin-triplet wave functions

$$
\begin{aligned}
|1+1\rangle &= \left\langle 1+1 \middle| \frac{1}{2}\frac{1}{2}\frac{1}{2}\frac{1}{2} \right\rangle \left| \frac{1}{2}\frac{1}{2} \right\rangle, \\
|1\;\;0\rangle &= \left\langle 1\,0 \middle| \frac{1}{2}\frac{1}{2}\frac{1}{2} - \frac{1}{2} \right\rangle \left| \frac{1}{2} - \frac{1}{2} \right\rangle + \left\langle 10 \middle| \frac{1}{2}\frac{1}{2} - \frac{1}{2}\frac{1}{2} \right\rangle \left| -\frac{1}{2}\frac{1}{2} \right\rangle, \\
|1-1\rangle &= \left\langle 1-1 \middle| \frac{1}{2}\frac{1}{2} - \frac{1}{2} - \frac{1}{2} \right\rangle \left| -\frac{1}{2} - \frac{1}{2} \right\rangle,
\end{aligned}
\tag{11.44}
$$

or by using the table in figure 11.6:

$$
\begin{aligned}
|1+1\rangle &= \left| \frac{1}{2}\frac{1}{2} \right\rangle, \\
|1\;\;0\rangle &= \frac{1}{\sqrt{2}} \left(\left| \frac{1}{2} - \frac{1}{2} \right\rangle + \left| -\frac{1}{2}\frac{1}{2} \right\rangle \right), \\
|1-1\rangle &= \left| -\frac{1}{2} - \frac{1}{2} \right\rangle.
\end{aligned}
\tag{11.45}
$$

For $j \equiv s = 0$ the spin-singlet wave function is

$$|00\rangle = \left\langle 00 \middle| \frac{1}{2}\frac{1}{2}\frac{1}{2} - \frac{1}{2} \right\rangle \left| \frac{1}{2} - \frac{1}{2} \right\rangle + \left\langle 00 \middle| \frac{1}{2}\frac{1}{2} - \frac{1}{2}\frac{1}{2} \right\rangle \left| -\frac{1}{2}\frac{1}{2} \right\rangle, \qquad (11.46)$$

or

$$|00\rangle = \frac{1}{\sqrt{2}}\left(\left| \frac{1}{2} - \frac{1}{2} \right\rangle - \left| -\frac{1}{2}\frac{1}{2} \right\rangle \right). \qquad (11.47)$$

The triplet wave functions are therefore *symmetric* under the permutation of the two neutrons, and the singlet wave function is *antisymmetric*. For the orbital wave function $\phi_L(n, n)$ in (11.40) the permutation amounts to a space inversion (parity transformation) under which

$$P\phi(n, n) = (-1)^L \phi(n, n). \qquad (11.48)$$

Hence ϕ is symmetric for even L, and antisymmetric for odd L. Since with identical fermions the total wave function (11.40) must be antisymmetric one obtains the selection rule

$$L \text{ even} \Rightarrow s = 0, \qquad L \text{ odd} \Rightarrow s = 1. \qquad (11.49)$$

The first possibility ($s = 0$) is excluded by angular momentum conservation $J_f = J_i = 1$ in reaction (11.38). Hence L is odd ($L = 1$) and the parity of the final state is

$$P_f = -1. \qquad (11.50)$$

Parity conservation imposes that $P_f = P_i = -1$. According to (11.39) the parity of the pion is negative:

$$\boxed{P(\pi) = -1}. \qquad (11.51)$$

In section 12.2.1 we show experimentally that the spin of the pion is $J(\pi) = 0$. With negative parity (11.37) then implies that the angular momentum of the $q\bar{q}$ pair must be $\ell = 0$, and that the $q\bar{q}$ pair is in the singlet state. The pion is therefore an ${}^1S_0$ $q\bar{q}$ state. In section 11.6 we show experimentally that the internal parity of the π^0 meson is also negative.

As a second example of parity conservation, let us determine experimentally the internal parity of the negative kaon. Consider the capture of a stopped K^- in helium:

$$K^- \, {}^4\text{He} \to {}^4\text{H}_\Lambda \, \pi^0, \qquad {}^4\text{H}_\Lambda \to {}^4\text{He} \, \pi^-. \qquad (11.52)$$

A proton is replaced by a Λ hyperon in the ${}^4\text{H}_\Lambda$ nucleus ($K^- p \to \Lambda$). This kind of *hypernucleus* is indeed observed [8]. From the isotropic angular distribution of the emitted π^- in the ${}^4\text{H}_\Lambda$ decay one infers that no angular momentum is carried away. Since the spin of ^{4}He is zero, the total angular momentum in the decay is also zero, hence the hypernucleus ${}^4\text{H}_\Lambda$ is spinless. In section 12.3.1 we will show experimentally that the spin of the kaon is also zero. Since all particles have spin zero, the orbital angular momentum L is equal to the total angular momentum which is conserved between the initial and final states. The parity of the initial state (K^{-4}He) is

$$P_i = (-1)^{L_i} P(K) \underbrace{P(^4\text{He})}_{+1}, \tag{11.53}$$

and that of the final state ($^4\text{H}_\Lambda \pi^0$)

$$P_f = (-1)^{L_f} \underbrace{P(^4\text{H}_\Lambda)}_{+1} \underbrace{P(\pi^0)}_{-1}, \tag{11.54}$$

since $P(\Lambda) = +1$. Therefore with $P_i = P_f$ and $L_i = L_f$ the internal parity of the kaon is

$$\boxed{P(K) = -1}, \tag{11.55}$$

independent of L_i. According to (11.37) the kaon is also a $^1S_0$ $q\bar{q}$ state.

Hadrons are classified according to their spins and parities. The $^1S_0$ mesons (π, η, K, etc) with $J^P = 0^-$ are dubbed *pseudoscalar* mesons and the $^3S_1$ mesons (ρ, ω, ϕ, etc) with $J^P = 1^-$ are called *vector* mesons[5]. We have seen in section 10.3 that vector mesons are produced in e^+e^- annihilation by virtual photons. As a consequence of parity conservation in the electromagnetic interaction the internal parity of the photon is negative: $J^P(\gamma) = 1^-$.

11.5 Gauge invariance

Let $\psi(\vec{r}, t)$ be the wave function of a charged particle with mass m and charge $q = Qe$. Because of charge conservation the wave function

$$\psi'(\vec{r}, t) = e^{iq\chi}\psi(\vec{r}, t) \tag{11.56}$$

(with χ = constant) qualifies equally to describe the physical system. We now extend this statement to a *local* phase transformation:

$$\psi'(\vec{r}, t) = e^{iq\chi(\vec{r}, t)}\psi(\vec{r}, t). \tag{11.57}$$

In general, in the absence of electromagnetic fields $\psi'(\vec{r}, t)$ does not describe the physical system, that is, the free Schrödinger equation is not invariant under the transformation (11.57). However, the *field-dependent* Schrödinger equation

$$\frac{(\vec{p} - q\vec{A})^2}{2m}\psi(\vec{r}, t) = (H - q\phi)\psi(\vec{r}, t) \tag{11.58}$$

[5] The orbital excitations with $\ell > 0$ (figure 10.4) are the scalar ($^3P_0$: $J^P = 0^+$), axial vector ($^{1,3}P_1$: $J^P = 1^+$) and tensor ($^3P_2$: $J^P = 2^+$) mesons.

($\vec{A}$ = vector potential, ϕ = scalar potential of the electromagnetic field) stays invariant under the transformation (11.57), provided that the *gauge transformation*

$$\vec{A} \to \vec{A}' = \vec{A} + \mathrm{grad}\chi, \tag{11.59}$$

$$\phi \to \phi' = \phi - \frac{\partial\chi}{\partial t}, \tag{11.60}$$

is performed simultaneously in the Schrödinger equation (the proof is left as an exercise for the reader). Invariance therefore requires the presence of an electromagnetic field (that of the photon) to which the charged particle couples. With the definitions

$$\vec{B} \equiv \mathrm{rot}\vec{A}, \tag{11.61}$$

$$\vec{E} \equiv -\mathrm{grad}\phi - \frac{\partial\vec{A}}{\partial t}, \tag{11.62}$$

Maxwell's field equations are (in Heaviside–Lorentz units with $\epsilon_0 = \mu_0 = 1$):

$$\mathrm{div}\vec{B} = 0, \tag{11.63}$$

$$\mathrm{rot}\vec{E} = -\frac{\partial\vec{B}}{\partial t}, \tag{11.64}$$

$$\mathrm{div}\vec{E} = \rho, \tag{11.65}$$

$$\mathrm{rot}\vec{B} = \vec{j} + \frac{\partial\vec{E}}{\partial t}. \tag{11.66}$$

These equations are invariant under the gauge transformation (11.59), (11.60) since $\vec{E}' = \vec{E}$ and $\vec{B}' = \vec{B}$. (Note that Maxwell's equations apply to massless photons while Proca's equations, introduced in section 16.5, apply to massive photons and are not gauge invariant.) Summarizing:

> A global phase transformation is related to a symmetry. A *gauge theory* is a theory in which the physical system is also invariant under the corresponding local phase transformation, while the simultaneous gauge transformation of the fields leaves the field equations invariant. The gauge bosons associated with the fields are massless.

QED (electromagnetic interaction) and QCD (strong interaction) are gauge theories. The compensating fields are the photon for the former, and the eight massless gluons for the latter. The gauge theory of the weak interaction will be discussed in chapter 16. The problem with the massive W and Z^0 gauge bosons is circumvented by introducing the Higgs field (section 16.7).

Gauge theories, especially QED, are extremely successful in describing physical phenomena. Therefore, the conservation of baryon number B (invariance under the global phase transformation $\exp[\mathrm{i}B\chi]$) also suggests invariance under the corresponding local phase transformation, which then requires compensating fields coupling to the baryon number. Such a *fifth force*, which is sensitive to the mass number A, has been sought so far without success (e.g. by Eötvös-type torsion balance experiments). Hence baryon number conservation may not be an exact symmetry after all. Consequently, the proton may have a finite lifetime. Similar arguments apply to the conservation of lepton number.

11.6 The helicity of the photon

We now assume that the mass of the photon is zero and choose a particular gauge in which

$$\mathrm{div}\vec{A} \equiv 0. \tag{11.67}$$

This can be achieved by choosing $\chi(\vec{r}, t)$ in (11.59) so that

$$\Delta\chi = -\mathrm{div}\vec{A} \Rightarrow \mathrm{div}\vec{A}' \equiv 0. \tag{11.68}$$

In this *Coulomb gauge* one obtains using (11.62) and Maxwell's third equation (11.65)

$$\mathrm{div}\vec{E} = -\Delta\phi = \rho \Rightarrow \phi = \int \frac{\rho(\vec{r}', t)}{|\vec{r} - \vec{r}'|}\mathrm{d}V', \tag{11.69}$$

namely Coulomb's law. On the other hand, Maxwell's fourth equation (11.66) leads with (11.61) to

$$\begin{aligned}\mathrm{rot}\vec{B} &= \mathrm{rot}\ \mathrm{rot}\vec{A} = \overbrace{\mathrm{grad}\ \mathrm{div}\ \vec{A}}^{=0} - \mathrm{div}\ \mathrm{grad}\vec{A} = \vec{j} + \frac{\partial\vec{E}}{\partial t} \\ &= \vec{j} - \mathrm{grad}\left(\frac{\partial\phi}{\partial t}\right) - \frac{\partial^2\vec{A}}{\partial t^2}.\end{aligned} \tag{11.70}$$

For a *free* photon ($\phi \equiv 0$ in the absence of charge and currents) one obtains the wave equation

$$\Delta\vec{A} = \frac{\partial^2\vec{A}}{\partial t^2}, \tag{11.71}$$

the solution of which is

$$\vec{A} = \vec{A}_0\, \mathrm{e}^{\mathrm{i}(kz-\omega t)}, \tag{11.72}$$

when choosing the z-axis along the flight direction of the photon with energy ω and momentum k. Therefore

$$\mathrm{div}\vec{A} = A_{0z}\mathrm{i}k\mathrm{e}^{\mathrm{i}(kz-\omega t)} = 0 \Rightarrow A_{0z} = 0 \Rightarrow A_z = 0. \tag{11.73}$$

The vector potential is *perpendicular* to the direction of propagation of the photon. The amplitudes A_{0x} and A_{0y} in (11.72) are complex. The axes can be chosen without loss of generality so that $A_{0x} \equiv A_0$ is real and $A_{0y} \equiv A_0' \exp(\mathrm{i}\delta)$, where A_0' is real, hence

$$\begin{aligned} A_x &= A_0 \mathrm{e}^{\mathrm{i}(kz-\omega t)}, \\ A_y &= A_0' \mathrm{e}^{\mathrm{i}(kz-\omega t+\delta)}. \end{aligned} \tag{11.74}$$

For $\delta = 0$ the wave is linearly polarized, otherwise it is elliptically polarized. Consider the particular case $A_0 = A_0'$ and $\delta = \pm\pi/2$ (*circularly* polarized wave). At $z = 0$ the components of the polarization vector $\vec{A}$ develop as

$$\begin{aligned} \delta = +\frac{\pi}{2} : A_x &= A_0 \cos \omega t, \\ A_y &= A_0 \sin \omega t, \end{aligned} \tag{11.75}$$

$$\begin{aligned} \delta = -\frac{\pi}{2} : A_x &= A_0 \cos \omega t, \\ A_y &= -A_0 \sin \omega t, \end{aligned} \tag{11.76}$$

where we have taken the real parts in (11.74). Equation (11.75) describes a right-circularly polarized photon ($\vec{A}$ is a right-handed screw in the direction of propagation), while (11.76) describes a left-circularly polarized photon (left-handed screw). From

$$\begin{aligned} A_x &= A_0 \cos(kz - \omega t) \qquad \text{and} \\ A_y &= A_0 \cos(kz - \omega t \pm \pi/2) = \mp A_0 \sin(kz - \omega t), \end{aligned} \tag{11.77}$$

it follows with (11.61) and (11.62) that

$$\begin{aligned} \vec{E} &= -\partial \vec{A}/\partial t = (-A_0\omega \sin(kz - \omega t), \mp A_0\omega \cos(kz - \omega t), 0) \perp \vec{A}, \\ \vec{B} &= \mathrm{rot}\vec{A} = (\pm A_0 k \cos(kz - \omega t), -A_0 k \sin(kz - \omega t), 0) \,//\, \vec{A}. \end{aligned} \tag{11.78}$$

The fields $\vec{B}$ and $\vec{E}$ are orthogonal to the flight direction of the photon. This is true in *any* gauge but not in the presence of charges or currents, as in a wave guide where the photon acquires an effective mass, or for the virtual γ exchanged between two charges in figure 7.9 where the electric field lies *parallel* to the flight direction of the γ.

The fields of the free photon also rotate, $\vec{E}$ staying perpendicular and $\vec{B}$ parallel (or antiparallel) to $\vec{A}$ (figure 11.7).

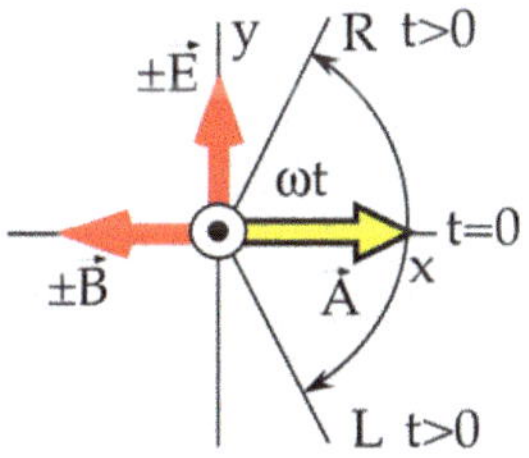

Figure 11.7. Time dependence of the polarization vector $\vec{A}$ for left-circularly polarized (L) and right-circularly polarized (R) photons at $z = 0$. The photon flies towards the reader.

Let us define the x-axis in the direction of $\vec{E}$ at $t = 0$ and consider right-circular and left-circular waves. We introduce the unit vectors in the direction of $\vec{E}(t)$ in the complex plane:

$$\begin{aligned} e_R &= \overbrace{\cos \omega t}^{e_x} + \mathrm{i} \overbrace{\sin \omega t}^{e_y}, \\ e_L &= \cos \omega t - \mathrm{i} \sin \omega t, \\ e_z &= 1, \end{aligned} \tag{11.79}$$

for right-circular (e_R) and left-circular (e_L) polarization, respectively. Under a rotation of the electric field with angle ϕ about the z-axis, the components transform as

$$\begin{aligned} e'_x &= e_x \cos \phi - e_y \sin \phi, \\ e'_y &= e_x \sin \phi + e_y \cos \phi, \\ e'_z &= e_z, \end{aligned} \tag{11.80}$$

whence

$$\begin{aligned} e'_R &\equiv e'_x + \mathrm{i}e'_y = \overbrace{\left(e_x + \mathrm{i}e_y\right)}^{e_R}\mathrm{e}^{\mathrm{i}\phi}, \\ e'_L &\equiv e'_x - \mathrm{i}e'_y = \underbrace{\left(e_x - \mathrm{i}e_y\right)}_{e_L}\mathrm{e}^{-\mathrm{i}\phi}. \end{aligned} \tag{11.81}$$

Rotations about z are generated by the operator $\mathrm{e}^{\mathrm{i}J_z\phi}$:

$$\begin{aligned} e'_R &= \mathrm{e}^{\mathrm{i}J_z\phi} e_R = \mathrm{e}^{\mathrm{i}\phi} e_R, \\ e'_L &= \mathrm{e}^{\mathrm{i}J_z\phi} e_L = \mathrm{e}^{-\mathrm{i}\phi} e_L, \\ e'_z &= \mathrm{e}^{\mathrm{i}J_z\phi} e_z = e_z. \end{aligned} \tag{11.82}$$

One obtains the important result that

> e_R is an eigenstate of J_z with eigenvalue $M = 1$,
> e_L is an eigenstate of J_z with eigenvalue $M = -1$,
> e_z is an eigenstate of J_z with eigenvalue $M = 0$,

where M is the projection of the total angular momentum,

$$M = m_\ell + m_s. \tag{11.83}$$

The projection m_ℓ of the orbital angular momentum is zero when the z-axis points in the flight direction of the photon. M is thus equal to the projection m_s of the photon spin (*helicity*). Since $\vec{E}$ is orthogonal to the z-axis the eigenstate e_z is not populated and one obtains (figure 11.8) the helicity -1 for a left-circularly polarized photon (left-handed γ) and $+1$ for a right-circularly polarized one (right-handed γ).

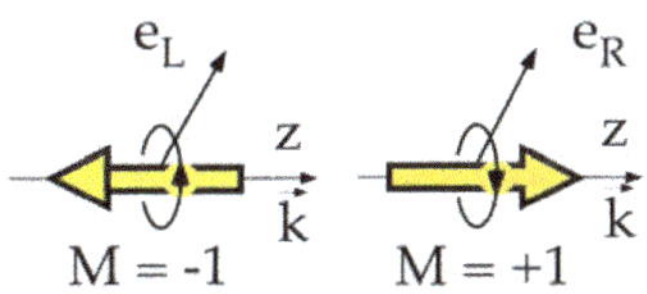

Figure 11.8. The helicities of left-circularly and right-circularly polarized photons which propagate in the $+z$-direction.

One finds that a free or massless photon cannot have the helicity $M = 0$, and is therefore *transverse* (meaning that the fields are transverse with respect to the direction of propagation). As a consequence, the *Landau–Yang theorem* states that

a spin-1 system does not decay into two photons

(nor into two massless gluons).

To prove the theorem, consider the wave function ψ of the two-photon system in the rest frame of the decaying spin-1 particle. The wave function must be a vector (or an axial vector) constructed from the products of the available vectors, $\vec{e}_1$, $\vec{e}_2$ and $\vec{k} \equiv \vec{k}_2 - \vec{k}_1 = 2\vec{k}_2$. The following combinations fulfil this condition: $\vec{k}\,(\vec{e}_1 \cdot \vec{e}_2)$, $(\vec{e}_1 \times \vec{e}_2)$, $\vec{k}$, and $\vec{k} \times (\vec{e}_1 \times \vec{e}_2)$. The last wave function vanishes due to the transversality of the photons. In addition, the wave functions have to be symmetric under the permutation of the two photons (Bose–Einstein symmetry for bosons). Therefore the first three wave functions also vanish:

$$\begin{aligned} \psi &\propto \overbrace{\left(\vec{k}_2 - \vec{k}_1\right)}^{=\vec{k}}\left(\vec{e}_1 \cdot \vec{e}_2\right) + \left(\vec{k}_1 - \vec{k}_2\right)\left(\vec{e}_2 \cdot \vec{e}_1\right) = 0, \\ \psi &\propto \left(\vec{e}_1 \times \vec{e}_2\right) + \left(\vec{e}_2 \times \vec{e}_1\right) = 0, \\ \psi &\propto \left(\vec{k}_2 - \vec{k}_1\right) + \left(\vec{k}_1 - \vec{k}_2\right) = 0. \end{aligned} \tag{11.84}$$

A suitable wave function cannot be found and therefore the Landau–Yang theorem is validated.

As a first application, let us determine the spin and parity of the neutral pion. The decay $\pi^0 \to 2\gamma$ already excludes $J(\pi^0) = 1$ by virtue of the Landau–Yang theorem. We consider $J(\pi^0) = 0$ as the simplest possibility and prove that the angle ϕ between the polarizations remains constant as a function of time by constructing the 2γ wave function directly from (11.79). Let $e_{L,R}$ be the direction of the $\vec{E}$-field for the photon propagating in the $-z$-direction and $E_{L,R}$ that of the photon flying in the $+z$-direction. Since we have assumed $J(\pi^0) = 0$, the photons are either both left-circularly ($e_L E_L$) or both right-circularly ($e_R E_R$) polarized (figure 11.9). These two states are related through parity transformation (mirroring):

$$e_R E_R \overset{P}{\longleftrightarrow} e_L E_L. \tag{11.85}$$

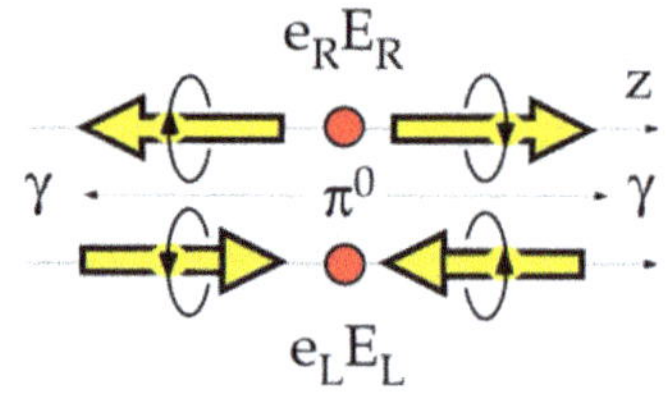

Figure 11.9. In the 2γ decay of a spinless π^0 the photons are both right-handed or both left-handed.

Because of parity conservation, the physical wave functions $\psi_\pm$ must be eigenstates of the parity operator. One guesses that

$$\begin{aligned} \psi_+ &= e_R E_R + e_L E_L, \quad \text{with} \quad P\psi_+ = +\psi_+, \\ \psi_- &= e_R E_R - e_L E_L, \quad \text{with} \quad P\psi_- = -\psi_-. \end{aligned} \tag{11.86}$$

The polarization vectors for the photons propagating towards $-z$ are $e_R = e_x - \mathrm{i}e_y$ and $e_L = e_x + \mathrm{i}e_y$. The wave function of even parity is then with (11.79) and $\omega \equiv \frac{m_{\pi^0}}{2}$

$$\begin{aligned} \psi_+ &= \left(e_x - \mathrm{i}e_y\right)\left(E_x + \mathrm{i}E_y\right) + \left(e_x + \mathrm{i}e_y\right)\left(E_x - \mathrm{i}E_y\right) = 2\left(e_x E_x + e_y E_y\right) \\ &= 2[\cos \omega t \cos(\omega t + \phi) + \sin \omega t \sin(\omega t + \phi)] = 2 \cos \phi, \end{aligned} \tag{11.87}$$

independent of t. Similarly, for the odd parity[6]

$$\psi_- = 2\mathrm{i} \sin \phi. \tag{11.90}$$

The angular distribution between the directions of polarization is then

$$\left|\psi_+\right|^2 \propto \cos^2 \phi = \frac{1}{2}(1 + \cos 2\phi) \quad \text{if } J^P\left(\pi^0\right) = 0^+, \tag{11.91}$$

$$\left|\psi_-\right|^2 \propto \sin^2 \phi = \frac{1}{2}(1 - \cos 2\phi) \quad \text{if } J^P\left(\pi^0\right) = 0^-. \tag{11.92}$$

Although the polarization vectors rotate in opposite directions (figure 11.9), the angle ϕ remains constant over time. This apparent contradiction is due to the fact that the 2γ system is a quantum superposition of right-circularly and left-circularly

[6] The wave functions can also be constructed directly from the kinematic variables as follows: for positive parity the wave function of a spin-0 state is a scalar (the sign does not change under space inversion), and is symmetric:

$$\psi_+ \propto \vec{e}_1 \cdot \vec{e}_2 = \cos \phi, \tag{11.88}$$

where ϕ is the angle between the directions of polarization. For negative parity the wave function is pseudo-scalar (the sign changes under space inversion) and is symmetric:

$$\psi_- \propto \overbrace{\left(\vec{k}_2 - \vec{k}_1\right)}^{=\vec{k}}\left(\vec{e}_1 \times \vec{e}_2\right) = k\left|\vec{e}_1 \times \vec{e}_2\right| \propto \sin \phi. \tag{11.89}$$

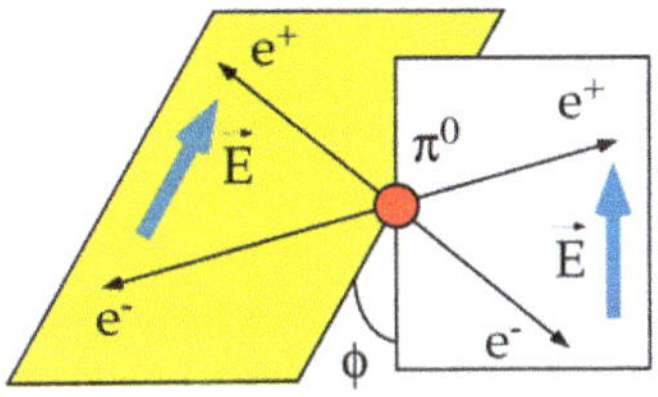

Figure 11.10. The planes spanned by the e^+e^- pairs in the decay $\pi^0 \to (e^+e^-)(e^+e^-)$ and the angle ϕ. The leptons are accelerated along the $\vec{E}$-field of the virtual photon.

polarized waves, see (11.86). For a spinless π^0 of *positive* parity the $\vec{E}$-fields of the photons prefer to be *parallel* ($\phi = 0$), while for *negative* parity they are preferably *orthogonal* ($\phi = 90°$).

The π^0 decay into two Dalitz pairs

$$\pi^0 \to \left(e^+e^-\right)\left(e^+e^-\right) \tag{11.93}$$

($f = 3.3 \times 10^{-5}$) was investigated in a hydrogen bubble chamber [9]. The neutral pions were produced by the reaction $\pi^- p \to \pi^0 n$. One can interpret reaction (11.93) as being due to the *internal* production of two e^+e^- pairs in $\pi^0 \to 2\gamma$ (rate $\propto (\sqrt{\alpha}\sqrt{\alpha})^2 \approx 5 \times 10^{-5}$). In the π^0 rest frame the e^+e^- pairs are emitted in opposite directions under small opening angles. Since the leptons are accelerated along the $\vec{E}$-field of the photon the pairs span two planes, each lying preferably parallel to the polarization of the photon. One measures the angle ϕ between the e^+e^- planes (figure 11.10).

However, in the decay (11.93) the photons are virtual and (11.92) is not directly applicable. A detailed calculation gives the angular distribution

$$w(\phi) \propto 1 \pm \chi \cos 2\phi, \tag{11.94}$$

with the plus sign for $P = +1$ (and the minus sign for $P = -1$), which agrees with (11.92) when $\chi \equiv 1$. The parameter χ depends on the opening angle and energy sharing in each lepton pair. The data are shown in figure 11.11 (64 decays). The preferred angle ϕ is 90° and thus the π^0 is a *pseudoscalar* meson.

As a second application, consider the ground states ($\ell = 0$) of positronium, $^1S_0$ and $^3S_1$. Observed are two decay components, a fast one ($\tau = 124$ ps) coupling to 2γ, and a much slower one ($\tau = 142$ ns) leading to 3γ. The additional coupling factor α and the reduced phase space are responsible for the longer lifetime of the latter. According to the Landau–Yang theorem, the e^+e^- pair in the spin triplet case cannot decay into 2γ. We therefore find for parapositronium ($^1S_0$), and orthopositronium ($^3S_1$):

$$^1S_0\left(e^+e^-\right) \to 2\gamma, \qquad {}^3S_1\left(e^+e^-\right) \to 3\gamma. \tag{11.95}$$

For both states the internal parity should be negative, $P = -(-1)^{\ell=0} = -1$ if fermions and antifermions have a relative negative parity. Therefore, one concludes that $J^P = 0^-$ for parapositronium and $J^P = 1^-$ for orthopositronium.

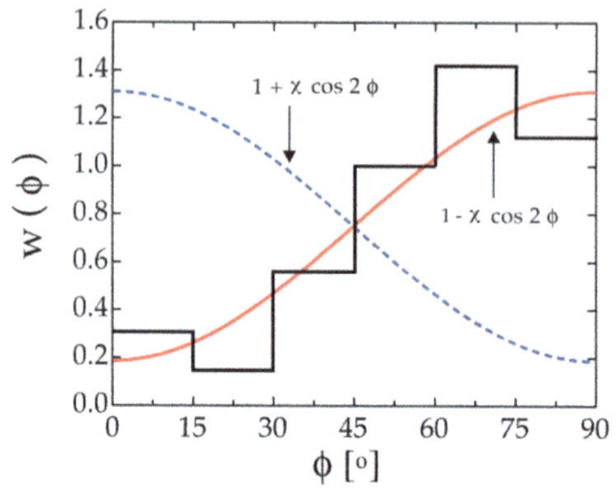

Figure 11.11. Measured ϕ distribution. The solid curve shows the fit for a $J^P = 0^-$ neutral pion, while the dashed curve shows the distribution for $J^P = 0^+$ ($\chi = 0.75$). After [9].

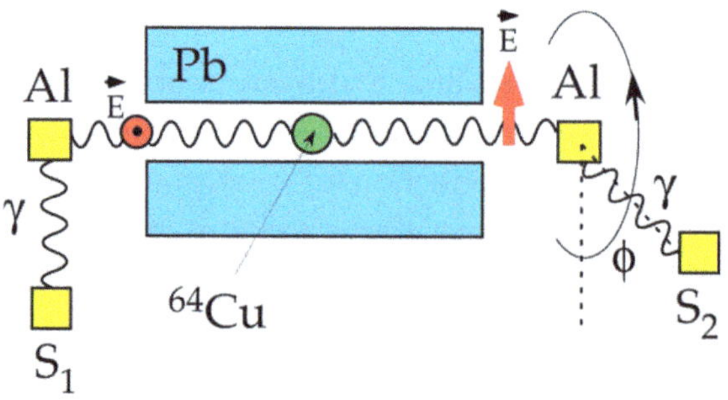

Figure 11.12. Diagram of the apparatus used to determine the relative parity between the fermion and anti-fermion. After [10].

The experiment of Wu and Shaknov [10] proves that the directions of polarization in parapositronium decay are preferably perpendicular, in analogy to π^0 decay. Figure 11.12 shows a diagram of the apparatus. Positronium is generated using a β^+ emitter (^{64}Cu). The two 511 keV annihilation quanta are emitted back-to-back in the source and are scattered by the aluminium blocks before being detected by the scintillation counters S_1 and S_2.

The Compton scattering cross section of polarized photons depends on the azimuthal angle ϕ: the photons scattered by the aluminium targets are emitted preferably in the direction orthogonal to the incident polarization. This is comparable to a classical linear antenna in which an excited electron oscillates along the applied $\vec{E}$-field. The re-emitted radiation reaches its maximum in the plane perpendicular to the oscillating dipole. One measures the coincidence rate S_1S_2 as a function of azimuthal angle ϕ between the two detectors. The result

$$\frac{S_1S_2(\phi = \pi/2)}{S_1S_2(\phi = 0)} = 2.04 \pm 0.08 \tag{11.96}$$

shows clearly that the polarizations are preferably orthogonal so that $P = -1$ for $^1S_0\,(e^+e^-)$. The relative parity between fermions and antifermions is therefore negative.

11.7 *C* parity

Charge conjugation (or *C* parity) transforms a particle into its corresponding antiparticle. The operation flips the sign of the electric charge, magnetic moment, baryon and lepton numbers, as well as all flavour quantum numbers such as strangeness, charm, etc. Space coordinates and kinematic variables (momentum, angular momentum, spin) are not affected. Examples of charge conjugation are:

$$\begin{array}{ccc} \pi^+ & \overset{C}{\rightarrow} & \pi^-, \\ n & \rightarrow & \bar{n}, \\ \nu_e & \rightarrow & \bar{\nu}_e. \end{array} \tag{11.97}$$

Particles with non-vanishing Q, B, flavour or lepton number are not eigenstates of C parity, nor are gluons (e.g. $r\bar{b}$ is transformed into $\bar{r}b$). In contrast, self-conjugated particles such as

$$\begin{array}{ccc} \gamma & \overset{C}{\rightarrow} & \gamma, \\ \pi^0 & \rightarrow & \pi^0, \\ \omega & \rightarrow & \omega, \end{array} \tag{11.98}$$

are eigenstates of C. The eigenvalue of C parity is ± 1 since

$$C^2|\ \rangle = |\ \rangle \Rightarrow C|\ \rangle = \pm 1|\ \rangle. \tag{11.99}$$

To determine the sign, consider first the photon. Since the electric charge changes under C, so does the sign of the electromagnetic potential ($\phi \rightarrow -\phi$, $\vec{A} \rightarrow -\vec{A}$), and correspondingly that of the fields $\vec{E}$ and $\vec{B}$. Hence

$$C|\gamma\rangle = -|\gamma\rangle \Rightarrow \boxed{J^{PC}(\gamma) = 1^{--}}. \tag{11.100}$$

C parity is, like parity, a *multiplicative* quantum number:

$$C|n\gamma\rangle = (-1)^n|n\gamma\rangle. \tag{11.101}$$

Consequently, assuming C conservation in $\pi^0 \rightarrow \gamma\gamma$, the C parity of the π^0 meson is $C = +1$:

$$C|\pi^0\rangle = +|\pi^0\rangle \Rightarrow J^{PC}(\pi^0) = 0^{-+}. \tag{11.102}$$

Indeed, the C parity violating decay into 3γ ($C = -1$) has not been observed. The measured ratio of branching ratios is

$$\frac{f\left(\pi^0 \rightarrow 3\gamma\ [C = -1]\right)}{f\left(\pi^0 \rightarrow 2\gamma\ [C = +1]\right)} < 3.1 \times 10^{-8}, \tag{11.103}$$

from which one concludes that C is conserved in electromagnetic interactions. From the electromagnetic decay $\omega \to \pi^0\gamma$ ($f = 8.3\%$) one finds that

$$C|\omega\rangle = -|\omega\rangle \Rightarrow J^{PC}(\omega) = 1^{--}. \tag{11.104}$$

Conservation of C parity in a reaction implies that the charge conjugated reaction runs with the same cross section, that is

$$\frac{\mathrm{d}\sigma}{\mathrm{d}\Omega}(ab \to cd) = \frac{\mathrm{d}\sigma}{\mathrm{d}\Omega}(\bar{a}\bar{b} \to \bar{c}\bar{d}). \tag{11.105}$$

In particular, the momentum and angular distributions of hadrons are equal to those of the antihadrons in the conjugated reaction. Likewise, conservation of C parity in a decay means that the partial width is equal to that of the charge conjugated decay. For example, the partial widths for the decays $\rho^+ \to \pi^+\pi^0$ and $\rho^- \to \pi^-\pi^0$ are equal. As another example, consider the self-conjugate decay

$$\eta \to \pi^+\pi^-\pi^0 \; (f = 22.9\%). \tag{11.106}$$

This process is *electromagnetic*, since the decay branching ratio is comparable to that of the (obviously) electromagnetic $\eta \to 2\gamma$ ($f = 39.4\%$). As we shall see in section 12.4 reaction (11.106) also violates isospin symmetry, which is another conservation law of strong interactions.

C invariance in electromagnetic interactions requires the momentum distributions of the positive and negative pions to be identical. This was tested with the apparatus shown in figure 11.13. Incident pions of 718 MeV c^{-1} are used to produce η mesons from the reaction $\pi^-p \to \eta n$ in a hydrogen target. Reactions such as $\pi^-p \to \pi^+\pi^-n$ and $\pi^+\pi^-\pi^0n$ contribute to most of the background. Therefore, it is important to reconstruct the kinematics of the reaction with sufficient constraints. The neutron is detected by a circular array of thick plastic scintillation counters (signal from the recoil proton). The momentum of the η meson is determined by the neutron angle

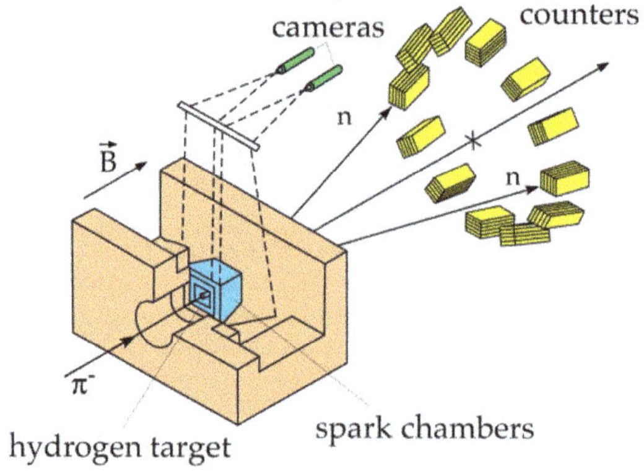

Figure 11.13. The apparatus used to test C invariance in η decay. After [11].

(two-body kinematics). The emerging two charged pions are registered by spark chambers Fukui and Miyamoto 1959 [12] monitored with cameras, and their momentum vectors are measured in the longitudinal magnetic field. One can then calculate the momentum of the invisible π^0 in the decay (11.106) and thus the invariant mass of the $\pi^+\pi^-\pi^0$ system, which must agree with the η mass. The decay is conveniently studied in the rest frame of the η meson by analysing the $\eta \to 3\pi$ Dalitz plot (section 12.3). The kinetic energies $T(\pi^+)$ and $T(\pi^-)$ of the charged pions are then calculated in the η rest frame by a suitable Lorentz transformation. If C is a symmetry one expects equal numbers $N(\pi^+)$ of events with $T(\pi^+) > T(\pi^-)$ as $N(\pi^-)$ with $T(\pi^-) > T(\pi^+)$. The asymmetry

$$A = \frac{N(\pi^+) - N(\pi^-)}{N(\pi^+) + N(\pi^-)} = (0.28 \pm 0.26) \times 10^{-2} \tag{11.107}$$

was measured. The main difficulty of the experiment lies in the fact that positive and negative pions are deflected in opposite directions by the magnetic field, so that small imperfections in the cylindrical symmetry affect the detection probability and lead to artificial asymmetries. Such systematic errors can be partially corrected by reversing the polarity of the field. The current best value for the asymmetry A is $(0.09\,^{+0.13}_{-0.17}) \times 10^{-2}$ [13], in good agreement with C invariance in electromagnetic interactions.

Similar experiments can be carried out for strong interaction processes. C parity is also conserved in *strong* interactions. However, in *weak* interactions C (like P) is not conserved (chapter 13).

We emphasize that C parity is not defined for particles with non-vanishing Q, B or flavour numbers. However, particle–antiparticle pairs $a\bar{a}$ are eigenstates of C. Consider for example a particle with spin s_a and its corresponding antiparticle. The C parity of the $a\bar{a}$ pair with relative angular momentum ℓ and spin s (where $s = 0, 1, \ldots, 2s_a$) is

$$\boxed{C(a\bar{a}) = (-1)^{\ell+s}}. \tag{11.108}$$

This can be seen as follows:

$$\begin{aligned} C|a\bar{a}\rangle &= K|\bar{a}a\rangle = K\phi_\ell(\bar{a}, a)\chi_s(\bar{a}, a) \\ &= K\begin{bmatrix} -(-1)^\ell(-1)^{s+1} & \text{for fermions} \\ (-1)^\ell(-1)^s & \text{for bosons} \end{bmatrix}\phi_\ell(a, \bar{a})\chi_s(a, \bar{a}) \\ &= K(-1)^{\ell+s}|a\bar{a}\rangle, \end{aligned} \tag{11.109}$$

with $K = \pm 1$. We have used the parity of the orbital wave function ϕ_ℓ and the symmetry properties of the Clebsch–Gordan coefficients for the spin wave function χ_s (see the spin wave function of the neutron pair in the reaction (11.38), $\pi^- d \to nn$). Under permutation

$$\begin{aligned} \chi_s &= A(s=0), \qquad S(s=1)\ldots \text{ for fermions, and} \\ \chi_s &= S(s=0), \qquad A(s=1)\ldots \text{ for bosons,} \end{aligned} \tag{11.110}$$

where S = symmetric and A = antisymmetric. The sign of K in (11.109) can be determined for bosons and fermions. For example, a $2\pi^0$ system has $C = +1$ according to (11.102), while $s = 0$ and ℓ is even due to Bose–Einstein symmetry for identical bosons, hence $K = +1$. For fermions we use the ground state of para-positronium: ${}^1S_0(e^+e^-) \rightarrow 2\gamma$ requires $C = C(\gamma)^2 = +1$, hence with $\ell = 0$ and $s = 0$ we have again $K = +1$. This is also true for orthopositronium ${}^3S_1(e^+e^-) \rightarrow 3\gamma$ $(s = 1)$ with $C = C(\gamma)^3 = -1$. Formula (11.108) therefore applies to both bosons *and* fermions. Let us discuss a few useful applications of this formula.

1. The quantum numbers of parapositronium and orthopositronium are $J^{PC} = 0^{-+}$, respectively 1^{--}.
2. In the quark model, the C parity of the π^0 meson ($d\bar{d} - u\bar{u}$ with $s = 0$) is $C = (-1)^{\ell=0} = +1$ and that of the ω meson ($u\bar{u} + d\bar{d}$ with $s = 1$) $C = (-1)^{(\ell=0)+1} = -1$, in accordance with (11.102) and (11.104). Likewise, the ρ^0 meson ($d\bar{d} - u\bar{u}$ with $\ell = 0$, $s = 1$) has $C = -1$.
3. The conservation of C in strong interactions requires the ρ^0 to decay into $\pi^+\pi^-$ with $C(\pi^+\pi^-) = (-1)^L = -1$, hence with odd relative angular momentum L between the pions ($L = 1$, as $J(\rho) = 1$). On the other hand, ρ^0 *does not* decay into $\pi^0\pi^0$ ($C = +1$) since L would have to be even. This decay would violate C as well as the conservation of total angular momentum and is therefore excluded by all three interactions.

We will show experimentally in section 14.4 that neutrinos are left-handed and antineutrinos right-handed. This leads to the violation of both P and C invariance in the weak interaction, because neither right-handed neutrinos (P transformation or mirroring) nor left-handed antineutrinos (C transformation) exist. However, CP conservation requires the existence of *both* left-handed and right-handed antineutrinos (figure 11.14). The composite operation CP is, with few exceptions (in the $K^0 - \overline{K}^0$ system and in hadronic B decays) a symmetry of the weak interaction, and therefore the product of eigenvalues $C \cdot P$ is conserved.

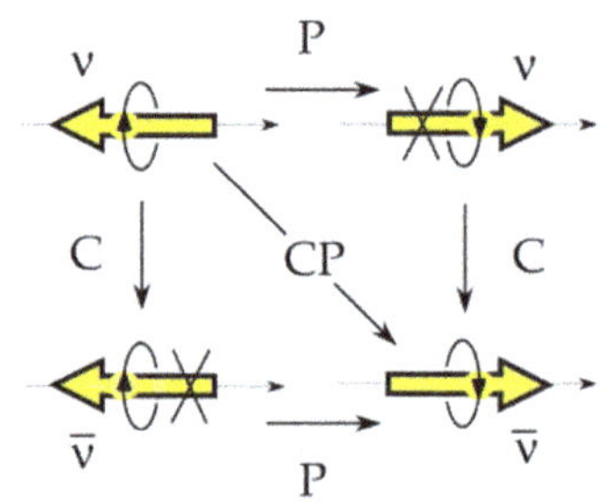

Figure 11.14. P, C and CP transformation for neutrinos. Right-handed neutrinos and left-handed antineutrinos do not exist. The neutrinos fly from left to right.

11.8 Time reversal invariance

Elementary processes are invariant under time reversal transformation T if they evolve with the 'same' probability in the forward and backward time directions. In particular, the differential cross section for the reaction $ab \to cd$ is equal to that of the time-reversed reaction $cd \to ab$ at equal centre-of-mass energies, and up to kinematic factors such as the density of final states. This statement will be refined in section 12.1 with a concrete example. More precisely, time reversal invariance requires

$$|\mathcal{M}(ab \to cd)|^2 = |\mathcal{M}(cd \to ab)|^2 \tag{11.111}$$

for the transition matrix $\mathcal{M}$ at the *same* total centre-of-mass energy $\sqrt{s}$. The three fundamental interactions are invariant under time reversal (except in the $K^0 - \overline{K}^0$ system and in hadronic B decays, see chapters 19 and 20).

Figure 11.15 shows the differential cross sections for the reactions

$$\alpha\ ^{24}\text{Mg} \to p\ ^{27}\text{Al} \qquad \text{and} \qquad p\ ^{27}\text{Al} \to \alpha\ ^{24}\text{Mg} \tag{11.112}$$

as a function of incident kinetic energy in the laboratory, for protons or α particles scattered under 168.1° in the centre-of-mass system. The two measured values at the same total energies in the centre-of-mass are superimposed. The cross sections are not equal due to the different kinematic factors (phase space in the final state), but they show exactly the same energy dependence, once normalized to each other. The excellent agreement proves that the transition matrix (11.111) of the strong interaction is invariant under time reversal. T invariance will be used in section 12.2.1 to derive the spin of the pion.

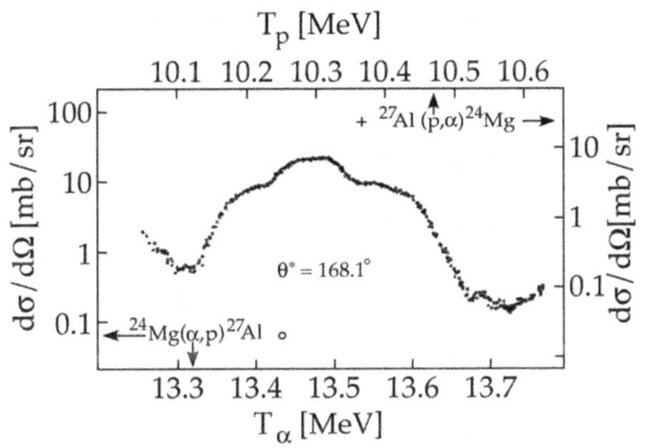

Figure 11.15. Differential cross sections for $\alpha\ ^{24}\text{Mg} \to p\ ^{27}\text{Al}$ (left ordinate) and for the reversed reaction (right ordinate) at $\theta^* = 168.1°$. The kinetic energies (bottom and top abscissae, respectively) correspond to the same centre-of-mass energy. After [14].

11.8.1 Magnetic and electric dipole moments

Let us first recall the transformation properties of various physical quantities under $P(\vec{r} \to -\vec{r})$ and $T\,(t \to -t)$.

$$P: \begin{bmatrix} \vec{r} & \to & -\vec{r} \\ \vec{p} & \to & -\vec{p} \\ \vec{s} & \to & \vec{s} \\ \vec{E} & \to & -\vec{E} \\ \vec{B} & \to & \vec{B} \end{bmatrix} \qquad T: \begin{bmatrix} \vec{r} & \to & \vec{r} \\ \vec{p} & \to & -\vec{p} \\ \vec{s} & \to & -\vec{s} \\ \vec{E} & \to & \vec{E} \\ \vec{B} & \to & -\vec{B} \end{bmatrix}. \tag{11.113}$$

The sign flip $\vec{s} \to -\vec{s}$ under T follows from the definition $\vec{r} \times \vec{p}$ of an angular momentum. The magnetic field $\vec{B}$ ($\mathrm{rot}\vec{B} = \vec{j}$) also changes sign with $\vec{j}$ under T, but not under P (being an axial vector), while the opposite holds for the vector $\vec{E}$ ($\mathrm{rot}\vec{E} = -\frac{\partial \vec{B}}{\partial t}$).

The energy $H = -\vec{\mu} \cdot \vec{B}$ of a *magnetic* dipole $\vec{\mu} \propto \vec{s}$ contains a term of the form

$$\vec{s} \cdot \vec{B} \overset{P \text{ or } T}{\rightleftharpoons} \vec{s} \cdot \vec{B}, \tag{11.114}$$

and is invariant under P and T. For an elementary particle, there is no other preferred direction than that of its spin, hence the *electric* dipole moment $\vec{d}$ must lie parallel to $\vec{s}$. The energy $H = -\vec{d} \cdot \vec{E}$ of an electric dipole therefore includes a term of the form

$$\vec{s} \cdot \vec{E} \overset{P \text{ or } T}{\rightleftharpoons} -\vec{s} \cdot \vec{E} \tag{11.115}$$

and is neither invariant under P nor T. The electric dipole moment leads to an asymmetrical charge distribution along the spin direction. If T is a symmetry of the fundamental interactions then elementary particles do not possess any electric dipole moment[7].

Finite electric dipole moments have been sought so far without success for the electron and the muon. However, given that the energy perturbation from the dipole moment is very small compared to the acceleration in the $\vec{E}$-field, neutral particles such as the neutron may be more suitable for electric dipole searches. The best upper limits on the electric dipole moment have been achieved with the technique of nuclear magnetic resonance (NMR). In an homogeneous $\vec{B}$-field the magnetic moment $\vec{\mu}$ of the nucleon precesses around the direction of the field with the *Larmor*

[7] Atoms and molecules possess electric dipole moments, although T is not violated. The electric dipole moment $\vec{d}$ lies along the asymmetrical charge distribution and does not change sign under T. Furthermore, as mentioned already, the induced electric dipole moment in the hydrogen atom originates from a mixture of degenerate states with different orbital angular momenta and opposite parities. For elementary particles electric dipole moments cannot be induced that way since particles of equal mass but with opposite parity do not exist (e.g. there is no neutron with internal parity $P = -1$).

frequency ω_L. For the proton the time evolution of the spin vector is classically (principle of angular momentum)

$$\frac{d\vec{s}}{dt} = \vec{\mu} \times \vec{B} \quad \text{with} \quad \vec{\mu} = g_p \overbrace{\frac{e}{2m_p}}^{\mu_N} \vec{s}, \tag{11.116}$$

where μ_N is the nuclear magneton (section 2.4). Expressed in components,

$$\begin{aligned} \frac{ds_x}{dt} &= g_p \mu_N s_y B, \\ \frac{ds_y}{dt} &= -g_p \mu_N s_x B, \\ \frac{ds_z}{dt} &= 0. \end{aligned} \tag{11.117}$$

The solution of these coupled differential equations is

$$\vec{s} = (s_\perp \sin \omega_L t, \, s_\perp \cos \omega_L t, \, s_z) \tag{11.118}$$

where ω_L is the Larmor frequency

$$\boxed{\omega_L = \frac{geB}{2m_p}}, \tag{11.119}$$

and

$$s_\perp \equiv \sqrt{s_x^2 + s_y^2}. \tag{11.120}$$

From the quantum mechanical viewpoint the perturbation

$$H = -\vec{\mu} \cdot \vec{B} \tag{11.121}$$

lifts the degeneracy of the levels with different spin projections m_s. For a spin-$\frac{1}{2}$ particle with mass M, the energy difference between spin 'up' and spin 'down' is

$$\Delta E = g\mu \left(m_s = \frac{1}{2} \right) B - g\mu \left(m_s = -\frac{1}{2} \right) B = \frac{geB}{2M} = \omega_L. \tag{11.122}$$

A radio-frequency wave with magnetic field $\vec{B}_{RF}$ irradiated *perpendicularly* to the static $\vec{B}$-field causes a spin flip when the *resonance* condition

$$\omega_{RF} = \omega_L \tag{11.123}$$

is fulfilled (figure 11.16). The g-factor can be determined from the resonance frequency, provided that the ratio of charge/mass is known.

Assume now that the particle also possesses an electric dipole moment $\vec{d}$. Equation (11.121) would then read

$$H = -\vec{\mu} \cdot \vec{B} - \vec{d} \cdot \vec{E}, \tag{11.124}$$

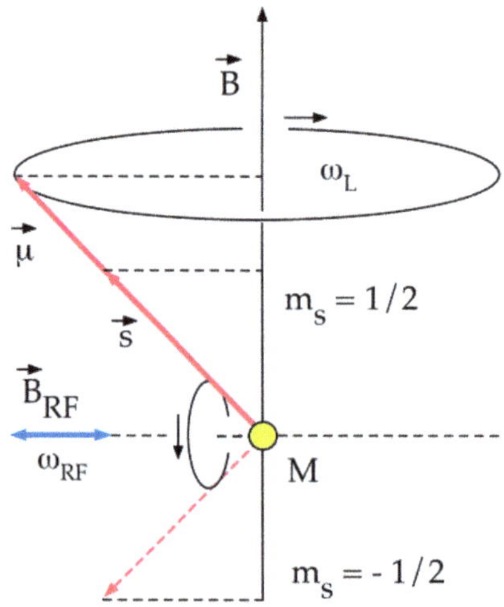

Figure 11.16. Spin precession in the magnetic field. The spin flips when the applied frequency coincides with the Larmor frequency, $\omega_{RF} = \omega_L$.

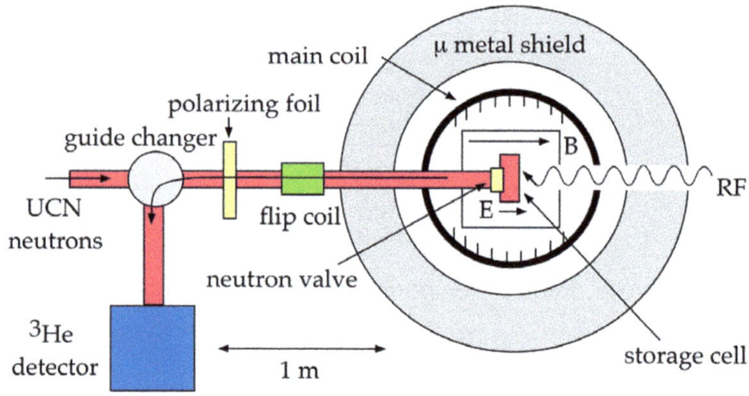

Figure 11.17. The magnetic bottle used to measure the electric dipole moment of the neutron. Adapted from [15].

with the additional perturbation induced by the applied electric field $\vec{E}$. The resonance condition, for parallel or antiparallel $\vec{B}$ and $\vec{E}$-fields, becomes:

$$\omega_{RF} = \omega_L \pm 2Ed. \tag{11.125}$$

Experimentally, slow neutrons from a nuclear reactor are stored in a magnetic bottle. One looks for a shift in the resonance frequency when an electric field is switched on. In the experiment by Pendlebury *et al* [15] cold neutrons from the reactor at the Institut Laue-Langevin (ILL) in Grenoble are moderated by a D_2O absorber before being transferred in a highly polished tube to a turbine, where they are further slowed down by rotating blades. The resulting ultra-cold neutrons ($v \sim 5\,\mathrm{m\,s^{-1}}$) then fly through a polarizing foil (0.5 μm Fe–Co) and are stored in a 5ℓ beryllium storage cell, confined by a longitudinal magnetic field of 10 mG (figure 11.17). Polarized neutrons (e.g. with spins in the +z-direction) are reflected by

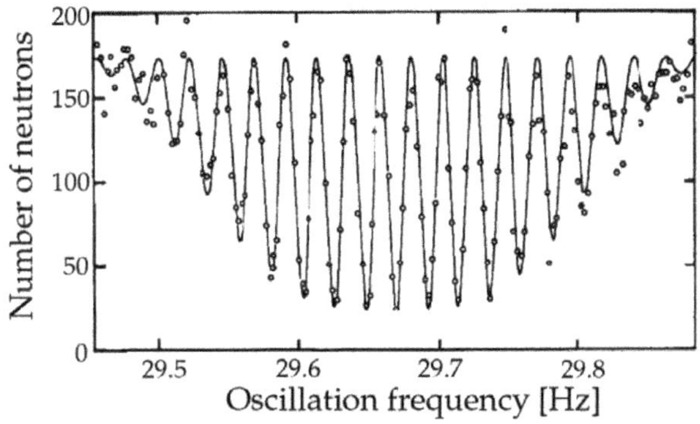

Figure 11.18. Resonance curve for ultra-cold neutrons in a magnetic field of 10 mG. After [15].

the walls of the bottle and survive for about 100 s before being absorbed. A 30 Hz pulse is applied to rotate the polarization of the neutrons into the direction perpendicular to the z-axis. The spins then precess in the magnetic field until a second high-frequency pulse is applied. If the second pulse is in phase with the first the polarization resumes its rotation towards the $-z$-direction. The bottle is then opened and the polarizing foil now acts as analyser. Neutrons with spin projections in the $+z$-direction are allowed to pass through and are registered in a ^{3}He gas counter (reaction $n^3\text{He} \to p^3\text{H}$), while those with spin in the $-z$-direction are reflected. The counting rate reaches a minimum when the applied high-frequency agrees with the Larmor frequency.

If instead the second high-frequency pulse is applied with a phase shift of 90° with respect to the first, the shape of the resonance curve becomes dispersive (figure 11.18). One then determines the frequency at which the slope of the counting rate is largest and looks for a shift in the frequency when a longitudinal electric field ($\pm 10\,\text{kV cm}^{-1}$) is applied.

The result for the electric dipole moment of the neutron is $d = (-3 \pm 5) \times 10^{-26}$ cm $\times\, e$ [16]. The most accurate measurement so far is [17]

$$d < 2.9 \times 10^{-26}\ \text{cm} \times e, \tag{11.126}$$

in agreement with T conservation. Compare the length of such a mini-antenna to the neutron size of $\sim 2 \times 10^{-13}$ cm!

In the following section we deal with CPT symmetry which holds for all three interactions. If CPT is conserved, T violation immediately follows from CP violation. In the standard model CP violation in the $K^0 - \overline{K}^0$ system and in hadronic B decays arises from complex terms in the Cabibbo–Kobayashi–Maskawa (CKM) mixing matrix (section 16.3). The standard model then predicts the electric dipole moment of the neutron to be around $d \sim 10^{-32}$ cm $\times\, e$. A more accurate measurement is therefore of great interest. Difficulties lie in obtaining the required homogeneity and stability of the magnetic field and a perfect parallelism between the $\vec{B}$ and $\vec{E}$-fields. For instance, a small perpendicular component of $\vec{E}$ appears in the rest frame of the neutron as a contribution to $\vec{B}$ which affects the resonance frequency.

11.9 *CPT* invariance

All interactions are symmetric under the composite operation *CPT*: processes between elementary particles run identical to the *mirrored*, time *reversed* processes between the corresponding *antiparticles. CPT* invariance ($[CPT, H] = 0$) implies that the masses, charges, magnetic moments and mean lifetimes of particles and antiparticles are equal:

$$m = \overline{m}, \qquad Q = -\overline{Q}, \qquad \mu = -\overline{\mu}, \qquad \tau = \overline{\tau}. \qquad (11.127)$$

These identities actually already follow from *C* invariance in strong and electromagnetic interactions. Any violation of *CPT* is therefore to be searched for in weak interactions. Table 11.1 shows the experimental upper limits for the differences between various measured quantities, relative to their mean values (see also [7]).

The magnetic moments of the μ^+ and μ^- are measured using $g - 2$ experiments (section 13.3.3). The extremely accurate result on the mass difference between K^0 and $\overline{K}^0$ is achieved from a measurement of the mass difference between the K_1 and K_2 states (chapter 19).

The mass of the antiproton was measured using slow antiprotons stored in a Penning trap [18]. Antiprotons of 6 MeV from the Low Energy Antiproton Ring (LEAR) at CERN are moderated with thin foils and captured in a suitable arrangement of $\vec{B}$ and $\vec{E}$-fields (a solenoidal $\vec{B}$-field with a superimposed $\vec{E}$ quadrupole field, figure 11.19) at the temperature of liquid helium (4.2 K). The antiprotons are cooled down to 0.3 meV ($\simeq$ 4.2 K) by collisions with an electron gas. The trapped antiprotons perform three types of motion: axial oscillation, cyclotron rotation and the much slower magnetron rotation (figure 11.19, left). The *cyclotron* frequency

$$\omega_c = \frac{eB}{M} \qquad (11.128)$$

of the orbiting antiprotons is measured with an oscillator and compared with the cyclotron frequency of protons in the same trap. (More precisely, one measures ω_c

Table 11.1. Some experimental results on *CPT* symmetry.

Measurement	Value
$(m_{e^+} - m_{e^-})$/mean value	$<8 \times 10^{-9}$
$\|m_p - m_{\bar{p}}\|/m_p$	$<7 \times 10^{-10}$
$\|m_{K^0} - m_{\overline{K}^0}\|$/mean value	$<6 \times 10^{-19}$
$(\tau_{\mu^+} - \tau_{\mu^-})$/mean value	$(2 \pm 8) \times 10^{-5}$
$(\tau_{\tau^+} - \tau_{\tau^-})$/mean value	$<7 \times 10^{-3}$
$\|Q_{e^+} + Q_{e^-}\|/e$	$<4 \times 10^{-8}$
$(g_{e^+} - g_{e^-})$/mean value	$(-0.5 \pm 2.1) \times 10^{-12}$
$(g_{\mu^+} - g_{\mu^-})$/mean value	$(-0.11 \pm 0.12) \times 10^{-8}$

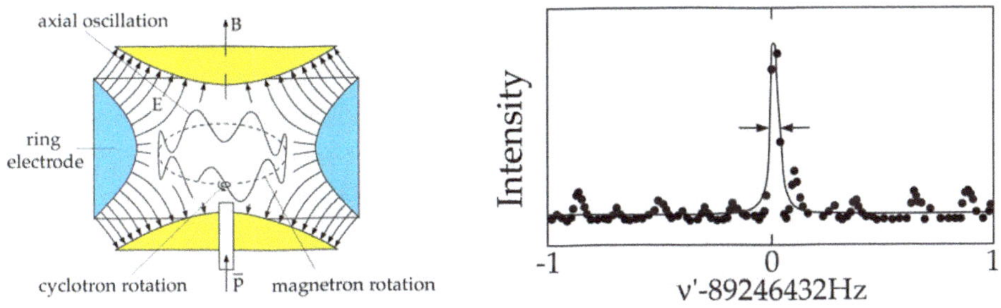

Figure 11.19. Left: principle of a Penning trap (used here for the storage of antiprotons). Right: resonance signal from a single antiproton. $\nu' = \omega_c'/2\pi$ is plotted with the cyclotron frequency modified by the axial motion. The resolution is better than 2×10^{-10}. After [18].

modified by the axial oscillation and the axial oscillation frequency which depends on the electric field.) H^- ions are used instead of protons to avoid introducing systematic biases when reversing the signs of the fields. This is actually a comparative measurement of the ratio e/M for a proton and antiproton. Figure 11.19 (right) shows the resonance line from a single stored antiproton.

Current experiments at CERN's Antiproton Decelerator (AD) are studying the properties of antihydrogen. Figure 11.20 (left) shows the ATHENA apparatus used to demonstrate the production of slow antihydrogen ($e^+\bar{p}$ atom). The antiprotons produced by the CERN Proton Synchrotron (see figure 9.8) are decelerated in the AD, extracted and finally captured in a Penning trap. Positrons from a radioactive ^{22}Na source are stored in a second trap. To induce positron capture by the antiprotons, the antiproton and positron clouds are transferred to a third trap and made to overlap. The electrically neutral antihydrogen atoms then migrate to the cryostat walls where they annihilate. The antihydrogen detector consists of two layers of silicon strips to track the charged annihilation pions. Simultaneously, the two 511 keV photons from positron–electron annihilation are detected in an array of 192 CsI scintillation crystals (figure 11.20, right) which also measures their energies [19, 20].

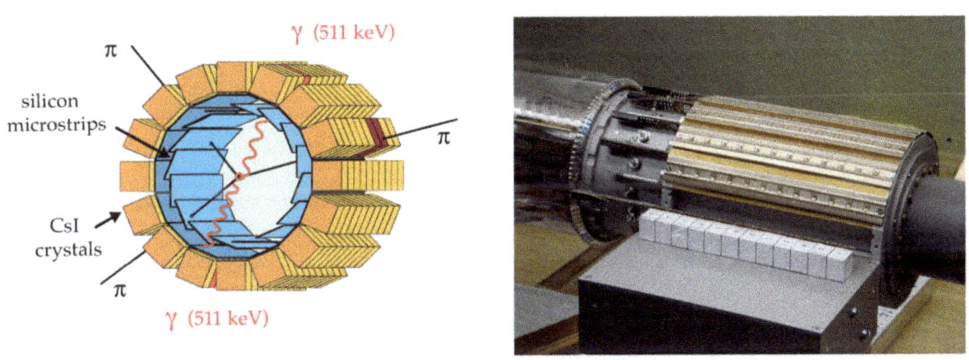

Figure 11.20. Left: diagram of the ATHENA antihydrogen detector. Right: the detector with one row of CsI crystals prior to installation.

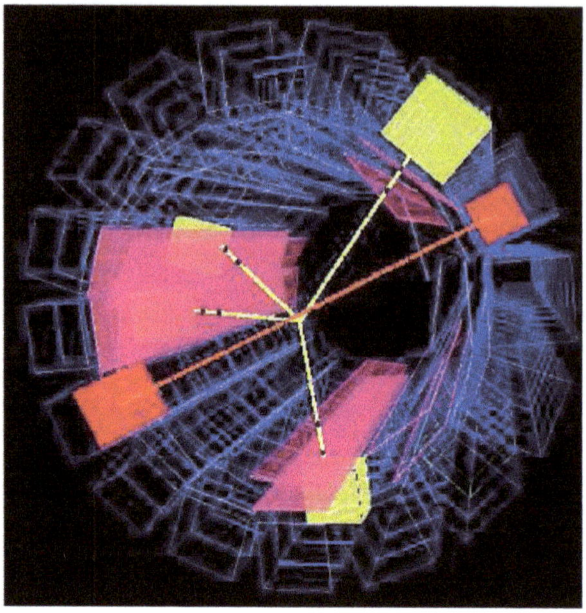

Figure 11.21. The electronic reconstruction of an antihydrogen annihilation on the cryostat wall of the ATHENA detector. The yellow lines show the tracks from four charged pions and the red lines show the directions of the two 511 keV annihilation photons which are emitted in opposite directions. Image credit: CERN.

Figure 11.21 shows an online annihilation event into four charged pions and two 511 keV photons from e^+e^- annihilation.

Since antihydrogen atoms possess a magnetic dipole moment (that of the positron), they can, in principle, be captured in an inhomogeneous magnetic field. The first attempts were carried out successfully at CERN [21].

The aim of such experiments is to test *CPT* invariance by comparing atomic transitions in hydrogen and antihydrogen. In principle, since the *CPT* theorem plays a central role in quantum field theories, one does not expect to see any violation of *CPT* in physical phenomena. However, experience shows, as with the surprising 1957 discovery of parity non-conservation in weak interactions, that, ultimately, the experiments will have the last word.

References

[1] See, e.g., Cohen-Tannoudgi C, Diu B and Laloë F 1997 *Quantum Mechanics* (New York: Wiley) p 140

[2] See, e.g., Bransden B H and Jochain C J 1983 *Physics of Atoms and Molecules* (London: Longman) p 220

[3] Amsler C *et al* 1975 *Phys. Lett.* B **57** 289

[4] Mango S, Runolfsson Ö and Borghini M 1969 *Nucl. Instrum. Methods* **72** 45

[5] Abragam A 1955 *Phys. Rev.* **98** 1729

[6] Adelberger E G *et al* 1975 *Phys. Rev. Lett.* **34** 402
[7] Olive K *et al* (Particle Data Group) 2014 *Chin. Phys.* C **38** 090001
[8] Pochodzalla J and Walcher T H 2007 *Proc. of the IX Int. Conf. on Hypernuclear and Strange Particle Physics* (Berlin: Springer)
[9] Plano R *et al* 1959 *Phys. Rev. Lett.* **3** 525
[10] Wu C S and Shaknov I 1950 *Phys. Rev.* **77** 136
[11] Jane M R *et al* 1974 *Phys. Lett.* B **48** 260
[12] Fukui S and Miyamoto S 1959 *Nuovo Cim.* **11** 113
[13] Ambrosino F *et al* 2008 *J. High Energy Phys.* **2008** JHEP05(2008)006
[14] von Witsch W, Richter A and von Brentano P 1968 *Phys. Rev.* **169** 923
[15] Pendlebury J M *et al* 1984 *Phys. Lett.* B **136** 327
[16] Smith K F *et al* 1990 *Phys. Lett.* B **234** 191
[17] Baker C A *et al* 2006 *Phys. Rev. Lett.* **97** 131801
[18] Gabrielse G *et al* 1995 *Phys. Rev. Lett.* **74** 3544
[19] Amoretti M *et al* 2002 *Nature* **419** 456
[20] Amsler C *et al* 2014 *Int. J. Mod. Phys.* A **29** 1430
[21] Andresen G B *et al* 2010 *Nature* **468** 673

IOP Publishing

Nuclear and Particle Physics

Claude Amsler

Chapter 12

Hadronic interactions

12.1 Cross section

Consider a reaction between two particles 1 and 2 in the initial state i leading to the final state f with $n - 2$ particles:

$$1 + 2 \to 3 + 4 + \cdots + n. \tag{12.1}$$

The 4-momenta are denoted by $\mathcal{P}_j = (E_j, \vec{k}_j)$, the masses by m_1, m_2 and the spins by s_1, s_2. The incident particles are assumed to be unpolarized. The general formula for the differential cross section reads

$$\boxed{\mathrm{d}\sigma = (2\pi)^4 \delta\left(\mathcal{P}_f - \mathcal{P}_i\right) \frac{1}{FK} |\mathcal{M}|^2 \mathrm{d}LIPS}, \tag{12.2}$$

where

$$\begin{aligned} \delta\left(\mathcal{P}_f - \mathcal{P}_i\right) &= \delta\left(\mathcal{P}_3 + \mathcal{P}_4 + \cdots + \mathcal{P}_n - \mathcal{P}_1 - \mathcal{P}_2\right) \\ &= \delta\left(\vec{k}_3 + \cdots + \vec{k}_n - \vec{k}_1 - \vec{k}_2\right)\delta\left(E_3 + \cdots + E_n - E_1 - E_2\right), \end{aligned} \tag{12.3}$$

$$F = 4 \times \sqrt{\left(\mathcal{P}_1 \cdot \mathcal{P}_2\right)^2 - m_1^2 m_2^2}, \tag{12.4}$$

$$K = \left(2s_1 + 1\right)\left(2s_2 + 1\right), \tag{12.5}$$

$$\mathrm{d}LIPS = \frac{\mathrm{d}^3\vec{k}_3}{(2\pi)^3 2E_3} \frac{\mathrm{d}^3\vec{k}_4}{(2\pi)^3 2E_4} \cdots \frac{\mathrm{d}^3\vec{k}_n}{(2\pi)^3 2E_n}. \tag{12.6}$$

The δ-function ensures conservation of total energy and total momentum. The factor F describes the flux of incident particles. The factor $\mathrm{d}LIPS$ is the Lorentz invariant *phase space*.

doi:10.1088/978-0-7503-1140-3ch12

© IOP Publishing Ltd 2015

The factor $\mathcal{M}$ stands for the transition amplitude describing the actual physics of the process. In first order perturbation theory the transition amplitude is given by

$$V_{fi} = -\mathrm{i} \int \psi_f^* V \psi_i \mathrm{d}^4 x = -\mathrm{i}(2\pi)^4 \delta\left(\mathcal{P}_f - \mathcal{P}_i\right)\mathcal{M}, \tag{12.7}$$

where V is the perturbing potential. The wave functions have to be expressed in Lorentz invariant form and should be normalized so that σ has the dimension of a surface. $\mathcal{M}$ is usually unknown for strong interaction processes since perturbation theory is not applicable. The term $|\mathcal{M}|^2$ assumes implicitly a sum over all spins in the initial and final states.

The multiplicity factor K is the spin multiplicity in the initial state. The expression (12.2) consists of a series of Lorentz invariant terms and is thus independent of the reference system (and each term could be evaluated in different reference frames). In particular, the terms $\mathrm{d}^3\vec{k}_j/E_j$ in $\mathrm{d}LIPS$ are Lorentz invariant and take the same values in any reference frame (the proof is left as an exercise for the reader).

Similarly, the partial width for the decay of a particle with mass M into $n-2$ particles is given by:

$$\boxed{\mathrm{d}\Gamma = (2\pi)^4 \delta\left(\mathcal{P}_f - \mathcal{P}_i\right) \frac{1}{2M} |\mathcal{M}|^2 \mathrm{d}LIPS}. \tag{12.8}$$

12.2 Two-body final state

Let us calculate from (12.2) the total cross section for the two-body reaction

$$1 + 2 \rightarrow 3 + 4 \tag{12.9}$$

with kinematic quantities given in the *centre-of-mass system* of the reaction. With $E \equiv E_1 + E_2$ and $\vec{k}_1 = -\vec{k}_2$ one obtains for the total cross section

$$\sigma = \frac{1}{4\pi^2 FK} \int \frac{\mathrm{d}^3\vec{k}_3}{2E_3} \frac{\mathrm{d}^3\vec{k}_4}{2E_4} |\mathcal{M}|^2 \delta\left(\vec{k}_3 + \vec{k}_4\right) \delta\left(E_3 + E_4 - E\right). \tag{12.10}$$

Integrating over $\vec{k}_4$ ($\vec{k}_4 = -\vec{k}_3$) gives

$$\sigma = \frac{1}{4\pi^2 FK} \int \frac{1}{4E_3(k_3)E_4(k_3)}$$

$$|\mathcal{M}|^2 k_3^2 \mathrm{d}k_3 \mathrm{d}\cos\theta_3 \mathrm{d}\phi_3\ \delta\left(E_3(k_3) + E_4(k_3) - E\right), \tag{12.11}$$

where θ_3 and ϕ_3 are the polar and azimuthal emission angles of particle 3, and where

$$E_3(k_3) = \sqrt{k_3^2 + m_3^2}, \qquad E_4(k_3) = \sqrt{k_3^2 + m_4^2}. \tag{12.12}$$

We use the property of the δ-distribution

$$\int f(x)\delta(g(x))\mathrm{d}x = f(x_0)\frac{1}{|g'(x_0)|} \quad \text{with} \quad x_0, \text{ so that } g(x_0) = 0, \tag{12.13}$$

and integrate over $k_3 \equiv x$:

$$\sigma = \frac{1}{4\pi^2 FK} \frac{1}{4E_3(k_3)E_4(k_3)} \frac{k_3^2}{\frac{dE_3}{dk_3} + \frac{dE_4}{dk_3}} \int |\mathcal{M}|^2 d\cos\theta_3 d\phi_3. \tag{12.14}$$

The derivatives are obtained from (12.12):

$$\frac{dE_3}{dk_3} = \frac{k_3}{E_3} \quad \text{and} \quad \frac{dE_4}{dk_3} = \frac{k_3}{E_4}. \tag{12.15}$$

Hence

$$\sigma = \frac{1}{4\pi^2 FK} \frac{1}{4E_3 E_4} \frac{k_3^2}{k_3} \frac{E_3 E_4}{E_3 + E_4} \int |\mathcal{M}|^2 d\cos\theta_3 d\phi_3$$

$$= \frac{1}{16\pi^2 FK} \frac{k_3}{E} \int |\mathcal{M}|^2 d\cos\theta_3 d\phi_3, \tag{12.16}$$

since $E = E_3 + E_4$. The flux factor is given by

$$F = 4 \times \sqrt{(\mathcal{P}_1 \cdot \mathcal{P}_2)^2 - m_1^2 m_2^2} = 4 \times \sqrt{\left(E_1 E_2 - \vec{k}_1 \cdot \vec{k}_2\right)^2 - m_1^2 m_2^2}. \tag{12.17}$$

Since in the centre-of-mass system $\vec{k}_2 = -\vec{k}_1$ and $E_{1,2}^2 = k_1^2 + m_{1,2}^2$ one obtains

$$F = 4 \times \sqrt{\left(k_1^2 + m_1^2\right)\left(k_1^2 + m_2^2\right) + 2k_1^2 E_1 E_2 + k_1^4 - m_1^2 m_2^2}$$

$$= 4 \times \sqrt{2k_1^4 + k_1^2\left(m_1^2 + m_2^2\right) + 2k_1^2 E_1 E_2}$$

$$= 4k_1 \sqrt{k_1^2 + m_1^2 + k_1^2 + m_2^2 + 2E_1 E_2} = 4k_1 \sqrt{\left(E_1 + E_2\right)^2} = 4k_1 E. \tag{12.18}$$

The (Lorentz invariant) total cross section is then

$$\sigma = \frac{1}{64\pi^2 KE^2} \frac{k_3}{k_1} \int |\mathcal{M}|^2 d\cos\theta_3 d\phi_3, \tag{12.19}$$

where all kinematic variables refer to the centre-of-mass system, and θ_3, ϕ_3 are the scattering and azimuthal angles of one of the two outgoing particles. Integrating over ϕ_3 gives

$$\sigma = \frac{1}{64\pi^2 KE^2} \frac{k_3}{k_1} \int |\mathcal{M}|^2 2\pi d\cos\theta_3. \tag{12.20}$$

With $d\Omega = 2\pi d\cos\theta_3$ one obtains the differential cross section (angular distribution) in the *centre-of-mass* frame

$$\boxed{\frac{d\sigma}{d\Omega} = \frac{1}{64\pi^2 KE^2} \frac{k_3}{k_1} |\mathcal{M}|^2}. \tag{12.21}$$

By comparison one also obtains the differential partial decay width for a particle of mass M. Replacing the flux factor $4k_1E$ by $2M$ and comparing (12.2) and (12.8) leads to

$$\frac{\mathrm{d}\Gamma}{\mathrm{d}\Omega} = \frac{k}{32\pi^2 M^2}\,|\mathcal{M}|^2, \tag{12.22}$$

where k is the daughter momentum in the rest frame of the mother particle. The term $|\mathcal{M}|^2$ assumes implicitly a sum over all initial and final state spins.

12.2.1 Detailed balance

As an illustration, let us determine the spin of the charged pion by measuring the total cross section σ_A of

$$pp \to d\pi^+ \quad \text{(reaction } A\text{)} \tag{12.23}$$

and the total cross section σ_B of the time-reversed reaction

$$\pi^+ d \to pp \quad \text{(reaction } B\text{)} \tag{12.24}$$

at the *same* centre-of-mass energy E. The differential cross section is measured at a fixed incident energy and is integrated over all scattering angles. The kinematics of this two-body reaction is fully determined by detecting only one of the two outgoing particles: the four momenta in the centre-of-mass frame $\vec{k}_1$, $\vec{k}_2$, $\vec{k}_3$, $\vec{k}_4$ are coplanar (they define the scattering plane) and $\vec{k}_2 = -\vec{k}_1$ is given by the centre-of-mass energy E. In the scattering plane four parameters describe the components of $\vec{k}_3$ and $\vec{k}_4$, of which three are constrained by energy and momentum conservation, thus leaving only one degree of freedom, for example θ_3. However, in practice both outgoing particles have to be identified to suppress background (here for instance from $pp \to pp$). Hence both final state particles need to be detected and the expected angular correlation verified.

Since reactions A and B have the same centre-of-mass energy

$$E = 2\sqrt{m_p^2 + k_p^2} = \sqrt{m_d^2 + k_d^2} + \sqrt{m_\pi^2 + k_\pi^2\left(= k_d^2\right)}, \tag{12.25}$$

the momenta of the incoming (outgoing) particles in reaction A are equal to the momenta of the outgoing (incoming) particles in reaction B. For A one expects from (12.19) with $K = (2s_p + 1)^2 = 4$

$$\sigma_A = \frac{1}{4 \times 64\pi^2 E^2}\frac{k_d}{k_p}\int |\mathcal{M}_A|^2 \mathrm{d}\cos\theta_d \mathrm{d}\phi_d, \tag{12.26}$$

and for B with $K = (2s_d + 1)(2s_\pi + 1) = 3(2s_\pi + 1)$ and $s_d = 1$

$$\sigma_B = \left(\frac{1}{2}\right)\frac{1}{3(2s_\pi + 1)64\pi^2 E^2}\frac{k_p}{k_d}\int |\mathcal{M}_B|^2 \mathrm{d}\cos\theta_p \mathrm{d}\phi_p. \tag{12.27}$$

The integrations over θ_d and θ_p run from 0 to π. However, integrating over θ_p from 0 to π leads for σ_B to a differential cross section that is two times too large because the process in which one of the protons is emitted into the forward hemisphere with respect to the incident pion cannot be distinguished from the backward case (figure 12.1), hence the factor $\frac{1}{2}$ in (12.27).

The transition amplitudes are related by *time reversal invariance*. According to (11.111)

$$|\mathcal{M}_A|^2 = |\mathcal{M}_B|^2. \tag{12.28}$$

The spin of the pion is then determined by building the ratio of cross sections

$$\frac{\sigma_A}{\sigma_B} = \frac{k_d}{4k_p} \times \frac{6k_d(2s_\pi + 1)}{k_p} = \frac{3}{2}(2s_\pi + 1)\frac{k_d^2}{k_p^2}. \tag{12.29}$$

Figure 12.2 shows the measured cross sections as a function of incident energy in the laboratory. The data points for reactions A and B, corresponding to the same centre-of-mass energies, are displayed in the same plot. For instance, for the lab kinetic energy $T_p = 500$ MeV (reaction A)

$$T_p + m_p = \frac{E^2 - 2m_p^2}{2m_p}, \tag{12.30}$$

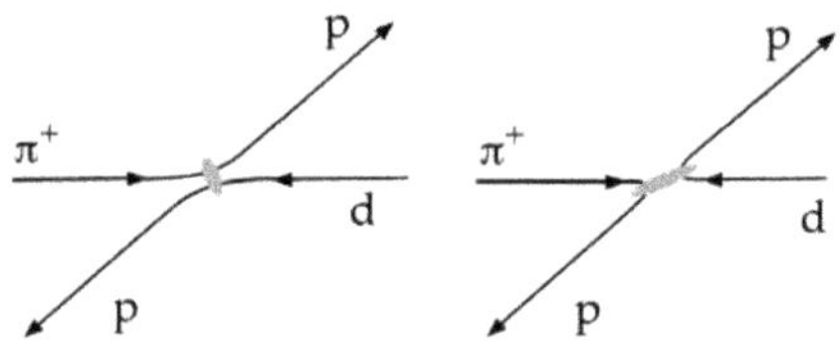

Figure 12.1. Forward protons (left) cannot be distinguished from backward protons (right).

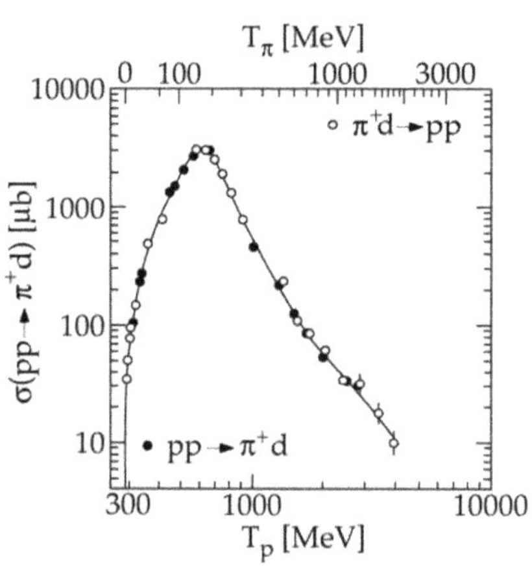

Figure 12.2. Total cross section σ_A for $pp \to d\pi^+$ (black dots) as a function of laboratory kinetic energy (bottom scale). The measured cross section σ_B for $\pi^+ d \to pp$ (open dots) as a function of kinetic energy (top scale) has been converted to σ_A using (12.29), assuming $s_\pi = 0$ for the spin of the pion. After [1].

see (8.61), hence

$$E = \sqrt{2m_p(T_p + m_p) + 2m_p^2} = 2111 \text{ MeV}. \tag{12.31}$$

The corresponding laboratory kinetic energy for reaction B is

$$T_\pi = \frac{E^2 - m_d^2 - m_\pi^2}{2m_d} - m_\pi = 105 \text{ MeV}. \tag{12.32}$$

The momenta in the centre-of-mass systems, calculated from (12.25), are $k_p = 484$ MeV c^{-1} and $k_d = 179$ MeV c^{-1}.

In figure 12.2 the measured cross section σ_B is converted to σ_A using (12.29) and assuming $s_\pi = 0$. The excellent agreement between σ_A and σ_B follows from time reversal invariance and shows unambiguously that the spin of the pion is zero.

12.3 Three-body final state

Starting from (12.8) we now calculate the partial decay width of a state R with mass M into three daughter particles with masses m_j ($j = 1, 2, 3$):

$$R \to 1 + 2 + 3. \tag{12.33}$$

The kinematics of this three-body system is determined by *two* parameters in the rest frame of R, e.g. the kinetic energies T_1 and T_2: the three outgoing momenta $\vec{k}_j$ spanning the decay plane are described by $3 \times 2 = 6$ parameters which are constrained by energy and momentum conservation (three equations) thus leaving $6 - 3 = 3$ free parameters. Moreover, the orientation of the reference axes of the rest frame can be chosen arbitrarily in the decay plane, hence two kinematical variables suffice to determine the final state completely.

We now treat the important special case of a constant transition amplitude $\mathcal{M}$ which does not depend on these two variables. The partial width is then proportional to the three-body *phase space*

$$\mathrm{d}\Gamma \propto \mathrm{d}^9 R_3 \equiv \frac{\mathrm{d}^3\vec{k}_1}{2E_1}\frac{\mathrm{d}^3\vec{k}_2}{2E_2}\frac{\mathrm{d}^3\vec{k}_3}{2E_3}\delta\left(\vec{k}_1 + \vec{k}_2 + \vec{k}_3\right)\delta\left(E_1 + E_2 + E_3 - M\right). \tag{12.34}$$

Integrating over $\vec{k}_3 = -\vec{k}_1 - \vec{k}_2$ gives

$$\mathrm{d}^6 R_3 = \frac{1}{8}\frac{\mathrm{d}^3\vec{k}_1}{E_1}\frac{\mathrm{d}^3\vec{k}_2}{E_2}\frac{1}{E_3}\delta\left(E_1 + E_2 + E_3 - M\right), \tag{12.35}$$

with

$$E_3 = \sqrt{\left(\vec{k}_1 + \vec{k}_2\right)^2 + m_3^2}. \tag{12.36}$$

We choose as the provisional z-axis the direction along $\vec{k}_1$, so that for particle 2

$$\mathrm{d}^3\vec{k}_2 = k_2^2 \mathrm{d}k_2 d \cos\theta \mathrm{d}\phi. \tag{12.37}$$

Equation (12.36) gives for fixed $\vec{k}_1$ and $\vec{k}_2$

$$E_3^2 = m_3^2 + k_1^2 + k_2^2 + 2k_1k_2 \cos\theta \Rightarrow \mathrm{d}\cos\theta = \frac{E_3\mathrm{d}E_3}{k_1k_2}. \tag{12.38}$$

Integrating (12.35) over θ and ϕ gives together with (12.37) and (12.38)

$$\begin{aligned} \mathrm{d}^4R_3 &= \frac{\pi}{4}\frac{\mathrm{d}^3\vec{k}_1}{E_1E_2}\frac{k_2^2\mathrm{d}k_2}{k_1k_2}\int_{E_3^{\min}}^{E_3^{\max}} \mathrm{d}E_3\delta(E_1+E_2+E_3-M) \\ &= \frac{\pi}{4}\frac{\mathrm{d}^3\vec{k}_1}{E_1E_2}\frac{k_2^2\mathrm{d}k_2}{k_1k_2}. \end{aligned} \tag{12.39}$$

Choosing as the z-axis the flight direction of the laboratory one can write for particle 1

$$\mathrm{d}^3\vec{k}_1 = k_1^2\mathrm{d}k_1\mathrm{d}\cos\theta_z\mathrm{d}\phi_z, \tag{12.40}$$

which leads to

$$\mathrm{d}^4R_3 = \frac{\pi}{4}\frac{k_1k_2\mathrm{d}k_1\mathrm{d}k_2}{E_1E_2}\mathrm{d}\cos\theta_z\mathrm{d}\phi_z. \tag{12.41}$$

Since furthermore

$$k_i = \sqrt{E_i^2 - m_i^2} \Rightarrow \mathrm{d}k_i = \frac{E_i}{k_i}\mathrm{d}E_i, \tag{12.42}$$

one finds after integrating over θ_z and ϕ_z ($\rightarrow 4\pi$):

$$\mathrm{d}^2R_3 = \pi^2\mathrm{d}E_1\mathrm{d}E_2 = \pi^2\mathrm{d}T_1\mathrm{d}T_2 = w(T_1, T_1)\mathrm{d}T_1\mathrm{d}T_2, \tag{12.43}$$

where T_1 and T_2 are kinetic energies. The probability that the kinetic energies T_1 and T_2 lie between T_1 and $T_1 + \mathrm{d}T_1$ and between T_2 and $T_2 + \mathrm{d}T_2$, respectively, is *constant*, that is, independent of T_1 and T_2. The events are uniformly distributed in the two-dimensional representation T_1 versus T_2 (*Dalitz plot*). Such events are called *phase space distributed.* Let us examine the boundary of the physically allowed region. The maximum and minimum energies of particle 3 are

$$E_3^{\max} = \sqrt{m_3^2 + (k_1+k_2)^2}, \qquad E_3^{\min} = \sqrt{m_3^2 + (k_1-k_2)^2}. \tag{12.44}$$

Since

$$T_1 + T_2 + T_3 = M - m_1 - m_2 - m_3 \equiv Q, \tag{12.45}$$

one usually plots the events within an equilateral triangle of height Q (figure 12.3) in which phase space distributed events should be uniformly distributed. However, since in general the amplitude $\mathcal{M}$ depends on T_1 and T_2 the resulting non-uniformity of the Dalitz plot yields information on the dynamics of the decay, such as the spin of the decaying state. We shall treat several examples in the following sections.

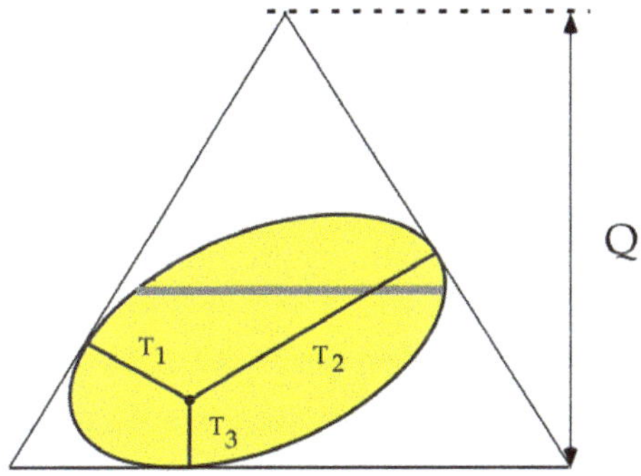

Figure 12.3. Dalitz plot of the three-body decay in the rest frame of the decaying state R. The kinematically allowed range, shown by the yellow area within the equilateral triangle of height Q, is uniformly populated for phase space distributed events. In contrast, an intermediate resonance Z decaying into particles 1 and 2 populates the dark band at fixed T_3.

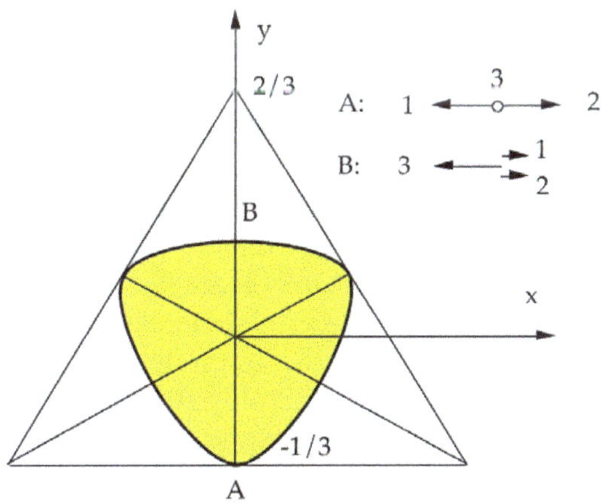

Figure 12.4. Dalitz plot of the three-body decay when the three daughter particles have identical masses.

The state R may also decay into an intermediate state Z with mass M and width Γ_z,

$$R \to Z + 3, \quad \text{where} \quad Z \to 1 + 2, \tag{12.46}$$

in which case the amplitude $\mathcal{M}$ depends on the kinematical parameters since T_3 is fixed by the two-body kinematics $R \to Z + 3$. Our integration above is no longer valid. The events appear in a *band* of width Γ_z around the line $T_3 = \text{constant}$ in the Dalitz plot (figure 12.3). This property is used in three-body decays to identify intermediate states. Furthermore, the directions of the particles are *collinear* for events located at the edges of the Dalitz plot. On the right edge particle 2 flies backwards in the rest frame of Z while flying forwards on the left edge. The event distribution of events along the band in figure 12.3 then describes the angular distribution of particles 1 and 2 in the rest frame of Z, from which its spin can be determined [2].

The Dalitz plot becomes symmetric in the special case $m_1 = m_2 = m_3 = m$ (figure 12.4). Let us introduce the variables

$$x \equiv \frac{1}{\sqrt{3}Q}(T_1 - T_2) \qquad \text{and} \qquad y \equiv \frac{T_3}{Q} - \frac{1}{3}. \tag{12.47}$$

At point A particle 3 is at rest while 1 and 2 are emitted in opposite directions. On the other hand, particle 3 reaches its maximum possible energy at point B, recoiling against 1 and 2. In the non-relativistic approximation ($Q \ll m$) we can write for point B that

$$T_1 + T_2 + T_3 = \frac{k_1^2}{2m} + \frac{k_2^2}{2m} + T_3 = Q. \tag{12.48}$$

Since $k_1 = k_2 = \frac{k3}{2}$ at point B, (12.48) gives

$$\frac{k_3^2}{4m} + T_3 = \frac{3}{2}T_3 = Q. \tag{12.49}$$

From (12.47) the ordinate of point B is then

$$y_B = \frac{1}{3}. \tag{12.50}$$

The ordinate of point A being $y_A = -\frac{1}{3}$, the Dalitz plot (figure 12.4) reduces to a *circle* for $Q \ll m$. In contrast, in the highly relativistic case ($Q \gg m$) the Dalitz plot becomes a *triangle*. Figure 12.5 shows for example the Dalitz plot of the decay $\omega \to 3\gamma$ which proceeds through the intermediate states $\pi^0\gamma$ and $\eta\gamma$. A triangular Dalitz plot is also obtained for orthopositronium decay into 3γ.

We now deal with two important Dalitz plot analyses.

12.3.1 The spin of the charged kaon

The spin of the charged kaon was determined using the three-body decay

$$K^\pm \to \pi^\pm\pi^+\pi^- \ (Q = 75 \text{ MeV}) \tag{12.51}$$

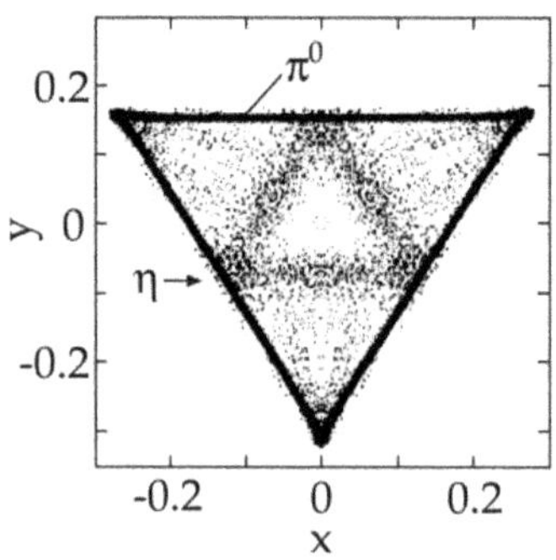

Figure 12.5. Dalitz plot of $\omega \to 3\gamma$. The ω meson was produced in $\bar{p}p$ annihilation into $\omega\pi^0$. From [3].

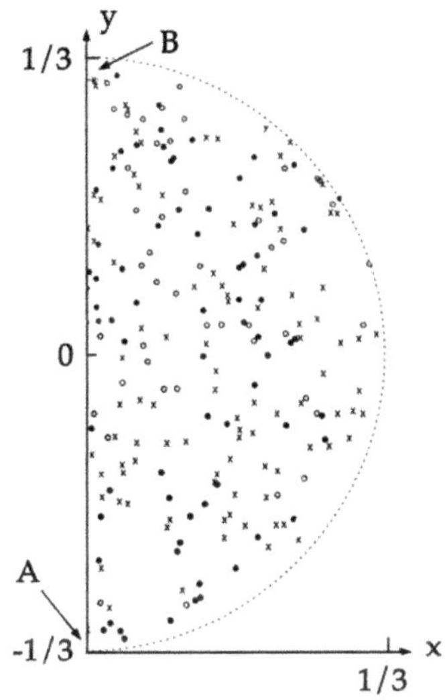

Figure 12.6. Dalitz plot of the decay $K^{\pm} \rightarrow \pi^{\pm}\pi^{+}\pi^{-}$ (219 decays from three different experiments). The variable $x > 0$ (see (12.47)) refers to the two like-charge pions. The dotted curve shows the boundary of the physically allowed area (a circle in the non-relativistic limit). After [4].

(figure 12.6). At point B the angular momentum ℓ between the two like-charge pions vanishes since their relative momentum is zero (point B in figure 12.4). Hence for $\ell > 0$ no events should be observed in the vicinity, in contrast to the data. At point A the $\pi^{\mp}$ is at rest with respect to the $\pi^{\pm}\,\pi^{\pm}$ dipion and the relative angular momentum is $L = 0$. No events should be observed for $L > 0$ around point A either, again in contradiction with the data. Therefore $\ell = L = 0$, leading with spinless pions to the conclusion that $J(K) = 0$.

12.3.2 The spin of the ω meson

The spin and parity of the ω meson were determined by analysing the Dalitz plot of the strong decay

$$\omega \rightarrow \pi^{+}\pi^{-}\pi^{0} \ (Q = 368 \text{ MeV}). \tag{12.52}$$

The isospin of the ω meson is $i(\omega) = 0$. In the final state the isospin of a dipion is $i(2\pi) = 1$ since the isospin of the third pion is 1 and the decay conserves isospin. The total wave function of the dipion must be symmetric, which then demands the angular momentum $\ell(2\pi)$ to be odd. Let L be the angular momentum between the dipion and the third pion. The parity $P = P(\pi)^{3}(-1)^{\ell+L}$ is conserved in strong interactions. Hence the simplest possibilities for the spin and parity of the ω meson are

$$\begin{aligned} \ell &= 1, \qquad L = 0 \Rightarrow J^{P} = 1^{+}, \\ \ell &= 1, \qquad L = 1 \Rightarrow J^{P} = 0^{-},\ 1^{-} \text{ and } 2^{-}. \end{aligned} \tag{12.53}$$

Looking at figure 10.4 we may ignore the 2^{-} case which is unlikely for such a light meson. With ℓ odd the transition amplitude $\mathcal{M}$ is *antisymmetric* with respect to the

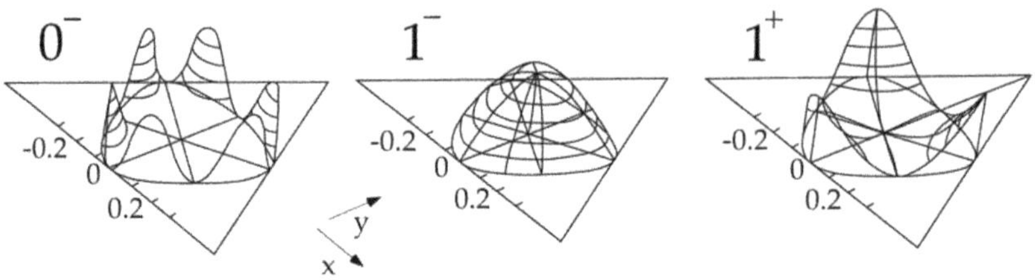

Figure 12.7. Population of the $\omega \to \pi^+\pi^-\pi^0$ Dalitz plot assuming $J^P(\omega) = 0^-$, 1^-, and 1^+. The variables x and y are defined by (12.47). After [5].

permutation of the two pions (antisymmetry of the orbital wave function). The intrinsic parity in the final state being negative ($(P(\pi))^3 = -1$), $\mathcal{M}$ is a scalar for 0^-, an axial vector for 1^- and a vector for 1^+. We can construct the following antisymmetric amplitudes with the help of the momenta $\vec{k}_j$ and kinetic energies T_j:

$$\begin{aligned}
0^- &: \mathcal{M} \propto \left(T_1 - T_2\right)\left(T_2 - T_3\right)\left(T_3 - T_1\right), \\
1^- &: \mathcal{M} \propto \left(\vec{k}_1 \times \vec{k}_2\right) + \left(\vec{k}_2 \times \vec{k}_3\right) + \left(\vec{k}_3 \times \vec{k}_1\right), \\
1^+ &: \mathcal{M} \propto E_3\left(\vec{k}_1 - \vec{k}_2\right) + E_1\left(\vec{k}_2 - \vec{k}_3\right) + E_2\left(\vec{k}_3 - \vec{k}_1\right).
\end{aligned} \tag{12.54}$$

We make the approximation $m_{\pi^\pm} = m_{\pi^0}$. (A description of the Zemach formalism to express transition amplitudes as a function of kinematics variables is given in [6].)

For 1^+ the population of the Dalitz plot vanishes in the centre and at the points B (figure 12.7). For 1^- the intensity vanishes on the boundaries and reaches a maximum in the centre, while for 0^- the intensity vanishes on the symmetry axes. Let us assume that $J^P(\omega) = 1^-$. Since $\vec{k}_3 = -\vec{k}_1 - \vec{k}_2$ one obtains

$$|\mathcal{M}|^2 \propto |\vec{k}_1 \times \vec{k}_2|^2 = |\vec{k}(\pi^+) \times \vec{k}(\pi^-)|^2. \tag{12.55}$$

Plotting the measured intensity as a function of the parameter

$$\xi = \frac{|\vec{k}(\pi^+) \times \vec{k}(\pi^-)|^2}{|\mathcal{M}|^2_{\max}} \tag{12.56}$$

leads to a linear distribution in ξ. The population reaches its maximum for $\xi = 1$ (the centre of the Dalitz plot) and vanishes for $\xi = 0$ at the boundaries. Figure 12.8 shows the distribution of ξ [7]. The ω meson was produced in $\bar{p}p$ annihilation into $\pi^+\pi^-(\omega \to \pi^+\pi^-\pi^0) \to 2\pi^+2\pi^-\pi^0$. The flat background stems from the combinatorial ambiguity of reconstructing the ω meson in the five-pion final state. The distribution is indeed linear in ξ, which establishes that $J^P(\omega) = 1^-$.

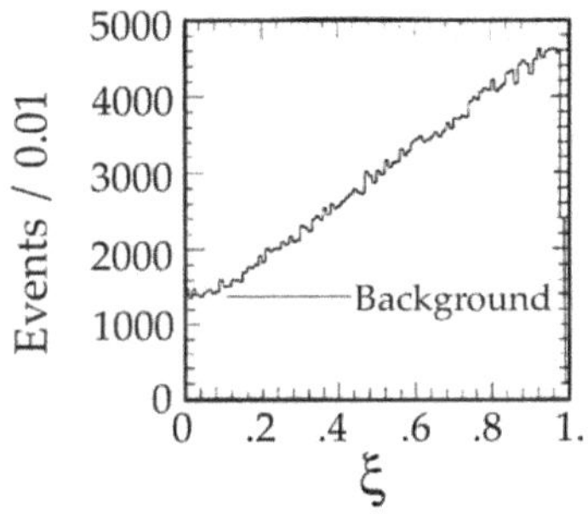

Figure 12.8. ξ distribution for $\omega \to \pi^+\pi^-\pi^0$ (see the text for details). After [7].

12.3.3 Invariant mass distribution

Invariant masses (introduced in section 8.2) are often used as Dalitz plot variables, rather than kinetic energies. Let us define for three-body processes, such as the decay $R \to 1 + 2 + 3$, the two-body invariant masses

$$\begin{aligned} m_{12}^2 &\equiv (\mathcal{P}_1 + \mathcal{P}_2)^2, \\ m_{13}^2 &\equiv (\mathcal{P}_1 + \mathcal{P}_3)^2, \\ m_{23}^2 &\equiv (\mathcal{P}_2 + \mathcal{P}_3)^2. \end{aligned} \tag{12.57}$$

It is straightforward to show that

$$\begin{aligned} m_{12}^2 + m_{13}^2 + m_{23}^2 &= m_1^2 + m_2^2 + m_3^2 + (\mathcal{P}_1 + \mathcal{P}_2 + \mathcal{P}_3)^2 \\ &= m_1^2 + m_2^2 + m_3^2 + M^2, \end{aligned} \tag{12.58}$$

where M is the mass of the state R. As expected, only two of the three invariant masses can be chosen independently.

We now prove that the population of phase space distributed events—expressed as a function of m_{12}^2 versus m_{13}^2—is homogeneous over the Dalitz plot. Evaluating (12.57) in the rest frame of R (the centre-of-mass system of the three-body final state) leads to

$$\begin{aligned} m_{12}^2 &= (E_1 + E_2)^2 - \overbrace{(\vec{k}_1 + \vec{k}_2)}^{-\vec{k}_3}{}^2 = (E_1 + E_2)^2 - (\overbrace{E_3}^{M-E_1-E_2})^2 + m_3^2, \\ m_{13}^2 &= \underbrace{(E_1 + E_3)}_{M-E_2}{}^2 - \underbrace{(\vec{k}_1 + \vec{k}_3)}_{-\vec{k}_2}{}^2 = (M - E_2)^2 - E_2^2 + m_2^2. \end{aligned} \tag{12.59}$$

The homogeneous distribution (12.43) becomes, when expressed in the new variables,

$$d^2R_3 = \pi^2 dE_1 dE_2 = \pi^2 \left| F\left(m_{12}^2, m_{13}^2\right) \right| dm_{12}^2 dm_{13}^2. \tag{12.60}$$

The Jacobi determinant is given by

$$\frac{1}{F\left(m_{12}^2, m_{13}^2\right)} = \frac{\partial\left(m_{12}^2, m_{13}^2\right)}{\partial\left(E_1, E_2\right)}$$

$$= \begin{vmatrix} 2\left(E_1 + E_2\right) + 2\left(M - E_1 - E_2\right) & \cdots \\ 0 & -2\left(M - E_2\right) - 2E_2 \end{vmatrix}$$

$$= -4M^2, \tag{12.61}$$

so that

$$\boxed{\mathrm{d}^2 R_3 = \frac{\pi^2}{4M^2} \mathrm{d}m_{12}^2 \mathrm{d}m_{13}^2}. \tag{12.62}$$

The Dalitz plot density m_{12}^2 versus m_{13}^2 is indeed uniform.

Figure 12.9 shows, as a first example, the Dalitz plot of the reaction

$$K^- p \to \Lambda\, \pi^+ \pi^-, \tag{12.63}$$

with 1.22 GeV c^{-1} kaons, and where the Λ decays into $\pi^- p$ [8]. The (Lorentz) invariant masses (12.57) can be calculated directly by inserting the π^+, π^- and Λ momenta measured in the *laboratory* system. The accumulation of events at $m_{12} \equiv m(\Lambda\pi^-)$ and $m_{13} \equiv m(\Lambda\pi^+) \simeq 1.4$ GeV stems from the decay of the intermediate resonance

$$\Sigma^{\pm}(1385) \to \Lambda\pi^{\pm}. \tag{12.64}$$

Note that, due to the linear relationship (12.58) between the squares of the invariant mass (M being here the centre-of-mass energy), a resonance in the 2π system with mass $m_{23} \equiv m(\pi^+\pi^-)$, would appear as a diagonal band in the Dalitz plot.

Figure 12.10 (left) shows, as a second example, the Dalitz plot of $\bar{p}p$ annihilation into $3\pi^0$ measured by the Crystal Barrel experiment at CERN [9]. The antiprotons are stopped in a liquid hydrogen target and the π^0 mesons are detected through their 2γ decays in a barrel-shaped assembly of 1380 CsI (Tl) crystals (figure 12.10 right, a photograph of the apparatus is shown in figure 12.11). The non-uniform distribution of events in the Dalitz plot signals the presence of intermediate resonances which decay

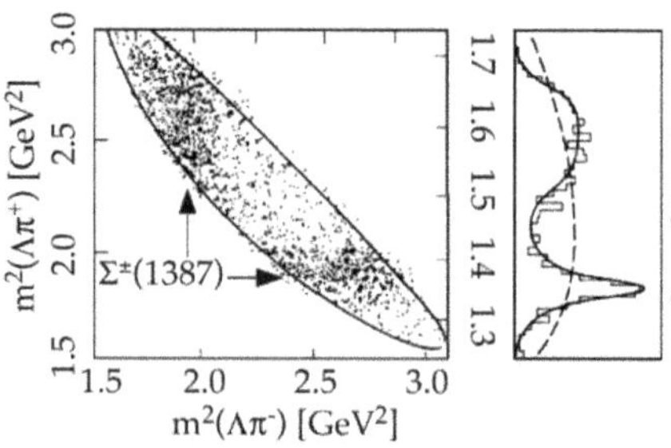

Figure 12.9. Dalitz plot of the reaction $K^- p \to \pi^+\pi^-\Lambda$. The dashed curve in the projection (right) shows the expected distribution for phase space distributed events. After [8].

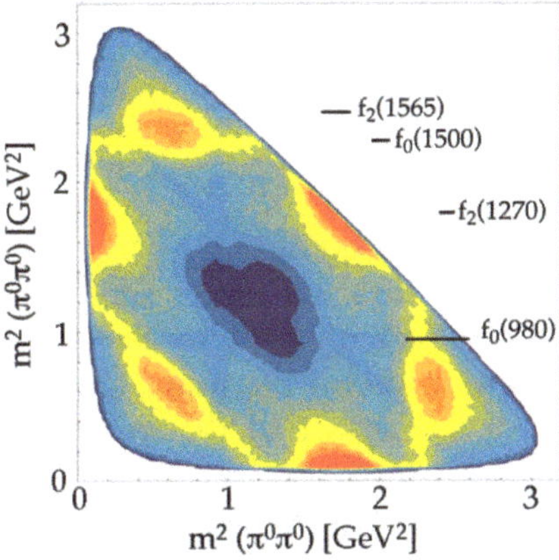

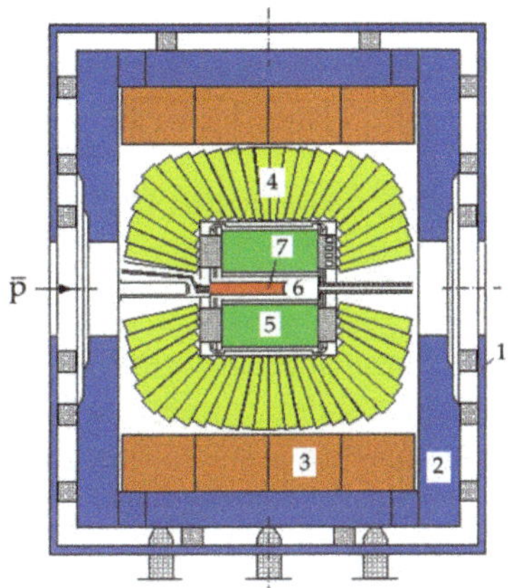

Figure 12.10. Left: experimental Dalitz plot of the annihilation $\bar{p}p \rightarrow 3\pi^0$. The bright (dark) zones correspond to high (low) event density. For symmetry reasons, there are six entries per event. After [9] and [10]. Right: the Crystal Barrel detector at CERN was used to study $\bar{p}p$ annihilation into charged and neutral mesons: magnet yoke (1,2), coil (3), CsI crystals (4), drift chamber (5), wire chamber (6) and hydrogen target (7). After [11].

Figure 12.11. The Crystal Barrel studied light meson spectroscopy from 1989 to 1996. The detector was operated with stopping antiprotons at CERN's Low Energy Antiproton Ring (LEAR). Image credit: CERN.

into $2\pi^0$, such as the scalar mesons $f_0(980)$ and $f_0(1500)$, and the tensor mesons $f_2(1270)$ and $f_2(1565)$. Note the uniform distribution of events in the $f_0(1500)$ band which points to a spin-0 state. The $f_0(1500)$ state has been interpreted as a glueball state [12].

12.4 Charge independence and isospin invariance

Mirror nuclei are pairs of nuclei with the same mass number but with $Z_1 = N_2$ and $Z_2 = N_1$, such as ${}^{13}_{6}C$ and ${}^{13}_{7}N$, or ${}^{17}_{8}O$ and ${}^{17}_{9}F$. The observed energy spectra of excited mirror nuclei are similar (figure 12.12). We have seen in chapter 6 that the unpaired neutron or proton determines the excitation spectrum. Furthermore, the fact that the proton and neutron masses are almost equal indicates that the *strong* interaction does not distinguish between the proton and neutron.

As already discussed, hadrons are grouped in isospin multiplets with nearly degenerate masses. For instance, the nucleon is a doublet of *isospin* $\vec{I}$ with quantum number $i = \frac{1}{2}$ and projection $i_3 = \frac{1}{2}$ (proton) or $-\frac{1}{2}$ (neutron). Since the strong interaction does not distinguish between the $2i + 1$ multiplet states, H commutes with $\vec{I}$:

$$\left[H, \vec{I}\right] = 0. \tag{12.65}$$

This implies spherical symmetry in the three-dimensional isospin space, in analogy to angular momentum in coordinate space for the hydrogen atom: $[H, \vec{L}] = 0$. For a given i the levels are $2i + 1$ times degenerate, e.g. $m_p = m_n$ for $i = \frac{1}{2}$, and the strong interaction is *invariant* under *rotations* in isospin space. For example, both the two-dimensional *isospinor*

$$\psi' = e^{iI_2\alpha}\psi \tag{12.66}$$

and ψ itself qualify to describe the nucleon. The rotation angle $\alpha = \pi$ flips the sign of i_3 and the representation (12.66) causes the transition

$$|p\rangle = \begin{pmatrix} 1 \\ 0 \end{pmatrix} \leftrightarrow |n\rangle = \begin{pmatrix} 0 \\ 1 \end{pmatrix} \tag{12.67}$$

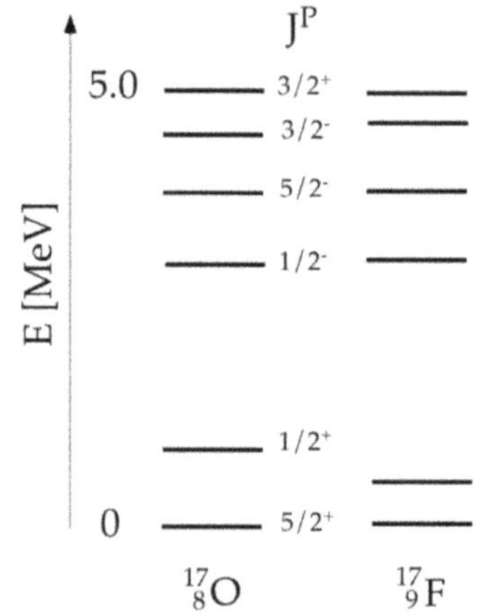

Figure 12.12. The excitation spectrum of ${}^{17}_{8}O$ and ${}^{17}_{9}F$.

(figure 12.13). This invariance under rotations in isospin space is called *charge independence* (sometimes also called 'charge symmetry') and (12.65) implies that i_3 is conserved in strong interactions. Furthermore, one finds experimentally (see the following section) that i itself is conserved in strong interactions (*isospin invariance*). Hence $\vec{I}^2$ commutes with H, as does $\vec{L}^2$ for the hydrogen atom. Switching the electromagnetic interaction as a perturbation, which distinguishes between p and n, one obtains

$$[H, I_1] \neq 0, \qquad [H, I_2] \neq 0, \qquad \text{whereas } [H, I_3] = 0. \tag{12.68}$$

Spherical symmetry is broken in isospin space and $m_p \neq m_n$, but i_3 remains conserved. The analogous Zeeman effect lifts the degeneracy of the hydrogen levels in the presence of a magnetic field.

However, isospin i is not conserved in electromagnetic interactions, as can be seen from the following two allowed decays:

$$\omega \to \pi^0\gamma, \qquad \omega \to \eta\gamma. \tag{12.69}$$

In the first case, i conservation would imply that $i(\gamma) = 1$ since $i(\omega) = 0$ and $i(\pi^0) = 1$. However, the second decay would be in contradiction: $i(\gamma) = 0$ since $i(\eta) = 0$. Nonetheless, i_3 is conserved in both decays if we define $i_3(\gamma) = 0$. A further example of charge independence in electromagnetic interactions is the radiative decay $\rho^\pm \to \pi^\pm\gamma$ for which $i_3 = \pm 1$. We shall see in chapter 13 that neither i nor i_3 are conserved in weak interactions. Note that radiative decays involving W exchange, such as the hyperon decay $\Sigma^+ \to p\gamma$, are in fact weak processes that do not conserve i_3. Table 12.1 summarizes the conservation laws for i and i_3 in the three fundamental interactions.

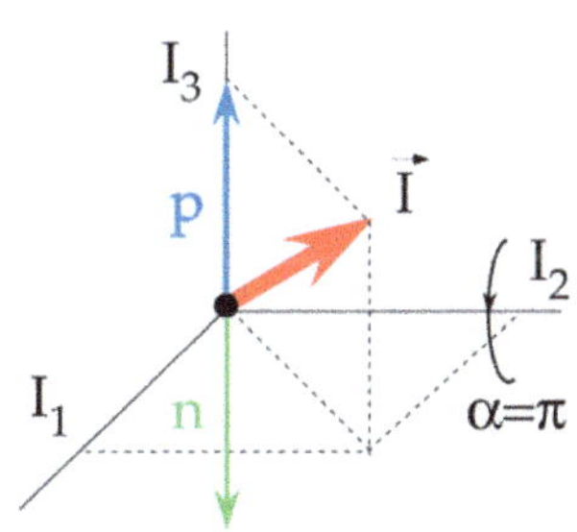

Figure 12.13. Rotation in isospin space around the I_2-axis. For $\alpha = \pi$ the proton becomes a neutron.

Table 12.1. Isospin conservation and charge independence in strong, weak and electromagnetic interactions.

	Strong	Electromagnetic	Weak
i	yes	no	no
i_3	yes	yes	no

12.4.1 Nucleon–nucleon scattering

As a first example of charge independence, let us consider the strong interaction between two nucleons, *pp*, *np* or *nn*. The wave function is composed of orbital and spin parts (11.40):

$$\psi(N_1, N_2) = \phi_L(N_1, N_2)|sm\rangle \equiv \phi_L(N_1, N_2)\chi_s(N_1, N_2), \tag{12.70}$$

where $s = 0$ or 1. We append the isospinor $\alpha_i(N_1, N_2)$:

$$\psi(N_1, N_2) = \phi_L(N_1, N_2)\chi_s(N_1, N_2)\alpha_i(N_1, N_2), \tag{12.71}$$

and distinguish between three possible configurations:

$$\begin{aligned} \alpha_i(N_1, N_2) &= |pp\rangle \Rightarrow i = 1 \text{ since } i_3 = 1, \\ &= |nn\rangle \Rightarrow i = 1 \text{ since } i_3 = -1, \\ &= |np\rangle \Rightarrow i = 0 \text{ or } 1 \text{ since } i_3 = 0. \end{aligned} \tag{12.72}$$

Charge independence requires the strong force F between *pp* and *nn* to be equal:

$$F(pp) = F(nn). \tag{12.73}$$

However, for *np* only the $i = 1$ component should be comparable to *pp*:

$$F(np) \neq F(pp) \quad \text{but} \quad F(np)_{i=1} = F(pp). \tag{12.74}$$

The total wave function $\psi(N_1, N_2)$ is *antisymmetric* under the permutation of the two nucleons since the system consists of two identical fermions (nucleons). Therefore,

$$\psi(N_2, N_1) = (-1)^L(-1)^{s+1}(-1)^{i+1}\psi(N_1, N_2) = -\psi(N_1, N_2), \tag{12.75}$$

where we have used the Clebsch–Gordan coefficients for coupling two spin- or isospin-$\frac{1}{2}$ particles, see (11.45) and (11.47). From (12.75) it follows that

$$L + s + i = \text{odd}. \tag{12.76}$$

For the ground state ($L = 0$) one finds that

$$\begin{aligned} s = 0 &\Rightarrow i = 1 \Rightarrow {}^1S_0(pp, np \text{ or } nn), \\ s = 1 &\Rightarrow i = 0 \Rightarrow {}^3S_1(np). \end{aligned} \tag{12.77}$$

The ${}^3S_1$ state occurs for $i = 0$ only and binds to form the isospin singlet spin-1 deuteron d with charge $Q = +1$. Following charge independence, the interactions between *pp*, *np* and *nn* are equal in the ${}^1S_0$ system which does not bind. A spin-0 'deuteron' would exist as an isospin triplet with the three possible charges, $\tilde{d}^{++}$, $\tilde{d}^{+}$

and $\tilde{d}^0$. Nuclear fusion in the Sun would then proceed through the electromagnetic interaction

$$pp \to \tilde{d}^{++}\gamma, \tag{12.78}$$

instead of the much slower weak process

$$pp \to de^+\nu_e \tag{12.79}$$

(section 14.3), and the Sun would have extinguished a long time ago. We owe life on Earth to the fact that the spin of the deuteron is equal to 1!

Ratios between cross sections can be derived from charge independence and isospin conservation in strong interactions. As an example, consider the reactions

$$pp \to d\pi^+ \qquad \text{and} \qquad np \to d\pi^0. \tag{12.80}$$

The Clebsch–Gordan decompositions (figure 11.6) give with $i(\pi) = 1$ and $i(d) = 0$ the isospinors

$$\begin{aligned} |np\rangle &= \frac{1}{\sqrt{2}}(|10\rangle - |00\rangle), \\ |pp\rangle &= |11\rangle, \\ \langle d\pi^0| &= \langle 10|, \\ \langle d\pi^+| &= \langle 11|. \end{aligned} \tag{12.81}$$

The matrix elements of the transition operator H are

$$\begin{aligned} \left\langle d\pi^+|H|pp\right\rangle &= \langle 11|H|11\rangle, \\ \left\langle d\pi^0|H|np\right\rangle &= \frac{1}{\sqrt{2}}(\langle 10|H|10\rangle - \langle 10|H|00\rangle). \end{aligned} \tag{12.82}$$

The terms $\langle 11|H|11\rangle$ and $\langle 10|H|10\rangle$ are equal, due to charge independence, while the second term in the latter equation vanishes by virtue of isospin conservation. Squaring the matrix elements one obtains the ratio of total cross sections at equal incident energies

$$\frac{\sigma\left(pp \to d\pi^+\right)}{\sigma\left(np \to d\pi^0\right)} = 2, \tag{12.83}$$

which agrees with the data. We have ignored the small mass differences between p and n and between π^+ and π^0, which lead to slightly different phase space factors.

Figure 12.14 shows a measurement of the differential cross section for the reaction $np \to d\pi^0$ with 675 MeV neutrons, as a function of centre-of-mass angle θ [13].

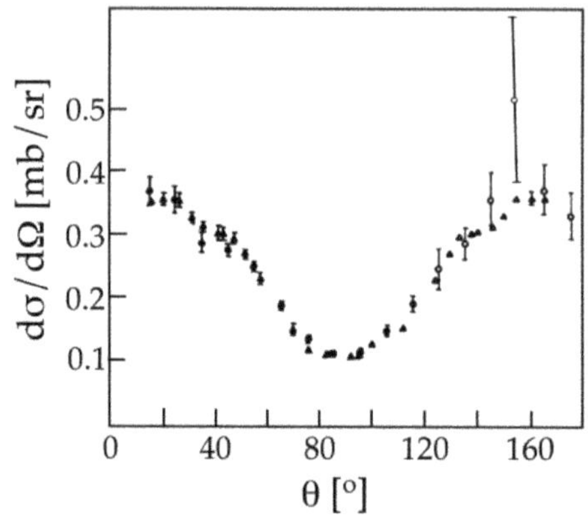

Figure 12.14. Differential cross section in the centre-of-mass system for $np \to d\pi^0$ with 675 MeV neutrons and for $pp \to d\pi^+$ at the same centre-of-mass energy. After [13].

The neutrons were produced by the circulating proton beam in the Berkeley 184-inch cyclotron, traversing a beryllium target. The neutral pions were not detected. The deuterons were detected and their momenta measured with spark chambers located upstream and downstream of a dipole magnet. Neutrons were not monoenergetic and their momenta could not be measured. Nevertheless, measuring the deuteron momentum and scattering angle is sufficient to fully determine the kinematics of the reaction. The angular distribution was compared to $pp \to d\pi^+$ data, obtained from $\pi^+ d \to pp$ data, using detailed balance (section 12.2.1). The data for both reactions are plotted in figure 12.14. For $pp \to d\pi^+$ the cross section was divided by two, following (12.83). Perfect agreement is observed, in accord with isospin invariance and charge independence in strong interactions.

12.4.2 Pion–nucleon scattering

As a second example of charge independence let us discuss the πN system which couples $i = \frac{1}{2}$ to $i = 1$. The combined isospin is either $i = \frac{1}{2}$ or $i = \frac{3}{2}$:

$$\begin{aligned}
\left|\pi^+ p\right\rangle &: i_3 = \frac{3}{2} \Rightarrow i = \frac{3}{2}, \\
\left|\pi^- p, \pi^0 n\right\rangle &: i_3 = -\frac{1}{2} \Rightarrow i = \frac{1}{2}, \frac{3}{2}, \\
\left|\pi^- n\right\rangle &: i_3 = -\frac{3}{2} \Rightarrow i = \frac{3}{2}, \\
\left|\pi^+ n, \pi^0 p\right\rangle &: i_3 = \frac{1}{2} \Rightarrow i = \frac{1}{2}, \frac{3}{2}.
\end{aligned} \tag{12.84}$$

The isospin decomposition gives for the first three states (with $j_1 = 1$ and $j_2 = \frac{1}{2}$, replace J by i and M by i_3 in the Clebsch–Gordan table, figure 11.6):

$$
\begin{aligned}
|\pi^+ p\rangle &= \Big|\overbrace{\tfrac{3}{2}\,\tfrac{3}{2}}^{i\ i_3}\Big\rangle, \\
|\pi^- p\rangle &= \left\langle \frac{3}{2} - \frac{1}{2}\middle| 1\frac{1}{2} - 1\frac{1}{2}\right\rangle \left|\frac{3}{2} - \frac{1}{2}\right\rangle + \left\langle \frac{1}{2} - \frac{1}{2}\middle| 1\frac{1}{2} - 1\frac{1}{2}\right\rangle \left|\frac{1}{2} - \frac{1}{2}\right\rangle \\
&= \sqrt{\frac{1}{3}}\left|\frac{3}{2} - \frac{1}{2}\right\rangle - \sqrt{\frac{2}{3}}\left|\frac{1}{2} - \frac{1}{2}\right\rangle, \\
|\pi^0 n\rangle &= \sqrt{\frac{2}{3}}\left|\frac{3}{2} - \frac{1}{2}\right\rangle + \sqrt{\frac{1}{3}}\left|\frac{1}{2} - \frac{1}{2}\right\rangle.
\end{aligned}
\tag{12.85}
$$

The transition matrix elements

$$
\mathcal{M}_{\frac{1}{2}} \equiv \left\langle i = \frac{1}{2}\middle| H \middle| i = \frac{1}{2}\right\rangle \qquad \text{and} \qquad \mathcal{M}_{\frac{3}{2}} \equiv \left\langle i = \frac{3}{2}\middle| H \middle| i = \frac{3}{2}\right\rangle \tag{12.86}
$$

do not depend on i_3 due to charge independence. Isospin invariance also forbids transitions from $i = \frac{1}{2}$ to $i = \frac{3}{2}$:

$$
\left\langle i = \frac{1}{2}\middle| H \middle| i = \frac{3}{2}\right\rangle = 0. \tag{12.87}
$$

The total cross sections for $\pi^+ p$ and $\pi^- p$ scattering are

$$
\begin{aligned}
\sigma_1 &\equiv \sigma(\pi^+ p \to \pi^+ p) \propto |\langle \pi^+ p|H|\pi^+ p\rangle|^2 \equiv \left|\mathcal{M}_{\frac{3}{2}}\right|^2, \\
\sigma_2 &\equiv \sigma(\pi^- p \to \pi^- p) \propto |\langle \pi^- p|H|\pi^- p\rangle|^2 \equiv \left|\frac{1}{3}\mathcal{M}_{\frac{3}{2}} + \frac{2}{3}\mathcal{M}_{\frac{1}{2}}\right|^2, \\
\sigma_3 &\equiv \sigma(\pi^- p \to \pi^0 n) \propto |\langle \pi^0 n|H|\pi^- p\rangle|^2 \equiv \left|\frac{\sqrt{2}}{3}\mathcal{M}_{\frac{3}{2}} - \frac{\sqrt{2}}{3}\mathcal{M}_{\frac{1}{2}}\right|^2.
\end{aligned}
\tag{12.88}
$$

Figure 12.15 shows the measured total cross sections as a function of pion kinetic energy in the laboratory. Inelastic scattering (pion production such as $\pi N \to \pi\pi N$) also contributes to the cross section above ~500 MeV. Below 500 MeV the excitation of the $\Delta(1232)$ resonance dominates. The resonance is produced in $\pi^+ p$, hence its isospin is $i = \frac{3}{2}$, and it exists in four charges with degenerate masses: Δ^{++}, Δ^+, Δ^0 and Δ^-.

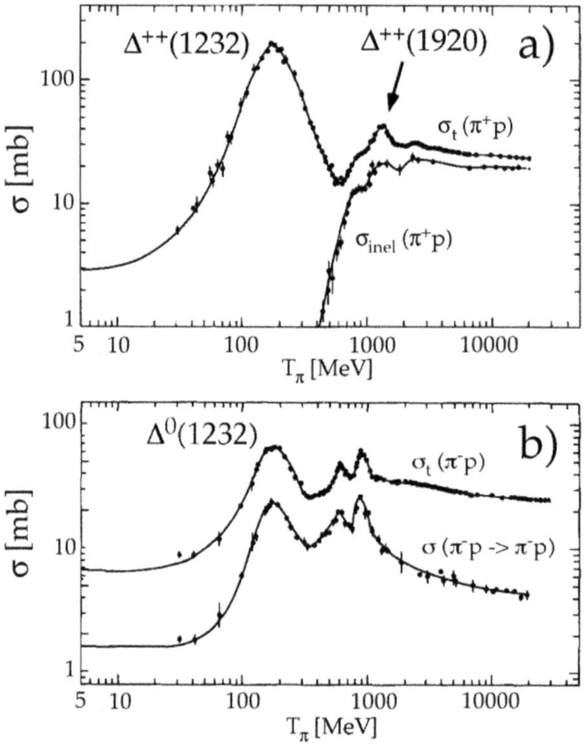

Figure 12.15. (*a*) Total cross section $\sigma(\pi^+p) = \sigma_1 + \sigma_{\text{inel}}$ as a function of pion kinetic energy (top curve). The bottom curve shows the inelastic contribution σ_{inel} from pion production. The $\Delta^{++}(1232)$ and $\Delta^{++}(1920)$ baryon resonances have $J^P = \frac{3}{2}^+$ and $\frac{7}{2}^+$, respectively. (*b*) Total cross sections $\sigma(\pi^-p) = \sigma_2 + \sigma_3 + \sigma_{\text{inel}}$ (top curve) and σ_2 (bottom curve). The structures around 600 and 900 MeV (mainly from the $J^P = \frac{3}{2}^-$ $N(1520)$ and $\frac{5}{2}^+$ $N(1680)$ excited baryons) have isospin $\frac{1}{2}$ since they are not excited in π^+p scattering. After [1].

Since this resonance dominates low energy πN scattering one can assume that $\mathcal{M}_{\frac{1}{2}} \simeq 0$ and predict in good approximation from (12.88) the ratios of the total cross sections

$$\sigma_1 : \sigma_2 : \sigma_3 \propto 9 : 1 : 2, \tag{12.89}$$

which are in excellent agreement with the data (figure 12.15).

From measurements of the differential cross section and analysing power (see figure 11.4) one finds that the Δ resonance is excited in a P wave of πN scattering ($L = 1$). Its parity is $P_\Delta = P_\pi P_N (-1)^L = +1$. Its spin $J = \frac{3}{2}$ can be inferred directly from the data in figure 12.15(*a*), by describing the enhancement in π^+p with the Breit–Wigner function

$$\boxed{\sigma_R(E) = \frac{2J+1}{(2s_\pi + 1)(2s_p + 1)} \frac{\pi}{k^2} \left[\frac{\Gamma^2}{(E - m_R)^2 + \Gamma^2/4} \right]}, \tag{12.90}$$

where k and E are evaluated in the centre-of-mass and $m_R = 1232$ MeV.

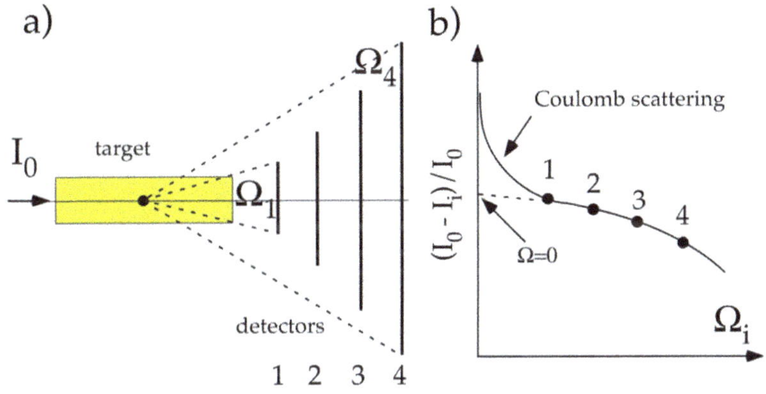

Figure 12.16. (*a*) Principle of a transmission experiment to measure total cross sections. (*b*) The total cross section is obtained by extrapolating the scattered intensities outside the detectors to zero solid angle.

Above 500 MeV the total cross section cannot be obtained simply by integrating the measured differential cross section, because of the contribution from inelastic channels. Total cross sections are typically measured with the transmission method. Consider a beam of intensity I_0 impinging on a target of length d, density ρ, and atomic weight A. One measures the scattered intensities I_i into a series of forward counters, each covering the solid angle Ω_i (figure 12.16(*a*)). The intensity $I_{\Omega_i} = I_0 - I_i$ of particles scattered *outside* the solid angle Ω_i is then calculated. The total cross section σ_T is obtained by extrapolating to zero solid angle $I_{(\Omega\to 0)}$ since

$$I_{(\Omega\to 0)} = I_0\left(1 - \exp\left(-\sigma_T nd\right)\right) \quad \text{with} \quad n = \rho\frac{L}{A} \tag{12.91}$$

(see figure 12.16(*b*)). Solving for σ_T gives the total cross section

$$\sigma_T = \frac{1}{nd}\ln\left[\frac{I_0}{I_0 - I_{(\Omega\to 0)}}\right]. \tag{12.92}$$

Note that the *differential* cross section does not necessarily vanish at $\theta = 0$, even in the absence of Coulomb interaction.

12.5 *G* parity

We have seen that mesons with charge $Q = 0$ and flavour quantum numbers $S = C = B' = 0$ are eigenstates of the C parity (section 11.7). Let us now include the electrically *charged* mesons with $S = C = B' = 0$ (the $i = 1$ $q\bar{q}$ mesons such as the pion), and define the combined operation of charge conjugation and rotation of 180° around the I_2-axis in isospin space. The G parity operation is defined as

$$\boxed{G = C\mathrm{e}^{\mathrm{i}\pi I_2}}. \tag{12.93}$$

For instance, this rotation transforms a π^+ into a π^- and C conjugation recovers the π^+. Hence the charged pion is an eigenstate of G parity. Since isospin is here an integer number, the isospinor behaves like an orbital wave function under rotations in three-dimensional space:

$$e^{i\pi L_y} Y_\ell^m(\theta, \phi) = Y_\ell^m(\theta + \pi, \pi - \phi) = (-1)^\ell Y_\ell^m(\theta, \phi). \tag{12.94}$$

Hence for such mesons

$$G|\ \rangle = (-1)^i C|\ \rangle. \tag{12.95}$$

The G parity of the π^0 is -1:

$$G|\pi^0\rangle = (-1)^{i=1} C|\pi^0\rangle = -|\pi^0\rangle. \tag{12.96}$$

By charge independence this is also true for the charged pion:

$$G|\pi^\pm\rangle = -|\pi^\pm\rangle. \tag{12.97}$$

The G parities of the ρ^0 and ω mesons are

$$\begin{aligned} G|\rho\rangle &= (-1)^1 C|\rho\rangle = +|\rho\rangle, \\ G|\omega\rangle &= (-1)^0 C|\omega\rangle = -|\omega\rangle. \end{aligned} \tag{12.98}$$

By charge independence this is also true for the charged ρ:

$$G|\rho^\pm\rangle = |\rho^\pm\rangle. \tag{12.99}$$

The conservation of G parity in strong interactions follows from C invariance and charge independence. Although not an additional symmetry, G parity is very handy, as illustrated by the following examples.

- The G parity of a system made of n pions is

  $$G|n\pi\rangle = (-1)^n |n\pi\rangle, \tag{12.100}$$

  from which the Furry theorem follows: the number of outgoing pions in $\pi\pi$ scattering is even (figure 12.17).
- Due to G parity conservation, the ρ decays into 2π and the ω into 3π. However, the ρ^0 does not decay into $\pi^0\pi^0$ (J, C and I conservations) and ω does not decay into $3\pi^0$ (C and I conservations).

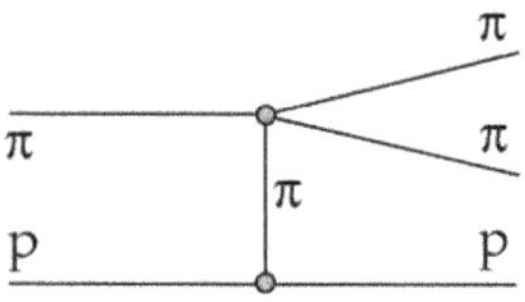

Figure 12.17. Inelastic πp scattering with π exchange in which a further pion is produced. The total number of pions at the vertex must be even.

- The G parity of the η meson is

$$G|\eta\rangle = (-1)^0 C|\eta\rangle = +|\eta\rangle. \tag{12.101}$$

The η cannot decay through the G-conserving strong process $\eta \to \pi\pi$ since $J^P(\eta) = 0^-$, which violates either P or J conservation. Therefore the η decays into 3π (with $f = 55.6\%$). As we have seen in section 11.7 the (obviously) electromagnetic decay $\eta \to 2\gamma$ occurs with a comparable branching ratio ($f = 39.4\%$). Thus, $\eta \to 3\pi$ is an electromagnetic decay which does not conserve isospin nor G.

- The G parity of a particle–antiparticle pair is, using (11.108) and (12.95),

$$G\left|a\bar{a}\right\rangle = (-1)^i C\left|a\bar{a}\right\rangle = (-1)^{\ell+s+i}\left|a\bar{a}\right\rangle. \tag{12.102}$$

For example, for a pion pair,

$$G\,|\pi^+\pi^-\rangle = (-1)^{\ell+i}|\pi^+\pi^-\rangle = +|\pi^+\pi^-\rangle. \tag{12.103}$$

Therefore, the isospin of the pair depends on the angular momentum between the pions. For ℓ even, $i = 0$ (or 2), while for ℓ odd, $i = 1$.

G parity is not defined for strange mesons ($S \neq 0$) nor for mesons with open charm or bottomness ($C \neq 0$, $B' \neq 0$). For example, the K^+ transforms under G parity as follows:

$$K^+ \overbrace{\Rightarrow}^{\exp(\mathrm{i}\pi I_2)} K^0 \overset{C}{\Rightarrow} \overline{K}^0. \tag{12.104}$$

Table 12.2. Conservation laws: [1]conserved for neutral currents; [2]conserved for hadrons; [3]conserved except in the $K^0 - \overline{K}^0$ system and in hadronic B decays.

Symmetry	Strong	Electromagnetic	Weak
Baryon number	√	√	√
Total angular momentum (J)	√	√	√
CPT	√	√	√
Flavour	√	√	×[1]
Colour	√	√[2]	√[2]
Lepton number	–	√	√
CP	√	√	√[3]
Time reversal (T)	√	√	√[3]
P parity	√	√	×
C parity	√	√	×
Isospin (i)	√	×	×
Isospin (i_3)	√	√	×
G parity	√	×	×

Hence K^+ is not an eigenstate of G. Kaon pairs such as K^+K^- and $K^0\overline{K}^0$ are not eigenstates of G either. The eigenstates $|a\bar{a}\rangle$ of G are the eigenstates of the isospin I:

$$\begin{aligned} i = 0 : |K\overline{K}\rangle &= \frac{1}{\sqrt{2}}\left(\left|K^0\,\overline{K}^0\right\rangle + |K^+\,K^-\rangle\right), \\ i = 1 : |K\overline{K}\rangle &= \frac{1}{\sqrt{2}}\left(\left|K^0\,\overline{K}^0\right\rangle - |K^+\,K^-\rangle\right). \end{aligned} \tag{12.105}$$

We conclude this chapter by summarizing in table 12.2 the conservation laws for the three fundamental interactions.

References

[1] Lock W O and Measday D F 1970 *Intermediate Energy Nuclear Physics* (London: Methuen)
[2] Amsler C and Bizot J C 1983 *Comput. Phys. Commun.* **30** 21
[3] Abele A *et al* 1997 *Phys. Lett.* B **411** 361
[4] Orear J, Harris G and Taylor S 1956 *Phys. Rev.* **102** 1676
[5] Stevenson M L *et al* 1962 *Phys. Rev.* **125** 687
[6] Zemach C 1964 *Phys. Rev.* B **133** 1201
[7] Weidenauer P *et al* 1993 *Z. Phys.* C **59** 387
[8] Shafer J *et al* 1963 *Phys. Rev. Lett.* **10** 179
[9] Amsler C *et al* 1995 *Phys. Lett.* B **342** 433
[10] Amsler C 1998 *Rev. Mod. Phys.* **70** 1293
[11] Aker E *et al* 1992 *Nucl. Instrum. Methods* A **321** 69
[12] Amsler C and Close F E 1996 *Phys. Rev.* D **53** 295
[13] Wilson S S *et al* 1971 *Phys. Lett.* B **35** 83

IOP Publishing

Nuclear and Particle Physics

Claude Amsler

Chapter 13

Weak interactions

Weak decays are classified into three categories:

1. Purely *leptonic* interactions involving only leptons such as

$$\mu^+ \to e^+\nu_e\bar{\nu}_\mu \qquad \text{or} \qquad \nu_e e^- \to \nu_e e^-. \tag{13.1}$$

2. *Semileptonic* interactions involving leptons and hadrons, e.g.

$$n \to pe^-\bar{\nu}_e, \tag{13.2}$$

or flavour changing decays such as

$$K^+ \to \pi^0 e^+\nu_e. \tag{13.3}$$

3. *Non-leptonic* interactions among hadrons only, e.g.

$$K^+ \to \pi^+\pi^0 \qquad \text{or} \qquad \Lambda \to p\pi^-. \tag{13.4}$$

The rule

$$\boxed{\Delta S = \Delta Q} \tag{13.5}$$

with $|\Delta S| = 1$ applies to *semileptonic* decays of hadrons with strangeness $S \neq 0$, where Q is the charge of the *hadrons*. For example, the $(S = -1)$ Σ^- hyperon decays semileptonically into a neutron ($f = 10^{-3}$):

$$\Sigma^- \to ne^-\bar{\nu}_e. \tag{13.6}$$

The strangeness S and charge Q of the hadron both increase by one unit. In contrast, the decay of the $(S = -1)$ Σ^+ hyperon

$$\Sigma^+ \to ne^+\nu_e \tag{13.7}$$

doi:10.1088/978-0-7503-1140-3ch13 © IOP Publishing Ltd 2015

is forbidden by the $\Delta S = \Delta Q$ rule since the charge decreases by 1. As further examples, the semileptonic decays

$$K^0(S=+1) \to \pi^- e^+ \nu_e \qquad \text{and} \qquad \overline{K}^0(S=-1) \to \pi^+ e^- \bar{\nu}_e \tag{13.8}$$

are allowed ($f = 7 \times 10^{-4}$) while the decays

$$K^0 \to \pi^+ e^- \bar{\nu}_e \qquad \text{and} \qquad \overline{K}^0 \to \pi^- e^+ \nu_e \tag{13.9}$$

are forbidden.

Before the discovery of quarks the $\Delta S = \Delta Q$ rule was purely empirical. The rule can be understood in the quark model from the underlying quark dynamics, namely by the quark transitions $s \to W^- u$ (with $W \to$ leptons) in which $\Delta S = \Delta Q = +1$.

Isospin i is not conserved in weak interactions, as the examples (13.4) of non-leptonic decays with strangeness $S \neq 0$ show. The empirical rule

$$\boxed{|\Delta i| = \frac{1}{2}} \tag{13.10}$$

holds for these decays. Transitions with $|\Delta i| = \frac{3}{2}$ are also possible, but much less probable. As an example, consider the decays $\Lambda \to \pi^- p$ and $\Lambda \to \pi^0 n$, and decompose the isospin wave function of the πN system in its $i = \frac{1}{2}$ and $i = \frac{3}{2}$ components, see (12.85):

$$\begin{aligned} \left|\pi^- p\right\rangle &= -\sqrt{\frac{2}{3}}\left|\frac{1}{2}-\frac{1}{2}\right\rangle + \sqrt{\frac{1}{3}}\left|\frac{3}{2}-\frac{1}{2}\right\rangle, \\ \left|\pi^0 n\right\rangle &= \sqrt{\frac{1}{3}}\left|\frac{1}{2}-\frac{1}{2}\right\rangle + \sqrt{\frac{2}{3}}\left|\frac{3}{2}-\frac{1}{2}\right\rangle. \end{aligned} \tag{13.11}$$

The partial decay width is proportional to $|\langle \pi N | H | \Lambda \rangle|^2$, that is to the square of the Clebsch–Gordan coefficients. Neglecting the slightly larger phase space for $\Lambda \to \pi^0 n$, one predicts the ratio of partial widths

$$R = \frac{\Gamma(\Lambda \to \pi^0 n)}{\Gamma(\Lambda \to \pi^- p)} = \begin{bmatrix} \frac{1}{2} \text{ if } i = \frac{1}{2}, & \text{hence } |\Delta i| = \frac{1}{2} \\ 2 \text{ if } i = \frac{3}{2}, & \text{hence } |\Delta i| = \frac{3}{2} \end{bmatrix}, \tag{13.12}$$

since $i(\Lambda) = 0$. The experimental value $R = 0.560 \pm 0.016$ agrees with the $|\Delta i| = \frac{1}{2}$ rule.

As a further example, consider the decay

$$K^+ \to \pi^+ \pi^0 \tag{13.13}$$

($f = 21\%$). The isospin of the pion pair is $i = 1$ or 2 while $i = 0$ is excluded since $i_3 = +1$. For $i = 1$, the symmetry of the total wave function for bosons requires the relative angular momentum ℓ to be odd. However, since the spin of the kaon is zero, angular momentum conservation requires ℓ to be zero, thus excluding $i = 1$.

Consequently, the transition proceeds over $|\Delta i| = \frac{3}{2}$ from $i(K^+) = \frac{1}{2}$ to $i(\pi^+\pi^0) = 2$. The relatively long mean life of the $K^\pm$ meson ($\tau = 12$ ns) is due to the low probability of $|\Delta i| = \frac{3}{2}$ transitions.

For heavy quarks one can also introduce the rules $\Delta C = \Delta Q$ and $\Delta B' = \Delta Q$ which apply to the semileptonic decays of charm and bottom hadrons respectively. For instance, $B^+(C = 0,\ B' = +1) \to \overline{D}^0(C = -1,\ B' = 0)\ e^+\nu_e$, while $B^+ \nrightarrow D^0 e^- \bar{\nu}_e$, see (10.44) for the definitions of the flavour quantum numbers C and B'.

13.1 The discovery of the neutron

In 1932 James Chadwick observed a penetrating neutral radiation with α particles impinging on a beryllium target. The reaction was

$$\alpha\ {}^9_4\mathrm{Be} \to n\ {}^{12}_6\mathrm{C}. \tag{13.14}$$

The α particles were obtained from a polonium source. The neutral radiation n (the neutron) triggered a reaction in the nearby cloud chamber. The mass m of the neutron was determined by measuring the maximum recoil energy that could be observed with hydrogen, helium, or nitrogen gas in the cloud chamber: assume that a neutron with velocity v from reaction (13.14) collides elastically with a nucleus of mass M in the gas of the cloud chamber. In a *central* collision the nucleus recoils with maximum velocity V in the forward direction. If v' is the speed of the neutron after the collision, momentum and energy conservation requires that

$$\begin{aligned} mv &= MV - mv', \\ \frac{1}{2}mv^2 &= \frac{1}{2}MV^2 + \frac{1}{2}mv'^2. \end{aligned} \tag{13.15}$$

The maximum recoil velocity is

$$V = \frac{2mv}{M + m}. \tag{13.16}$$

V can be calculated from the range of the longest tracks in the chamber. If one builds the ratio of maximum velocities for two different gases, for instance hydrogen H and nitrogen N, one finds

$$\frac{V_H}{V_N} = \frac{M_N + m}{M_H + m}, \tag{13.17}$$

from which m can be determined. Chadwick found for the neutron mass $m = (1.16 \pm 0.12)\ m_p$.

Recent accurate measurements of the neutron mass (at the level of some 20 ppb) were obtained with the reaction $np \to d\gamma$, by using the cold neutrons available at the Institut Laue-Langevin (ILL) reactor in Grenoble. The γ energy ($\simeq$2.2 MeV) is essentially the binding energy of the deuteron. A precise measurement is achieved by

Bragg-scattering the γ quanta off bent Si or Ge crystals [1]. The neutron mass can then be calculated from the well-known proton and deuteron masses, taking into account the deuteron recoil energy.

13.1.1 Neutron lifetime

The neutron mean lifetime is about 15 min. A precise measurement is of great interest, for example to determine the coupling between u and d quarks (see section 13.2.4 below). There are basically two methods to measure the lifetime τ_n. In the first approach one measures the activity of a neutron beam by detecting the decay protons. This method makes the (reasonable) assumption that β-decay is the only possible decay process, otherwise the measured decay constant would be too small and correspondingly, τ_n too large. In the second approach, the number of cold neutrons stored in a trap is monitored as a function of time. It is interesting to note that current activity measurements lead to a lifetime about 7 s longer than trap experiments [2].

Figure 13.1 shows the neutron trap used at the ILL reactor in Grenoble [3]. Ultra-cold neutrons with typical velocities of a few m/s are stored in a trap with the inner wall covered by Fomblin oil, a hydrogen-free organic compound with which neutron absorption is minimized[1]. The neutrons are released, detected in a ^{3}He gas detector by the reaction $n^3\text{He} \rightarrow p^3\text{H}$, and counted as a function of storage time. The energy spectrum is influenced by gravity. This can be studied by moving the absorbing polyethylene roof up and down. The lifetime of bottled neutrons is smaller than τ_n,

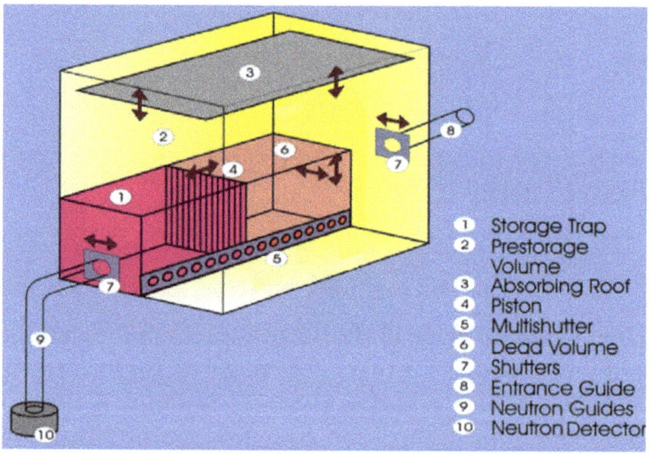

Figure 13.1. Trap for measuring the neutron lifetime with ultra-cold neutrons. The volume of the storage trap is about 1 m^3. The length of the measuring trap (in red) can be varied between 0 and 75 cm (see the text for details). After [3].

[1] The de Broglie wavelength of ultra-cold neutrons is much larger than the typical lattice constant. Neutrons are repelled from the bottle walls by coherent elastic scattering off surface atoms if their velocity perpendicular to the surface is smaller than a critical value v_c. Beryllium is a popular coating with $v_c = 6.9$ m s^{-1}. For Fomblin oil, $v_c = 4.6$ m s^{-1}.

due to absorption by the trap walls. With monoenergetic neutrons, τ_n could be determined from measurements with different trap lengths L, extrapolated to $L \to \infty$. However, the survival probability depends on the neutron velocity, which varies with the number of wall collisions. This is circumvented by performing two sets of measurements, first with the trap length L and storage times t_1 and t_2, then repeating with the trap length L' and storage times $t_i' = (L'/L)\, t_i$ to equalize the number of wall collisions [3]. The result is $\tau_n = (880.7 \pm 1.8)$ s.

13.2 β-decay

We have learned in chapter 7 that β-decay is mediated by the exchange of the heavy W boson. In β^--decay a d quark turns into a u quark by emitting a W^-, while in β^+-decay a u quark transforms into a d quark by emitting a W^+. Thus the weak interaction is extremely short-range, due to the large W mass ($1/M_W \sim 2 \times 10^{-3}$ fm), and one can in good approximation consider the process as a point-like *contact interaction* (figure 13.2). The four-fermion interaction theory was formulated by Enrico Fermi in 1934, shortly after the discovery of the neutron. The exchange of a heavy W boson was introduced later, when it was realized that the cross sections for neutrino scattering would diverge with increasing neutrino energy, leading to a violation of unitarity at very high energies.

Consider the β-decay of a nucleus of mass M into a daughter nucleus of mass M' (with recoil energy E and recoil momentum $\vec{p}$). The partial decay width is, according to (12.8):

$$\mathrm{d}\Gamma = (2\pi)^4 \delta\left(\mathcal{P}_f - \mathcal{P}_i\right) \frac{1}{2M} |\mathcal{M}|^2 \frac{\mathrm{d}^3\vec{p}}{(2\pi)^3 2E} \cdot \frac{\mathrm{d}^3\vec{p}_e}{(2\pi)^3 2E_e} \cdot \frac{\mathrm{d}^3\vec{p}_\nu}{(2\pi)^3 2E_\nu}. \tag{13.18}$$

We absorb the factors $2M$, $2E$, $2E_e$ and $2E_\nu$ in the transition amplitude. The amplitude $\mathcal{M}$ contains the four wave functions of the mother and daughter nuclei, and of the electron and the antineutrino. These wave functions are normalized so that $\int \psi^\dagger \psi \mathrm{d}V = 2E$, where E is the energy of the particle with wave function ψ. Since we have absorbed the $2E$ factors in the transition amplitude, ψ is now normalized to unity: $\int \psi^\dagger \psi \mathrm{d}V = 1$.

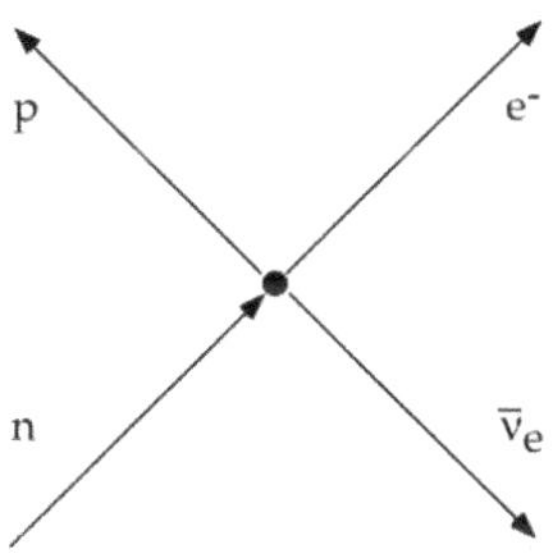

Figure 13.2. Contact interaction between four fermions in the decay $n \to pe^- \bar{\nu}_e$.

Integrating over $\vec{p}$ gives

$$\mathrm{d}\Gamma = \frac{(2\pi)^4}{(2\pi)^3}\delta(\underbrace{M - E}_{\equiv E_0} - E_e - E_\nu)|\mathcal{M}|^2 \frac{\mathrm{d}^3\vec{p}_e}{(2\pi)^3} \cdot \frac{\mathrm{d}^3\vec{p}_\nu}{(2\pi)^3}. \tag{13.19}$$

When neglecting the recoil energy, we have that

$$E_0 = M - E \simeq M - M' = Q + m_e \left(+m_\nu \simeq 0\right). \tag{13.20}$$

With

$$\mathrm{d}^3\vec{p}_e = 4\pi p_e^2\,\mathrm{d}p_e \qquad \text{and} \qquad \mathrm{d}^3\vec{p}_\nu = 4\pi p_\nu^2\,\mathrm{d}p_\nu \tag{13.21}$$

one finds that

$$\mathrm{d}\Gamma = 2\pi\delta\left(E_0 - E_e - E_\nu\right)|\mathcal{M}|^2 \overbrace{\frac{p_e^2\,\mathrm{d}p_e}{2\pi^2}}^{\mathrm{d}n_e} \cdot \overbrace{\frac{p_\nu^2\,\mathrm{d}p_\nu}{2\pi^2}}^{\mathrm{d}n_\nu}, \tag{13.22}$$

where E_e and E_ν are the *total* energies of the electron and the neutrino. The two factors after the transition amplitude are the density of electron and neutrino states, that is the number of states with momenta between p_e and $p_e + \mathrm{d}p_e$ and between p_ν and $p_\nu + \mathrm{d}p_\nu$ respectively. These factors follow from (3.16). For example

$$\mathrm{d}n_e = \frac{1}{8}4\pi\rho^2\mathrm{d}\rho \quad \text{with} \quad \rho = \frac{p_e d}{\pi}, \tag{13.23}$$

hence

$$\mathrm{d}n_e = \frac{p_e^2 d^3}{2\pi^2}\mathrm{d}p_e = \frac{p_e^2 \tau}{2\pi^2}\mathrm{d}p_e. \tag{13.24}$$

The electron wave function is normalized over the volume $\tau = d^3$. The factor τ is absorbed in the transition amplitude and the wave function renormalized accordingly.

Fermi's *golden rule* for β-decay reads

$$\boxed{\mathrm{d}\Gamma = 2\pi\delta\left(E_0 - E_e - E_\nu\right)|\mathcal{M}|^2\mathrm{d}n}, \tag{13.25}$$

where

$$\mathrm{d}n \equiv \frac{p_e^2\,\mathrm{d}p_e}{2\pi^2} \cdot \frac{p_\nu^2\,\mathrm{d}p_\nu}{2\pi^2} \tag{13.26}$$

is the *phase space* factor.

13.2.1 Allowed and forbidden transitions

The transition amplitude for the decay $|i\rangle \to |f\rangle e\bar{\nu}_e$ takes the form

$$\mathcal{M} \propto \int \psi_f^* \psi_e^* \psi_{\nu e}^* \Omega \psi_i \mathrm{d}V, \tag{13.27}$$

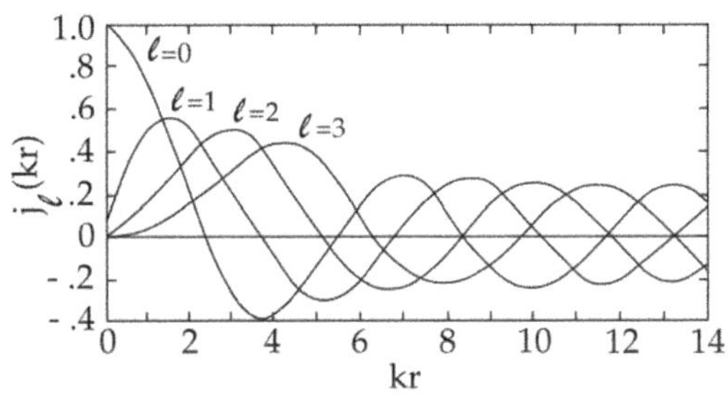

Figure 13.3. Spherical Bessel functions $j_\ell(kr)$.

where the transition operator Ω acts on the spins. The wave function of the lepton with momentum k is a plane wave of the form

$$\psi \propto \exp\left(i\vec{k}\cdot\vec{r}\right) = \exp(ikr\cos\theta) = \sum_\ell (2\ell+1) i^\ell j_\ell(kr) P_\ell(\cos\theta). \quad (13.28)$$

The quantum number ℓ denotes the orbital angular momentum carried away by the lepton. The spherical Bessel functions $j_\ell(kr)$ are shown in figure 13.3. The integral in (13.27) extends over the volume of the nucleus. At large distances (typically $r > 7$ fm) the nuclear wave functions ψ_i and ψ_f vanish. In β-decay the energies of the leptons lie in the range of at most a few MeV. For example, with

$$k < 3\text{ MeV } c^{-1} = 1.5\times 10^{-2}\text{ fm}^{-1} \Rightarrow kr < 0.1 \quad (13.29)$$

for the largest nuclei. Decays in which neither lepton removes angular momentum ($\ell = 0$) are called '*allowed*' and do not change the internal parity of the nucleus. For such decays $j_0(kr) \simeq 1$ (figure 13.3) for both leptons when performing the integration (13.27). One obtains the *energy-independent* transition amplitude

$$\mathcal{M} = g\int \psi_f^* \Omega \psi_i \, dV, \quad (13.30)$$

with coupling constant g.

For $\ell > 0$ the transition amplitude (13.27) is much smaller, being reduced by the Bessel functions in the range $kr < 0.1$ (figure 13.3). Such β-decays are suppressed, leading to long mean lives. Transitions are dubbed *first-forbidden*, *second-forbidden*, etc, depending on the total angular momentum L carried away by the leptons. The partial width decreases by typically 10^2–10^4 with each additional unit of L. Furthermore, the amplitude depends on the lepton energies through the Bessel functions. As an illustration consider the β-decay of the (4^-) ^{40}K nucleus (figure 5.7). The transition via electron capture (EC) into the (2^+) excited ^{40}Ar state is first-forbidden ($L = 1$ and parallel lepton spins, $s = 1$) and is further suppressed by the small final state phase space. The electron capture to the ($J^P = 0^+$) ground state of ^{40}Ar and the β^--decay into the ($J^P = 0^+$) ground state of ^{40}Ca are third-forbidden ($L = 3$ and $s = 1$). This explains the extremely long half-life of ^{40}K ($t_{1/2} = 1.3\times 10^9$ years).

Let us return to *allowed* decays ($\ell = 0$ for both leptons, i.e. $L = 0$) and distinguish between two cases:

1. In *Fermi* transitions the lepton spins are antiparallel ($s = 0$). The total angular momentum j of the leptons is therefore equal to $j = L = 0$ (figure 13.4(a)).

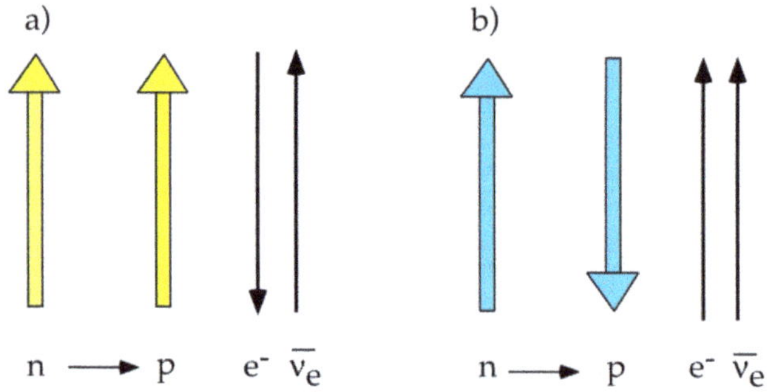

Figure 13.4. Spin configurations in (*a*) Fermi transitions and (*b*) Gamow–Teller transitions.

The spin of the nucleus is conserved between the initial (i) and final state (f),

$$\boxed{J_f = J_i}, \tag{13.31}$$

and thus the operator Ω is equal to unity (no nuclear spin flip), leading to the transition amplitude

$$\mathcal{M} = g_V \int \psi_f^* \psi_i \mathrm{d}V = g_V M_F, \tag{13.32}$$

where g_V is the *vector* coupling constant.

2. In *Gamow–Teller* transitions the leptons are emitted with parallel spins ($s = 1$), and thus remove the total angular momentum $j = 1$ (figure 13.4(*b*)). Hence the spin of the decaying nucleon flips. Therefore,

$$\left|J_f - J_i\right| \leqslant j(=1) \leqslant J_f + J_i. \tag{13.33}$$

Gamow–Teller transitions with $J_f = J_i = 0$ are forbidden. The selection rule is

$$\boxed{J_f = J_i \qquad \text{or} \qquad J_f = J_i \pm 1}. \tag{13.34}$$

The spin operator Ω is the Pauli spin vector $\vec{s} = \frac{1}{2}\vec{\sigma}$ and the transition amplitude is

$$\vec{\mathcal{M}} = g_A \int \psi_f^* \vec{\sigma} \psi_i \mathrm{d}V = g_A \vec{M}_{GT}, \tag{13.35}$$

where g_A is the *axial vector* coupling constant.

Parity does not change in both types of transitions since $L = 0$. Allowed β-decays with $|\Delta J| = 1$ are pure Gamow–Teller, while β-decays with $J_i = 0 \rightarrow J_f = 0$ are pure

Fermi transitions. Often both types contribute in *mixed* transitions. The partial decay width (13.25) is then proportional to the amplitude squared

$$|\mathcal{M}|^2 = g_V^2 \left|M_F\right|^2 + g_A^2 \left|\vec{M}_{GT}\right|^2 \equiv g_V^2 \left[\left|M_F\right|^2 + \frac{g_A^2}{g_V^2} \left|\vec{M}_{GT}\right|^2 \right]. \tag{13.36}$$

The ratio g_A^2/g_V^2 is equal to = 1.61, see (13.56) below. In nuclear β-decays the vector coupling constant is

$$\boxed{g_V = G_F \cos\theta_C}, \tag{13.37}$$

where G_F is the *Fermi coupling constant* and θ_C the *Cabibbo* angle which describes the coupling between u and d quarks. Since $\cos\theta_C \simeq 0.97$ a fair approximation is $g_V \simeq G_F$.

For example, consider the three isobars ^{14}C, ^{14}N and ^{14}O. They have the same excitation spectrum (figure 13.5), apart from level shifts due to Coulomb repulsion. They consist of ^{12}C cores with two additional nucleons,

$$\begin{aligned} ^{14}O &= {}^{12}C + 2p \left(i_3 = 1\right), \\ ^{14}N &= {}^{12}C + np \left(i_3 = 0\right), \\ ^{14}C &= {}^{12}C + 2n \left(i_3 = -1\right). \end{aligned} \tag{13.38}$$

They build an isospin triplet. The nuclear wave functions are therefore identical for levels with equal spin. The ground state of the ^{14}O nucleus ($J_i^P = 0^+$) decays by β^+ emission into the first excited state of ^{14}N ($J_f^P = 0^+$),

$$^{14}O \to {}^{14}N^* e^+ \nu_e. \tag{13.39}$$

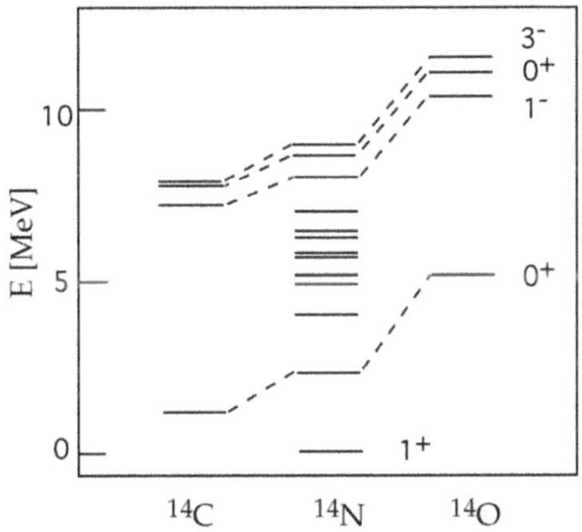

Figure 13.5. The excitation spectrum of $A = 14$ isobars. After [4].

This decay is a pure Fermi transition. Since the nuclear wave functions of mother and daughter nuclei are identical, they fully overlap in (13.32) and one obtains

$$|M_F|^2 = \left|\int \psi_f^* \psi_i \mathrm{d}V\right|^2 = 2 \Rightarrow |\mathcal{M}|^2 = 2\, G_F^2 \cos^2\theta_C \simeq 2\, G_F^2. \tag{13.40}$$

The factor 2 arises from the fact that either proton in ^{14}O can decay. The Fermi coupling constant G_F can be estimated from the half-life $t_{1/2} = 71$ s.

The half-lives of such *superallowed* transitions within isospin multiplets are short. The decay of the ^{14}N excited state (0^+) into the ^{14}C ground state (figure 13.5) is also a superallowed Fermi transition. The subsequent β^--decay to the ground state of the ^{14}N nucleus

$$^{14}C \rightarrow {}^{14}Ne^-\bar{\nu}_e \tag{13.41}$$

is a pure Gamow–Teller transition ($0^+ \rightarrow 1^+$) with the half-life $t_{1/2} = 5730$ years. This decay is used for absolute dating in archaeology. The ^{14}C isotope stems from the reaction $n^{14}N \rightarrow p^{14}C$ with neutrons produced by cosmic rays in the upper atmosphere.

The β-decay of the free neutron

$$n \rightarrow pe^-\bar{\nu}_e \tag{13.42}$$

($\frac{1}{2}^+ \rightarrow \frac{1}{2}^+$) is a mixed superallowed transition, as the proton and neutron form an isospin doublet, and are therefore described by the same wave function. The contribution from the Fermi part is

$$|M_F|^2 = \left|\int \psi_p^* \psi_n \mathrm{d}V\right|^2 = 1 \quad \text{with} \quad \psi_p = \psi_n, \tag{13.43}$$

while that from the Gamow–Teller part is

$$|\vec{M}_{GT}|^2 = \left|\int \psi_p^* \vec{\sigma} \psi_n \mathrm{d}V\right|^2 = 3 \quad \text{with} \quad \psi_p = \psi_n, \tag{13.44}$$

since

$$|\langle \sigma_x \rangle|^2 = |\langle \sigma_y \rangle|^2 = |\langle \sigma_z \rangle|^2 = 1. \tag{13.45}$$

With (13.36) one obtains for the decay of the neutron

$$|\mathcal{M}|^2 = G_F^2 \cos^2\theta_C \left[1 + 3\frac{g_A^2}{g_V^2}\right]. \tag{13.46}$$

Let us now calculate the energy distribution of the electron, starting from Fermi's golden rule (13.25), first for nuclear β-decay, then for the free neutron. For *massless* neutrinos, $p_\nu^2 = E_\nu^2 = (E_0 - E_e)^2$ from (13.20). We integrate (13.25) over $p_\nu = E_\nu$ and obtain the continuous momentum distribution of the electron:

$$\frac{d\Gamma}{dp_e} = \frac{1}{2\pi^3} |\mathcal{M}|^2 p_e^2 \left(E_0 - E_e\right)^2, \tag{13.47}$$

with $E_0 = Q + m_e$.

Due to the Coulomb attraction in β^--decay the overlap of the electron wave function with the nucleus is slightly enhanced, which increases the transition probability. The reverse holds for β^+-decay: Coulomb repulsion slightly de-localizes the positron wave function and therefore reduces the decay width. This effect is particularly pronounced at low electron or positron energies. The Coulomb correction is taken into account by the Fermi function $F(E_e, Z)$ which depends on the energy E_e of the electron or positron and on the atomic number Z (figure 13.6). The corrected momentum distribution of the electron or positron with kinetic energy T_e is, using (13.20),

$$\boxed{N(p_e) \equiv \frac{d\Gamma}{dp_e} = \frac{1}{2\pi^3} F\left(E_e, Z\right) |\mathcal{M}|^2 p_e^2 \underbrace{\left(E_0 - E_e\right)^2}_{(Q - T_e)^2}}. \tag{13.48}$$

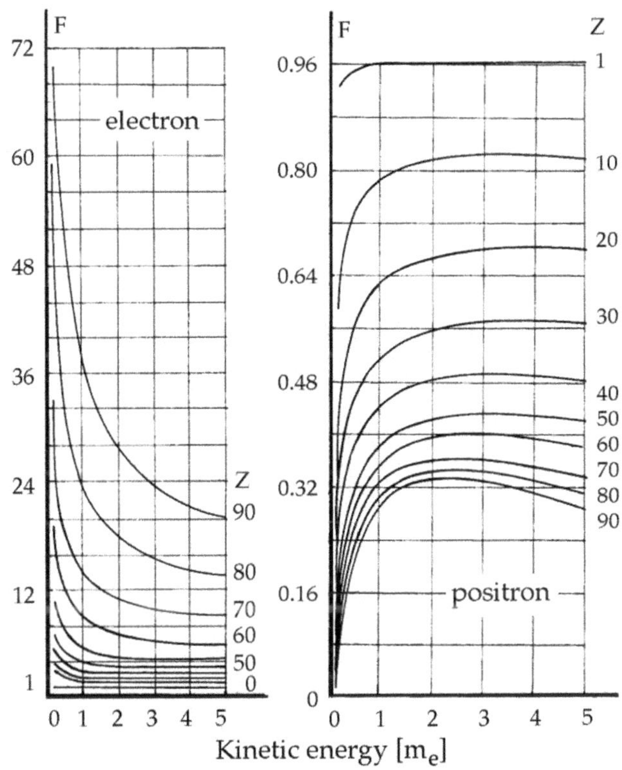

Figure 13.6. Left: Fermi function for β^--decay ($Z = 0$ and $F = 1$ for neutron decay). Right: Fermi function for β^+-decay. After [5].

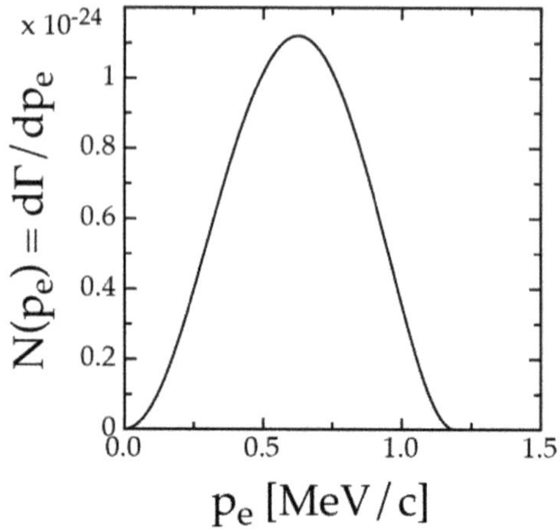

Figure 13.7. Momentum distribution of the electron in the decay of the free neutron, calculated with $G_F \cos\theta_C = 1.135 \times 10^{-5}\,\mathrm{GeV}^{-2}$, $g_A^2/g_V^2 = 1.61$ and $F = 1$.

For allowed decays the transition amplitude does not depend on the energy of the electron, therefore the shape of the momentum spectrum is fully determined by the three-body phase space. Figure 13.7 shows as an example the momentum distribution of the electron in the free neutron decay. The mean lifetime of the neutron ($\tau_n = \frac{1}{\Gamma} \simeq 880$ s) is obtained by integrating over p_e. In section 14.2.1 we will show how the mass of the neutrino can be deduced from the shape of the β spectrum.

13.2.2 *ft*-value

The width Γ (and hence the mean lifetime) of *allowed* β-decays is obtained by integrating (13.48) over p_e:

$$\begin{aligned}\Gamma &= \frac{1}{2\pi^3}|\mathcal{M}|^2 \int_0^{p_{\max}} F(E_e, Z)p_e^2(E_0 - E_e)^2 \mathrm{d}p_e \\ &= \frac{m_e^5\,|\mathcal{M}|^2}{2\pi^3} \int_0^{\eta_{\max}} F(\epsilon, Z)\eta^2(\epsilon_0 - \epsilon)^2 \mathrm{d}\eta \\ &= \frac{m_e^5\,|\mathcal{M}|^2}{2\pi^3} \overbrace{\int_1^{\epsilon_0} F(\epsilon, Z)\epsilon\sqrt{\epsilon^2 - 1}(\epsilon_0 - \epsilon)^2 \mathrm{d}\epsilon}^{f(\epsilon_0, Z)},\end{aligned} \tag{13.49}$$

where

$$\eta \equiv \frac{p_e}{m_e}, \qquad \epsilon \equiv \frac{E_e}{m_e}, \qquad \epsilon_0 \equiv \frac{E_0}{m_e}. \tag{13.50}$$

The function $f(\epsilon_0, Z)$ has been computed numerically and is tabulated [6]. Since width and half-life are related by

$$\Gamma = \frac{\ln 2}{t_{\frac{1}{2}}} \tag{13.51}$$

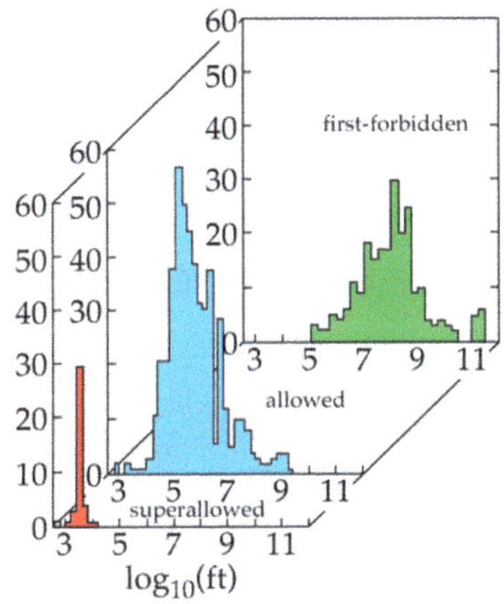

Figure 13.8. $\log_{10} ft$ distributions for superallowed, allowed and first-forbidden β-decays. After [8].

one obtains a relation between the half-life and the nuclear transition amplitude:

$$\boxed{ft \equiv t_{\frac{1}{2}} f\left(\epsilon_0, Z\right)} = \frac{2\pi^3 \ln 2}{m_e^5 |\mathcal{M}|^2}. \tag{13.52}$$

The ft-value, an estimate of the interaction strength, can be computed from the half-life and by using the tabulated function $f(\epsilon_0, Z)$. Superallowed transitions have the smallest ft-values, in the range $\log_{10} ft = 3$–4. For example, for the ^{14}O decay (13.39), $\log_{10} ft \simeq 3.5$ [7]. The ft-value increases with angular momentum L. However, L cannot be determined exactly from the ft-value since $|\mathcal{M}|^2$ varies substantially within a given class (figure 13.8).

13.2.3 Coupling constants G_F and g_A/g_V

The Fermi coupling constant G_F can be determined from a pure Fermi transition with well-known nuclear transition amplitude, such as the superallowed decay (13.39). Using the ft-value calculated from the half-life and E_0 one typically finds with $|\mathcal{M}|^2 = 2\ g_V^2$

$$G_F \simeq g_V = 1.14 \times 10^{-5}\,\mathrm{GeV}^{-2}. \tag{13.53}$$

To determine the axial vector coupling g_A, a pure Gamow–Teller decay such as $^6\mathrm{He}(0^+) \rightarrow {}^6\mathrm{Li}(1^+)\, e^- \bar{\nu}_e$ could be chosen. Unfortunately the calculation of the transition amplitude $|\vec{M}_{GT}|$ depends on a model for the nuclear wave function. Instead, one resorts to cold neutrons stored in a bottle and measures their lifetime τ_n. The ratio $|g_A/g_V|$ can be calculated from τ_n using (13.46). Precise measurements are obtained by measuring the angular distribution of the electron from polarized neutrons. The angular distribution of the electron is given by [9]

$$w\left(E_e, \theta\right) = N\left(E_e\right)(1 + A\beta \cos \theta), \tag{13.54}$$

for 100% polarized neutrons and where β is the electron velocity, θ the angle between the neutron spin and the electron momentum and A the asymmetry

$$A = -\frac{2\lambda(\lambda + 1)}{1 + 3\lambda^2} \simeq -0.12, \tag{13.55}$$

from which $\lambda \equiv g_A/g_V$ can be determined, including its sign. The recommended value is [2]

$$\lambda = \frac{g_A}{g_V} = -1.2723 \pm 0.0023. \tag{13.56}$$

13.2.4 Sargent rule

A useful approximation for the total width can be derived from (13.49) by setting $F = 1$ and assuming that the electron is highly relativistic ($p_e \simeq E_e \simeq T_e$):

$$\Gamma \simeq \frac{1}{2\pi^3}|\mathcal{M}|^2 \int_0^Q T_e^2(Q - T_e)^2 \mathrm{d}T_e = \frac{1}{2\pi^3}|\mathcal{M}|^2 \int_0^Q \left(T_e^2 Q^2 - 2T_e^3 Q + T_e^4\right)\mathrm{d}T_e$$
$$= \frac{1}{2\pi^3}|\mathcal{M}|^2 \frac{Q^5}{30}. \tag{13.57}$$

The Sargent rule

$$\boxed{\Gamma \simeq \frac{|\mathcal{M}|^2 Q^5}{60\pi^3}} \tag{13.58}$$

agrees surprisingly well with the data (figure 13.9): for allowed processes the total width increases with the fifth power of the Q-value. Small deviations are to be expected since the nuclear transition amplitude depends on the decay. For small Q-values the electron becomes non-relativistic and the approximation breaks down. This is for example the case for ${}^3_1$H decay (tritium decay with $Q = 18.6$ keV).

For strangeness conserving β-decays (such as neutron decay) G_F is multiplied by the *cosine* of the Cabibbo angle to take the coupling between d and u quarks into account, see (13.37). As detailed in section 16.2 the transition amplitude for semi-leptonic $|\Delta S| = 1$ decays must be multiplied by the *sine* of the *Cabibbo* angle θ_C, to account for the smaller coupling between s and u quarks, that is $\cos\theta_C$ must be replaced by $\sin\theta_C$. Transitions from the second quark generation to the first (or vice versa), such as $\Sigma^- \to ne^-\bar{\nu}_e$, are then hindered by the $\sin^2\theta_C$ factor. This explains the large deviations for hyperon decays in figure 13.9 (top right).

The most accurate value for G_F is derived from the purely leptonic muon β-decay, $\mu \to e\nu\bar{\nu}$ [10]:

$$G_F = (1.1663787 \pm 0.0000006) \times 10^{-5}\ \mathrm{GeV}^{-2}. \tag{13.59}$$

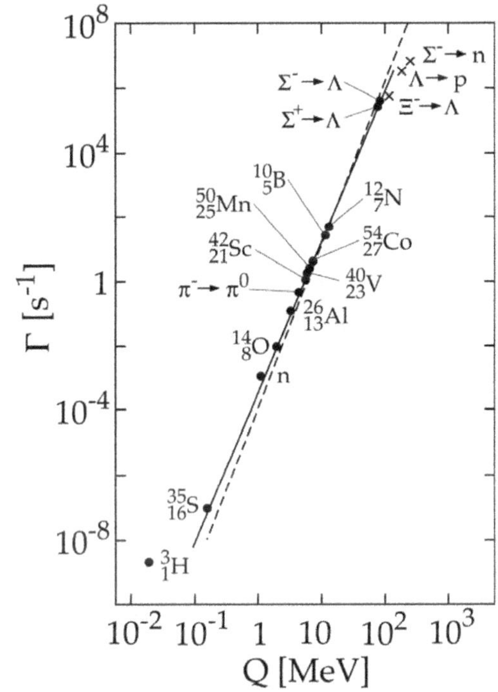

Figure 13.9. Partial widths for various β-decays as a function of Q-value. The Sargent rule predicts a slope $\log \Gamma/\log Q = 5$ (dashed line). The solid straight line corresponds to a slope of 4.67. After [8].

This value is to be compared with the one obtained from β-decays (e.g. (13.53)). Using the results from 20 superallowed $0^+ \to 0^+$ nuclear β-decays, as well as branching ratios for leptonic and semileptonic kaon decays, one obtains [11]

$$\sin\theta_C = 0.22534 \pm 0.00093. \tag{13.60}$$

The Q^5 dependence in the Sargent rule also applies to pure leptonic β-decays. Compare for example the muon and tauon β-decays:

$$\begin{aligned} \mu^+ &\to e^+\nu_e\bar{\nu}_\mu \left(f_\mu \simeq 1\right), \\ \tau^+ &\to e^+\nu_e\bar{\nu}_\tau \left(f_\tau = 17.8\%\right). \end{aligned} \tag{13.61}$$

The partial widths should scale with the fifth power of $Q \simeq m_{\mu,\tau}$:

$$\Gamma_\tau\left(e^+\nu_e\bar{\nu}_\tau\right) = \Gamma_\mu\left(e^+\nu_e\bar{\nu}_\mu\right)\frac{m_\tau^5}{m_\mu^5} = \frac{1}{\tau_\mu}\left(\frac{m_\tau}{m_\mu}\right)^5. \tag{13.62}$$

The mean life of the tauon can be predicted from that of the muon, $\tau_\mu = 2.197$ µs:

$$\tau_\tau = \frac{1}{\Gamma_\tau(\text{tot})} = \frac{f_\tau}{\Gamma_\tau\left(e^+\nu_e\bar{\nu}_\tau\right)} = f_\tau\,\tau_\mu\left(\frac{m_\mu}{m_\tau}\right)^5 = 0.291 \text{ ps}, \tag{13.63}$$

in agreement with the measured value, $\tau_\tau = 0.291 \pm 0.001$ ps.

Alternatively, the decay $\tau^+ \to \mu^+ \nu_\mu \bar{\nu}_\tau$ ($f_\tau = 17.4\%$) could be used. The Fermi coupling constant has been tested to be equal in $\mu \to e\nu\bar{\nu}$ and $\tau \to e\nu\bar{\nu}$ (*lepton universality*): $G_F^2(\tau)/G_F^2(\mu) = 1.0016 \pm 0.0042$ [12].

13.3 Parity violation

The first indication of parity non-conservation in weak interactions came from the observation of the two kaon decays

$$K^+ \to \pi^+\pi^0 \; (f = 21\%) \qquad \text{and} \qquad K^+ \to \pi^+\pi^-\pi^+ \; (f = 6\%). \quad (13.64)$$

In the first decay the angular momentum is zero since the kaon is spinless. Assuming parity conservation the intrinsic parity of the kaon is $P(K) = P(\pi)^2 = +1$. In the second decay all angular momenta are zero (section 12.3.1) and therefore the internal parity is $P(K) = P(\pi)^3 = -1$. This contradiction was initially solved unsatisfactorily by postulating the existence of two different kaons with the same mass, but opposite parities, the 'θ' $\to 2\pi$ and the 'τ' $\to 3\pi$.

The hypothesis of parity violation in weak interactions was formulated by Tsung-Dao Lee and Chen Ning Yang in 1956 [13], soon followed by direct proof from the experiments of Chien-Shiung Wu [14] and Richard Garwin [15].

13.3.1 Parity violation in nuclear β-decay

Wu and collaborators [14] investigated the (Gamow–Teller) decay

$$^{60}\mathrm{Co}\left(5^+\right) \to {}^{60}\mathrm{Ni}^*\left(4^+\right) e^-\bar{\nu}_e, \quad (13.65)$$

with polarized ^{60}Co-nuclei, where the daughter nucleus decays as

$$^{60}\mathrm{Ni}^*\left(4^+\right) \to {}^{60}\mathrm{Ni}\left(2^+\right)\gamma \to {}^{60}\mathrm{Ni}\left(0^+\right) 2\gamma. \quad (13.66)$$

They observed that the electron from the decay (13.65) is emitted preferably in the direction opposite to the spin of the ^{60}Co nucleus. This forward–backward asymmetry shows that parity is not conserved in β-decays (figure 13.10(*a*)).

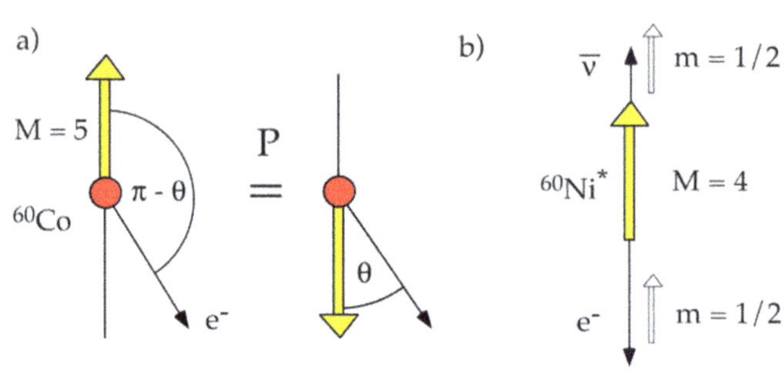

Figure 13.10. ^{60}Co β-decay. (*a*) The number of emitted electrons is larger under $\pi - \theta$ than under θ while they should be equal under parity conservation (mirroring). (*b*) Since the electron is preferably emitted in the direction opposite to the nuclear spin, angular momentum conservation requires its helicity to be negative. The antineutrino flies preferably in the direction of the nuclear spin with positive helicity.

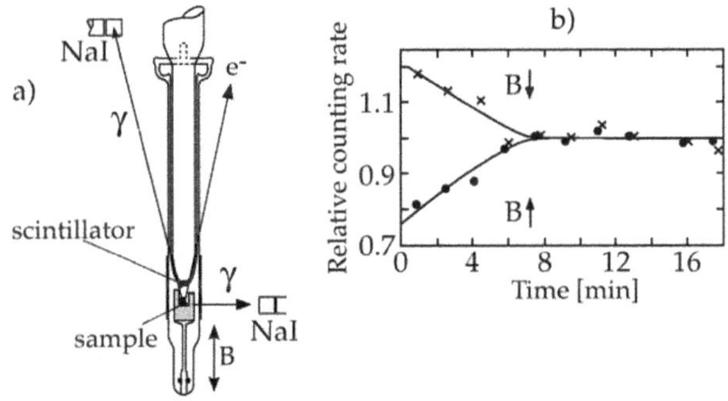

Figure 13.11. (*a*) The apparatus used to demonstrate parity violation in nuclear β-decay. (*b*) e^- rate relative to $B = 0$ as a function of time. After [14].

Figure 13.11 shows the experimental arrangement. The spins of the radioactive ^{60}Co nuclei are aligned in a cryostat by adiabatic demagnetization [16]. The temperature of the sample is 10 mK and the magnetic field is oriented along the axis of the cryostat. The electrons are detected by a scintillation counter located above the target. The polarization of the sample is determined by measuring with two NaI crystals the anisotropy of the γ emission from ^{60}Ni* decay. Figure 13.11(*b*) shows the relative electron rate in the scintillation counter for the two field orientations, as the polarization decreases during the sample warm-up. An excess of counts is observed in the direction opposite to the nuclear spin.

Since the electrons prefer to be emitted into the backward hemisphere with respect to the nuclear spin, conservation of angular momentum ($J_i = 5 \rightarrow J_f = 4$) requires them to be left-handed, while the antineutrinos fly into the forward hemisphere with positive helicity (right-handed antineutrinos, see figure 13.10(*b*)). This fact is of central importance in weak interactions:

> The *weak* interaction between charged currents (hence with $W^{\pm}$ exchange) involves only *left-handed* fermions and *right-handed* antifermions.

This rule, which follows from the $V - A$ structure of weak interactions (chapter 16), applies when the mass of the fermion becomes negligible compared to its energy, that is for highly relativistic particles. The rule is always correct, independent of mass and energy, provided that 'handedness' is replaced by 'chirality'. The chirality of a fermion (antifermion) is -1 ($+1$) in weak interactions, see section 15.3.

Since neutrinos are (nearly) *massless*, the rule implies that *neutrinos* are *left-handed* and *antineutrinos right-handed*. All experiments are indeed compatible with left-handed neutrinos (a prominent experimental proof will be described in section 14.4). Therefore the right-handed mirror version of the massless neutrino does not exist (figure 11.14) and the amount of parity violation is maximum.

13.3.2 Parity violation in muon production and decay

The μ^+ from the decay $\pi^+ \to \mu^+\nu_\mu$ is 100% polarized (left-handed, i.e. negative helicity) in the rest frame of the π^+. This follows from the conservation of angular momentum for a spinless pion, and from the negative helicity of the neutrino (figure 13.12(*a*)). In this decay the muon is non-relativistic ($T = 4$ MeV), so that the 'wrong' negative helicity of the μ^+ does not violate the right-handedness rule for antifermions. In contrast, the decay $\pi^+ \to e^+\nu_e$ is strongly suppressed ($f \simeq 10^{-4}$) since the electron is extremely relativistic ($T \simeq 70$ MeV, see chapter 16).

Let us now consider the β-decay of the positive muon, $\mu^+ \to e^+\nu_e\bar{\nu}_\mu$, in the muon rest frame. The energy distribution of the positron reaches its maximum value at $E_e \simeq m_\mu/2 = 52.8$ MeV while the two neutrinos fly in the direction opposite to the positron (figure 13.12(*b*)). One expects from angular momentum conservation that the (extremely relativistic and hence right-handed) positron flies in the direction of the muon spin while the opposite direction is forbidden. This forward–backward asymmetry violates parity conservation.

The angular distribution of the positron with respect to the spin direction depends on the positron energy and on the four independent parameters ρ, η, δ and ξ which in the $V - A$ theory take the values $\frac{3}{4}$, 0, $\frac{3}{4}$ and 1, respectively [17]. These parameters have been measured accurately by using positive muons produced from pions decaying at the surface of a target irradiated with 500 MeV protons [18]. They agree within 0.1% with the $V - A$ predictions.

Assuming $V - A$ one obtains the following angular distribution of the positron in the rest frame of the μ^+:

$$\boxed{\frac{d^2\Gamma_\mu}{d\Omega d\epsilon} = \frac{1}{4\pi\tau_\mu} n(\epsilon)\left[1 + \alpha_+(\epsilon)\cos\theta\right]}. \tag{13.67}$$

The angle θ refers to the direction of the positron with respect to the muon spin. The function $n(\epsilon)$ describes the energy distribution (*Michel* spectrum) of the positron, given by

$$n(\epsilon) = 2\epsilon^2(3 - 2\epsilon), \tag{13.68}$$

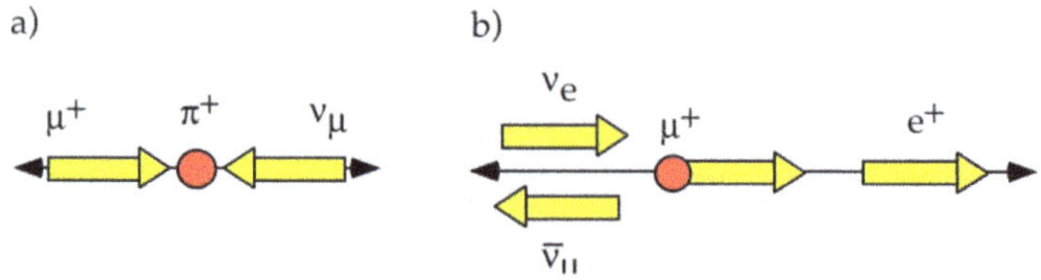

Figure 13.12. (*a*) π^+ decay. (*b*) μ^+ decay for the positron with maximum energy. The neutrinos fly in the direction opposite to the positron.

with $\int_0^1 n(\epsilon)\mathrm{d}\epsilon = 1$. The parameter ϵ is the reduced positron energy

$$\epsilon = \frac{2E_e}{m_\mu} \tag{13.69}$$

with maximum value $\epsilon = 1$. The function

$$\alpha_+(\epsilon) = \frac{2\epsilon - 1}{3 - 2\epsilon} \tag{13.70}$$

describes the parity violating asymmetry. The functions $n(\epsilon)$, $\alpha_+(\epsilon)$ and the angular distribution of the positron are shown in figure 13.13 (left).

When α_+ is equal to 1 (maximum energy $\epsilon = 1$) the positron cannot fly backwards, as expected from figure 13.12(*b*). The mean angular distribution (averaged over ϵ) is

$$\begin{aligned} w(\cos\theta) = \frac{\mathrm{d}\Gamma_\mu}{\mathrm{d}\Omega} &= \frac{1}{4\pi\tau_\mu}\int_0^1 n(\epsilon)\big[1 + \alpha_+(\epsilon)\cos\theta\big]\mathrm{d}\epsilon \\ &= \frac{1}{4\pi\tau_\mu}\big[1 + \langle\alpha_+\rangle\cos\theta\big], \end{aligned} \tag{13.71}$$

with the mean asymmetry

$$\langle\alpha_+\rangle = \int_0^1 \alpha_+(\epsilon)n(\epsilon)\mathrm{d}\epsilon = +\frac{1}{3}. \tag{13.72}$$

One sees that the positron flies preferably *in the direction* of the muon spin. Figure 13.13 (right) illustrates the angular distribution for $\alpha_+ = 1$ and $\frac{1}{3}$.

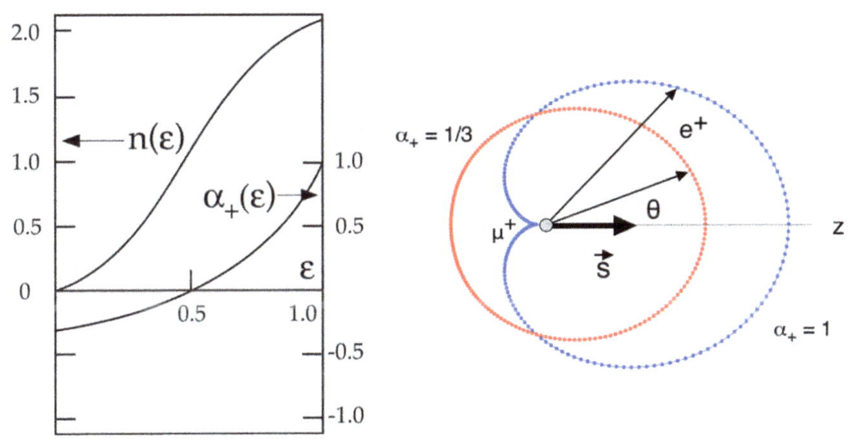

Figure 13.13. Left: energy distribution $n(\epsilon)$ of the positron and asymmetry $\alpha_+(\epsilon)$ in μ^+ decay. Right: angular distribution $1 + \alpha_+(\epsilon)\cos\theta$ of the positron for $\alpha_+ = 1$ and $\alpha_+ = \frac{1}{3}$.

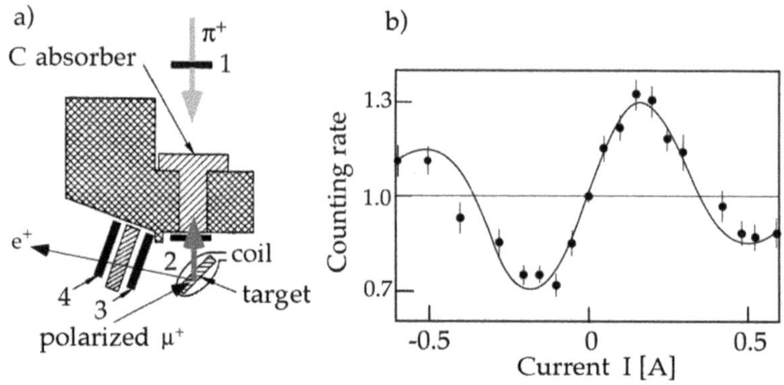

Figure 13.14. (*a*) Diagram of the equipment used to demonstrate parity violation in the decay chain $\pi \to \mu \to e$ (see the text for details). (*b*) Counting rate as a function of coil current I (relative to $I = 0$). The curve describes the expected behaviour. After [15].

Similar arguments apply to the μ^- in the decay of the negative pion $\pi^- \to \mu^- \bar{\nu}_\mu$. The μ^- is 100% polarized (right-handed, i.e. positive helicity) in the rest frame of the pion, so that the electron prefers to be emitted in the direction *opposite* to the μ^- spin with the asymmetry $\alpha_-(\epsilon) = -\alpha_+(\epsilon)$. On average $\langle \alpha_- \rangle = -\frac{1}{3}$.

Figure 13.14(*a*) shows a diagram of the historical apparatus used to demonstrate the asymmetric angular distribution of the positron [15]. Positive muons from a 85 MeV π^+ beam traverse a carbon absorber. Consider a forward-emitted and a backward-emitted muon in the rest frame of the pion, with respect to the pion beam direction. In the laboratory both muons fly into the absorber. However, the range of the forward-emitted muon is larger than that of the backward-emitted one. Therefore, provided that the absorber thickness is chosen correctly, the former traverses the absorber and reaches the target, while the latter stops in the absorber. The forward muon reaching the target has its spins pointing upwards in figure 13.14(*a*). However, the sample of forward muons can be contaminated by backward muons (due to the finite solid angle and range straggling) so that the polarization P is smaller than 1, and hence the asymmetry α_+ is reduced to $\alpha_+ P$.

The clock is started by the scintillation counters (coincidence 12). The muon spin precesses in the magnetic field B with the *Larmor* frequency

$$\boxed{\omega_L = \frac{g_\mu e B}{2m_\mu}}, \tag{13.73}$$

where g_μ is the g-factor of the muon. One measures with the scintillation counters 3 and 4 the rate of decay positrons, as a function of the magnetic field and for a given delay time Δt between the coincidences 34 and 12. By adding an absorber between counters 3 and 4 one can select high energy positrons for which the asymmetry α_+ is larger (figure 13.13, left). Figure 13.14(*b*) shows the rate as a function of coil current. The peaks correspond to the situation in which the spins point in the direction of the counters 3 and 4 after the delay time Δt. With unpolarized muons (or for $\alpha_+ \equiv 0$) the angular distribution would be isotropic (a parity conserving *anisotropy* is not possible for polarized spin-$\frac{1}{2}$ particles) and hence the counting rate would be independent of the magnetic field. This experiment proves that leptons prefer negative helicities, and thus that parity is violated in weak interactions.

A similar but simpler experiment can be set up in the student lab with stopping cosmic ray muons (which turn out to be polarized) to verify that parity is violated in muon decay, and that $g_\mu \simeq 2$ [19].

13.3.3 The anomalous magnetic moment of the muon

The magnetic dipole moment of the muon, hence the factor g_μ, is determined by measuring the Larmor frequency (13.73) in a known magnetic field[2]. Radiative corrections, such as the ones shown in figure 13.15, lead to a g-factor which deviates slightly from the value $g_\mu = 2$ predicted by the Dirac equation for spin-$\frac{1}{2}$ fermions (chapter 15).

Consider the circular path of a polarized non-relativistic muon in the homogeneous magnetic field B. During the time interval t the momentum vector $\vec{p}$ rotates by the angle

$$\theta_C = \omega_C t = \frac{eBt}{m_\mu}. \tag{13.74}$$

The spin vector $\vec{s}$ rotates in the *same* direction by the angle

$$\theta_L = \omega_L t = \frac{g_\mu eBt}{2m_\mu}. \tag{13.75}$$

The angle between $\vec{p}$ and $\vec{s}$ varies with time as

$$\Delta\theta \equiv \theta_L - \theta_C = \frac{eBt}{m_\mu}\left(\frac{g_\mu - 2}{2}\right), \tag{13.76}$$

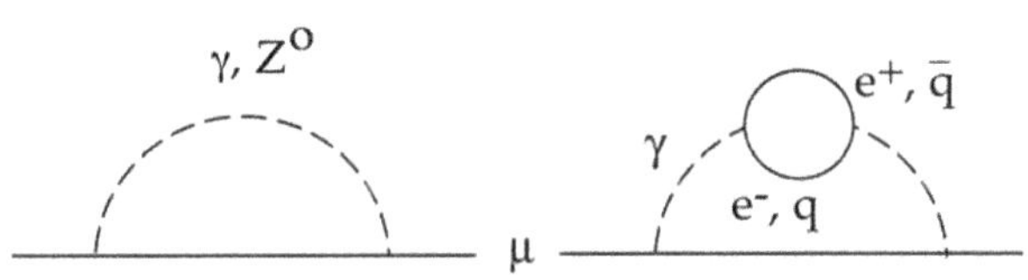

Figure 13.15. Sample of radiative corrections contributing to the anomalous ($g_\mu \neq 2$) magnetic moment of the muon.

---

[2] Conversely, assuming $g_\mu = 2$, the magnetization in solids can be studied with stopping muons (muon spin rotation experiments (μSR)).

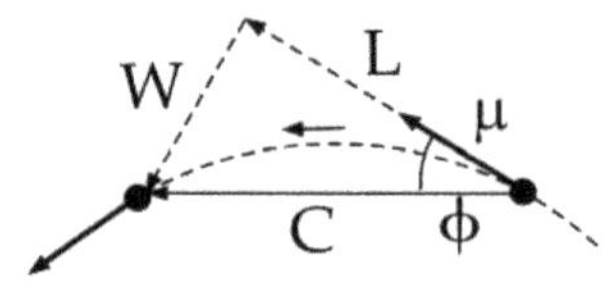

Figure 13.16. Thomas spin precession along a curved path.

and the corresponding angular velocity is

$$\omega_a = \frac{eB}{m_\mu}\overbrace{\left(\frac{g_\mu - 2}{2}\right)}^{a_\mu}. \tag{13.77}$$

The *anomalous* magnetic moment a_μ of the muon vanishes for $g_\mu = 2$, in which case the spin $\vec{s}$ remains always parallel to the momentum $\vec{p}$ ($\omega_a = 0$).

The relativistic treatment leads to the same result for ω_a in the laboratory. Let us denote the kinematic quantities in the rest frame of the muon by an asterisk. The momentum vector of the relativistic muon rotates in the laboratory with the cyclotron frequency $\omega_C = eB/\gamma m_\mu$ by the angle $\theta_C = (eB/\gamma m_\mu)t$. In the muon rest frame the rotation angle of the laboratory is $\theta_C^* = \gamma\theta_C$. This can be seen as follows (figure 13.16): let us approximate the muon circular path by the chord C and consider a small rotation $\theta_C = \phi \simeq \frac{W}{L}$ in the laboratory. In the muon rest frame the segment parallel to the momentum direction is contracted by the relativistic factor γ, hence $\phi^* \simeq \frac{W}{L/\gamma} = \gamma\frac{W}{L} = \gamma\phi$. The difference $\Delta\phi = \phi(\gamma - 1)$ is the *Thomas precession.*

The cyclotron frequency is

$$\omega_C^* = \frac{\theta_C^*}{t^*} = \frac{e\gamma B}{m_\mu}. \tag{13.78}$$

with $t^* = t/\gamma$. The precession frequency in the rest frame is

$$\omega_L^* = \frac{g_\mu eB^*}{2m_\mu} = \frac{g_\mu e\gamma B}{2m_\mu}, \tag{13.79}$$

therefore

$$\omega_a^* = \omega_L^* - \omega_C^* = \frac{e\gamma B a_\mu}{m_\mu}, \tag{13.80}$$

which gives the angular velocity in the laboratory

$$\omega_a = \frac{\omega_a^*}{\gamma} = \frac{eB}{m_\mu}a_\mu. \tag{13.81}$$

The deviation from the value $g_\mu = 2$ can be obtained directly from ω_a in so-called g–2 experiments. Figure 13.17(a) shows a diagram of the apparatus used at CERN to perform such a measurement [20].

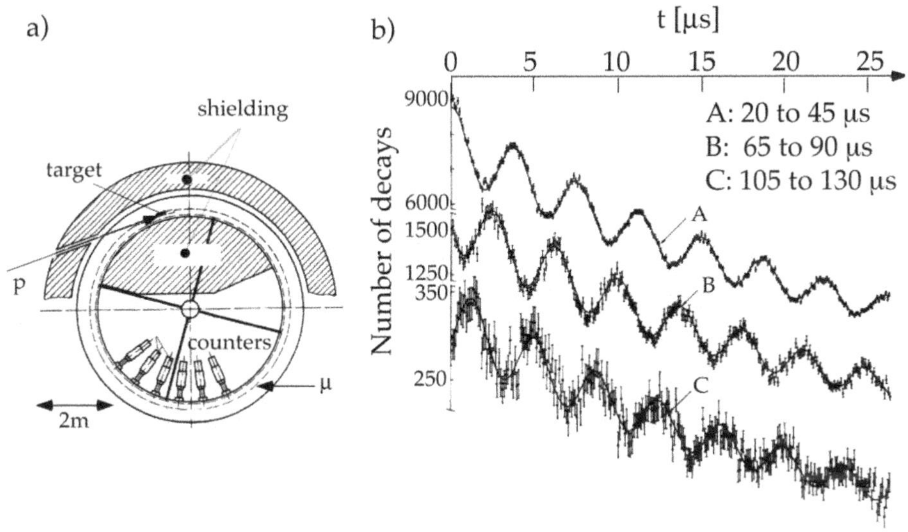

Figure 13.17. (*a*) Diagram of the equipment used to measure *g*–2 of the muon. (*b*) Time distribution of the decay electron. After [20].

A pulsed 10 GeV proton beam strikes a target and produces (mainly) pions. Some of the decay muons are captured in the magnetic storage ring ($B = 1.7\,\mathrm{T}$). A series of lead glass counters detect the electrons from muon decays which are emitted perpendicularly to the muon beam. One measures the counting rate as a function of time between the proton pulse and the lead glass signal. Figure 13.17(*b*) shows the measured time distribution. The cosine oscillation due to the angular distribution (13.67) is superimposed on the exponential muon decay curve (lifetime $\gamma\tau_\mu$). The period of the oscillation $T = 2\pi/\omega_a \sim 3.8\ \mu\mathrm{s}$, which corresponds to about 80 revolutions around the ring, leads with (13.81) to $a_\mu \simeq 10^{-3}$. The correction $g_\mu - 2 = 2a_\mu$ of about 2×10^{-3} stems from radiative corrections (figure 13.15). The experiment can be performed with negative or positive muons by reversing the polarity of the magnetic field.

Considerable improvements have been achieved over the years in *g*–2 experiments. The goal of high precision measurements is to test the *CPT* invariance between μ^+ and μ^- and to compare the results with QED predictions. The current values for the anomalous magnetic moments of the positive and negative muons are [21]:

$$\begin{aligned} a(\mu^+) &= (11\,659\,204 \pm 8) \times 10^{-10}, \\ a(\mu^-) &= (11\,659\,215 \pm 9) \times 10^{-10}. \end{aligned} \tag{13.82}$$

The values for μ^- and μ^+ are consistent within errors, as expected from *CPT* conservation (table 11.1). However, there is currently a discrepancy of 3.6 standard deviations between the data and predictions [2] and further, more precise measurements are under way (figure 13.18).

Figure 13.18. The superconducting coil of the most recent *g*–2 experiment performed in Brookhaven being moved to Fermilab. The red support structure is for transportation. The outer and two inner coils, when enclosed in the iron yoke, provide a homogeneous field of 1.45 T. Image courtesy of Brookhaven National Laboratory.

References

[1] Kessler E G *et al* 1999 *Phys. Lett.* A **255** 221
[2] Olive K *et al* (Particle Data Group) 2014 *Chin. Phys.* C **38** 090001
[3] Pichlmaier A *et al* 2010 *Phys. Lett.* B **693** 221
[4] Povh B, Rith K, Scholz C and Zetsche F 2004 *Particles and Nuclei* (Berlin: Springer) p 22
[5] Marmier P 1960 *Kernphysik* (Zurich: Verlag des Vereins der Mathematiker und Physiker an der ETH Zürich)
[6] Feenberg E and Trigg G 1950 *Rev. Mod. Phys.* **22** 399
[7] Bardin R K *et al* 1962 *Phys. Rev.* **127** 583
[8] Williams W S C 1991 *Nuclear and Particle Physics* (Oxford: Oxford University Press)
[9] Abele H *et al* 2002 *Phys. Rev. Lett.* **88** 211801
[10] Webber D M *et al* 2011 *Phys. Rev. Lett.* **106** 041803
[11] Hardy J C and Towner I S 2009 *Phys. Rev.* C **79** 055502
[12] Ablikim M *et al* 2014 *Phys. Rev.* D **90** 012001
[13] Lee T D and Yang C N 1956 *Phys. Rev.* **104** 254
[14] Wu C S *et al* 1957 *Phys. Rev.* **105** 1413
[15] Garwin R L *et al* 1957 *Phys. Rev.* **105** 1415
[16] Kittel C 1988 *Einführung in die Festkörperphysik* (Berlin: Oldenbourg)
[17] Michel L 1950 *Proc. Phys. Soc. London* A **63** 514
[18] Bayes R *et al* 2011 *Phys. Rev. Lett.* **106** 041804
[19] Amsler C 1974 *Am. J. Phys.* **42** 1067
[20] Bailey J W *et al* 1968 *Phys. Lett.* B **28** 287
[21] Bennett G W *et al* 2006 *Phys. Rev.* D **73** 072003

IOP Publishing

Nuclear and Particle Physics

Claude Amsler

Chapter 14

Neutrinos

14.1 The discovery of the neutrino

In a nuclear reactor the antineutrinos are produced by the fission products which undergo β^--decay. Figure 14.1 shows the energy spectrum of reactor neutrinos. About six antineutrinos are emitted on average for every nuclear fission.

The electron antineutrino ($\bar{\nu}_e$) was observed for the first time in 1959 at the Savannah River power plant by Frederick Reines and collaborators [2] who used the *inverse* β-decay

$$\bar{\nu}_e p \rightarrow e^+ n. \tag{14.1}$$

The Q-value for this reaction is $Q = -1.8$ MeV. Figure 14.2 shows a sketch of the experiment. Antineutrinos from the reactor interact with protons in a 200ℓ water target. The emitted positrons stop in the target and annihilate with electrons into two 511 keV photons which are detected in liquid scintillators located above and below the target. The latter is thin (7.6 cm) to minimize photon absorption. Neutrons from reaction (14.1) drift through the water target and are moderated to thermal energies after a few collisions. A small admixture of cadmium (in the form of $CdCl_2$) is added to the water to absorb thermal neutrons. The mean lifetime of the neutrons between emission and absorption is a few μs. The excited Cd nuclei decay through the emission of photons (total energy of 9.1 MeV) which are also detected by the scintillators. One searches for two prompt photons from e^+e^- annihilation, coincident with a delayed γ signal of 9.1 MeV.

Figure 14.3(a) shows the experimental arrangement. Two water targets are used to increase the counting rate. Paraffin and lead bricks surround the apparatus to shield it from direct neutrons and photons emitted by the reactor. About three events per day are recorded from reaction (14.1) with a neutrino flux of 10^{13} s^{-1} cm^{-2}. This rate corresponds to a cross section of about 10^{-19} b. Figure 14.3(b) shows the time

doi:10.1088/978-0-7503-1140-3ch14

© IOP Publishing Ltd 2015

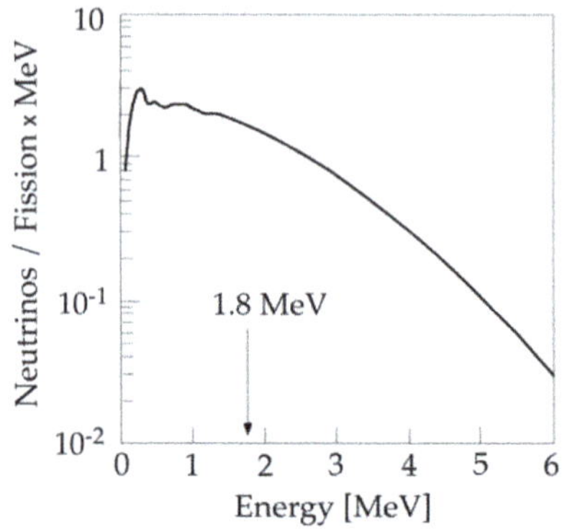

Figure 14.1. Calculated energy distribution of antineutrinos from a ^{235}U nuclear reactor. After [1].

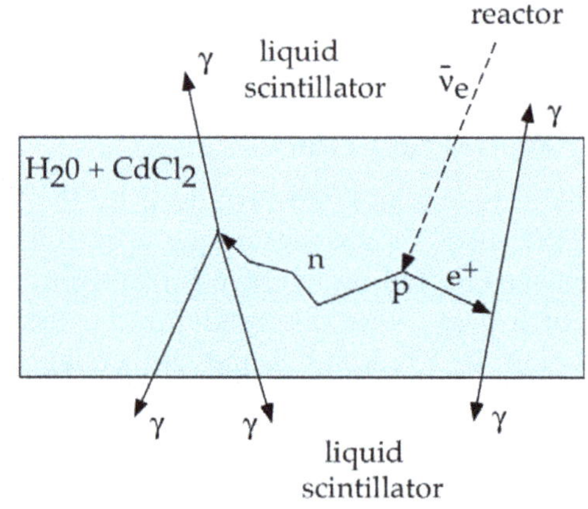

Figure 14.2. Diagram of the experiment that discovered the electron antineutrino.

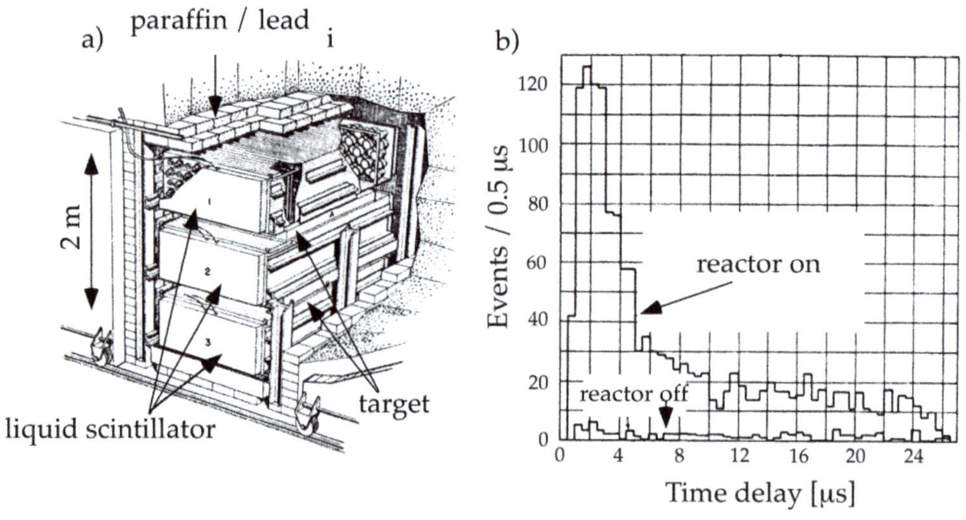

Figure 14.3. (*a*) The experimental set up used to detect the $\bar{\nu}_e$. (*b*) Time distribution between the positron signal and the delayed γ emission with the reactor on and off. After [2].

distribution between prompt and delayed photons with the reactor on and off. The peak at short delay times is due to antineutrinos.

The ν_μ and $\bar{\nu}_\mu$ were discovered in 1962 using the decays of energetic π^+ and π^- mesons [3] and searching for the reactions

$$\begin{aligned} \nu_\mu n &\rightarrow \mu^- p, \\ \bar{\nu}_\mu p &\rightarrow \mu^+ n. \end{aligned} \tag{14.2}$$

Since only muons and no electrons were produced, the electron neutrino and the muon neutrino had to be different particles.

We now describe the observation of the tau neutrino ν_τ from the decay of the τ lepton (or tauon). The latter was discovered in 1975 at the SPEAR e^+e^- storage ring [4]. Above a collider energy of $\sqrt{s} \simeq 3.6\,\text{GeV}$ peculiar events were observed, in which only $e^\pm\mu^\mp$ pairs were produced. The underlying reaction was the pairwise production of a new heavy charged lepton, the $\tau^\pm$, with a mass of about 1.8 GeV: $e^+e^- \rightarrow \tau^+\tau^-$ [5]. The two tauons then disintegrated into $e^+\nu_e\bar{\nu}_\tau$ (or $\mu^+\nu_\mu\bar{\nu}_\tau$) and $\mu^-\bar{\nu}_\mu\nu_\tau$ (or $e^-\bar{\nu}_e\nu_\tau$), respectively. Because of lepton number conservation, another type of neutrino had to be emitted together with the ν_e and the ν_μ accompanying the electron and the muon: the tau neutrino ν_τ.

This neutrino was eventually detected directly by the DONUT experiment at Fermilab in 2001 [6]. Protons of 800 GeV impinge on a target, generating secondary particles such as D_s^+ (D_s^-) mesons, which then decay into $\tau^+\nu_\tau$ ($\tau^-\bar{\nu}_\tau$). The charged particles are stopped in a 36 m long absorber, while the ν_τ (or $\bar{\nu}_\tau$) flying through is converted in iron foils by the reaction $\nu_\tau N \rightarrow \tau^- X$ (or $\bar{\nu}_\tau N \rightarrow \tau^+ X$). The tauon then decays leptonically or semileptonically (table 14.1) within a few millimeters with a mean lifetime of 0.29 ps. The track of the tauon, its decay point and charged decay products are observed in a stack of emulsion detectors. Figure 14.4 shows one of the four events observed by DONUT. The particle is identified as an electron (or a positron) and its momentum ($\simeq 2.9\,\text{GeV}\,c^{-1}$) is estimated by sampling the induced electromagnetic shower.

Table 14.1. Main decay modes of the τ lepton [7].

$\tau^+ \rightarrow$	f [%]
$e^+\nu_e\bar{\nu}_\tau$	17.8
$\mu^+\nu_\mu\bar{\nu}_\tau$	17.4
$\pi^+\bar{\nu}_\tau$	10.8
$\pi^+\pi^0\bar{\nu}_\tau$	25.9
$\pi^+ \geqslant 2\pi^0\bar{\nu}_\tau$	10.9
$2\pi^+\pi^-\bar{\nu}_\tau$	9.8
$2\pi^+\pi^-\pi^0\bar{\nu}_\tau$	4.8

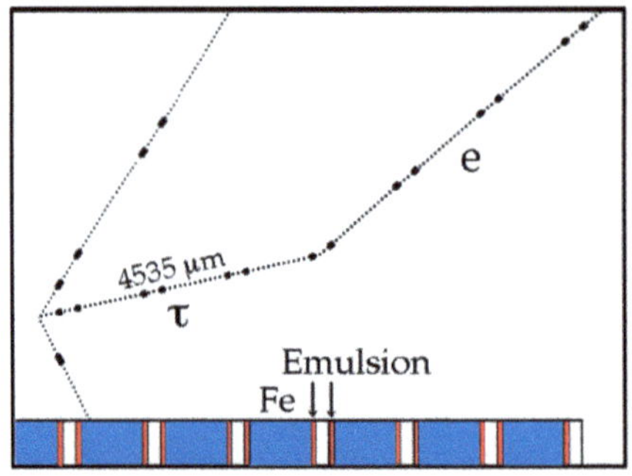

Figure 14.4. A $\nu_\tau N \rightarrow \tau^- X$ event in the DONUT experiment. The (invisible) ν_τ enters the apparatus from the left and the τ lepton decays into an electron ($\tau^- \rightarrow e^- \bar{\nu}_e \nu_\tau$). The dotted lines interpolate the tracks between the measured coordinates in the emulsions. After [6].

14.2 Mass measurements

14.2.1 ν_e mass

The mass of the electron neutrino is measured with nuclear β-decays. The *Kurie* function is obtained from the momentum distribution (13.48) of the electron for massless neutrinos ($p_\nu = E_\nu$):

$$\sqrt{\frac{N(p_e)}{F(E_e, Z)p_e^2}} = \frac{|\mathcal{M}|}{\sqrt{2\pi^3}}(Q - T_e). \tag{14.3}$$

When histogramming the data (*Kurie* plot) one expects a linearly decreasing behaviour as a function of the kinetic energy T_e of the electron, with endpoint at $T_e(\text{max}) = Q$ (figure 14.5).

Let us now assume a finite neutrino mass m_ν. One obtains for the energy of the neutrino, neglecting again the recoil energy,

$$E_\nu = M - M' - T_e - m_e = Q - T_e, \tag{14.4}$$

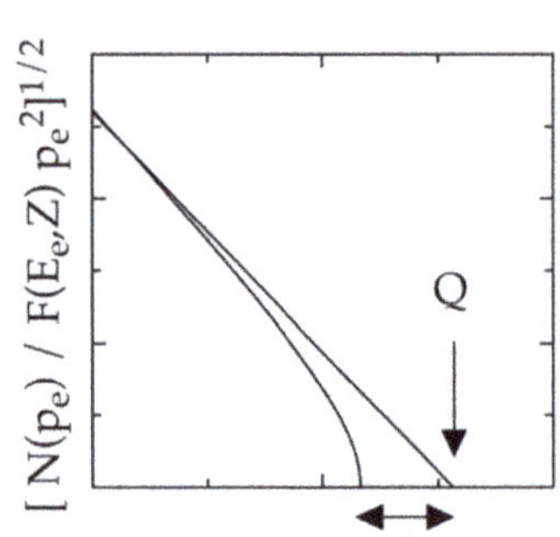

Figure 14.5. Kurie plot for $m_\nu = 0$ (straight line) and for $m_\nu > 0$.

where M and M' are the masses of the parent and daughter nuclei, and where we have defined $Q \equiv M - M' - m_e$, that is the Q-value for a vanishing neutrino mass. The momentum of the neutrino is

$$p_\nu = \sqrt{E_\nu^2 - m_\nu^2} = \sqrt{(Q - T_e)^2 - m_\nu^2}. \tag{14.5}$$

Integrating (13.22) over p_ν, using the property (12.13) of the δ-function leads to

$$\begin{aligned} \frac{\mathrm{d}\Gamma}{\mathrm{d}p_e} &= \int \frac{1}{2\pi^3} \delta(E_0 - E_e - E_\nu)\, |\mathcal{M}|^2 p_e^2 p_\nu^2 \,\mathrm{d}p_\nu \\ &= \frac{1}{2\pi^3} |\mathcal{M}|^2 p_e^2 p_\nu^2 \frac{1}{\left|\frac{\mathrm{d}E_\nu}{\mathrm{d}p_\nu}\right|} = \frac{1}{2\pi^3} |\mathcal{M}|^2 p_e^2 p_\nu E_\nu. \end{aligned} \tag{14.6}$$

The Kurie function now reads

$$\boxed{\sqrt{\frac{N(p_e)}{F(E_e, Z) p_e^2}} = \frac{|\mathcal{M}|}{\sqrt{2\pi^3}} \left[(Q - T_e)^2 - m_\nu^2 \right]^{\frac{1}{4}} \sqrt{Q - T_e}}. \tag{14.7}$$

The mass term modifies the shape of the energy spectrum at the endpoint which now lies at the energy $T_e(\max) = Q - m_\nu$ (figure 14.5).

The electron spectrum in tritium β-decay

$$^3\mathrm{H}\left(\frac{1}{2}^+\right) \rightarrow {}^3\mathrm{He}\left(\frac{1}{2}^+\right) e^- \bar{\nu}_e \;\; (Q = 18.57\,\mathrm{keV}) \tag{14.8}$$

was studied using several experiments. Tritium is one of the β emitters with the lowest endpoints. The half-life of this mixed (Fermi/Gamow–Teller) transition is 12.3 years. In one of the pioneering experiments (figure 14.6(*a*)) the tritium ions are implanted in

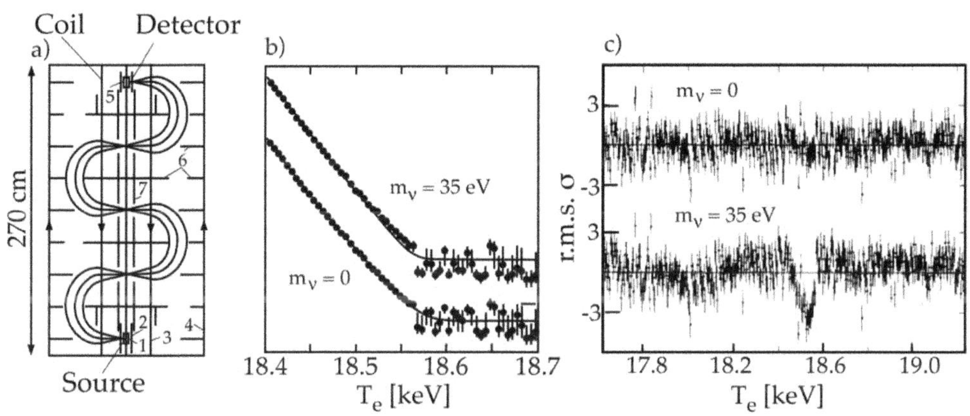

Figure 14.6. (*a*) The spectrometer used to study the energy distribution of the electron in the β-decay of tritium. (*b*) Kurie plot around the endpoint. The curves show calculations for two neutrino masses. For clarity the data for $m_\nu = 35\,\mathrm{eV}$ have been shifted upwards. (*c*) Residuals between the fits and data in rms standard deviations. Adapted from [8].

a thin carbon film and the decay electrons are analysed in a toroidal magnetic spectrometer [8]. Electrons near the endpoint are detected in a proportional counter by applying a (variable) positive high-voltage to the radioactive source. The precise impact point of the electron on the detector is determined by comparing the amplitudes measured at both ends of the anode wire. With a dispersion of $1\,\text{eV}\,\text{mm}^{-1}$ at the detector, an rms energy resolution of 12 eV can be achieved. Figure 14.6(*b*) shows the Kurie spectrum with fits for two different neutrino masses. Measured is the square of the neutrino mass, see (14.7): $m^2_{\nu e} = -11 \pm 63\,\text{eV}^2$. The corresponding upper limit is 18 eV (95% confidence level). The neutrino mass of ~35 eV that had been reported earlier [9] is therefore excluded. As discussed in the following section, the current upper limit for the electron neutrino is $m_{\nu_e} = 2.3\,\text{eV}$ [10].

14.2.2 Solenoid retarding spectrometers

Recent, more accurate measurements of the ν_e mass have been performed using solenoid retarding spectrometers (see for example [10]). Let us recall the motion of a particle with mass m and charge q in a *slowly* decreasing magnetic field (figure 14.7). In the plane perpendicular to the magnetic field B_z the motion is a circle of radius R with velocity $v_\perp$. However, the slowly decreasing field with transverse component $B_r > 0$ induces a Lorentz force F which accelerates the particles along the magnetic field lines (for either sign of the charge q). Conversely, in an increasing field the particle is decelerated along the field lines and eventually reverses its direction (principle of the magnetic bottle).

From the Maxwell equation div $\vec{B} = 0$ one obtains in cylindrical coordinates:

$$\frac{1}{r}\frac{\partial}{\partial r} r B_r + \frac{\partial B_z}{\partial z} = 0. \tag{14.9}$$

For a slowly varying field we can assume that $\partial B_z/\partial z$ does not change within the disk of radius R. Integrating over r gives

$$\int_0^R \frac{\partial}{\partial r} r B_r \mathrm{d}r \simeq -\frac{\partial B_z}{\partial z}\int_0^R r\mathrm{d}r \Rightarrow B_r \simeq -\frac{R}{2}\left(\frac{\partial B_z}{\partial z}\right). \tag{14.10}$$

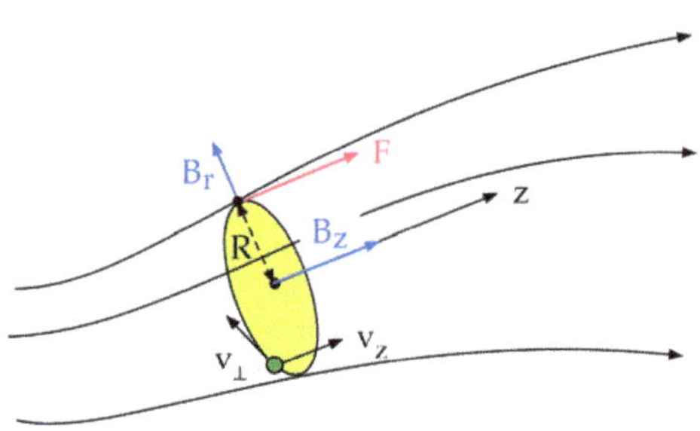

Figure 14.7. The charged particle rotates around the magnetic field lines. The slowly decreasing field with transverse component $B_r > 0$ induces a force which accelerates the particles along the z-direction.

The magnetic moment M of the current loop is (current × surface)

$$M = \left(\frac{qv_\perp}{2\pi R}\right)\pi R^2 = \frac{qv_\perp R}{2}. \tag{14.11}$$

With (14.10) the longitudinal force induced by B_r is

$$F = qv_\perp B_r = -\left(\frac{qv_\perp R}{2}\right)\frac{\partial B_z}{\partial z} = -M\frac{\partial B_z}{\partial z}. \tag{14.12}$$

Since kinetic energy T is conserved, part of the transverse energy must be transferred to the longitudinal component $T_z = \frac{1}{2}mv_z^2$. Then

$$F = m\frac{\mathrm{d}v_z}{\mathrm{d}t} = \frac{1}{v_z}\frac{\mathrm{d}T_z}{\mathrm{d}t} = \frac{\mathrm{d}t}{\mathrm{d}z}\cdot\frac{\mathrm{d}T_z}{\mathrm{d}t}, \tag{14.13}$$

and therefore with (14.12)

$$\frac{\mathrm{d}T_z}{\mathrm{d}t} = -M\frac{\partial B_z}{\partial z}\cdot\frac{\mathrm{d}z}{\mathrm{d}t} = -M\frac{\partial B_z}{\partial t}. \tag{14.14}$$

On the other hand, we have that

$$\frac{\mathrm{d}T_z}{\mathrm{d}t} = \overbrace{\frac{\mathrm{d}}{\mathrm{d}t}}^{0}\left(T - \frac{mv_\perp^2}{2}\right). \tag{14.15}$$

From the relation

$$mv_\perp = qRB_z \tag{14.16}$$

one obtains by inserting $qR = \frac{2M}{v_\perp}$ from (14.11)

$$v_\perp = \frac{2MB_z}{mv_\perp} \Rightarrow v_\perp^2 = \frac{2MB_z}{m}. \tag{14.17}$$

Introducing into (14.15) leads to

$$\frac{\mathrm{d}T_z}{\mathrm{d}t} = \frac{\mathrm{d}}{\mathrm{d}t}(-MB_z). \tag{14.18}$$

By comparing (14.14) and (14.18) one finds that the magnetic moment M induced by the loop is a constant of motion. It is easy to show that the particle flies along the surface of a magnetic flux tube since the flux ϕ through the loop is constant:

$$\phi = B_z \pi R^2 = B_z \pi \cdot \underbrace{\frac{4M^2}{q^2 v_\perp^2}}_{R^2 \text{ from (14.11)}} = B_z \pi \frac{4M^2}{q^2} \cdot \underbrace{\frac{m}{2MB_z}}_{1/v_\perp^2 \text{ from (14.17)}} = \frac{2\pi m M}{q^2}. \tag{14.19}$$

One concludes from (14.11) that

$$v_\perp R = \frac{2M}{q} = \text{constant} = v_\perp \cdot \underbrace{\frac{mv_\perp}{qB_z}}_{R \text{ from (14.16)}}, \tag{14.20}$$

and therefore that

$$\boxed{v_\perp^2 / B_z = \text{constant}}. \tag{14.21}$$

A diagram of the solenoid retarding spectrometer used in [10] to measure the energy distribution of the electron in tritium β-decay, is shown in figure 14.8 (left). Electrons emitted from the source (field $B_z(0)$) curl around the magnetic field lines and are accelerated towards the centre of the spectrometer, where the magnetic field reaches its minimum value ($B_z(1)$). Cylindrical electrodes provide a retarding electric field which transmits only electrons with longitudinal kinetic energies T_z sufficient to overcome the potential step U, that is for $T_z > eU$. The value of the retarding energy eU is chosen just below the endpoint energy of the β spectrum. The advantage of such a device is its large acceptance at the source (2π sr) which increases the counting rate at the endpoint. Technical details can be found in [11].

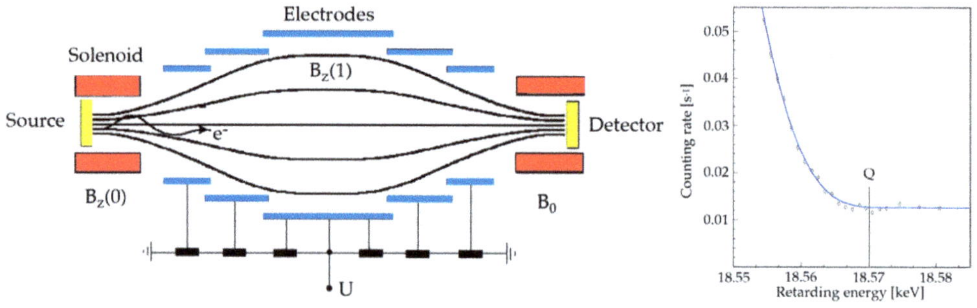

Figure 14.8. Left: diagram of a solenoid retarding spectrometer. Right: energy distribution of the electron around the endpoint (18.57 keV) in tritium β-decay. Adapted from [12].

Let us calculate the energy resolution for an electron of kinetic energy T. According to (14.21), the perpendicular kinetic energy $T_\perp(1)$ at the centre of the spectrometer is

$$T_\perp(1) = T_\perp(0)\frac{B_z(1)}{B_z(0)}, \tag{14.22}$$

for an electron emitted with transverse energy $T_\perp(0)$ at the source. The longitudinal energy at the centre is then

$$T_z(1) = T - T_\perp(1) = T - T_\perp(0)\frac{B_z(1)}{B_z(0)}. \tag{14.23}$$

At the centre the minimum and maximum longitudinal energies of an electron with kinetic energy T are therefore

$$T_{\text{min}} = T\left[1 - \frac{B_z(1)}{B_z(0)}\right] \qquad \text{and} \qquad T_{\text{max}} = T, \tag{14.24}$$

corresponding to the perpendicular energies $T_\perp(0) = T$ and $T_\perp(0) = 0$ at the source, respectively. The retarding potential is chosen such that $eU = T_{\text{min}}$ so that electrons with energies below T_{min} are not transmitted. The energy resolution of the spectrometer at the energy T is then given by

$$\boxed{\frac{\Delta T}{T} = \frac{B_z(1)}{B_z(0)}}, \tag{14.25}$$

which is typically 2×10^{-4} [11]. Note that the spectrometer measures only the integrated rate above the retarding energy, which is nonetheless sufficient around the endpoint energy to determine an upper limit for the neutrino mass.

Figure 14.8 (right) shows the energy distribution around the endpoint from a recent experiment with tritium [12]. The fit leads to the (95% confidence level) upper limit $m_{\nu e} = 2.3\,\text{eV}$. One knows from the width of the Z^0 resonance that only three light neutrinos exist (section 17.2.2). A fourth light neutrino would therefore be 'sterile' (meaning that it would not interact weakly). The existence of such a neutrino would lead to a kink in the energy distribution of the electron. The experiment excludes such a neutrino in the mass range between 1.7 and 191 eV [12].

The KATRIN experiment (figure 14.9) will study the endpoint region with much higher statistics and a better energy resolution, with the goal of reaching the 0.2 eV upper limit [13].

14.2.3 ν_μ and ν_τ masses

An upper limit for the mass m_{ν_μ} of the muon neutrino can be obtained from π^+ decay at rest into $\mu^+\nu_\mu$. From energy and momentum conservation one obtains

$$m_{\nu_\mu}^2 = \mathcal{P}_{\nu_\mu}^2 = \left(\mathcal{P}_\pi - \mathcal{P}_\mu\right)^2 = \left(m_\pi - E_\mu\right)^2 - \vec{p}_\mu^{\,2} = m_\pi^2 + m_\mu^2 - 2m_\pi\sqrt{m_\mu^2 + \vec{p}_\mu^{\,2}}. \tag{14.26}$$

Figure 14.9. The KATRIN solenoid retarding spectrometer on its way to its data taking site. Image courtesy of Karlsruhe Institute of Technology.

By measuring the momentum $|\vec{p}_\mu|$ of the muon in a homogeneous magnetic field an upper limit of 170 keV (90% confidence level) could be determined for m_{ν_μ}, using the well-known masses m_π and m_μ [14].

An upper limit for τ neutrino mass m_{ν_τ} has been obtained by measuring the missing mass in τ decays. The CLEO-II experiment at the e^+e^- storage ring in Cornell studied τ pairs produced by the reaction $e^+e^- \to \tau^+\tau^-$ [15]. The energies of the colliding leptons (and hence of the outgoing τ leptons) are $E = 5.3$ GeV. One of the τ leptons is identified through its leptonic decay or through the decay $\tau^\pm \to \pi^\pm \nu_\tau$. The second τ lepton then has to be emitted with the same momentum, but in the opposite direction. Its momentum $\vec{p}_\tau$ and its decay into pionic channels are used to calculate m_{ν_τ}. Decay channels with high pion multiplicities, such as $\tau^\pm \to \pi^\pm\pi^+\pi^-\pi^0\nu_\tau$ ($f = 4.8\%$, see table 14.1) are chosen, since the contribution of a finite neutrino mass to the neutrino energy is larger than for low pion multiplicities. The total energy E_H and the total momentum $\vec{p}_H$ of the pions are measured. The momenta of charged pions are determined with drift chambers in a solenoidal magnet, and the momenta of the two photons from π^0 decay are obtained from a matrix of CsI crystals. The mass of the ν_τ is given by

$$\begin{aligned} m_{\nu_\tau}^2 &= \left(E - E_H\right)^2 - \left(\vec{p}_\tau - \vec{p}_H\right)^2 \\ &= m_\tau^2 + m_H^2 - 2EE_H + 2\sqrt{E^2 - m_\tau^2}\sqrt{E_H^2 - m_H^2}\cos\theta. \end{aligned} \tag{14.27}$$

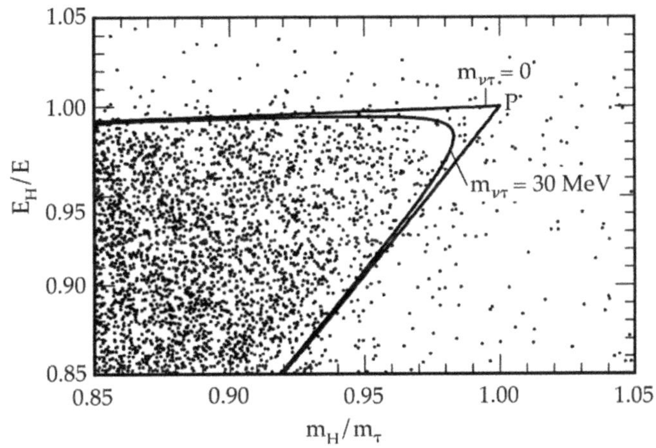

Figure 14.10. E_H/E versus m_H/m_τ for $m_{\nu_\tau} = 0$ and 30 MeV. Note the suppressed horizontal and vertical scales. The lines limit the kinematically allowed range. At point P a (massless) ν_τ is at rest in the laboratory system. After [15].

The invariant mass of the pions is denoted by m_H and θ is the angle between $\vec{p}_\tau$ and $\vec{p}_H$. Figure 14.10 shows the kinematically allowed range in a diagram E_H/E versus m_H/m_τ, for $m_{\nu_\tau} = 0$ and $m_{\nu_\tau} = 30$ MeV. An upper limit of 28 MeV (95% confidence level) is determined for m_{ν_τ} after background subtraction.

In section 14.6.1, which is about neutrino oscillations, we will show that neutrino masses are finite, but that the mass difference between the lightest and heaviest neutrino is at most ~50 meV. Given the upper limit of ~2 eV for the ν_e mass obtained from β-decay (see the previous section), the measured ν_μ and ν_τ masses just described are no longer competitive.

14.3 Solar neutrinos

The weak interaction plays a central role for life on Earth, being responsible for generating solar energy. The main process in the Sun is the production of α particles from proton–proton fusion. The *pp main cycle* begins with the weak interaction

$$pp \rightarrow de^+\nu_e \; (Q = 0.42 \text{ MeV}). \tag{14.28}$$

The extremely small cross section for this reaction, $\sigma \sim 10^{-23}$ b at a typical temperature of 1.5×10^7 K in the centre of the Sun, is compensated by the large number of available protons. The deuterium nuclei then generate α particles through the strong interaction

$$\begin{aligned} dp &\rightarrow {}^3\text{He}\,\gamma \; (Q = 5.49 \text{ MeV}), \\ {}^3\text{He}\,{}^3\text{He} &\rightarrow \alpha\, 2p \; (Q = 12.86 \text{ MeV}), \end{aligned} \tag{14.29}$$

while the positrons annihilate:

$$e^+e^- \to 2\gamma \; (Q = 1.02 \text{ MeV}). \tag{14.30}$$

The energy released by the main cycle is $Q = 26.7$ MeV.

The following strong interactions are used instead in *controlled fusion* ($t \equiv {}^3$H):

$$\begin{aligned} dd &\to tp \; (Q = 4.0 \text{ MeV}) \qquad \text{or} \\ &\to {}^3\text{He}\, n \; (Q = 3.3 \text{ MeV}) \qquad \text{and} \\ dt &\to \alpha n \; (Q = 17.6 \text{ MeV}). \end{aligned} \tag{14.31}$$

The *calculated* energy spectrum of solar *pp* neutrinos is shown in figure 14.11. The maximum energy corresponds to the Q-value of 420 keV. The neutrino flux at the surface of the Earth can be calculated from the measured solar power above the atmosphere ($L \simeq 3.9 \times 10^{26}$ W):

$$\phi = \frac{2L}{4\pi R^2 Q} = 6.2 \times 10^{10} \text{ cm}^{-2} \text{ s}^{-1}, \tag{14.32}$$

where R is the distance of the Sun, and for two neutrinos per main cycle.

Apart from the main cycle, other reactions also contribute to the production of solar neutrinos, albeit with lower probabilities (figure 14.12). Further ^{4}He is produced by the ppII and ppIII reactions. The pep reaction and the electron capture on ^{7}Be produce monoenergetic neutrinos. The β^+-decay of ^{8}B leads to a small (but experimentally essential) contribution from neutrinos of energies up to 14.6 MeV (figure 14.11).

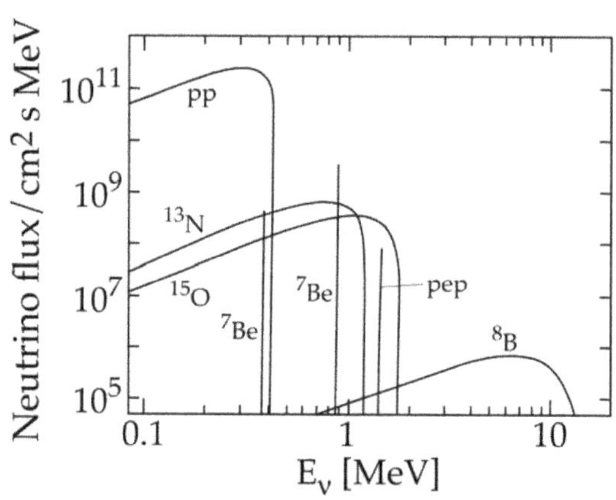

Figure 14.11. Calculated neutrino spectrum from the Sun at the surface of the Earth. The main contribution comes from the *pp* cycle. Note the monoenergetic neutrinos from the *pep* and ^{7}Be reactions (see also figure 14.12). The latter fluxes are given in cm^{-2} s^{-1}.

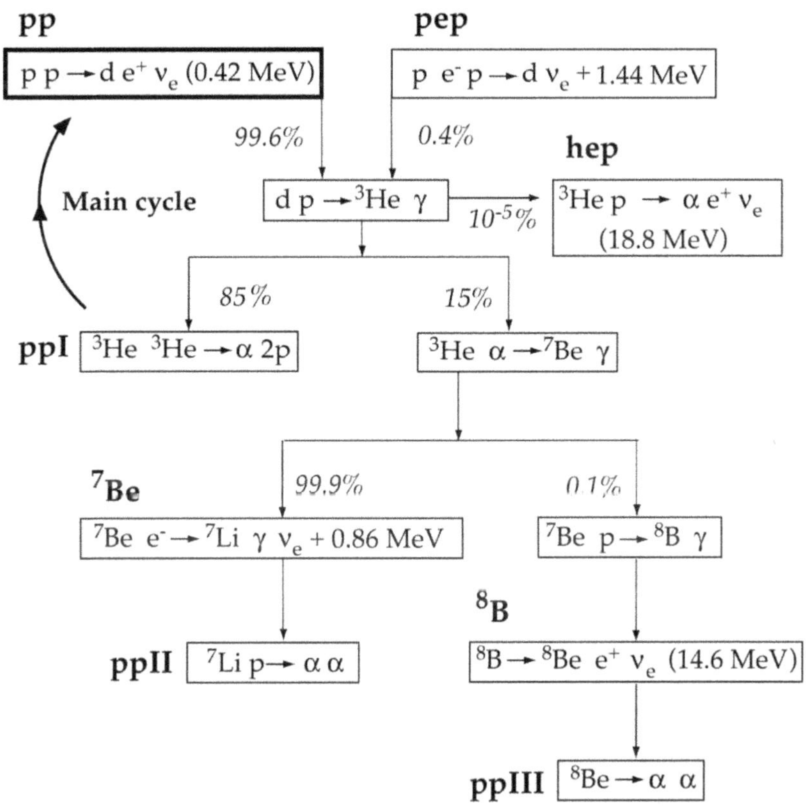

Figure 14.12. In the *pp* cycle (ppI) hydrogen is converted into helium. Neutrinos are produced from the *pp* reaction, from ^{8}B decay, and to a lesser extent from the *hep* reaction (the maximum energies are given in parentheses). Monoenergetic neutrinos are generated (1.44, 0.38 MeV and 0.86 MeV) by the *pep* reaction and by ^{7}Be decay (electron capture).

Organic elements are produced in the interior of stars by the CNO cycle:

$$\begin{aligned}
p^{12}\mathrm{C} &\rightarrow {}^{13}\mathrm{N}\,\gamma \\
{}^{13}\mathrm{N} &\rightarrow {}^{13}\mathrm{C}\,e^{+}\,\nu_e\ (+1.20\ \mathrm{MeV}) \\
p^{13}\mathrm{C} &\rightarrow {}^{14}\mathrm{N}\,\gamma \\
p^{14}\mathrm{N} &\rightarrow {}^{15}\mathrm{O}\,\gamma \\
{}^{15}\mathrm{O} &\rightarrow {}^{15}\mathrm{N}\,e^{+}\,\nu_e\ (+1.73\ \mathrm{MeV}) \\
p^{15}\mathrm{N} &\rightarrow {}^{12}\mathrm{C}\,\alpha.
\end{aligned} \tag{14.33}$$

The element ^{12}C is produced by α^{8}Be fusion. However, even with large α densities the production of carbon with this reaction would be unlikely since the ^{8}Be isotope decays quickly into 2α with a lifetime of 10^{-16} s (ppIII in figure 14.12). However, carbon production is enhanced in α^{8}Be fusion by the excitation of the Hoyle resonance $^{12}\mathrm{C}^{*}(0^{+})$ which requires 287 keV, supplied thermally in the star interior.

The resonance decays by γ emission into the ^{12}C ground state: $^{12}C^*(0^+) \rightarrow {}^{12}C(0^+)\ 2\gamma$ ($Q = 7.4$ MeV). The fraction of α particles produced by the CNO cycle in the Sun is only about 1.7%, but rises rapidly in more massive stars with core temperatures above 1.5×10^7 K.

The ^{8}B neutrinos were the first to be observed from the Sun. Conservation of lepton number forbids the reaction $\nu_e p \rightarrow n e^+$. However, absorption on the neutron $\nu_e n \rightarrow p e^-$ is allowed. The chlorine isotope ^{37}Cl was used as a neutron target by Raymond Davis and collaborators [16], employing the neutrino capture process

$$\nu_e\ {}^{37}_{17}\mathrm{Cl} \rightarrow {}^{37}_{18}\mathrm{Ar}^*\ e^-, \tag{14.34}$$

with $Q = -810$ keV. The cross section reaches its maximum value for neutrinos from ^{8}B decay. A container with 390 000ℓ tetrachloroethylene (C_2Cl_4) was used as a target (the ^{37}Cl isotopic fraction in chlorine is 24%). To reduce the production of protons by cosmic ray muons and suppress the background from the reaction $p^{37}\mathrm{Cl} \rightarrow n^{37}\mathrm{Ar}^*$, the experiment was performed in an old 1.5 km deep gold mine (Homestake in South Dakota). The Ar* atoms were periodically flushed with helium gas and detected in an argon proportional counter. The excited ^{37}Ar* nucleus decays by electron capture with a mean life of 49 d:

$$^{37}\mathrm{Ar}^*\ e^- \rightarrow {}^{37}\mathrm{Cl}\ \nu_e + \text{Auger electron}. \tag{14.35}$$

The observation of the 2.8 keV Auger electron in the proportional counter was the evidence for reaction (14.34). The first results yielded only upper limits for the neutrino flux [16]. However, over the years the experiment observed on average 0.5 Ar atom per day, while solar models predicted three times more, namely about 1.5 atom per day. Whether these models described the fusion mechanisms occurring in the Sun reliably was not obvious at that time. A deficit in the solar flux for neutrinos from the *pp* cycle was also observed with the reaction

$$\nu_e\ {}^{71}_{31}\mathrm{Ga} \rightarrow {}^{71}_{32}\mathrm{Ge}^*\ e^- \tag{14.36}$$

using 30 t of gallium [17]. We now know that the deficit is due to ν_e oscillations to ν_μ and ν_τ during their journey to Earth[1], as explained in section 14.6. Note that a finite and large enough magnetic moment of the neutrino could also contribute to the observed neutrino deficit [18]. The spin of the neutrino would precess in the solar magnetic field, leading to 'sterile' right-handed neutrinos which would not interact weakly, and therefore would remain invisible. However, this requires new physics: in the standard model the spin precession is negligible since the magnetic moment of the neutrino is very small, $\mu_\nu = 3 \times 10^{-19} m_\nu$ [eV] μ_B (Bohr magnetons). The experimental upper limit for the neutrino magnetic moment is $2.9 \times 10^{-11} \mu_B$ [7].

[1] The conversion of ^{8}B neutrinos is massively enhanced by interactions with the solar medium. This effect is negligible for the lower energy *pp* neutrinos, see section 14.6.6.

The reaction (14.34) was also sought with reactor antineutrinos:

$$\bar{\nu}_e\,{}^{37}_{17}\mathrm{Cl} \rightarrow {}^{37}_{18}\mathrm{Ar}^* e^-. \tag{14.37}$$

The negative result [19] shows that neutrinos and antineutrinos are different particles (see section 14.5).

14.4 The helicity of the neutrino

That neutrinos are left-handed was first demonstrated by the experiment of Maurice Goldhaber and colleagues [20]. Let us consider the level scheme of the (0^-) ^{152}Eu isomeric state (figure 14.13(a)). In the electron capture

$$pe^- \rightarrow n\nu_e \tag{14.38}$$

the neutrino is emitted in the direction opposite to the recoiling daughter, the (1^-) excited $^{152}\mathrm{Sm}^*$ isotope (figure 14.13(b)). Let us assume that the neutrino is left-handed. Since the spin of ^{152}Eu is zero, angular momentum conservation implies that the spin of $^{152}\mathrm{Sm}^*$ must be antiparallel to its flight direction (the spin of the captured electron is parallel to the spin of the neutrino). The *forward* emitted 960 keV γ in the decay to the ^{152}Sm ground state

$$^{152}\mathrm{Sm}^*(1^-) \rightarrow {}^{152}\mathrm{Sm}(0^+)\,\gamma \tag{14.39}$$

is therefore left-circularly polarized (negative helicity), while the backward emitted γ is right-circularly polarized (positive helicity). The opposite reasoning would apply to a right-handed neutrino: right-circularly polarized forward γ and left-circularly polarized backward γ.

Now, since the energy of the forward γ is larger in the laboratory than the 960 keV transition energy, this photon can perform nuclear resonance scattering:

$$\gamma^{152}\mathrm{Sm} \rightarrow {}^{152}\mathrm{Sm}^* \rightarrow \gamma^{152}\mathrm{Sm}. \tag{14.40}$$

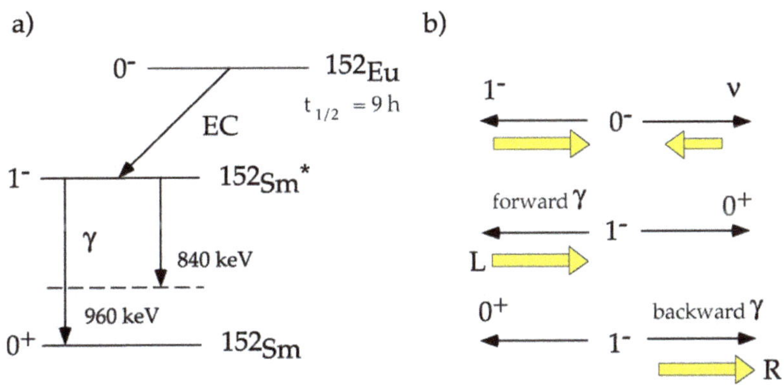

Figure 14.13. Measuring the neutrino helicity. (a) Level scheme of the ^{152}Eu isotope. (b) For a left-handed neutrino the 960 keV γ emitted in the direction of the $^{152}\mathrm{Sm}^*$ nucleus is left-circularly polarized.

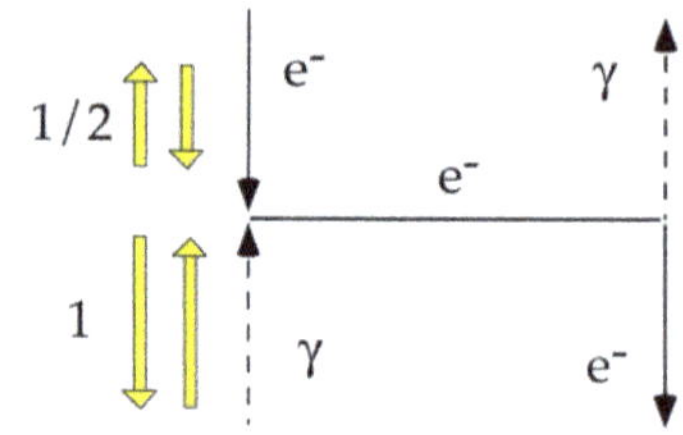

Figure 14.14. For Compton scattering the spins of the γ and of the electron are antiparallel since the exchanged electron has spin $\frac{1}{2}$.

This reaction cannot occur for the backward γ since its energy in the laboratory is smaller than 960 keV. Hence, a measurement of the helicity of the forward γ determines the helicity of the neutrino. Note that the experiment is only possible thanks to the very short lifetime (3×10^{-14} s) of $^{152}Sm^*$, otherwise its spin could be reversed by collisions with neighbouring nuclei.

The helicity of the photon was measured using Compton scattering on polarized electrons. In the first approximation, a right-handed (left-handed) photon does not scatter on a left-handed (right-handed) electron since the projection of the spin of the exchanged virtual electron along the quantization axis would be $\frac{3}{2}$ (figure 14.14). Thus photons are more easily scattered on polarized electrons if the spins are antiparallel. This leads to an asymmetry in the scattering of polarized photons in a magnetized iron target when the direction of the magnetic field is reversed. With iron an electron polarization of only about 5% can be achieved, hence the asymmetry is expected to be small.

Figure 14.15(*a*) shows the experimental arrangement. The γ rays from Sm* decay traverse a magnetized iron absorber where they scatter off polarized electrons. Some of the photons reach the Sm_2O_3 target below, where they undergo nuclear resonance scattering before reaching the NaI detector. Since these photons are emitted forwards, the neutrinos from Eu decay are emitted upwards in figure 14.15(*a*).

Figure 14.15(*b*) shows the energy spectrum. One measures the energy spectrum for both polarities of the *B*-field (see figure 14.15(*a*)), that is with electron spins in the iron absorber pointing upwards ($N_\uparrow$ events) and downwards ($N_\downarrow$ events), and calculates the asymmetry for the 960 keV line:

$$\delta = \frac{N_\uparrow - N_\downarrow}{N_\uparrow + N_\downarrow}. \tag{14.41}$$

The asymmetry is positive for left-handed photons since the losses through Compton scattering in the iron absorber are reduced for parallel spins. The result is

$$\delta = +0.034 \pm 0.006. \tag{14.42}$$

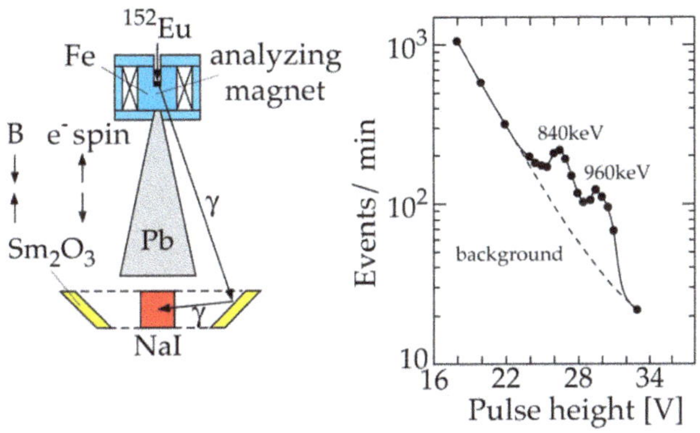

Figure 14.15. (*a*) The apparatus used to measure the helicity of the neutrino. (*b*) Energy spectrum of the forward γ. After [20].

The expected value is slightly higher for 100% polarized left-handed photons: $\delta = +0.050$. However, taking into account the solid angle acceptance one expects a photon polarization of only 75%, thus bringing result (14.42) in full agreement with the expectation for left-handed neutrinos.

14.5 Dirac and Majorana neutrinos

In the standard model neutrinos and antineutrinos are *C*-conjugated *Dirac* particles and are therefore different:

$$\boxed{\nu_e \neq \bar{\nu}_e}. \tag{14.43}$$

Reaction (14.37) is then forbidden by lepton number conservation. We have just shown that neutrinos (for example from β^+-decay) are left-handed, while antineutrinos (e.g. from β^--decay) are right-handed. However, apart from their different helicities, neutrinos and antineutrinos could nonetheless be identical (self-conjugated) particles and reaction (14.37) would not occur either since the incident (anti)neutrino has the *wrong* helicity. In this case neutrinos would be of the *Majorana* type:

$$\boxed{\nu_e \equiv \bar{\nu}_e}. \tag{14.44}$$

Whether neutrinos are Dirac or Majorana particles is still an open issue. Double β-decay (first proposed by Maria Goeppert-Mayer in 1935 [21]) leads to the simultaneous emission of two neutrinos. However, the observation of *neutrinoless* double β-decay would solve the ambiguity between Dirac and Majorana neutrinos in favour of the latter: in this process the neutrino emitted by the β-decay of the first neutron would be reabsorbed by a neighbouring neutron (figure 14.16(*a*)). This is only possible for Majorana neutrinos. Since the spin of the neutrinos has to flip, the

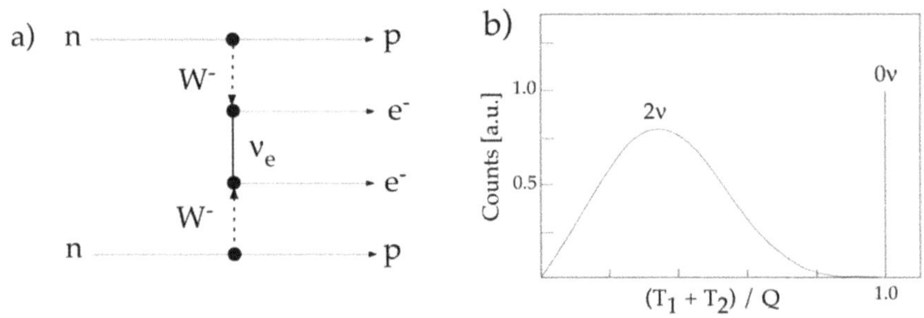

Figure 14.16. (*a*) Neutrinoless double *β*-decay. (*b*) The fraction of the energy Q carried by the two electrons in double *β*-decay, with and without neutrinos in the final state.

neutrino also has to be massive. The interest in double *β*-decay was revived in the 1990s with the observation of neutrino oscillations, implying that neutrinos are not massless (section 14.6).

We have seen, when dealing with the liquid drop model of the nucleus, that several even–even nuclei have stable isobars (lower parabola in figure 5.6(*b*)). Double β^--decay occurs via $Z \to (Z+2)\, 2e^-\, 2\bar{\nu}$, and double β^+-decay through $Z \to (Z-2)\, 2e^+ 2\nu$. For the latter nuclei the electron capture processes e^+ EC 2ν and 2EC 2ν are also possible but very difficult to observe. For β^+-decays the Q-values are smaller than $\Delta\mathcal{M}$ since $4m_e$ have to be subtracted from Q. Since the mean life increases rapidly with decreasing Q-value only decays with relatively large Q-values can, in practice, be measured. There are eleven double β^--decays with $Q > 2$ MeV, but only one β^+-decay with $Q > 800$ keV (^{124}Xe → ^{124}Te) [22]. Hence mostly double β^--decays have been investigated so far. The nuclei have spin 0 and positive parity. The transition is therefore not suppressed by nuclear matrix elements.

The energy distributions of the electrons in 0ν and 2ν double *β*-decays are very different. For 0ν the electron energies sum up to the constant value Q, while for 2ν the energy spectrum is continuous (figure 14.16(*b*)). The energies are measured, for example, with high-resolution germanium detectors. Natural germanium contains 7.7% ^{76}Ge which decays into $^{76}\mathrm{Se} + 2e^- 2\bar{\nu}_e$ ($Q = 2$ MeV). The background from ambient γ activity at the surface of the Earth increases rapidly below 2.6 MeV. To reduce the background from cosmic rays the detectors are operated underground. The 2ν decay of ^{76}Ge has been observed with a half-life $t_{1/2} \simeq 1.8 \times 10^{21}$ years [23]. The lower limit for the corresponding 0ν decay is $t_{1/2} = 2.1 \times 10^{25}$ years [24] which does not confirm the evidence reported earlier by another experiment [25].

The tracks of the two stopping electrons can also be observed directly using time projection chambers (TPCs) (figure 14.17). However, the energies can only be determined roughly from the lengths of the tracks. A xenon TPC enriched with ^{136}Xe was operated at 5 atm in the Gotthard tunnel in Switzerland. A lower limit of 3.4×10^{23} years could be derived for the neutrinoless double *β*-decay $^{136}\mathrm{Xe} \to {}^{136}\mathrm{Ba} + 2\mathrm{e}^-$ [27]. The 2ν decay was observed later with a half-life of $(2.17 \pm 0.06) \times 10^{21}$ years [28].

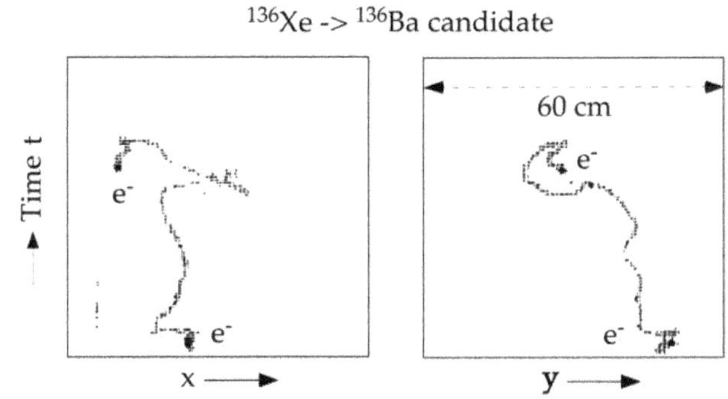

Figure 14.17. x and y projections of a $2e^-$ background event in a xenon TPC. Note the large ionization losses at the end of the trajectories. After [26].

14.6 Neutrino oscillations

The search for neutrino oscillations was first proposed by Bruno Pontecorvo in 1946 to test the conservation of lepton number [29]. Since 1998 we have known that the identity of neutrinos changes as a function of time. Lepton number conservation does not apply between neutrinos. This implies that neutrinos are not massless as we shall show in section 14.6.1.

The first evidence for neutrino oscillation came from neutrinos produced in the atmosphere. Atmospheric neutrinos (and antineutrinos) stem from the decay chain $\pi \to \mu\nu_\mu$ ($\mu \to e\nu_e\nu_\mu$) with pions generated in the atmosphere by high energy cosmic protons. Atmospheric neutrinos were studied using the Super-Kamiokande detector in Japan [30]. The apparatus consists of a 50 000 t water tank equipped with some 11 000 phototubes (50 cm diameter) in a 1000 m deep mine. The direction of the outgoing lepton from the reaction $\nu_e(\bar{\nu}_e)N \to e^-(e^+)X$ or $\nu_\mu(\bar{\nu}_\mu)$ $N \to \mu^-(\mu^+)X$ is determined by detecting the Čerenkov radiation. Muons and electrons produce light rings (see section 9.10 on RICH counters). While the rings from muons are sharp, the rings from electrons are smeared by electromagnetic showers (figure 14.18, right). Thus ν_e and ν_μ can be distinguished by analysing the shapes of the Čerenkov light rings, and their energies can be calculated from their diameters.

Figure 14.19 shows that the flux N_u of muon neutrinos from the antipodes is smaller than the flux N_d from the surface above the detector. Since neutrino absorption or scattering by the Earth is negligible one concludes that some of the muon neutrinos have changed flavour on their longer path from the antipodes to the detector. In contrast, this asymmetry is much smaller for electron neutrinos, and even compatible with zero (figure 14.19). It follows that ν_μ oscillates primarily into ν_τ. A ν_μ deficit was also reported for neutrinos produced by the KEK accelerator in Japan, 250 km from the Super-Kamiokande detector [31].

The solar neutrinos from the main (pp) cycle also oscillate along their flight path to the Earth. The Sudbury Neutrino Observatory (SNO) in Ontario (Canada) has

Figure 14.18. Left: the Super-Kamiokande detector in the 1000 m deep Mozumi Mine in Japan. Right: Čerenkov rings of a ν_μ-induced muon of 600 MeV (top) and smeared rings from a ν_e-induced electron of 490 MeV (bottom). The false colours indicate the photomultiplier pulse height or the light arrival time. Image credit: Kamioka-Observatory, ICRR, University of Tokyo.

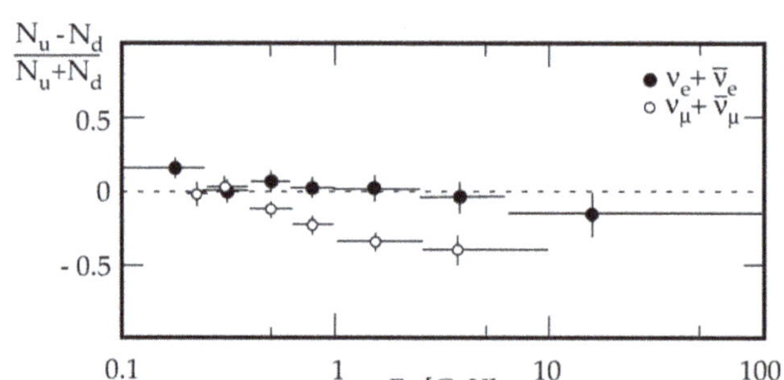

Figure 14.19. Asymmetry between the upward neutrino flux N_u and the downward flux N_d in the Super-Kamiokande detector, for ν_e (or $\bar{\nu}_e$) and ν_μ (or $\bar{\nu}_\mu$). Adapted from [30].

definitively demonstrated the deficit in solar neutrinos. The solar neutrino flux was measured from the observed rates of the two reactions

$$\nu_e n \rightarrow e^- p, \tag{14.45}$$

$$\nu n \rightarrow \nu n. \tag{14.46}$$

The first reaction (charged current with W exchange (CC)) is only possible with ν_e, while the second reaction runs with the same cross section for all neutrino flavours,

ν_e, ν_μ and ν_τ (neutral current with Z^0 exchange (NC)). The absolute neutrino flux can therefore be derived from the neutral current reaction, independent of oscillations. If the ν_e vanishes by oscillating into ν_μ or ν_τ, the calculated electron neutrino flux from the observed rate for reaction (14.45) would be smaller than expected from reaction (14.46).

The SNO detector is an acrylic sphere containing 1000 t of heavy water [32]. (D_2O is used in the Canadian CANDU nuclear reactors as a moderator.) The sphere (figure 14.20) is installed 2039 m below the surface in an old copper mine to reduce the background from cosmic rays. Since the binding energy of deuterium is 2.2 MeV, only neutrinos from ^8B decay can be observed (section 14.3). The Čerenkov light of the electron from reaction (14.45) is detected by 9438 photomultipliers. Salt (2 t of NaCl) is dissolved in the heavy water to detect the recoil neutron from reaction (14.46). Neutron capture in ^{35}Cl leads to a γ cascade with a total energy of about 8 MeV that can be measured by recording the Čerenkov light generated by the Compton recoil electrons. The latter can be distinguished from the light produced by the electrons from reaction (14.45) since the angular distribution of the γ from ^{35}Cl is isotropic. The D_2O sphere is embedded in a radiochemically clean water tank to suppress γ-rays from the surrounding rock (Th and U isotopes).

SNO finds [33] for the ratio of *neutrino fluxes* ϕ_{CC} from reaction (14.45) to ϕ_{NC} from reaction (14.46):

$$\frac{\phi_{CC}}{\phi_{NC}} = (30.6 \pm 3.5)\%, \tag{14.47}$$

thus proving that about $\frac{2}{3}$ of the electron neutrinos have disappeared, in accordance with the previous Homestake result (section 14.3). On the other hand, the measured

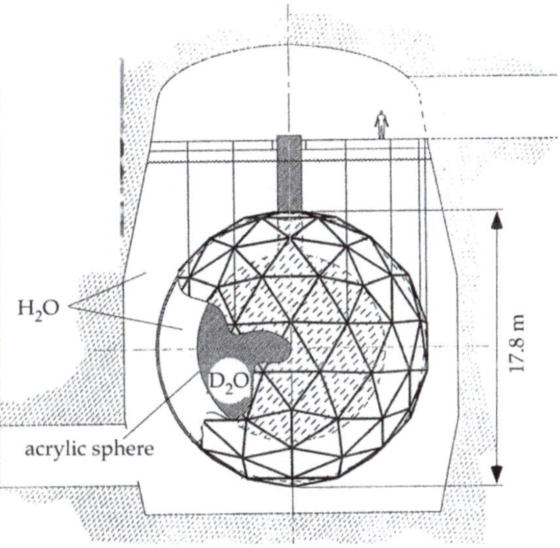

Figure 14.20. The solar neutrino detector SNO at Sudbury. After [32]. Image credit: Roy Kaltschmidt, Lawrence Berkeley National Laboratory.

flux ϕ_{NC} is in agreement with predictions from solar models. The long lasting puzzle of the solar neutrino deficit has finally been solved.

14.6.1 The oscillation mechanism

Since neutrino flavours oscillate, ν_e, ν_μ and ν_τ are not eigenstates of the (weak) interaction, but linear superpositions of the eigenstates $\nu_i \equiv \nu_1$, ν_2 and ν_3 with masses m_1, m_2 and m_3:

$$|\nu_\alpha\rangle = \sum_{i=1}^{3} U_{\alpha i}|\nu_i\rangle \qquad \text{with} \qquad \alpha = 1(e), 2(\mu), 3(\tau), \tag{14.48}$$

where the matrix U is unitary. For antineutrinos $U_{\alpha i}$ has to be replaced by its complex conjugate. We use Greek letters for flavour states and Roman letters for eigenstates. With the notations ν_e, ν_μ or ν_τ we mean the neutrinos that were generated in association with electrons, muons or tauons. For example, ν_μ is the neutrino emitted in the decay $\pi^+ \to \mu^+\nu_\mu$. Since ν_μ is a superposition of ν_1, ν_2 and ν_3 one would obtain with an infinitely precise measurement of the muon momentum (and the known π and μ masses) three possible mass values m_1, m_2 or m_3, averaging to

$$m_{\nu\mu} = \langle\nu_\mu|H|\nu_\mu\rangle = \sum_{i,j=1}^{3} U_{\mu i}U_{\mu j}^{*}\overbrace{\langle\nu_j|H|\nu_i\rangle}^{\delta_{ij}m_i} = \sum_{i=1}^{3}|U_{\mu i}|^2 m_i \text{ with } \sum_{i=1}^{3}|U_{\mu i}|^2 = 1. \tag{14.49}$$

The probability to measure m_i would be $|U_{\mu i}|^2$.

Consider now at time $t = 0$ a neutrino flavour ν_α with momentum $p \gg m_\alpha$. The eigenstates ν_i have the same momentum p, but different energies E_i (and velocities β_i) because the masses can be different. (A detailed quantum mechanical treatment can be found in [34].) The time evolution of the ν_i-component in its rest frame at *laboratory* time $t = \tau_i\gamma_i$ is

$$|\nu_i(\tau_i)\rangle = \mathrm{e}^{-\mathrm{i}m_i\tau_i}\,|\nu_i\rangle, \tag{14.50}$$

where τ_i is the proper time. By applying a Lorentz transformation into the neutrino rest frame one obtains the relation

$$m_i = \gamma_i E_i - \beta_i\gamma_i p \tag{14.51}$$

which inserted into (14.50) gives

$$|\nu_i(t)\rangle = \mathrm{e}^{-\mathrm{i}(E_i - \beta_i p)t}\,|\nu_i\rangle = \mathrm{e}^{-\mathrm{i}(E_i - p)L}|\nu_i\rangle. \tag{14.52}$$

Given their small masses, we have assumed neutrinos to be highly relativistic ($\beta_i \simeq 1$) and have denoted with $L \simeq t$ the distance travelled by the neutrino (in natural units). Furthermore, since $m_i \ll p$ we may write

$$E_i = \sqrt{p^2 + m_i^2} \simeq p\left(1 + \frac{m_i^2}{2p^2}\right) = p + \frac{m_i^2}{2p}. \tag{14.53}$$

Introducing the mean energy $E \simeq p$ one obtains

$$\left|\nu_i(L)\right\rangle = \mathrm{e}^{-\mathrm{i}\frac{m_i^2 L}{2E}}\left|\nu_i\right\rangle. \tag{14.54}$$

The composition of the ν_α beam is then

$$\left|\nu_\alpha(L)\right\rangle = \sum_{i=1}^{3} U_{\alpha i}\mathrm{e}^{-\mathrm{i}\frac{m_i^2 L}{2E}}\left|\nu_i\right\rangle \tag{14.55}$$

after the distance L. Note that energy is conserved during the oscillation process. Indeed, the energy of the neutrino with flavour α

$$E_\alpha = \langle \nu_\alpha(L)|H|\nu_\alpha(L)\rangle = \sum_{i=1}^{3} E_i |U_{\alpha i}|^2 \tag{14.56}$$

is independent of L (or t). The amplitude for finding a neutrino ν_β at the distance L is given by

$$\left\langle \nu_\beta \middle| \nu_\alpha(L)\right\rangle = \sum_{i,j=1}^{3}\left\langle \nu_j \middle| U_{\beta j}^{*} U_{\alpha i}\mathrm{e}^{-\mathrm{i}\frac{m_i^2 L}{2E}} \middle| \nu_i\right\rangle = \sum_{i=1}^{3} U_{\beta i}^{*} U_{\alpha i}\mathrm{e}^{-\mathrm{i}\frac{m_i^2 L}{2E}}, \tag{14.57}$$

since

$$\left\langle \nu_i \middle| \nu_j \right\rangle = \delta_{ij}. \tag{14.58}$$

Starting from a neutrino ν_α, the probability of finding a ν_β at the distance L is then

$$\boxed{I_{\alpha\beta} = \left|\left\langle \nu_\beta \middle| \nu_\alpha(L)\right\rangle\right|^2 = \left|\sum_{i=1}^{3} U_{\alpha i} U_{\beta i}^{*}\mathrm{e}^{-\mathrm{i}\frac{m_i^2 L}{2E}}\right|^2}. \tag{14.59}$$

Since the matrix U is unitary:

$$\left|\sum_{i=1}^{3} U_{\alpha i} U_{\beta i}^{*}\right|^2 = \delta_{\alpha\beta}. \tag{14.60}$$

In particular, it follows from (14.59) that $I_{\alpha\alpha} = 1$ for $m_1 = m_2 = m_3$. Therefore oscillation is not possible with three massless neutrinos.

Consider as a simple example the oscillation between $N = 2$ neutrino flavours. The $N \times N$ unitary matrix is described by N^2 real parameters ($2N^2$ real parameters minus the N^2 unitarity conditions $U^\dagger = U^{-1}$). $2N$ phases can be absorbed in the definition of $|\nu_\alpha\rangle$ (i.e. $|\nu_\alpha\rangle \to \mathrm{e}^{\mathrm{i}\phi_\alpha}|\nu_\alpha\rangle$) and $|\nu_i\rangle$ without changing $I_{\alpha\beta}$. A common phase can still be chosen, leaving the matrix $U_{\alpha i} = \langle \nu_i | \nu_\alpha \rangle$ invariant. Thus $2N - 1$ phases are unphysical and the number of free real parameters is

$$n_f = N^2 - (2N - 1) = (N - 1)^2. \tag{14.61}$$

For $N = 2$ there is one degree of freedom ($n_f = 1$) and the 2×2 mixing matrix U takes the simple form

$$U = \begin{pmatrix} \cos\theta & \sin\theta \\ -\sin\theta & \cos\theta \end{pmatrix}. \tag{14.62}$$

The probability of finding flavour 2 at distance L, when flavour 1 was produced at the source, is given by (14.59):

$$\begin{aligned} I_{12} &= \left| -\sin\theta\cos\theta \mathrm{e}^{-\mathrm{i}\frac{m_1^2 L}{2E}} + \cos\theta\sin\theta \mathrm{e}^{-\mathrm{i}\frac{m_2^2 L}{2E}} \right|^2 \\ &= (\sin\theta\cos\theta)^2\left[2 - 2\mathrm{Re}\,\mathrm{e}^{\mathrm{i}\frac{(m_1^2 - m_2^2)}{2E}L}\right] = 2(\sin\theta\cos\theta)^2\left[1 - \cos\frac{(m_2^2 - m_1^2)}{2E}L\right] \\ &= \sin^2 2\theta \sin^2\left(\frac{\Delta m_{12}^2 L}{4E}\right), \end{aligned} \tag{14.63}$$

with

$$\boxed{\Delta m_{12}^2 \equiv m_2^2 - m_1^2}. \tag{14.64}$$

With masses expressed in GeV the distance L is given in GeV^{-1}. As a widely accepted convention L is expressed in km, Δm_{12}^2 in eV^2, and E in GeV. One finds from (14.63) with $L\,[\mathrm{GeV}^{-1}] = L\,[\mathrm{km}] \times (10^{21}/197.3)$:

$$\boxed{I_{12} = \sin^2 2\theta \sin^2\left(\frac{1.267\,\Delta m_{12}^2\,[\mathrm{eV}^2]\,L\,[\mathrm{km}]}{E\,[\mathrm{GeV}]}\right)}. \tag{14.65}$$

Figure 14.21 shows as an example the transition probability I_{12} as a function of L/E for $\theta = 34°$ and $\Delta m_{12}^2 = 7.5 \times 10^{-5}\,\mathrm{eV}^2$. The period of oscillation is $\pi E/1.267\Delta m_{12}^2 = 33\,\mathrm{km\,MeV}^{-1}$.

14.6.2 Oscillations between three neutrino flavours

We shall now describe the data for atmospheric and solar neutrinos, as well as results from accelerators and nuclear reactors, through oscillations between three

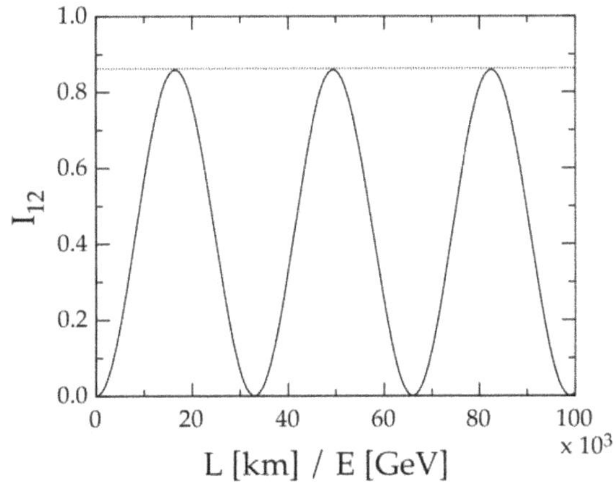

Figure 14.21. Probability of finding a neutrino of flavour 2 for a neutrino of flavour 1 produced at the source, as a function of distance/energy (see the text for details).

neutrino flavours. According to (14.61) the 3×3 matrix U contains $n_f = 4$ free parameters, which are denoted by the angles θ_{12}, θ_{13}, θ_{23}, and a phase δ. The usual parametrization is in terms of the Pontecorvo–Maki–Nagakawa–Sakata (PMNS) matrix, which has the same form as the Cabibbo–Kobayashi–Maskawa (CKM) matrix (16.67) in the quark sector[2]:

$$U = \begin{pmatrix} 1 & 0 & 0 \\ 0 & c_{23} & s_{23} \\ 0 & -s_{23} & c_{23} \end{pmatrix} \begin{pmatrix} c_{13} & 0 & e^{-i\delta} s_{13} \\ 0 & 1 & 0 \\ -e^{i\delta} s_{13} & 0 & c_{13} \end{pmatrix} \begin{pmatrix} c_{12} & s_{12} & 0 \\ -s_{12} & c_{12} & 0 \\ 0 & 0 & 1 \end{pmatrix}, \quad (14.66)$$

with $s_{ij} \equiv \sin\theta_{ij}$ and $c_{ij} \equiv \cos\theta_{ij}$. The matrix U represents a succession of three rotations (figure 14.22): a first rotation by θ_{12} around the 3-axis, a second rotation by θ_{13} around the new 2-axis, followed by a third rotation by θ_{23} around the new 1-axis. From the data (see the following sections) one obtains the set of angles [7]:

$$\theta_{12} \simeq 34°, \quad \theta_{23} \simeq 45° \quad \text{and} \quad \theta_{13} \simeq 9°. \quad (14.67)$$

These angles are much larger than the mixing angles in the quark sector (section 16.3). Experimentally one finds for the quadratic mass differences $\Delta m_{ij}^2 \equiv m_j^2 - m_i^2$ between the eigenstates ν_1, ν_2 and ν_3 [7]:

$$\left|\Delta m_{23}^2\right| \simeq 2.4 \times 10^{-3}\,\text{eV}^2 \quad \text{and} \quad \Delta m_{12}^2 \simeq 7.5 \times 10^{-5}\,\text{eV}^2. \quad (14.68)$$

[2] In the literature the superposition (14.48) is sometimes written with $U_{\alpha i}^*$ for neutrinos, in which case $\delta \to -\delta$ in (14.66).

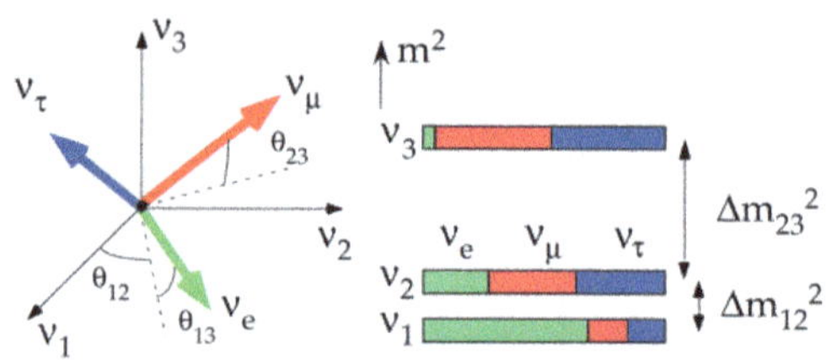

Figure 14.22. Left: definition of the Euler angles θ_{ij}. The dashed lines are projections into the 12-plane. Right: mass spectra and flavour contents of the neutrino eigenstates. The contribution from ν_e to ν_3 is small (~3%) and even vanishes in the limit $\theta_{13} \to 0$. For $\theta_{13} \neq 0$ the relative contributions from ν_μ and ν_τ to ν_1 and ν_2 depend on the phase δ. Shown here is the configuration for $\theta_{13} = 0$.

Therefore

$$\Delta m_{12}^2 + \Delta m_{23}^2 = \Delta m_{13}^2 \Rightarrow \Delta m_{13}^2 \simeq \Delta m_{23}^2. \tag{14.69}$$

Figure 14.22 shows the mass spectrum (the *normal* hierarchy with $\Delta m_{23}^2 > 0$). By convention the heavier of the two first states is labelled ν_2 ($m_2^2 > m_1^2$). The sign of Δm_{23}^2 is not known yet (the *inverted* hierarchy with ν_1 and ν_2 heavier than ν_3). For Majorana neutrinos this sign could be determined by measuring the rates for neutrinoless double β-decay [35]. Absolute masses cannot be measured with oscillations. However, it follows from the mass differences in figure 14.22 (right) and from (14.68) that the mass of the heaviest neutrino is at least 50 meV.

The ν_e contains a small ν_3-component since the angle θ_{13} is small, and therefore large samples of data will be required to determine the phase δ. For $0 < \delta < \pi$ the matrix U becomes complex, which induces CP violation $I(\nu_\alpha \to \nu_\beta) \neq I(\bar{\nu}_\alpha \to \bar{\nu}_\beta)$ and also T violation in the neutrino sector $I(\nu_\alpha \to \nu_\beta) \neq I(\nu_\beta \to \nu_\alpha)$, assuming that CPT is conserved ($I(\nu_\alpha \to \nu_\beta) = I(\bar{\nu}_\beta \to \bar{\nu}_\alpha)$) (see appendix D).

We now show that Δm_{23}^2 is determined using atmospheric neutrinos or neutrinos from accelerators, and that Δm_{12}^2 is obtained with neutrinos from the Sun or from nuclear reactors.

14.6.3 Δm_{23}^2 oscillations

To start with, let us assume that the small angle θ_{13} actually vanishes. The matrix U, (14.66), becomes with $\theta_{13} = 0$:

$$U = \begin{pmatrix} 1 & 0 & 0 \\ 0 & c_{23} & s_{23} \\ 0 & -s_{23} & c_{23} \end{pmatrix} \begin{pmatrix} c_{12} & s_{12} & 0 \\ -s_{12} & c_{12} & 0 \\ 0 & 0 & 1 \end{pmatrix} = \begin{pmatrix} c_{12} & s_{12} & 0 \\ -s_{12}c_{23} & c_{12}c_{23} & s_{23} \\ s_{12}s_{23} & -c_{12}s_{23} & c_{23} \end{pmatrix}. \tag{14.70}$$

The probability (14.59) of finding the neutrino flavour β, when the flavour α was produced, is

$$I_{\alpha\beta} = \left| U_{\alpha 1} U_{\beta 1}^* e^{-i\frac{m_1^2 L}{2E}} + U_{\alpha 2} U_{\beta 2}^* e^{-i\frac{m_2^2 L}{2E}} + U_{\alpha 3} U_{\beta 3}^* e^{-i\frac{m_3^2 L}{2E}} \right|^2. \tag{14.71}$$

Since $\Delta m_{12}^2 \ll \Delta m_{23}^2$ the term Δm_{12}^2 induces a slow modulation of the Δm_{23}^2 oscillation which can be neglected if the length L is not too large, that is, comparable to (or smaller than) the period of the Δm_{23}^2 oscillation. The term

$$\frac{\Delta m_{12}^2 L}{2E} \tag{14.72}$$

is then negligible and we can write

$$\begin{aligned} I_{\alpha\beta} = {} & \left|U_{\alpha 1}U_{\beta 1}^*\right|^2 + \left|U_{\alpha 2}U_{\beta 2}^*\right|^2 + \left|U_{\alpha 3}U_{\beta 3}^*\right|^2 + 2\mathrm{Re}\,U_{\alpha 1}U_{\beta 1}^*U_{\alpha 2}^*U_{\beta 2} \\ & + 2\mathrm{Re}\,U_{\alpha 1}U_{\beta 1}^*U_{\alpha 3}^*U_{\beta 3}\mathrm{e}^{\mathrm{i}\frac{\Delta m_{13}^2 L}{2E}} + 2\mathrm{Re}\,U_{\alpha 2}U_{\beta 2}^*U_{\alpha 3}^*U_{\beta 3}\mathrm{e}^{\mathrm{i}\frac{\Delta m_{23}^2 L}{2E}}. \end{aligned} \tag{14.73}$$

For $\alpha \neq \beta$ the unitarity conditions read

$$U_{\alpha 1}U_{\beta 1}^* + U_{\alpha 2}U_{\beta 2}^* + U_{\alpha 3}U_{\beta 3}^* = 0, \tag{14.74}$$

and therefore

$$\begin{aligned} & \left|U_{\alpha 1}U_{\beta 1}^*\right|^2 + \left|U_{\alpha 2}U_{\beta 2}^*\right|^2 + \left|U_{\alpha 3}U_{\beta 3}^*\right|^2 + 2\mathrm{Re}\,U_{\alpha 1}U_{\beta 1}^*U_{\alpha 2}^*U_{\beta 2} \\ & \quad = -2\mathrm{Re}\,U_{\alpha 1}U_{\beta 1}^*U_{\alpha 3}^*U_{\beta 3} - 2\mathrm{Re}\,U_{\alpha 2}U_{\beta 2}^*U_{\alpha 3}^*U_{\beta 3}. \end{aligned} \tag{14.75}$$

Introducing into (14.73) gives the probability

$$\begin{aligned} I_{\alpha\beta}(\alpha \neq \beta) = {} & 2\mathrm{Re}\,U_{\alpha 1}U_{\beta 1}^*U_{\alpha 3}^*U_{\beta 3}\left[\mathrm{e}^{\mathrm{i}\frac{\Delta m_{13}^2 L}{2E}} - 1\right] \\ & + 2\mathrm{Re}\,U_{\alpha 2}U_{\beta 2}^*U_{\alpha 3}^*U_{\beta 3}\left[\mathrm{e}^{\mathrm{i}\frac{\Delta m_{23}^2 L}{2E}} - 1\right], \end{aligned} \tag{14.76}$$

and with $\Delta m_{13}^2 \simeq \Delta m_{23}^2$:

$$\begin{aligned} I_{\alpha\beta}(\alpha \neq \beta) & = 2\mathrm{Re}\,U_{\alpha 3}^*U_{\beta 3}\overbrace{\left[U_{\alpha 1}U_{\beta 1}^* + U_{\alpha 2}U_{\beta 2}^*\right]}^{-U\alpha 3U_{\beta 3}^*} \times \underbrace{\left[\mathrm{e}^{\mathrm{i}\frac{\Delta m_{23}^2 L}{2E}} - 1\right]}_{\mathrm{e}^{\mathrm{i}\frac{\Delta m_{23}^2 L}{4E}}\underbrace{\left(\mathrm{e}^{\mathrm{i}\frac{\Delta m_{23}^2 L}{4E}} - \mathrm{e}^{-\mathrm{i}\frac{\Delta m_{23}^2 L}{4E}}\right)}_{2\mathrm{i}\sin\left[\frac{\Delta m_{23}^2 L}{4E}\right]}} \\ & = 4\left|U_{\alpha 3}^*U_{\beta 3}\right|^2 \mathrm{Re}\left(-\mathrm{i}\mathrm{e}^{\mathrm{i}\frac{\Delta m_{23}^2 L}{4E}}\sin\left[\frac{\Delta m_{23}^2 L}{4E}\right]\right). \end{aligned} \tag{14.77}$$

Therefore,

$$I_{\alpha\beta}(\alpha \neq \beta) = 4\left|U_{\alpha 3}^{*} U_{\beta 3}\right|^{2} \sin^{2}\left(\frac{\Delta m_{23}^{2} L}{4E}\right). \tag{14.78}$$

Introducing the matrix (14.70) with $U_{13} = 0$ leads to

$$\begin{aligned} I_{e\tau}(\nu_e \to \nu_\tau) &= I_{\tau e}(\nu_\tau \to \nu_e) = 0, \\ I_{e\mu}(\nu_e \to \nu_\mu) &= I_{\mu e}(\nu_\mu \to \nu_e) = 0. \end{aligned} \tag{14.79}$$

There is no ν_e-component in ν_3, and hence oscillation into ν_e is not possible. The oscillation occurs between ν_μ and ν_τ with probability

$$I_{\mu\tau}(\nu_\mu \to \nu_\tau) = I_{\tau\mu}(\nu_\tau \to \nu_\mu) \tag{14.80}$$

given by

$$\begin{aligned} I_{\mu\tau}(\nu_\mu \to \nu_\tau) &= \sin^2 2\theta_{23} \sin^2\left(\frac{\Delta m_{23}^2 L}{4E}\right) \\ &= \sin^2 2\theta_{23} \sin^2\left(\frac{1.267 \Delta m_{23}^2 \left[\mathrm{eV}^2\right] L[\mathrm{km}]}{E[\mathrm{GeV}]}\right). \end{aligned} \tag{14.81}$$

One recovers the oscillation formula for two neutrinos, (14.65). This applies to the atmospheric ν_μ which transforms into ν_τ. The data from figure 14.19 determine θ_{23} and Δm_{23}^2. The muon neutrinos transform entirely into tauon neutrinos with $\sin^2 2\theta_{23} \simeq 1$. For example, the period of the oscillation (14.81) is $\pi E/1.267\Delta m_{23}^2 \simeq$ 1033 km for 1 GeV neutrinos. For much larger distances the term (14.72) is not negligible. A slow modulation of the intensity (14.81) is observed and the intensities (14.79) do no longer vanish.

A 17 GeV mean energy ν_μ beam was sent from CERN to the Gran Sasso underground laboratory in the Apennine Mountains, where the OPERA detector is installed, to detect the oscillation of ν_μ into ν_τ along the 732 km long route through the Earth. The neutrino beam is produced with a parallel beam of pions and kaons obtained with a magnetic horn. This device consists of two axially symmetric conductors surrounding the beam pipe, with currents running in opposite directions, producing a toroidal magnetic field (see the photograph of a magnetic horn in figure 14.23).

The detector consists of 150 000 'bricks' of alternating (44 μm) thin AgBr emulsion films and 1 mm thick lead foils [36]. The conversion of a ν_τ into a τ lepton in the detector is sought, mediated by the reaction $\nu_\tau N \to \tau^- X$, $\tau^- \to \pi$'s $+ \nu_\tau$, see also figure 14.4 on the similar DONUT apparatus which discovered the ν_τ. The τ lepton is

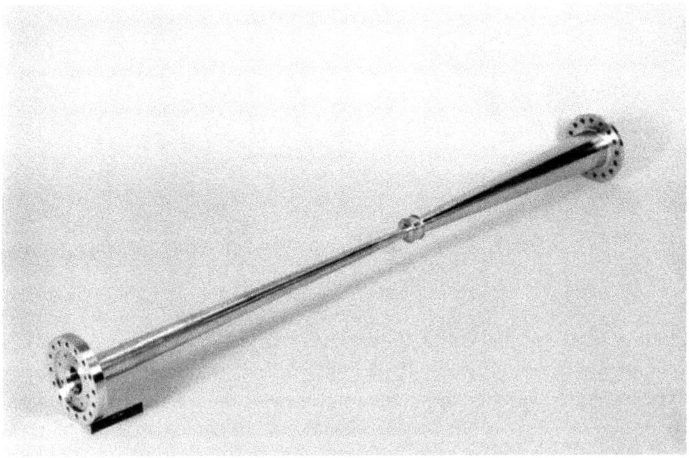

Figure 14.23. The inner conductor of a magnetic double-horn used for neutrino oscillation experiments. Image credit: CERN.

identified thanks to its relatively long mean decay length. The observation of about two τ decays yr^{-1} is expected.

Figure 14.24 shows the first event observed in 2010 [37]. Here the τ lepton decays into $\pi^-\pi^0$ and an invisible ν_τ. The two photons from π^0 decay are detected by their conversion into e^+e^- pairs. This event is the first experimental evidence for the *appearance* of neutrino flavour, while earlier experiments had observed only a flux deficit between production point and detection site.

14.6.4 Δm_{12}^2 oscillations

We now derive the formula for the antineutrino intensity I_{ee} ($\bar{\nu}_e \to \bar{\nu}_e$) from nuclear reactors, again assuming that $\theta_{13} = 0$, hence $U_{13} = 0$. In this case, the $\bar{\nu}_e$ does not contain any $\bar{\nu}_3$-component so that I_{ee} does not depend on Δm_{23}^2. A simple derivation starting from (14.59) uses, as before, the unitarity condition of the matrix U, giving the final result:

$$I_{ee} = \left| |U_{11}|^2 \, e^{-i\frac{m_1^2 L}{2E}} + |U_{12}|^2 \, e^{-i\frac{m_2^2 L}{2E}} \right|^2 = 1 - \sin^2 2\theta_{12} \sin^2\left(\frac{\Delta m_{12}^2 L}{4E}\right). \quad (14.82)$$

Figure 14.24 shows the measured antineutrino intensity as a function of energy E. The data were recorded in Japan with the KamLAND detector. The antineutrinos are detected by the inverse β-decay $\bar{\nu}_e p \to e^+ n$ in a vessel containing $1200\,m^3$ of liquid scintillator. The energy distributions measured from 53 nuclear reactors are added, after normalizing to the distance $L_0 = 180$ km (the average distance between the reactors and the detector). The detection threshold is 2.6 MeV. Reactor antineutrinos are emitted with the energy distribution shown in figure 14.1. One observes

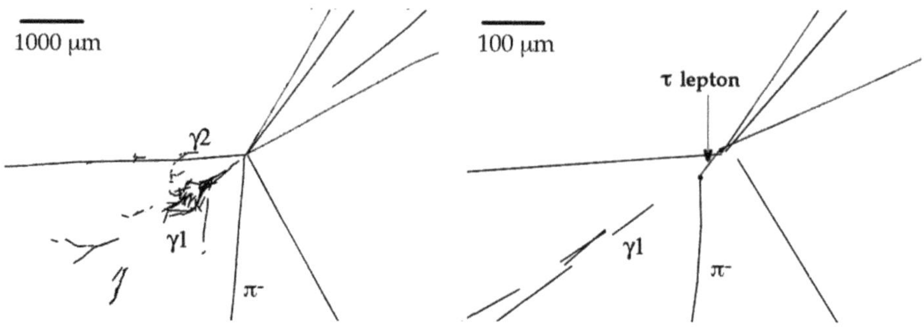

Figure 14.24. Detection of ν_τ by the OPERA experiment from the reaction $\nu_\tau N \to \tau^- X$, $\tau^- \to \pi$'s$+\nu_\tau$. The left panel shows the conversion of two photons from π^0 decay in the plane perpendicular to the neutrino beam. The magnified image on the right shows the projection of the tracks of the τ lepton and its π^- decay product, also in the plane transverse to the neutrino beam. The decay length is about 1.3 mm. After [37].

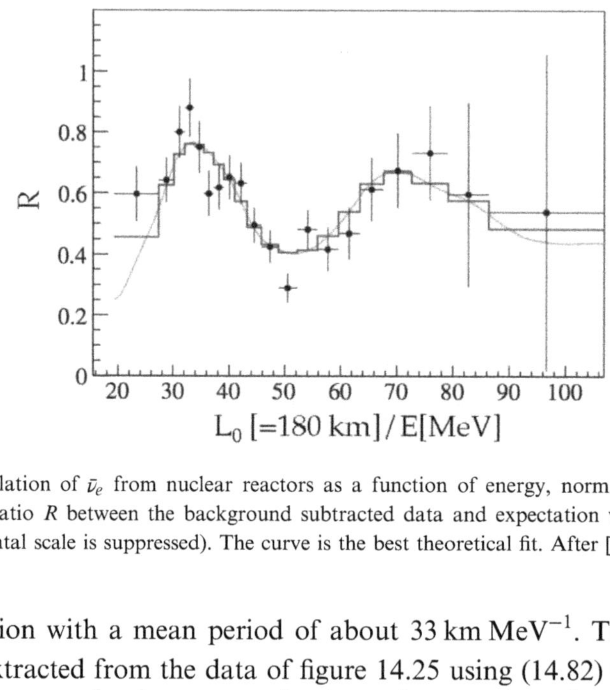

Figure 14.25. Oscillation of $\bar{\nu}_e$ from nuclear reactors as a function of energy, normalized to the distance $L_0 = 180$ km. The ratio R between the background subtracted data and expectation without oscillations is plotted (the horizontal scale is suppressed). The curve is the best theoretical fit. After [38].

a clear oscillation with a mean period of about 33 km MeV^{-1}. The angle θ_{12} and Δm_{12}^2 can be extracted from the data of figure 14.25 using (14.82) [38].

The angle θ_{12} can also be measured using solar neutrinos from the main cycle ($pp \to de^+\nu_e$). However, due to the very large size of the source, one cannot observe the oscillation, but only the intensity I_{ee} averaged over many oscillation periods.

14.6.5 Measurement of θ_{13}

The angle θ_{13} is a crucial parameter to determine. For very small values of θ_{13} the complex CP violating term, containing the phase δ in the matrix (14.66), becomes hard to measure. The eigenstate ν_3 does not contain any ν_e-component for $\theta_{13} = 0$ and early data had suggested that this angle might be rather small.

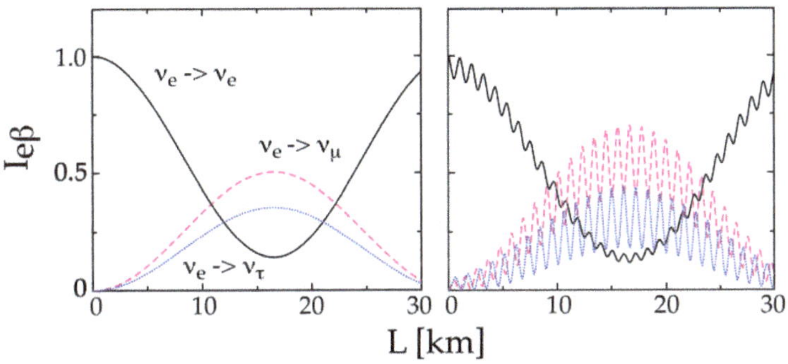

Figure 14.26. Oscillations of 1 MeV neutrinos when electron neutrinos are produced. The curves are computed with $\theta_{12} = 34°$, $\theta_{23} = 40°$, $\theta_{13} = 0$ (left) and $\theta_{13} = 9°$, $\delta = 0$ (right). Note that $I_{ee}(\nu_e \to \nu_e)$ does not depend on the phase δ.

One way to determine θ_{13} is to measure the fast Δm_{23}^2 modulation on the Δm_{12}^2 oscillation of the antineutrino flux from a nuclear reactor. Figure 14.26 shows the oscillation curves $\nu_e \to \nu_e, \nu_\mu, \nu_\tau$, calculated from (14.59) for 1 MeV neutrinos, with $\theta_{13} = 0$ (left) and $\theta_{13} = 9°$, $\delta = 0$ (right). With $\theta_{13} \neq 0$ one observes two oscillations, one with period $\pi E/1.267\Delta m_{12}^2 \simeq 33$ km (as in figure 14.21), the other with period $\pi E/1.267\Delta m_{23}^2 \simeq 1033$ m. The fast modulation vanishes for $\theta_{13} = 0$.

This fast modulation was observed by the Double Chooz reactor experiment in the Ardennes [39] as a deficit in the antineutrino intensity

$$I_{ee} \simeq 1 - \sin^2 2\theta_{13} \sin^2\left(\frac{\Delta m_{23}^2 L}{4E}\right), \tag{14.83}$$

which is readily obtained by replacing θ_{12} by θ_{13}, and Δm_{12}^2 by $\Delta m_{13}^2 \simeq \Delta m_{23}^2$ in (14.82). The neutrinos are detected via the inverse β-decay $\bar{\nu}_e p \to e^+ n$. The target is located 1050 m from the (dual) reactor. The vessel contains 10 m^3 of liquid scintillator to detect the positrons, mixed with 1 g/ℓ of gadolinium to capture the delayed neutrons. Figure 14.27 shows the energy distribution of antineutrinos at the detector. A clear deficit is observed when comparing the data with the expectation without oscillation. A total of 4121 events are observed, while 4344 ± 165 events are expected without oscillation. This result was the first indication of a finite value of about 9° for θ_{13} [39].

However, the intensity of the antineutrino source depends on fuel composition and on reactor power which need to be known to an accuracy of 1–2%. The antineutrino flux used in reactor experiments appears to have been underestimated by ~3%, following a reanalysis of the β-decay spectra of the fission products [40]. A better method is to measure the *relative* antineutrino fluxes with two (or more) detectors located at various distances from the reactor. Such an experiment is performed at the Daya Bay site near Shenzhen in China. The antineutrino flux from three pairs of dual reactors is measured in three underground caverns, using six detectors, each containing 20 t of gadolinium loaded liquid scintillator (figure 14.28).

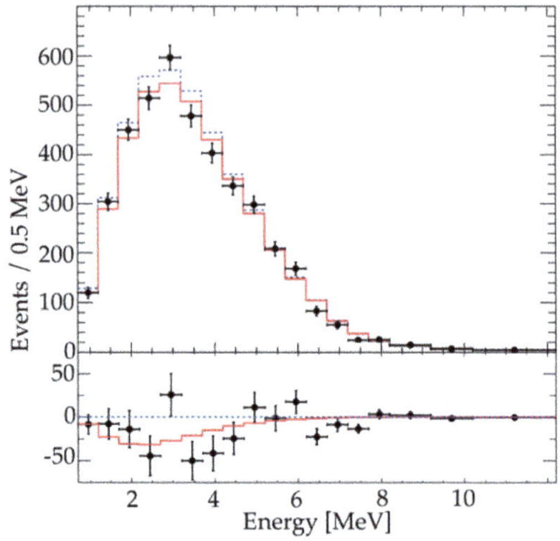

Figure 14.27. Disappearance of reactor antineutrinos. The top figure shows the energy distribution at the detector (crosses). Most of the data points lie below the dashed histogram (no oscillation hypothesis). The red histogram shows the best fit with oscillations. The bottom figure shows the difference between the data and prediction (no oscillation hypothesis) with the best fit shown by the red histogram. After [39].

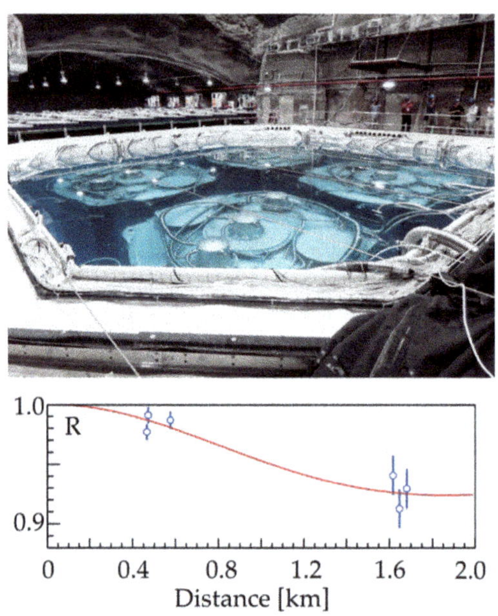

Figure 14.28. Top: one of the detector stations at the Daya Bay site photographed by the author (see the text for details). Bottom: disappearance of electron antineutrinos. The ratio R of the measured to expected flux assuming no oscillations is plotted as a function of distance for three detector stations. The line shows the fit. After [41].

The scintillator vessels are surrounded by liquid scintillator shields to suppress the background from the surrounding rocks, and are immersed in ultrapure water pools to detect the Čerenkov light emitted by residual cosmic ray muons. Figure 14.28 (bottom) shows the attenuation of the antineutrino flux at the three detector stations, as a function of (flux-averaged) distance. The observed attenuation leads to the first indisputable proof that θ_{13} is finite, $\theta_{13} = 8.8 \pm 0.8°$ [41].

Another way to measure θ_{13} is to search for the appearance probability $I_{\mu e}$ of ν_e in a ν_μ beam, with an oscillation period determined by Δm_{23}^2. This would also allow a measurement of the *CP* violating phase δ. The T2K collaboration [42] in Japan is searching for oscillations $\nu_\mu \to \nu_e$ with the frequency $\Delta m_{23}^2\ L/4E$ on the 295 km flight path between the J-PARC accelerator in Tokai and the Super-Kamiokande detector (figure 14.18). The neutrinos are detected slightly sideways (2.5°) from the direction of the primary pion beam. One then obtains a nearly monoenergetic neutrino beam with a mean energy of 600 MeV (section 8.3), and an oscillation period of about 620 km. The T2K collaboration has reported the observation of 8 $\nu_\mu \to \nu_e$ events, leading to $6.4° < \theta_{13} < 10.8°$ [43].

14.6.6 Matter effects on neutrino oscillations

So far we have been dealing with neutrino oscillations in vacuum. Neutrinos flying through matter—such as solar neutrinos travelling from the centre of the Sun to its surface—experience neutral and charged current interactions with the solar material. Neutral current interactions (mediated by Z^0 exchange) affect all neutrino flavours alike. However, for low energy neutrinos such as solar neutrinos, charged current interactions are only possible for the electron flavour via W exchange in $\nu_e e^- \to \nu_e e^-$. The ν_e then acquires an effective mass which departs from its mass in vacuum and, consequently, the eigenstates ν_i' differ from ν_i in vacuum. This is the so-called MSW effect formulated by Stanislav Mikheyev, Alexei Smirnov and Lincoln Wolfenstein on neutrino oscillations in matter [44]. Similar arguments apply to low energy $\bar{\nu}_e$.

Let us first consider two neutrinos, say ν_e and ν_μ, initially produced in vacuum. For the eigenstates we have according to (14.53), since H is diagonal:

$$E_i \simeq p + \frac{m_i^2}{2p} \Rightarrow H = \frac{1}{2p}\begin{pmatrix} m_1^2 & 0 \\ 0 & m_2^2 \end{pmatrix}. \tag{14.84}$$

We have dropped the additive term p which leads to the irrelevant common phase factor e^{-ipt} in the time evolution $|\nu_i(t)\rangle = e^{-iE_i t}|\nu_i\rangle$. The flavour states are, according to (14.62):

$$|\nu_\alpha\rangle = \begin{pmatrix} |\nu_e\rangle \\ |\nu_\mu\rangle \end{pmatrix} = \begin{pmatrix} \cos\theta & \sin\theta \\ -\sin\theta & \cos\theta \end{pmatrix}\begin{pmatrix} |\nu_1\rangle \\ |\nu_2\rangle \end{pmatrix} \equiv U|\nu_i\rangle. \tag{14.85}$$

The time evolution is given by

$$\mathrm{i}\frac{\partial}{\partial t}|\nu_i(t)\rangle = \mathrm{i}\frac{\partial}{\partial t}U^\dagger|\nu_\alpha(t)\rangle = H|\nu_i(t)\rangle = HU^\dagger|\nu_\alpha(t)\rangle, \tag{14.86}$$

hence

$$\mathrm{i}\frac{\partial}{\partial t}|\nu_\alpha(t)\rangle = \underbrace{UHU^\dagger}_{\equiv H\alpha}|\nu_\alpha(t)\rangle. \tag{14.87}$$

Inserting H and U gives for H_α

$$\begin{aligned} H_\alpha &= \frac{1}{2p}\begin{pmatrix} m_{\nu e}^2 & m_{e\mu}^2 \\ m_{\mu e}^2 & m_{\nu\mu}^2 \end{pmatrix} \\ &= \frac{1}{2p}\begin{pmatrix} \cos\theta & \sin\theta \\ -\sin\theta & \cos\theta \end{pmatrix}\begin{pmatrix} m_1^2 & 0 \\ 0 & m_2^2 \end{pmatrix}\begin{pmatrix} \cos\theta & -\sin\theta \\ \sin\theta & \cos\theta \end{pmatrix}, \end{aligned} \tag{14.88}$$

from which it is easy to show that

$$\boxed{\cos 2\theta = \frac{m_{\nu\mu}^2 - m_{\nu e}^2}{m_2^2 - m_1^2} = \frac{m_{\nu\mu}^2 - m_{\nu e}^2}{\Delta m_{12}^2}}, \tag{14.89}$$

with $\theta \equiv \theta_{12}$. We have seen earlier that measurements on oscillations in vacuum determine $\sin^2 2\theta_{12}$, see (14.65). An ambiguity arises with two possible values for the mixing angle, θ_{12} or $\frac{\pi}{2} - \theta_{12}$. For example, $\theta_{12} = 34°$ in Δm_{12}^2 oscillations but 56° is also possible. We have defined as m_1 the mass of the lightest of two eigenstates. Hence from (14.89) one finds that $m_{\nu_\mu} > m_{\nu_e}$ for $\theta_{12} < 45°$. Thus ν_e contains more ν_1, see (14.49) and also figure 14.22. In contrast, for $90° > \theta_{12} > 45°$ one would obtain $m_{\nu_\mu} < m_{\nu_e}$ and ν_μ would contain more ν_1. We shall see that the MSW effect on solar neutrinos lifts the ambiguity in favour of the lower value for θ_{12}.

We now move to oscillations in matter, in particular in the solar interior. The mass of the ν_e is modified by W exchange with solar electrons according to

$$m_{\nu_e}^2 \to m_{\nu_e}^2 + A \text{ with } A = 2\sqrt{2}G_F N_e p, \tag{14.90}$$

where N_e denotes the density of electrons and p the neutrino momentum [45]. Therefore H_α now reads

$$H_\alpha = \frac{1}{2p}\begin{pmatrix} m_{\nu_e}^2 + A & m_{e\mu}^2 \\ m_{\mu e}^2 & m_{\nu_\mu}^2 \end{pmatrix} \tag{14.91}$$

in the flavour basis. In the $\nu_1 - \nu_2$ basis the Hamiltonian $H = U^\dagger H_\alpha U$ becomes

$$H = \frac{1}{2p}\begin{pmatrix} \cos\theta & -\sin\theta \\ \sin\theta & \cos\theta \end{pmatrix}\begin{pmatrix} m_{\nu_e}^2 + A & m_{e\mu}^2 \\ m_{\mu e}^2 & m_{\nu_\mu}^2 \end{pmatrix}\begin{pmatrix} \cos\theta & \sin\theta \\ -\sin\theta & \cos\theta \end{pmatrix}$$

$$= \frac{1}{2p}\begin{pmatrix} m_1^2 + A\cos^2\theta & A\cos\theta\sin\theta \\ A\cos\theta\sin\theta & m_2^2 + A\sin^2\theta \end{pmatrix}. \quad (14.92)$$

The eigenvalues and eigenstates are obtained by diagonalizing H:

$$(m_1^2 + A\cos^2\theta - \lambda)(m_2^2 + A\sin^2\theta - \lambda) - A^2\cos^2\theta\sin^2\theta = 0. \quad (14.93)$$

The eigenvalues of the diagonalized Hamiltonian

$$H' = \frac{1}{2p}\begin{pmatrix} \lambda_1 & 0 \\ 0 & \lambda_2 \end{pmatrix} \quad (14.94)$$

are

$$\boxed{\lambda_{1,2} \equiv m'^2_{1,2} = \frac{1}{2}\left[m_1^2 + m_2^2 + A \mp \sqrt{(A - \Delta m_{12}^2\cos 2\theta)^2 + (\Delta m_{12}^2\sin 2\theta)^2}\right]}. \quad (14.95)$$

With $A = 0$ one recovers the masses

$$m'^2_{1,2} = m_{1,2}^2 \quad (14.96)$$

as expected. Note that the oscillation frequency, determined by the mass difference

$$\Delta m'^2_{12} = \sqrt{\left(A - \Delta m_{12}^2\cos 2\theta\right)^2 + \left(\Delta m_{12}^2\sin 2\theta\right)^2}, \quad (14.97)$$

reaches its minimum value when $A = \Delta m_{12}^2$ and even vanishes for $\theta = 0$.

The flavour states are related to the eigenstates ν_1' and ν_2' in matter through the transformation

$$\begin{pmatrix} |\nu_e\rangle \\ |\nu_\mu\rangle \end{pmatrix} = U'\begin{pmatrix} |\nu_1'\rangle \\ |\nu_2'\rangle \end{pmatrix} = \begin{pmatrix} \cos\theta' & \sin\theta' \\ -\sin\theta' & \cos\theta' \end{pmatrix}\begin{pmatrix} |\nu_1'\rangle \\ |\nu_2'\rangle \end{pmatrix}, \quad (14.98)$$

where θ' is the MSW mixing angle. This angle can be calculated from the Hamiltonian (14.91) in the flavour basis:

$$H_\alpha = \frac{1}{2p}\begin{pmatrix} m_{\nu_e}^2 + A & m_{e\mu}^2 \\ m_{\mu e}^2 & m_{\nu_\mu}^2 \end{pmatrix} = U'H'U'^\dagger$$

$$= \frac{1}{2p}\begin{pmatrix} \cos\theta' & \sin\theta' \\ -\sin\theta' & \cos\theta' \end{pmatrix}\begin{pmatrix} m'^2_1 & 0 \\ 0 & m'^2_2 \end{pmatrix}\begin{pmatrix} \cos\theta' & -\sin\theta' \\ \sin\theta' & \cos\theta' \end{pmatrix}. \quad (14.99)$$

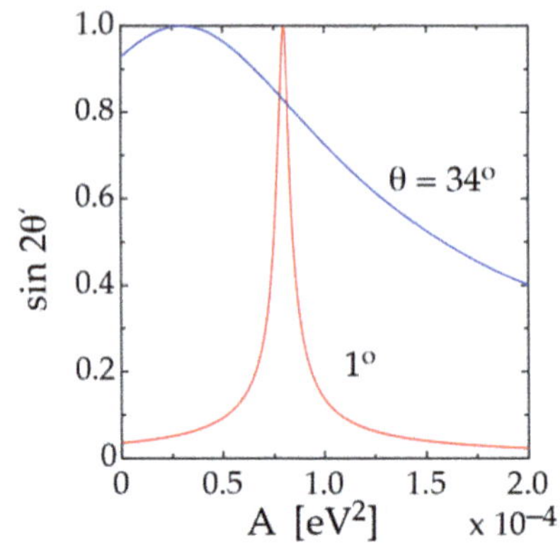

Figure 14.29. MSW angle θ' as a function of $A \propto N_e$ for a small mixing angle and for the realistic $\theta = 34°$.

The element $m^2_{e\mu}$, obtained by matrix multiplication,

$$m^2_{e\mu} = \Delta m'^2_{12} \sin\theta' \cos\theta'$$
$$= \sin\theta' \cos\theta' \sqrt{\left(A - \Delta m^2_{12} \cos 2\theta\right)^2 + \left(\Delta m^2_{12} \sin 2\theta\right)^2} \qquad (14.100)$$

can also be expressed as a function of θ by comparing with (14.88):

$$m^2_{e\mu} = \Delta m^2_{12} \sin\theta \cos\theta. \qquad (14.101)$$

The last two equations lead to the important relation

$$\boxed{\sin 2\theta' = \sin 2\theta \frac{1}{\sqrt{\left(A/\Delta m^2_{12} - \cos 2\theta\right)^2 + \sin^2 2\theta}}}. \qquad (14.102)$$

The function $\sin 2\theta'$, shown in figure 14.29 as a function of A, has the typical shape of a resonance curve, with a width proportional to $\sin 2\theta$. For $A \to 0$ (vacuum) we obtain that $\theta' \to \theta$, as expected, while for large A (high electron densities) $\theta' \to 90°$. At resonance

$$A = \Delta m^2_{12} \cdot \cos 2\theta \Rightarrow \theta' = 45°. \qquad (14.103)$$

According to (14.98), the ν_e and ν_μ are now equal mixtures of the eigenstates $|\nu_1'\rangle$ and $|\nu_2'\rangle$. Thus the MSW effect is able to massively enhance $\nu_e \to \nu_\mu$ oscillations even for a small mixing angle θ in vacuum.

14.6.7 The MSW effect in the Sun

The matter in the centre of the Sun is mostly composed of helium (64%) and hydrogen with the density $\rho \simeq 150\,\mathrm{g\,cm^{-3}}$. Charge neutrality requires the density of electrons to be equal to that of protons. The electron density is then given

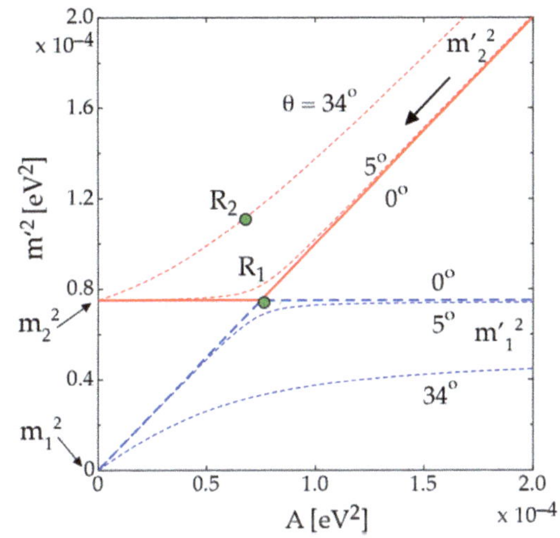

Figure 14.30. Mass eigenvalues as a function of $A \propto N_e$ for various mixing angles. The arrow shows the path of the ν_e flying from the centre to the surface of the Sun. Resonance occurs at the points R_1 and R_2.

by $N_e = \frac{\rho}{m_P}\left(\frac{0.64}{2} + 0.36\right) = 6.1 \times 10^{25}\,\text{cm}^{-3}$. According to (14.90) A [eV^2]= 1.5×10^{-5} p [MeV c^{-1}] while the measured value for Δm_{12}^2 is $\simeq 7.5 \times 10^{-5}$ eV^2. Resonance can therefore only occur for momenta larger than about 5 MeV c^{-1}, that is for neutrinos from ^{8}B decay (figure 14.11), but not for neutrinos from the main cycle (pp).

Let us now describe the time evolution of electron neutrinos produced in the centre of the Sun (figure 14.30), first assuming no $\nu_e - \nu_\mu$ mixing in vacuum, hence $\theta = 0$. According to (14.95) the eigenvalue $m_2'^2$ is

$$m_2'^2 = \frac{1}{2}\left(m_1^2 + m_2^2 + A + \left|A - \Delta m_{12}^2\right|\right) = \begin{bmatrix} m_2^2 & \text{for } A < \Delta m_{12}^2 \\ m_1^2 + A & \text{for } A > \Delta m_{12}^2 \end{bmatrix}. \tag{14.104}$$

In the centre of the Sun the ν_e is identical to ν_2' ($\theta' = 90°$ for large A). The neutrino then propagates to the surface in the ν_2' eigenstate while the density $N_e \propto A$ decreases. There are no transitions between ν_1' and ν_2', as long as A decreases slowly enough (adiabatic condition) [46]. At resonance we have that $A = \Delta m_{12}^2$ (point R_1 along the plain red curve in figure 14.30). At this point ν_2' becomes an equal mixture of ν_e and ν_μ and has the mass $m_2' = m_2$. When reaching the surface, where $A = 0$ ($\theta' = 0$, $|\nu_2'\rangle = |\nu_2\rangle$) the neutrino becomes entirely ν_2, and with $\theta = 0$ entirely ν_μ (see (14.85)). Full conversion from ν_e to ν_μ has occurred!

However, in reality $\theta = \theta_{12} \simeq 34°$ from reactor measurements or for solar pp neutrinos (which are not affected by the MSW effect). The ν_e state follows instead the dashed red curve in figure 14.30. Resonance occurs at the point R_2 and the ν_e reaches the surface in the state

$$\left|\nu_2'\right\rangle \to \left|\nu_2\right\rangle = \sin\theta_{12}\left|\nu_e\right\rangle + \cos\theta_{12}\left|\nu_\mu\right\rangle, \tag{14.105}$$

which then propagates as an eigenstate in vacuum to the Earth. From the ν_e produced in the Sun only a fraction $\sin^2\theta_{12} \simeq 31\%$ reaches the Earth. This is in excellent agreement with the flux deficit of about $\frac{1}{3}$ that was originally measured at the Homestake mine [16]. However, the complementary angle of 56° would lead to a fraction of 69% reaching the Earth. Hence the MSW effect brings agreement between the ^{8}B measurement and other data, and solves the ambiguity in the θ_{12} mixing angle.

References

[1] Ishimoto S *et al* 2002 *J. Nucl. Sci. Technol.* **39** 670
[2] Reines F *et al* 1960 *Phys. Rev.* **117** 159
[3] Danby G *et al* 1962 *Phys. Rev. Lett.* **9** 36
[4] Perl M *et al* 1975 *Phys. Rev. Lett.* **35** 1489
[5] Perl M *et al* 1977 *Phys. Lett.* B **70** 487
[6] Kodama K *et al* 2001 *Phys. Lett.* B **504** 218
[7] Olive K *et al* (Particle Data Group) 2014 *Chin. Phys.* C **38** 090001
[8] Fritschi M *et al* 1986 *Phys. Lett.* B **173** 485
[9] Lubimov V A *et al* 1980 *Phys. Lett.* B **94** 266
[10] Kraus C *et al* 2005 *Eur. Phys. J.* C **40** 447
[11] Picard A *et al* 1992 *Nucl. Instrum. Methods Phys. Res.* B **63** 345
[12] Kraus C *et al* 2013 *Eur. Phys. J.* C **73** 2323
[13] Weinheimer Ch 2002 *Prog. Part. Nucl. Phys.* **48** 141
[14] Assamagan K *et al* 1996 *Phys. Rev.* D **53** 6065
[15] Athanas M *et al* 2000 *Phys. Rev.* D **61** 052002
[16] Davis R *et al* 1968 *Phys. Rev. Lett.* **20** 1205
[17] Anselmann P *et al* 1994 *Phys. Lett.* B **327** 377
[18] Daraktchieva Z *et al* 2005 *Phys. Lett.* B **615** 153
[19] Davis R 1955 *Phys. Rev.* **97** 766
[20] Goldhaber M *et al* 1958 *Phys. Rev.* **109** 1015
[21] Goeppert-Mayer M 1935 *Phys. Rev.* **48** 512
[22] Boehm F and Vogel P 1992 *Physics of Massive Neutrinos* (Cambridge: Cambridge University Press)
[23] Guenther M *et al* 1997 *Phys. Rev.* D **55** 54
[24] Agostini M *et al* 2013 *Phys. Rev. Lett.* **111** 122503
[25] Klapdor-Kleingrothaus H V *et al* 2004 *Phys. Lett.* B **586** 198
Klapdor-Kleingrothaus H V and Krivosheina I V 2006 *Mod. Phys. Lett.* A **21** 1547
[26] Wong H T *et al* 1991 *Phys. Rev. Lett.* **67** 1218
[27] Vuilleumier J C *et al* 1993 *Phys. Rev.* D **48** 1009
[28] Albert J B *et al* 2014 *Phys. Rev.* C **89** 015502
[29] Pontecorvo B 1968 *Sov. Phys.–JETP* **26** 984
[30] Fukuda Y *et al* 1998 *Phys. Rev. Lett.* **81** 1562
[31] Aliu E *et al* 2005 *Phys. Rev. Lett.* **94** 081802
[32] Boger J *et al* 2000 *Nucl. Instrum. Methods Phys. Res.* A **449** 172
[33] Ahmed S N *et al* 2004 *Phys. Rev. Lett.* **92** 181301
[34] Beuthe M 2003 *Phys. Rep.* **375** 105

[35] Bilenky S M and Giunti C 2012 *Mod. Phys. Lett.* A **27** 1230015
[36] Acquafredda R *et al* 2009 *J. Instrum.* **4** P04018
[37] Agafonova N *et al* 2010 *Phys. Lett.* B **691** 138
[38] Abe S *et al* 2008 *Phys. Rev. Lett.* **100** 221803
[39] Abe Y *et al* 2012 *Phys. Rev. Lett.* **108** 131801
[40] Mention G *et al* 2011 *Phys. Rev.* D **83** 73006
[41] An F P *et al* 2012 *Phys. Rev. Lett.* **108** 171803
[42] Abe K *et al* 2011 *Nucl. Instrum. Methods Phys. Res.* A **659** 106
[43] Abe K *et al* 2013 *Phys. Rev.* D **88** 032002
[44] Wolfenstein S L 1978 *Phys. Rev.* D **17** 2369
Mikheyev S P and Smirnov A Yu 1989 *Prog. Part. Nucl. Phys.* **23** 41
[45] Halprin A 1986 *Phys. Rev.* D **34** 3462
[46] Schmitz N 1997 *Neutrinophysik* (Stuttgart: Teubner)

IOP Publishing

Nuclear and Particle Physics

Claude Amsler

Chapter 15

The Dirac equation

The relativistic wave equation introduced by Paul Adrien Maurice Dirac in 1928 describes the motion of a *free* particle with mass m and spin $\frac{1}{2}$, such as a free electron. For a comprehensive biography of Dirac see [1].

The relativistic relation between energy and momentum is

$$E^2 = \vec{p}^2 + m^2 \tag{15.1}$$

(expressed in natural units). Applying the correspondence principle

$$E \Rightarrow \mathrm{i}\frac{\partial}{\partial t}, \qquad \vec{p} \Rightarrow -\mathrm{i}\,\mathrm{grad} \tag{15.2}$$

leads to

$$\left(-\frac{\partial^2}{\partial t^2} + \Delta\right)\psi(\vec{x}, t) = m^2\psi(\vec{x}, t), \tag{15.3}$$

which can be rewritten with the d'Alembert operator $\Box \equiv \frac{\partial^2}{\partial t^2} - \Delta$ as

$$\boxed{\left(\Box + m^2\right)\psi(\vec{x}, t) = 0}. \tag{15.4}$$

This equation is known as the *Klein–Gordon* relativistic wave equation describing the motion of a free particle. Let us left-multiply (15.3) by ψ^*,

$$\psi^*\Delta\psi - \psi^*\frac{\partial^2\psi}{\partial t^2} = m^2\psi^*\psi, \tag{15.5}$$

doi:10.1088/978-0-7503-1140-3ch15 © IOP Publishing Ltd 2015

and subtract the complex conjugated (15.3), left-multiplied by ψ,

$$\psi^*\Delta\psi - \psi\Delta\psi^* + \psi\frac{\partial^2\psi^*}{\partial t^2} - \psi^*\frac{\partial^2\psi}{\partial t^2} = 0$$

$$\Rightarrow \text{div}\underbrace{\left(\psi^*\,\text{grad}\psi - \psi\,\text{grad}\psi^*\right)}_{\equiv\vec{j}} + \frac{\partial}{\partial t}\underbrace{\left(\psi\frac{\partial\psi^*}{\partial t} - \psi^*\frac{\partial\psi}{\partial t}\right)}_{\equiv\rho} = 0. \tag{15.6}$$

This is the well-known continuity equation:

$$\text{div}\,\vec{j} + \frac{\partial\rho}{\partial t} = 0. \tag{15.7}$$

Due to the second derivative of the wave function with respect to time t in (15.3), the probability density function ρ may be negative. A negative probability was not acceptable before the discovery of the positron in 1932. Since antiparticles are now known to exist, ρ and $\vec{j}$ in (15.7) are multiplied by the electric charge Qe of the particle and are interpreted as electric charge density and electric current density, respectively. In fact, a similar derivation of the continuity equation using the non-relativistic Schrödinger equation which contains only the first derivative $\partial/\partial t$, leads to the positive probability density function $\rho = \psi^*\psi = |\psi|^2$.

We therefore look for another relativistic wave equation that is linear in E and, for covariance reasons, also linear in $\vec{p}$. From the ansatz (Dirac, 1928)

$$E = \vec{\alpha}\vec{p} + \beta m \tag{15.8}$$

follows, together with (15.1), that

$$\left(\sum_{i=1}^{3}\alpha^i p^i + \beta m\right)\left(\sum_{j=1}^{3}\alpha^j p^j + \beta m\right) = \vec{p}^{\,2} + m^2. \tag{15.9}$$

The *operators* $\vec{\alpha}$ and β have to fulfil the equations

$$\begin{aligned} \alpha^i\alpha^j + \alpha^j\alpha^i &= 0 \quad (i \neq j), \\ \alpha^i\beta + \beta\alpha^i &= 0, \\ \left(\alpha^i\right)^2 &= I, \\ \beta^2 &= I, \end{aligned} \tag{15.10}$$

where I is the unit operator. In the lowest possible dimension, the operators α^1, α^2, α^3 and β are represented by 4×4 matrices which are traceless and Hermitian. The latter follows from the hermiticity of the Hamiltonian $\vec{\alpha}\vec{p} + \beta m$. The Hermitian conjugated matrix (denoted by †) is the complex conjugated transposed matrix. In components, $(\alpha^i_{kl})^\dagger = \alpha^{i*}_{lk}$ and $\beta^\dagger_{kl} = \beta^*_{lk}$. The eigenvalues are equal to ± 1, as can easily be seen from the last two equations of (15.10). Thus, by writing concisely $(\alpha^1, \alpha^2, \alpha^3)$ as $\vec{\alpha}$,

$$\begin{aligned} \vec{\alpha}^\dagger &= \vec{\alpha}, \\ \beta^\dagger &= \beta. \end{aligned} \tag{15.11}$$

The field-free Dirac equation reads

$$\boxed{(\vec{\alpha}\vec{p} + \beta m)\psi(\vec{x}, t) = E\psi(\vec{x}, t)} \tag{15.12}$$

or, with the correspondence principle,

$$(-\mathrm{i}\vec{\alpha} \cdot \mathrm{grad} + \beta m)\psi = \mathrm{i}\frac{\partial \psi}{\partial t}. \tag{15.13}$$

Left-multiplying by β and defining

$$\gamma^0 \equiv \beta, \qquad \vec{\gamma} \equiv \beta\vec{\alpha}, \tag{15.14}$$

leads to

$$(-\mathrm{i}\vec{\gamma} \cdot \mathrm{grad} + m)\psi = \mathrm{i}\gamma^0\frac{\partial \psi}{\partial t}. \tag{15.15}$$

With the definition

$$\boxed{\partial_\mu \equiv \frac{\partial}{\partial x^\mu} \equiv \left(\frac{\partial}{\partial t}, + \mathrm{grad}\right)} \tag{15.16}$$

the Dirac equation becomes

$$\left(\mathrm{i}\gamma^0\partial_0 + \mathrm{i}\gamma^k\partial_k - m\right)\psi(\vec{x}, t) = 0 \tag{15.17}$$

or in concise form:

$$\left(\mathrm{i}\gamma^\mu\partial_\mu - m\right)\psi(\vec{x}, t) = 0. \tag{15.18}$$

Summation over repeated indices k and μ is implicit: Roman indices run from 1 to 3, Greek indices from 0 to 3. The wave function $\psi(\vec{x}, t)$ is a column vector (*spinor*) with four components.

The following anti-commutation relations between γ matrices follow from the properties (15.10) and definitions (15.14):

$$\boxed{\gamma^\mu\gamma^\nu + \gamma^\nu\gamma^\mu = 2g^{\mu\nu}I}, \tag{15.19}$$

where $g^{\mu\nu}$ is the metric tensor

$$g^{00} = 1, \qquad g^{ii} = -1 \quad (i = 1, 2, 3), \qquad g^{\mu\nu} = 0 \quad (\mu \neq \nu). \tag{15.20}$$

Note the relations between the raised and lowered indices:

$$\begin{aligned} \gamma^\mu &\equiv g^{\mu\nu}\gamma_\nu \Rightarrow \gamma^0 = \gamma_0, & \gamma^i &= -\gamma_i, \\ \partial^\mu &\equiv g^{\mu\nu}\partial_\nu \Rightarrow \partial^0 = \partial_0, & \partial^i &= -\partial_i, \\ g_{\mu\nu} &\equiv g^{\mu\nu}. \end{aligned} \tag{15.21}$$

The following important properties of γ matrices follow from (15.10), (15.11) and:

$$\boxed{\gamma^{0\dagger} = \gamma^0, \qquad \vec{\gamma}^\dagger = -\vec{\gamma}}, \tag{15.22a}$$

$$\boxed{\left(\gamma^0\right)^2 = I, \qquad \left(\gamma^i\right)^2 = -I}. \tag{15.22b}$$

The solution of the free Dirac equation (15.18) is the plane wave

$$\psi(\vec{x}, t) = u(p)\mathrm{e}^{-\mathrm{i}px} = u(p)\mathrm{e}^{-\mathrm{i}p_\mu x^\mu}, \tag{15.23}$$

where

$$x^\mu \equiv (t, \vec{x}) \qquad \text{and} \qquad p^\mu \equiv (E, \vec{p}). \tag{15.24}$$

Substituting the derivatives

$$\frac{\partial}{\partial x^\mu}\mathrm{e}^{-\mathrm{i}p_\mu x^\mu} = -\mathrm{i}p_\mu \mathrm{e}^{-\mathrm{i}p_\mu x^\mu} \tag{15.25}$$

in (15.18) with $p_\mu = g_{\mu\nu}p^\nu = (E, -\vec{p})$, leads to the time-independent free Dirac equation

$$\left(\gamma^\mu p_\mu - m\right)u(p) = 0 \tag{15.26}$$

or in concise form:

$$\boxed{(\not{p} - m)u = 0}. \tag{15.27}$$

The abbreviation p 'slash' stands for the matrix sum

$$\not{p} \equiv \gamma^\mu p_\mu. \tag{15.28}$$

Conjugating (15.26) and right-multiplying by γ^0 gives

$$u^\dagger(p)\left(\gamma^{\mu\dagger}p_\mu - m\right)\gamma^0 = 0 \tag{15.29}$$

or, using the anti-commutation rules (15.19) and (15.22a),

$$\bar{u}(p)\left(\gamma^\mu p_\mu - m\right) = 0 \tag{15.30}$$

that is

$$\bar{u}(\not{p} - m) = 0, \tag{15.31}$$

the free Dirac equation for the *adjoint* spinor

$$\boxed{\bar{u}(p) \equiv u(p)^\dagger\gamma^0}. \tag{15.32}$$

15.1 Standard representation

The number of possible representations for the γ matrices is infinite. The simple and most used one is the Pauli–Dirac (or standard) representation which fulfils the anti-commutation relations (15.19):

$$\gamma^0 = \begin{pmatrix} I & 0 \\ 0 & -I \end{pmatrix}, \qquad \vec{\gamma} = \begin{pmatrix} 0 & \vec{\sigma} \\ -\vec{\sigma} & 0 \end{pmatrix}, \tag{15.33}$$

with the Pauli matrices

$$\sigma^1 = \begin{pmatrix} 0 & 1 \\ 1 & 0 \end{pmatrix}, \qquad \sigma^2 = \begin{pmatrix} 0 & -\mathrm{i} \\ \mathrm{i} & 0 \end{pmatrix}, \qquad \sigma^3 = \begin{pmatrix} 1 & 0 \\ 0 & -1 \end{pmatrix}. \tag{15.34}$$

In this representation

$$\vec{\alpha} = \begin{pmatrix} 0 & \vec{\sigma} \\ \vec{\sigma} & 0 \end{pmatrix} \qquad \text{and} \qquad \beta = \begin{pmatrix} I & 0 \\ 0 & -I \end{pmatrix}. \tag{15.35}$$

The Dirac equation (15.26) becomes

$$\left(\gamma^0 E - \vec{\gamma} \cdot \vec{p} - m\right) u(p) = 0 \tag{15.36}$$

or, in components,

$$\begin{pmatrix} E - m & -\vec{\sigma}\vec{p} \\ \vec{\sigma}\vec{p} & -[E + m] \end{pmatrix} u(p) = 0. \tag{15.37}$$

For $\vec{p} = 0$ one obtains

$$\begin{pmatrix} E - m & 0 \\ 0 & -[E + m] \end{pmatrix} u(0) = 0. \tag{15.38}$$

The solutions of the secular equation $E^2 - m^2 = 0$ and the eigenspinors are:

$$\begin{aligned} E = +m : u^1(\vec{p} = 0) &\propto \begin{pmatrix} 1 \\ 0 \\ 0 \\ 0 \end{pmatrix} \equiv \begin{pmatrix} \chi^1 \\ 0 \end{pmatrix}, \\ u^2(\vec{p} = 0) &\propto \begin{pmatrix} 0 \\ 1 \\ 0 \\ 0 \end{pmatrix} \equiv \begin{pmatrix} \chi^2 \\ 0 \end{pmatrix}, \\ E = -m : u^3(\vec{p} = 0) &\propto \begin{pmatrix} 0 \\ 0 \\ 1 \\ 0 \end{pmatrix} \equiv \begin{pmatrix} 0 \\ \chi^1 \end{pmatrix}, \\ u^4(\vec{p} = 0) &\propto \begin{pmatrix} 0 \\ 0 \\ 0 \\ 1 \end{pmatrix} \equiv \begin{pmatrix} 0 \\ \chi^2 \end{pmatrix}, \end{aligned} \tag{15.39}$$

where we have defined

$$\chi^1 \equiv \begin{pmatrix} 1 \\ 0 \end{pmatrix},$$
$$\chi^2 \equiv \begin{pmatrix} 0 \\ 1 \end{pmatrix}. \tag{15.40}$$

When $|\vec{p}\,| \geqslant 0$ the eigenspinors become

$$\boxed{u^s(p) = N\begin{pmatrix} \chi^s \\ \frac{\vec{\sigma}\vec{p}}{E+m}\chi^s \end{pmatrix}} \quad \text{for } E = \sqrt{\vec{p}^{\,2} + m^2} > 0 \tag{15.41}$$

and

$$u^{s+2}(p) = N\begin{pmatrix} \frac{-\vec{\sigma}\vec{p}}{-E+m}\chi^s \\ \chi^s \end{pmatrix} \quad \text{for } E = -\sqrt{\vec{p}^{\,2} + m^2} < 0, \tag{15.42}$$

with $s = 1$ or 2, and where the variable p denotes the 4-vector $(E, \vec{p}\,)$. For $\vec{p} \to 0$ one retrieves the spinors (15.39) up to an arbitrary normalization constant N. The proof follows directly by substituting into (15.37). For example, for $u^s(p)$:

$$(E - m)\chi^s - \overbrace{\vec{\sigma}\vec{p}\left(\frac{\vec{\sigma}\vec{p}}{E+m}\right)}^{\frac{p^2}{E+m}=\frac{E^2-m^2}{E+m}=E-m}\chi^s = 0, \tag{15.43}$$

where we have used the handy relation

$$(\vec{\sigma}\vec{A})(\vec{\sigma}\vec{B}) = \vec{A}\vec{B} + \mathrm{i}\vec{\sigma}(\vec{A} \times \vec{B}). \tag{15.44}$$

The negative energy solutions u^{s+2} in (15.42) for a particle with 4-momentum $-p$ were interpreted by Dirac as solutions with positive energy for a hypothetical antiparticle with 4-momentum $+p$. The spinors of the antiparticle are defined as

$$v^1(p) \equiv u^4(-p) = N\begin{pmatrix} \frac{\vec{\sigma}\vec{p}}{E+m}\chi^2 \\ \chi^2 \end{pmatrix}, \quad v^2(p) \equiv -\,u^3(-p) = -\,N\begin{pmatrix} \frac{\vec{\sigma}\vec{p}}{E+m}\chi^1 \\ \chi^1 \end{pmatrix} \tag{15.45}$$

or concisely,

$$\boxed{v^s(p) \equiv N\begin{pmatrix} \frac{\vec{\sigma}\vec{p}}{E+m}\eta\chi^s \\ \eta\chi^s \end{pmatrix}}, \tag{15.46}$$

with[1]

$$\eta \equiv \begin{pmatrix} 0 & -1 \\ 1 & 0 \end{pmatrix}. \tag{15.47}$$

The v spinors fulfil the Dirac equation (15.26) with $p \to -p$

$$\left(\gamma^\mu p_\mu + m\right)v(p) = 0, \tag{15.48}$$

or in other words

$$\boxed{(\not{p} + m)v = 0}. \tag{15.49}$$

The wave function of the antiparticle

$$\psi^c(\vec{x}, t) \equiv v(p)\mathrm{e}^{\mathrm{i}px} = v(p)\mathrm{e}^{\mathrm{i}p_\mu x^\mu} \tag{15.50}$$

satisfies the Dirac equation (15.18).

With 4×4 matrices representing the operators $\vec{\alpha}$ and β one obtains two pairs of solutions to describe the spin-$\frac{1}{2}$ particle and its antiparticle. It can be shown that:

1. $H = \vec{\alpha}\vec{p} + \beta m$ and $\vec{J} \equiv \vec{L} + \frac{1}{2}\vec{\sigma}$ commute with each other:

$$\left[H, \vec{J}\right] = \left[H, \vec{L}\begin{pmatrix} 1 & 0 \\ 0 & 1 \end{pmatrix} + \frac{1}{2}\begin{pmatrix} \vec{\sigma} & 0 \\ 0 & \vec{\sigma} \end{pmatrix}\right] = 0. \tag{15.51}$$

   The total angular momentum $\vec{J}$ is conserved, in contrast to the orbital angular momentum $\vec{L}$. The Dirac equation describes a spin-$\frac{1}{2}$ particle with spin $\vec{s} = \frac{1}{2}\vec{\sigma}$.

2. The Dirac equation predicts for the g-factor in the magnetic moment (11.116) the value $g = 2$. The field-dependent Dirac equation reads for the electron (charge $-e$):

$$\left[\vec{\alpha}\left(\vec{p} + e\vec{A}\right) + \beta m\right]\psi = [E + e\phi]\psi. \tag{15.52}$$

   At low energies (non-relativistic approximation $|\vec{p}\,| \ll m$) the two upper components of the u spinors and the two lower components of the v spinors dominate. The relation (15.52) can then be reduced to the field-dependent Schrödinger equation [2]:

$$\left[\frac{1}{2m}\left(\vec{p} + e\vec{A}\right)^2 + g\mu_B\,\vec{s}\cdot\vec{B} - e\phi\right]\psi = T\psi, \tag{15.53}$$

   with $g = 2$, $T \equiv E - m$ and for $m \gg \phi$, $m \gg T$. The magnetic field $\vec{B}$ is related to the vector potential $\vec{A}$ through $\vec{B} = \mathrm{rot}\,\vec{A}$.

[1] With this definition (note the minus sign in front of $u^3(-p)$), $v^s(p)$ can be constructed from $u^s(p)$ by applying the operation of charge conjugation, $\mathrm{i}\gamma^2 \times$ complex conjugation: $v^1(p) = \mathrm{i}\gamma^2 u^1(p)^*$ and $v^2(p) = \mathrm{i}\gamma^2 u^2(p)^*$, see section 15.5.

By choosing for the normalization constant N in (15.41) and (15.46) the value

$$N \equiv \sqrt{E+m}, \tag{15.54}$$

one obtains the important formulas

$$u^{r\dagger} u^s = v^{r\dagger} v^s = 2E\delta_{rs}, \tag{15.55a}$$

$$\bar{u}^r u^s = -\bar{v}^r v^s = 2m\delta_{rs}, \tag{15.55b}$$

and

$$\boxed{\sum_{s=1}^{2} u^s \bar{u}^s = \not{p} + m}, \tag{15.55c}$$

$$\boxed{\sum_{s=1}^{2} v^s \bar{v}^s = \not{p} - m}. \tag{15.55d}$$

These relationships, which are very useful for practical calculations, can be verified directly by inserting the free Dirac spinors (15.41) and (15.46). Note that the terms on the left-hand sides of (15.55c) and (15.55d) are *tensor products* of matrices. For example, (15.55c) gives

$$\sum_{s=1}^{2} u^s \bar{u}^s = N^2 \sum_s \begin{pmatrix} \chi^s \\ \frac{\vec{\sigma}\vec{p}}{E+m}\chi^s \end{pmatrix} \left(\chi^{s\dagger}, -\chi^{s\dagger} \frac{\vec{\sigma}\vec{p}}{E+m} \right). \tag{15.56}$$

We have used the free adjoint Dirac spinors

$$\bar{u}^s \equiv u^{s\dagger}\gamma^0 = N\left(\chi^{s\dagger}, -\chi^{s\dagger} \frac{\vec{\sigma}\vec{p}}{E+m} \right) \tag{15.57}$$

in the standard representation with $\vec{\sigma}^\dagger = \vec{\sigma}$. Since

$$\left(\frac{\vec{\sigma}\vec{p}}{E+m} \right)^2 = \frac{\vec{p}^{\,2}}{(E+m)^2} = \frac{E-m}{E+m}, \tag{15.58}$$

and

$$\sum_{s=1}^{2} \chi^s \chi^{s\dagger} = \underbrace{\begin{pmatrix} 1 \\ 0 \end{pmatrix}(1, 0)}_{\begin{pmatrix} 1 & 0 \\ 0 & 0 \end{pmatrix}} + \underbrace{\begin{pmatrix} 0 \\ 1 \end{pmatrix}(0, 1)}_{\begin{pmatrix} 0 & 0 \\ 0 & 1 \end{pmatrix}} = \begin{pmatrix} 1 & 0 \\ 0 & 1 \end{pmatrix} = I, \tag{15.59}$$

Table 15.1. Trace theorems and properties of γ matrices. The matrix γ^5 is defined as $\gamma^5 \equiv i\gamma^0\gamma^1\gamma^2\gamma^3$ (section 15.3) and ε is the usual antisymmetric unit tensor.

$\mathrm{Tr}(I) = 4$		
$\mathrm{Tr}(\not{a}\,\not{b}) = 4ab$	$\Rightarrow$	$\mathrm{Tr}(\gamma^\mu\gamma^\nu) = 4g^{\mu\nu}$
$\mathrm{Tr}(\not{a}\,\not{b}\,\not{c}\,\not{d}) = 4[(ab)(cd) + (ad)(bc) - (ac)(bd)]$	$\Rightarrow$	$\mathrm{Tr}(\gamma^\lambda\gamma^\mu\gamma^\nu\gamma^\sigma)$
	$=$	$4\left[g^{\lambda\mu}g^{\nu\sigma} + g^{\lambda\sigma}g^{\mu\nu} - g^{\lambda\nu}g^{\mu\sigma}\right]$
$\mathrm{Tr}\left(\gamma^5\right) = 0$		
Tr(product of an odd number of γ matrices) = 0		
$\mathrm{Tr}\left(\gamma^5\not{a}\,\not{b}\right) = 0$		
$\mathrm{Tr}\left(\gamma^5\not{a}\,\not{b}\,\not{c}\,\not{d}\right) = 4i\varepsilon^{\mu\nu\lambda\sigma}a_\mu b_\nu c_\lambda d_\sigma$	$\Rightarrow$	$\mathrm{Tr}\left(\gamma^5\gamma^\mu\gamma^\nu\gamma^\lambda\gamma^\sigma\right) = 4i\varepsilon^{\mu\nu\lambda\sigma}$
$\gamma^\mu\gamma_\mu = 4$	with	$\gamma_\mu = g_{\mu\nu}\gamma^\nu$
$\gamma_\mu\not{a}\gamma^\mu = -2\not{a}$		
$\gamma_\mu\not{a}\,\not{b}\gamma^\mu = 4ab$		
$\gamma_\mu\not{a}\,\not{b}\,\not{c}\gamma^\mu = -2\not{c}\,\not{b}\,\not{a}$		

one obtains

$$\sum_{s=1}^{2} u^s\bar{u}^s = N^2\begin{pmatrix} I & -\frac{\vec{\sigma}\vec{p}}{E+m} \\ \frac{\vec{\sigma}\vec{p}}{E+m} & -\frac{E-m}{E+m} \end{pmatrix} = \begin{pmatrix} E+m & -\vec{\sigma}\vec{p} \\ \vec{\sigma}\vec{p} & -E+m \end{pmatrix}$$
$$= \gamma^0 E - \vec{\gamma}\cdot\vec{p} + mI = \not{p} + m, \tag{15.60}$$

in accordance with (15.55c).

To avoid lengthy calculations with the explicit expressions (15.41) and (15.46) for the Dirac spinors and (15.33) for the γ matrices, one usually uses the relations (15.55a)–(15.55d) and the properties of γ matrices and 4-vectors given in table 15.1 which follow from the anti-commutation relations (15.19). For example, according to (15.19),

$$2\,\mathrm{Tr}\left(\gamma^\mu\gamma^\nu\right) = 8g^{\mu\nu} \tag{15.61}$$

and therefore

$$\mathrm{Tr}(\not{a}\,\not{b}) = \mathrm{Tr}\left(\gamma^\mu\gamma^\nu\right)a_\mu b_\nu = 4g^{\mu\nu}\,a_\mu b_\nu = 4a^\nu\,b_\nu = 4ab. \tag{15.62}$$

Several practical applications will be described in detail in the following chapters.

15.2 The discovery of the positron and of the antiproton

The positron was discovered in 1932 by Carl Anderson in cosmic rays, using a cloud chamber (figure 15.1) [3]. The larger curvature of the track above the 6 mm thick lead plate shows that the particle flies upwards and is positively charged (magnetic

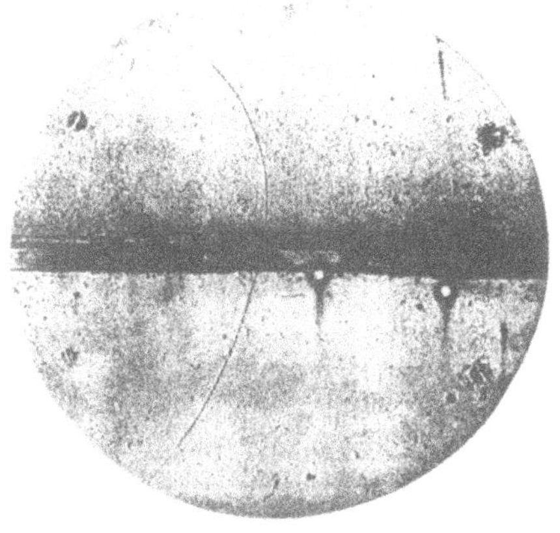

Figure 15.1. The discovery of the positron in a cloud chamber. The positively charged particle flies upwards through the 6 mm thick lead absorber. Reproduced with permission from [3]. Copyright 1933 the American Physical Society.

field pointing into the picture). The momentum can be measured from the curvature: 63 MeV c^{-1} before and 23 MeV c^{-1} after the lead absorber. A heavy particle with this incident momentum would have a much larger curvature after the absorber. A 63 MeV c^{-1} proton would not even cross the lead absorber.

The measurement of the magnetic moment of the proton by Otto Stern in 1933 gave a value ($g = 5.58$) at variance with that of the electron ($g = 2$). It was therefore not clear whether Dirac's theory could be applied to baryons, and thus whether a charge conjugate version of the proton would exist. One of the main motivations for the construction of the Bevatron at Berkeley was to search for the antiproton. Following (8.61), the threshold for antiproton production in the reaction $pp \to pp\bar{p}p$ is

$$E_1 = \frac{(4m_p)^2 - 2m_p^2}{2m_p} = 7m_p, \tag{15.63}$$

corresponding to a kinetic energy $T_1 = 6m_p = 5.6$ GeV. With a nucleus as the target the threshold is slightly lower than 5.6 GeV due to the Fermi motion of the protons. The Bevatron, which was constructed mainly to search for the antiproton, delivered a proton beam of 6.2 GeV, striking a copper target. A 1.2 GeV c^{-1} beam of secondary, negatively charged particles was set up to search for the antiproton which was discovered in 1955 [4]. The main difficulty was the overwhelming background from pion production (e.g. from $pn \to pp\pi^-$) with about 2 antiprotons for 10^5 pions. Antiprotons were distinguished from pions by Čerenkov counters and by measuring the time-of-flight over a distance of 12 m, and combining with the known beam momentum. The antiproton mass agreed within 5% with the proton mass. The antiproton yield was rather low, with 60 antiprotons observed during a 7 h long run. (In comparison, CERN's Antiproton Decelerator delivers 3×10^7 antiprotons of 5 MeV every 100 s.) The annihilation of the antiproton by matter was observed shortly thereafter with emulsion detectors (figure 15.2).

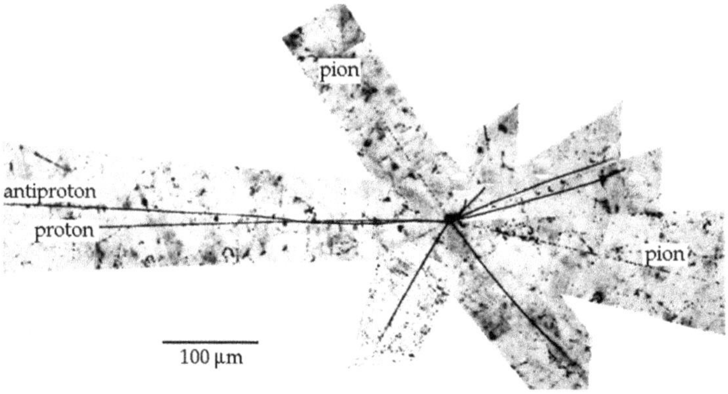

Figure 15.2. An antiproton annihilation observed by an emulsion detector. The antiproton enters from the left. The fat tracks are due to protons or nuclear fragments and the faint tracks to minimum ionizing particles (mostly pions). Reproduced with permission from [5]. Copyright 1956 Springer Science and Business Media.

15.3 Chirality and handedness

The γ^5 operator is defined as

$$\boxed{\gamma^5 \equiv \mathrm{i}\gamma^0\gamma^1\gamma^2\gamma^3}. \tag{15.64}$$

The following properties, easily derived from the anti-commutation relations (15.19),

$$\begin{aligned} \gamma^{5\dagger} &= \gamma^5, \\ (\gamma^5)^2 &= I, \\ \gamma^5\gamma^\mu + \gamma^\mu\gamma^5 &= 0, \end{aligned} \tag{15.65}$$

are independent of the representation. In the standard representation γ^5 takes the form

$$\gamma^5 = \begin{pmatrix} 0 & I \\ I & 0 \end{pmatrix}. \tag{15.66}$$

Let us investigate the action of the operators $\frac{1}{2}(1 \pm \gamma^5)$ on a Dirac u spinor. Consider the spinors

$$\begin{aligned} \phi_\ell = \frac{1}{2}\left(1 - \gamma^5\right)u &= \frac{N}{2}\begin{pmatrix} I & -I \\ -I & I \end{pmatrix}\begin{pmatrix} \chi^s \\ \frac{\vec{\sigma}\vec{p}}{E+m}\chi^s \end{pmatrix} \\ &= \frac{N}{2}\begin{pmatrix} \left[1 - \frac{\vec{\sigma}\vec{p}}{E+m}\right]\chi^s \\ -\left[1 - \frac{\vec{\sigma}\vec{p}}{E+m}\right]\chi^s \end{pmatrix} \equiv \begin{pmatrix} \Phi_\ell \\ -\Phi_\ell \end{pmatrix} \end{aligned} \tag{15.67}$$

and

$$\phi_r = \frac{1}{2}\left(1 + \gamma^5\right)u = \frac{N}{2}\begin{pmatrix} I & I \\ I & I \end{pmatrix}\begin{pmatrix} \chi^s \\ \frac{\vec{\sigma}\vec{p}}{E+m}\chi^s \end{pmatrix}$$

$$= \frac{N}{2}\begin{pmatrix} \left[1 + \frac{\vec{\sigma}\vec{p}}{E+m}\right]\chi^s \\ \left[1 + \frac{\vec{\sigma}\vec{p}}{E+m}\right]\chi^s \end{pmatrix} \equiv \begin{pmatrix} \Phi_r \\ \Phi_r \end{pmatrix}. \tag{15.68}$$

Inserting $u = \phi_\ell + \phi_r$ into the Dirac equation (15.37) gives

$$\begin{aligned} (E - m)\left(\Phi_\ell + \Phi_r\right) - \vec{\sigma}\vec{p}\left(-\Phi_\ell + \Phi_r\right) &= 0, \\ \vec{\sigma}\vec{p}\left(\Phi_\ell + \Phi_r\right) - (E + m)\left(-\Phi_\ell + \Phi_r\right) &= 0. \end{aligned} \tag{15.69}$$

Adding and subtracting the last two equations leads to the coupled (Weyl) equations

$$\begin{aligned} (E + \vec{\sigma}\vec{p})\Phi_\ell &= m\Phi_r, \\ (E - \vec{\sigma}\vec{p})\Phi_r &= m\Phi_\ell. \end{aligned} \tag{15.70}$$

For $m = 0$ (or in a reference frame in which the mass of the particle is negligible compared to its momentum) Φ_ℓ and Φ_r are eigenstates of the *helicity* $\vec{\sigma}\vec{p}/p$ with eigenvalues -1 and $+1$, respectively. Thus, in the highly relativistic limit ($m \ll E$), ϕ_ℓ and ϕ_r are spinors of a *left-handed* (L) and *right-handed* (R) fermion respectively:

$$\phi_\ell \to u_L \qquad \text{and} \qquad \phi_r \to u_R. \tag{15.71}$$

Let us now calculate the polarization of the fermion when its mass m is not negligible. With the definitions

$$a \equiv 1 - \frac{p}{E + m} \qquad \text{and} \qquad b \equiv 1 + \frac{p}{E + m} \tag{15.72}$$

(where $p \equiv |\vec{p}\,|$), the square brackets in (15.67) can be rewritten as

$$\left[1 - \frac{\vec{\sigma}\vec{p}}{E + m}\right] = \frac{a}{2}\left(1 + \frac{\vec{\sigma}\vec{p}}{p}\right) + \frac{b}{2}\left(1 - \frac{\vec{\sigma}\vec{p}}{p}\right). \tag{15.73}$$

Hence ϕ_ℓ becomes

$$\phi_\ell = \frac{N}{2}\begin{pmatrix} \frac{a}{2}\phi_R + \frac{b}{2}\phi_L \\ -\left[\frac{a}{2}\phi_R + \frac{b}{2}\phi_L\right] \end{pmatrix}, \tag{15.74}$$

where ϕ_R and ϕ_L are defined as

$$\phi_R \equiv \left(1 + \frac{\vec{\sigma}\vec{p}}{p}\right)\chi^s \qquad \text{and} \qquad \phi_L \equiv \left(1 - \frac{\vec{\sigma}\vec{p}}{p}\right)\chi^s. \tag{15.75}$$

The spinors ϕ_R and ϕ_L describe particles with positive and negative *helicity*, respectively, since

$$\begin{aligned} \frac{\vec{\sigma}\vec{p}}{p}\phi_R &= \left(\frac{\vec{\sigma}\vec{p}}{p} + 1\right)\chi^s = +\phi_R, \\ \frac{\vec{\sigma}\vec{p}}{p}\phi_L &= \left(\frac{\vec{\sigma}\vec{p}}{p} - 1\right)\chi^s = -\phi_L. \end{aligned} \tag{15.76}$$

The polarization of the state ϕ_ℓ is therefore

$$P \equiv \frac{\frac{a^2}{4} - \frac{b^2}{4}}{\frac{a^2}{4} + \frac{b^2}{4}} = -\frac{p}{E} = -\beta, \tag{15.77}$$

equal to the velocity of the particle. Thus for highly relativistic particles ($m \ll E$) $P = -1$ ($a = 0$ and $b = 1$), as expected. Similarly, from (15.68)

$$\phi_r = \frac{N}{2}\begin{pmatrix} \frac{b}{2}\phi_R + \frac{a}{2}\phi_L \\ \frac{b}{2}\phi_R + \frac{a}{2}\phi_L \end{pmatrix}, \tag{15.78}$$

and $P = +\beta$. For $m \ll E$ the polarization of the state ϕ_r is $P = +1$. Note that the spinors ϕ_ℓ and ϕ_r are eigenstates of γ^5 with *chiralities* –1 and +1, respectively:

$$\begin{aligned} \gamma^5\phi_\ell &= \frac{1}{2}\gamma^5\left(1 - \gamma^5\right)u = -\frac{1}{2}\left(1 - \gamma^5\right)u = -\phi_\ell, \\ \gamma^5\phi_r &= \frac{1}{2}\gamma^5\left(1 + \gamma^5\right)u = +\frac{1}{2}\left(1 + \gamma^5\right)u = +\phi_r. \end{aligned} \tag{15.79}$$

However, ϕ_ℓ and ϕ_r are not eigenstates of the Dirac equation (unless $m = 0$) and therefore chirality cannot be observed, in contrast to helicity and polarization. To summarize:

> In a reference frame, in which the mass of the fermion can be neglected, the operators $\frac{1}{2}(1 - \gamma^5)$ and $\frac{1}{2}(1 + \gamma^5)$ generate the eigenstates u_L and u_R of a left-handed and right-handed fermion, respectively.

Finally, one can show by using similar arguments that $\bar{u}_L = \bar{u}\frac{1}{2}(1 + \gamma^5)$ describes a *left-handed* fermion for $\beta \to 1$, while $\bar{u}_R = \bar{u}\frac{1}{2}(1 - \gamma^5)$ describes a *right-handed* one. In the following chapters we shall mostly deal with high energy fermions for which the words 'handedness' and 'chirality' become equivalent.

15.4 Parity

The momentum vector $\vec{p}$ changes sign under space inversion. Hence the Dirac spinors $u^s(E, \vec{p})$ transform into $u^s(E, -\vec{p})$ which do not fulfil the Dirac equation

(15.37), and are therefore not acceptable as physical states. However, appending the operator γ^0 to space inversion recovers the original spinors. The parity transformation is therefore defined as

$$Pu^s(E, \vec{p}) \equiv \gamma^0 u^s(E, -\vec{p}) = +u^s(E, \vec{p}). \tag{15.80}$$

Accordingly, parity transformation applied to the anti-spinors (15.46) gives

$$Pv^s(E, \vec{p}) \equiv \gamma^0 v^s(E, -\vec{p}) = -v^s(E, \vec{p}), \tag{15.81}$$

which shows that spin-$\frac{1}{2}$ particles and antiparticles have opposite parities.

15.5 Charge conjugation

The Dirac wave functions ψ and ψ^c of the particle and antiparticle are related by charge conjugation C. Let us prove that

$$\boxed{\psi^c = \overbrace{\mathrm{i}\gamma^2\gamma^0}^{C}\bar{\psi}^T = \mathrm{i}\gamma^2\psi^*} \tag{15.82}$$

is indeed the Dirac wave function of the antiparticle. Using the representation

$$\psi = \overbrace{\sqrt{E+m}}^{N}\begin{pmatrix} \chi^s \\ \frac{\vec{\sigma}\vec{p}}{E+m}\chi^s \end{pmatrix}\mathrm{e}^{-\mathrm{i}px} \tag{15.83}$$

one obtains

$$\begin{aligned} \psi^c &= \mathrm{i}\gamma^2\left(\gamma^0\right)^2\begin{pmatrix} \chi^s \\ \frac{\vec{\sigma}\vec{p}}{E+m}\chi^s \end{pmatrix}^* N\mathrm{e}^{\mathrm{i}px} \\ &= \mathrm{i}\gamma^2\begin{pmatrix} \chi^s \\ \frac{\vec{\sigma}^*\vec{p}}{E+m}\chi^s \end{pmatrix} N\mathrm{e}^{\mathrm{i}px} = \mathrm{i}\begin{pmatrix} 0 & \sigma_2 \\ -\sigma_2 & 0 \end{pmatrix}\begin{pmatrix} \chi^s \\ \frac{-\sigma_2\vec{\sigma}\sigma_2\vec{p}}{E+m}\chi^s \end{pmatrix} N\mathrm{e}^{\mathrm{i}px}, \end{aligned} \tag{15.84}$$

where we have written $\vec{\sigma}^*$ as $-\sigma_2\vec{\sigma}\sigma_2$, which can easily be verified by using the anti-commutation properties of Pauli matrices. With the η matrix defined in (15.47) ψ^c becomes

$$\psi^c = \mathrm{i}\begin{pmatrix} \frac{-\vec{\sigma}\vec{p}\,\sigma_2}{E+m}\chi^s \\ -\sigma_2\chi^s \end{pmatrix} N\mathrm{e}^{\mathrm{i}px} = \begin{pmatrix} \frac{\vec{\sigma}\vec{p}}{E+m}\eta\chi^s \\ \eta\chi^s \end{pmatrix} N\mathrm{e}^{\mathrm{i}px}, \tag{15.85}$$

in agreement with the wave functions (15.46) or (15.50) of the antiparticle.

15.6 Current

Let us take the Hermitian conjugate of the Dirac equation (15.18):

$$\left(\mathrm{i}\gamma^\mu\partial_\mu - m\right)\psi = 0 \;\Rightarrow\; \psi^\dagger\left(-\mathrm{i}\gamma^{\mu\dagger}\partial_\mu - m\right) = 0. \tag{15.86}$$

Right-multiplying by γ^0 and taking into account (15.19) and (15.22a) leads to

$$\psi^\dagger\gamma^0\left(-\mathrm{i}\gamma^\mu\partial_\mu - m\right) = 0 \Rightarrow \bar{\psi}\left(\mathrm{i}\gamma^\mu\partial_\mu + m\right) = 0. \tag{15.87}$$

Adding the left-hand side of (15.86), left-multiplied by $\bar{\psi}$, to (15.87) right-multiplied by ψ, gives

$$\bar{\psi}\left(\gamma^\mu\partial_\mu\psi\right) + \left(\bar{\psi}\gamma^\mu\partial_\mu\right)\psi = 0, \tag{15.88}$$

hence the continuity equation

$$\partial_\mu\underbrace{\left(\bar{\psi}\gamma^\mu\psi\right)}_{j^\mu} = 0. \tag{15.89}$$

The probability density

$$\rho = \bar{\psi}\gamma^0\psi = \psi^\dagger\psi = |\psi|^2 \tag{15.90}$$

is positive, as expected. The electric 4-*current* of the electron (charge $-e$) is

$$j^\mu = -e\bar{\psi}\gamma^\mu\psi = -e\bar{u}\gamma^\mu u = -e(\rho, \vec{j}\,) = -2e(E, \vec{p}\,), \tag{15.91}$$

where we have used the normalization (15.55a). Note that, with the transformation of the 4-vector $p \to -p$, the antiparticle current obtains the opposite sign: $j^\mu(-p) = 2e(E, \vec{p}\,)$.

15.6.1 Current–current interaction

We have given the formulas for cross sections (12.2) and partial widths (12.8) in section 12.1. We now calculate the transition amplitude $\mathcal{M}$ by assuming that first-order perturbation theory is applicable. Let us consider as a concrete example Møller scattering $e^-e^- \to e^-e^-$ (figure 15.3). Far away from the interaction point, the four leptons are described by *free* Dirac spinors ψ_1, ψ_2, ψ_3 and ψ_4, satisfying

$$(\vec{\alpha}\vec{p} + \beta m)\psi_k = E_k\psi_k \quad (k = 1, 2, 3, 4). \tag{15.92}$$

The incident electron 1 is scattered in the field of the photon emitted by the target electron 3. The transition amplitude is, following first order perturbation theory,

$$V_{fi} = -\mathrm{i}\int\psi_2^\dagger V\psi_1 \mathrm{d}^4x. \tag{15.93}$$

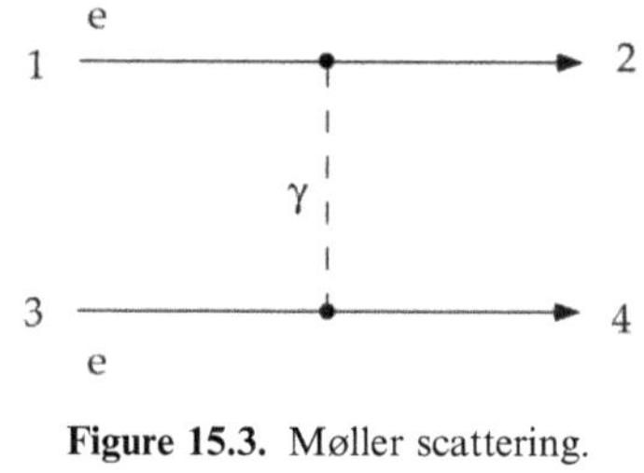

Figure 15.3. Møller scattering.

We now need an expression for the perturbation V. In the presence of the field $A^{\mu} = (\phi, \vec{A})$ the 4-momentum p^{μ} in the Dirac equation is to be replaced by $p^{\mu} - qA^{\mu}$, hence for electrons

$$\begin{aligned} \vec{p} &\to \vec{p} + e\vec{A}, \\ E &\to E + e\phi. \end{aligned} \tag{15.94}$$

The free Dirac equation is therefore modified in the vicinity of the interaction point to read

$$(\overbrace{\vec{\alpha}\vec{p} + \beta m}^{H0} + \overbrace{e[\vec{\alpha}\vec{A} - \phi]}^{V})\psi_k' = E_k \psi_k'. \tag{15.95}$$

Note that the spinors ψ_k' are no longer free Dirac spinors. The perturbing potential energy V can be read out:

$$V \equiv e[\vec{\alpha}\vec{A} - \phi]. \tag{15.96}$$

The transition amplitude (15.93) becomes

$$\begin{aligned} V_{fi} &= -\mathrm{i}e \int \psi_2^{\dagger} \overbrace{I}^{(\gamma^0)^2 = \beta^2} [\vec{\alpha}\vec{A} - \phi]\psi_1 \mathrm{d}^4x \\ &= \mathrm{i}e \int \bar{\psi}_2 [\overbrace{\gamma^0\phi - \vec{\gamma}\vec{A}}^{\gamma^{\mu}A_{\mu}}]\psi_1 \mathrm{d}^4x \\ &= -\mathrm{i} \int \left(-e\bar{\psi}_2\gamma^{\mu}\psi_1\right) A_{\mu} \mathrm{d}^4x \end{aligned} \tag{15.97}$$

with $A_{\mu} = (\phi, -\vec{A})$. Introducing the expression (15.23) for ψ_1 and ψ_2 gives

$$V_{fi} = -\mathrm{i} \int \left(-e\bar{u}_2\gamma^{\mu}u_1\right) A_{\mu} \mathrm{e}^{\mathrm{i}(p_2 - p_1)x} \mathrm{d}^4x. \tag{15.98}$$

Let us define as the electron *current* the 4-vectors

$$\boxed{j_{12}^{\mu} \equiv -e\bar{u}_2\gamma^{\mu}u_1, \qquad \text{and} \qquad J_{12}^{\mu} \equiv j_{12}^{\mu} \mathrm{e}^{\mathrm{i}(p_2 - p_1)x}}, \tag{15.99}$$

in analogy to the electromagnetic current (15.91). The field A^{μ} is produced by the target electron 3 which is the source current J_{34}^{μ}. The Maxwell equations (see (16.84) below) are in the Lorentz gauge:

$$\Box A^{\mu} = J_{34}^{\mu} = j_{34}^{\mu} \mathrm{e}^{\mathrm{i}(p_4 - p_3)x} = j_{34}^{\mu} \mathrm{e}^{\mathrm{i}qx}. \tag{15.100}$$

We have introduced the 4-momentum transfer

$$q \equiv p_4 - p_3 = p_1 - p_2, \tag{15.101}$$

which is the 4-momentum of the photon. Since

$$\Box \mathrm{e}^{\mathrm{i}qx} = -q^2 \mathrm{e}^{\mathrm{i}qx}, \tag{15.102}$$

the field A^μ can be expressed as

$$A^\mu = -\frac{1}{q^2} J^\mu_{34} \Rightarrow \Box A^\mu = J^\mu_{34}. \tag{15.103}$$

The transition amplitude (15.98) then becomes

$$\begin{aligned} V_{fi} &= -\mathrm{i} \int j^\mu_{12} \left[-\frac{1}{q^2} \right] j_{\mu 34}\, \mathrm{e}^{\mathrm{i}(p_2+p_4-p_1-p_3)x} \mathrm{d}^4 x \\ &= \overbrace{-\mathrm{i} j^\mu_{12} \left[-\frac{1}{q^2} \right] j_{\mu 34}}^{\mathcal{M}} \times (2\pi)^4 \delta(p_2 + p_4 - p_1 - p_3) \end{aligned} \tag{15.104}$$

(compare with expression (12.7) in section 12.1). The term in square brackets is the *propagator* of the photon. The transition amplitude then takes the form

$$\boxed{\mathcal{M} \propto \text{current} \times \text{propagator} \times \text{current}}, \tag{15.105}$$

where

$$\text{current} = -e\bar{u}(p')\gamma^\mu u(p) \tag{15.106}$$

is the electron current and

$$\text{propagator} = -\frac{1}{q^2} \tag{15.107}$$

is the photon propagator with $q = p - p'$. The incoming electron has 4-momentum p and the outgoing electron 4-momentum p'. The charge e is the coupling between the electron and photon. The δ-function ensures energy and momentum conservation. For a heavy $J^P = 1^-$ boson such as the $W^\pm$ boson exchanged in weak interactions, the Maxwell equations (15.100) contain an additional mass term (see (16.93) in section 16.5):

$$\left(\Box + M^2\right) W^\mu = J^\mu_{34} \Rightarrow W^\mu = -\frac{1}{q^2 - M^2} J^\mu_{34}, \tag{15.108}$$

where M is the boson mass and W^μ is the boson field. The propagator of the heavy boson then reads

$$\text{propagator} = -\frac{1}{q^2 - M^2}, \tag{15.109}$$

which reduces to the constant $1/M^2$ for relatively small momentum transfers $q^2 \ll M^2$. Then

$$\boxed{\mathcal{M} \propto j^\mu_{12} j_{\mu 34}} \tag{15.110}$$

describes a *point-like interaction*. This transition amplitude will be used to describe weak interactions, but the electric charge $-e$ in the current (15.106) will be replaced by the coupling constant of weak interactions. The precise expression for the weak current will be given in chapter 16.

References

[1] Farmelo G 2010 *The Strangest Man: The Hidden Life of Paul Dirac, Quantum Genius* (London: Faber and Faber)
[2] Schiff L 1955 *Quantum Mechanics* (London: McGraw-Hill) p 330
[3] Anderson C D 1933 *Phys. Rev.* **43** 491
[4] Chamberlain O *et al* 1955 *Phys. Rev.* **100** 947
[5] Chamberlain O *et al* 1956 *Nuovo Cimento* **3** 447

IOP Publishing

Nuclear and Particle Physics

Claude Amsler

Chapter 16

The electroweak interaction

16.1 $V - A$ interaction

We now deal with the weak interaction between four fermions. Consider for example nuclear β^+-decay, in which a proton transforms into a neutron, $p \to ne^+\nu_e$. Since the exchanged W boson is heavy the force is very short-range and one can in good approximation assume that the interaction is point-like. We first replace the antiparticle currents by the particle currents by considering the inverse β-decay $pe^- \to n\nu_e$, involving only particles (figure 16.1). According to (15.110) one expects a transition matrix of the form

$$\mathcal{M} \propto \left[\bar{u}(n)\gamma^\mu u(p)\right]\left[\bar{u}(\nu_e)\gamma_\mu u(e^-)\right]. \tag{16.1}$$

The u spinors are free Dirac spinors. The nucleon current describes an *incoming* proton (spinor $u(p)$) and an *outgoing* neutron (spinor $u(n)$), while the lepton current describes an *incoming* electron (spinor $u(e^-)$) and an *outgoing* neutrino (spinor $u(\nu_e)$). However, in the decay $p \to ne^+\nu_e$ an antiparticle (the positron) is emitted with 4-momentum p_e. Hence the spinor $u(e^-)$ in (16.1) describes an electron with 4-momentum $-p_e$. Therefore the transition matrix becomes

$$\mathcal{M} \propto \left[\bar{u}(n)\gamma^\mu u(p)\right]\left[\bar{u}(\nu_e)\gamma_\mu v(e^+)\right] \tag{16.2}$$

with

$$v(e^+)(p_e) = u(e^-)(-p_e), \tag{16.3}$$

the spinor of the antiparticle. The expression (16.1) was formulated in 1934 by Enrico Fermi to describe β-decay. However, Fermi but not Gamow–Teller transitions could be described. The next step was therefore to generalize the transition

doi:10.1088/978-0-7503-1140-3ch16 © IOP Publishing Ltd 2015

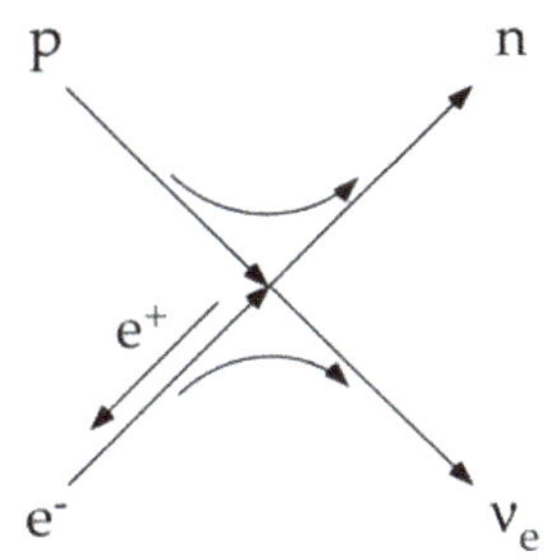

Figure 16.1. Interaction of the $p \to n$ and $e \to \nu_e$ weak charged currents.

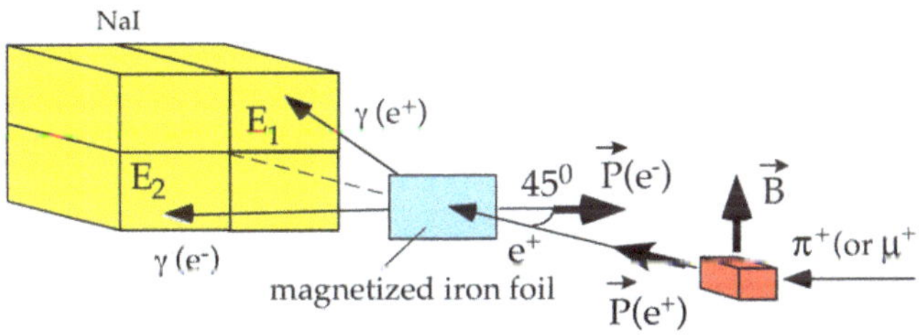

Figure 16.2. Measurement of the polarization of the positron in muon β-decay with a pion or a muon beam. The direction of the positron is measured with wire chambers. The B-field is used with the muon beam. After [1].

matrix (16.1) to include all independent combinations of γ matrices. In addition to γ^μ (vector V) the following four independent operators are possible:

$$I \ (S = \text{scalar}), \qquad \gamma^\mu\gamma^\nu - \gamma^\nu\gamma^\mu \ (T = \text{tensor}),$$
$$\gamma^\mu\gamma^5 \ (A = \text{axialvector}) \qquad \text{and} \qquad \gamma^5 \ (P = \text{pseudoscalar}). \tag{16.4}$$

All experiments show that only left-handed fermions or right-handed antifermions contribute to the interaction of weak charged currents. Consequently, only the V and A operators can occur in the particular $V - A$ combination, as we shall demonstrate below.

The experiment of Burkard *et al* [1] measured the longitudinal polarization of the positron in μ^+ decay at the Paul Scherrer Institute (PSI) (figure 16.2). Positive pions of 150 MeV c^{-1} are stopped and decay in a target. The net polarization of the emitted muons, which also stop in the target, is zero since they are distributed isotropically. The polarization of the e^+ from μ^+ decay is analysed with a magnetized iron foil using Bhabha scattering or e^+e^- annihilation:

$$e^+e^- \to e^+e^-, \qquad \text{and} \qquad e^+e^- \to \gamma\gamma. \tag{16.5}$$

The cross sections take the form

$$\sigma(E_1, E_2) \propto \left[1 + A(E_1, E_2)\vec{P}(e^+)\vec{P}(e^-)\right], \tag{16.6}$$

where $\vec{P}(e^+)$ is the polarization of the e^+ and $\vec{P}(e^-)$ that of the e^- in the magnetized foil, and E_1 and E_2 are the energies of the outgoing particles ($e^\pm$ or γ). The function $A(E_1, E_2) = A(E_2, E_1)$ is different for the two processes [2].

Not shown in figure 16.2 are scintillation counters and wire chambers behind the scattering foil to select scattered leptons or annihilation photons. The foil is tilted at an angle of 45° with respect to the positron direction. The asymmetry of rates is measured with the NaI counters, when the sign of the polarization $\vec{P}(e^-)$ is reversed by flipping the polarity of the magnetic field. The electron polarization in the magnetized foil is typically 4%. Thus the polarization $P(e^+)$ can be determined, the analysing power $A(E_1, E_2)$ being well known.

The polarization of the positron was also studied with a stopping polarized muon beam as a function of angle θ between the muon spin and the positron direction, see (13.67). The muon spin precession is achieved by applying a vertical B field (374 mT). The longitudinal polarization P_L of the positron was measured to be

$$P_L(e^+) = +0.998 \pm 0.045. \tag{16.7}$$

Thus the positron as an antiparticle is *right-handed.* We therefore assume that particles are *left-handed* and show that the weak charged current takes the form

$$j^\mu \propto \bar{u}'\gamma^\mu\left(1 - \gamma^5\right)u. \tag{16.8}$$

We have seen in section 15.3 that the spinors $(1 - \gamma^5)u$ and $\bar{u}(1 + \gamma^5)$ are left-handed. Hence the operator $(1 - \gamma^5)$ in (16.8) ensures that incoming fermions are left-handed. Decomposing $\bar{u}'$ into left- and right-handed components

$$\bar{u}' \equiv \bar{u}'\left[\frac{1}{2}\left(1 - \gamma^5\right) + \frac{1}{2}\left(1 + \gamma^5\right)\right], \tag{16.9}$$

leads to

$$j^\mu \propto \bar{u}'\frac{1}{2}\left(1 - \gamma^5\right)\gamma^\mu\left(1 - \gamma^5\right)u + \bar{u}'\frac{1}{2}\left(1 + \gamma^5\right)\gamma^\mu\left(1 - \gamma^5\right)u. \tag{16.10}$$

The first term (right-handed outgoing fermion) vanishes since

$$\left(1 - \gamma^5\right)\gamma^\mu\left(1 - \gamma^5\right) = \left(1 - \gamma^5\right)\left(1 + \gamma^5\right)\gamma^\mu = \Big(1 - \overbrace{\left(\gamma^5\right)^2}^{I}\Big)\gamma^\mu = 0. \tag{16.11}$$

Hence the factor $(1 - \gamma^5)$ in (16.8) ensures the left-handedness of *both* incoming and outgoing fermions. The generalization of the transition matrix (16.1) therefore reads

$$\boxed{\mathcal{M} = \frac{G_F}{\sqrt{2}}\left[\bar{u}(n)\gamma^\mu\left(1 - \gamma^5\right)u(p)\right]\left[\bar{u}(\nu_e)\gamma_\mu\left(1 - \gamma^5\right)u(e^-)\right]} \tag{16.12}$$

with the *Fermi coupling constant* G_F. A factor $1/\sqrt{2}$ is prepended by convention. The weak interaction is said to be of the $V - A$ type, due to the $\gamma^\mu - \gamma^\mu\gamma^5$ form of the operator. This form is valid for all interactions between weak charged currents,

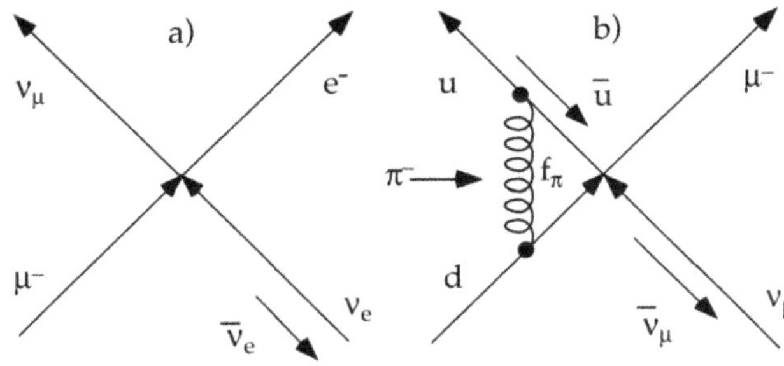

Figure 16.3. (*a*) μ^- decay. (*b*) π^- decay.

provided that the fermions are *point-like*. Due to the internal structure of the nucleon, the factor $\gamma^\mu(1 - \gamma^5)$ has to be replaced by $\gamma^\mu(1 + \lambda\gamma^5)$ for the nucleon β-decay, where $\lambda = g_A/g_V = -1.27$ (see (13.56)).

As a first example, let us outline the derivation of the muon lifetime. The transition matrix for the μ^- decay (figure 16.3(*a*)) reads

$$\mathcal{M} = \frac{G_F}{\sqrt{2}}\left[\bar{u}(\nu_\mu)\gamma^\mu\left(1 - \gamma^5\right)u(\mu^-)\right]\left[\bar{u}(e^-)\gamma_\mu\left(1 - \gamma^5\right)\underbrace{u\left(\nu_e[-p]\right)}_{=v(\bar{\nu}_e[p])}\right]. \tag{16.13}$$

The decay width is obtained [3] from the partial width dΓ, (12.8):

$$\Gamma = \frac{1}{\tau} = \frac{G_F^2 m_\mu^5}{192\pi^3}. \tag{16.14}$$

One recognizes here the Q^5 dependence of the Sargent rule (section 13.2.4). The constant G_F can be determined with a measurement of the muon mass using x-rays from muonic atoms [4] (section 2.3) and from a measurement of the muon lifetime [5]. The most precise value to date for τ_μ was obtained with the experiment of [6], leading to $G_F = (1.1663787 \pm 0.0000006) \times 10^{-5}$ GeV^{-2}.

As a second example, let us calculate the lifetime of the pion from the decay $\pi^- \to \mu^-\bar{\nu}_\mu$ (figure 16.3(*b*)). An expression such as (16.8) is not permitted for the quark current since quarks are bound, and therefore free Dirac spinors cannot be used. Since the pion is spinless the 4-momentum of the intermediate state

$$q^\mu = k^\mu + p^\mu \tag{16.15}$$

is the only available 4-vector to describe the quark current. Here k and p denote the 4-vectors of the antineutrino and the muon:

$$k \equiv \left(\omega, \vec{k}\right) \qquad \text{and} \qquad p \equiv (E, \vec{p}), \tag{16.16}$$

respectively. The ansatz for the hadronic current is (see (15.91)

$$j^\mu = f_\pi\left(k^\mu + p^\mu\right), \tag{16.17}$$

where the unknown pion decay constant f_π describes the strong interaction within the pion. The transition matrix is

$$\mathcal{M} = \frac{G_F f_\pi}{\sqrt{2}}(k^\mu + p^\mu)\Big[\bar{u}(p)\gamma_\mu\big(1-\gamma^5\big)\underbrace{u(-k)}_{=v(k)}\Big]$$
$$= \frac{G_F f_\pi}{\sqrt{2}}\Big[\bar{u}(p)k^\mu\gamma_\mu\big(1-\gamma^5\big)v(k) + p^\mu\bar{u}(p)\gamma_\mu\big(1-\gamma^5\big)v(k)\Big]. \quad (16.18)$$

From the Dirac equation

$$\bar{u}(p)\big(\gamma_\mu p^\mu - m\big) = 0 \quad (16.19)$$

it follows for the muon with mass m that

$$p^\mu\bar{u}(p)\gamma_\mu = m\bar{u}(p). \quad (16.20)$$

On the other hand, for a massless neutrino

$$k^\mu\gamma_\mu v(k) = 0, \quad (16.21)$$

so that also

$$k^\mu\gamma_\mu\gamma^5 v(k) = 0, \quad (16.22)$$

hence the first term in (16.18) vanishes, therefore

$$\mathcal{M} = \frac{mG_F f_\pi}{\sqrt{2}}\bar{u}(p)\big(1-\gamma^5\big)v(k). \quad (16.23)$$

According to (12.8) the partial decay width is

$$\mathrm{d}\Gamma = \frac{1}{2m_\pi}\sum_{s,s'=1}^{2}|\mathcal{M}|^2\frac{\mathrm{d}^3\vec{p}}{(2\pi)^3 2E}\frac{\mathrm{d}^3\vec{k}}{(2\pi)^3 2\omega}(2\pi)^4\delta(p+k-q). \quad (16.24)$$

The sum runs over the spins of the muon (s) and of the neutrino (s'). Let us write the components of the spinors and matrices explicitly and add over repeated indices. The contribution from the squares of the matrix is

$$\sum_{s,s'=1}^{2}|\mathcal{M}|^2 = \frac{G_F^2 f_\pi^2 m^2}{2}\sum_{s,s'=1}^{2}\bar{u}\big(1-\gamma^5\big)\overbrace{v v^\dagger\big(1-\gamma^5\big)\gamma^0 u}^{\bar{v}(1+\gamma^5)u}$$
$$= \frac{G_F^2 f_\pi^2 m^2}{2}\sum_{s,s'=1}^{2}\bar{u}_\alpha\big(1-\gamma^5\big)_{\alpha\beta}v_\beta\bar{v}_\gamma\big(1+\gamma^5\big)_{\gamma\delta}u_\delta$$
$$= \frac{G_F^2 f_\pi^2 m^2}{2}\underbrace{\sum_{s=1}^{2}u_\delta\bar{u}_\alpha}_{(\not{p}+m)_{\delta\alpha}}\big(1-\gamma^5\big)_{\alpha\beta}\underbrace{\sum_{s'=1}^{2}v_\beta\bar{v}_\gamma}_{\not{k}_{\beta\gamma}}\big(1+\gamma^5\big)_{\gamma\delta}$$
$$= \frac{G_F^2 f_\pi^2 m^2}{2}\mathrm{Tr}\Big[(\not{p}+m)\big(1-\gamma^5\big)\not{k}\big(1+\gamma^5\big)\Big]. \quad (16.25)$$

Since the trace of an odd product of γ matrices vanishes (table 15.1 in chapter 15) the trace in the final expression can be reduced to

$$\mathrm{Tr}\left[(\not{p} + m)\left(1 - \gamma^5\right)\not{k}\left(1 + \gamma^5\right)\right] = \mathrm{Tr}\left[\not{p}\left(1 - \gamma^5\right)\not{k}\left(1 + \gamma^5\right)\right] = \mathrm{Tr}\Big[\not{p}\underbrace{\left(1 - \gamma^5\right)^2}_{1-2\gamma^5+1}\not{k}\Big]$$

$$= 2\,\mathrm{Tr}\left[\not{p}\left(1 - \gamma^5\right)\not{k}\right]. \tag{16.26}$$

Furthermore, since

$$\begin{aligned} \mathrm{Tr}\left[\not{p}\gamma^5\not{k}\right] &= 0, \\ \mathrm{Tr}\left[\not{p}\not{k}\right] &= 4pk \end{aligned} \tag{16.27}$$

(table 15.1), the final result is $8pk$, hence

$$\sum|\mathcal{M}|^2 = 4G_F^2 f_\pi^2\, m^2 pk = 4G_F^2 f_\pi^2\, m^2\left(E\omega - \vec{p}\cdot\vec{k}\right). \tag{16.28}$$

When integrating the partial width (16.24) over $\vec{p}$ one obtains with momentum conservation $\vec{p} = -\vec{k}$, and assuming a massless neutrino so that $|\vec{k}| = \omega$:

$$\sum|\mathcal{M}|^2 = 4G_F^2 f_\pi^2\, m^2\omega(E + \omega). \tag{16.29}$$

The integration of (16.24) over $\vec{k}$ gives with $\mathrm{d}^3\vec{k} = 4\pi k^2\mathrm{d}k = 4\pi\omega^2\mathrm{d}\omega$

$$\Gamma = \frac{4G_F^2 f_\pi^2\, m^2}{(2\pi)^2 2m_\pi}4\pi\int\frac{\mathrm{d}\omega\omega^2}{4E\omega}\omega(E + \omega)\delta\underbrace{\left(E + \omega - m_\pi\right)}_{g(\omega)}. \tag{16.30}$$

From the property of the δ-function

$$\int f(\omega)\delta(g(\omega))\mathrm{d}\omega = f(\omega_0)\frac{1}{|g'(\omega_0)|} \quad \text{with} \quad g(\omega_0) = 0, \tag{16.31}$$

one obtains with $E = \sqrt{\omega^2 + m^2}$ the zeroes of the function $g(\omega) \equiv E + \omega - m_\pi$:

$$E_0 + \omega_0 - m_\pi = \sqrt{\omega_0^2 + m^2} + \omega_0 - m_\pi = 0 \Rightarrow \omega_0 = \frac{m_\pi^2 - m^2}{2m_\pi}. \tag{16.32}$$

Hence the integral in (16.30) becomes

$$\begin{aligned} \int\frac{\mathrm{d}\omega\omega^2}{4E}(E + \omega)\delta\left(E + \omega - m_\pi\right) &= \frac{\omega_0^2}{4E_0}\left(E_0 + \omega_0\right)\frac{1}{\frac{\omega_0}{E_0} + 1} = \frac{\omega_0^2}{4} \\ &= \frac{1}{4}\left(\frac{m_\pi^2 - m^2}{2m_\pi}\right)^2, \end{aligned} \tag{16.33}$$

leading to the final result for the partial width $\Gamma(\pi \to \mu\nu)$:

$$\Gamma = \frac{16\pi G_F^2 f_\pi^2 m^2}{(2\pi)^2 2m_\pi} \frac{\left(m_\pi^2 - m^2\right)^2}{16m_\pi^2} = \frac{G_F^2 f_\pi^2}{8\pi} \frac{m^2}{m_\pi^3}\left(m_\pi^2 - m^2\right)^2. \tag{16.34}$$

The pion decays with the branching ratio $f(\mu\nu_\mu) \simeq 1$ into $\mu\nu_\mu$, hence Γ is also the total width. By inserting the lifetime of the pion $\tau = 1/\Gamma = 26$ ns one obtains a numerical value for the pion decay constant

$$f_\pi \simeq m_\pi. \tag{16.35}$$

The same derivation is valid for $\pi^- \to e^-\bar{\nu}_e$ by simply replacing m by m_e. Comparing the decay branching ratios into $e\nu$ and $\mu\nu$ one expects that

$$\frac{f(e\nu)}{f(\mu\nu)} = \frac{\Gamma(e\nu)}{\Gamma(\mu\nu)} = \frac{m_e^2}{m_\mu^2}\left(\frac{m_\pi^2 - m_e^2}{m_\pi^2 - m_\mu^2}\right)^2 = 1.284 \times 10^{-4}. \tag{16.36}$$

The suppression of the electron mode stems from the fact that the probability for the emission of the ultrarelativistic electron with the 'wrong' (positive) helicity is very small compared to the muon mode for which the non-relativistic muon is emitted with $\beta = 0.27$ (section 8.3).

In the experiment by Britton *et al* [7] slow positive pions (83 MeV c^{-1}) are stopped in a scintillation counter and the energy of the positrons is measured using a NaI crystal. As the positron is monoenergetic one expects an accumulation of events around the energy $m_\pi/2 \sim 70$ MeV. However, the pion decays mostly into a positive muon which in turn decays into $e^+\nu_e\bar{\nu}_\mu$. The Michel spectrum of the positron, which reaches a maximum at ~53 MeV (see figure 13.13), introduces a large background, due to the finite energy resolution of the NaI counter. Because of the much longer lifetime of the muon ($\tau_\mu = 2.2$ μs compared to $\tau_\pi = 26$ ns) the Michel background can be reduced by a timing cut: figure 16.4(*a*) shows the energy spectrum of the positrons recorded within 30 ns of the pion stop signal. The additional 4 MeV energy deposit by stopping muons in the target scintillator can be used to further suppress the background (figure 16.4(*b*)).

The measured value of $(1.230 \pm 0.004) \times 10^{-4}$ [7] for the ratio (16.36) is in agreement with the expected one including radiative corrections: $(1.234 \pm 0.001) \times 10^{-4}$.

16.2 Weak isospin

Let us group the Dirac spinors into pairs and define for the leptons the flavour doublets

$$u \equiv \begin{pmatrix} \nu_e \\ e^- \end{pmatrix}, \begin{pmatrix} \nu_\mu \\ \mu^- \end{pmatrix}, \begin{pmatrix} \nu_\tau \\ \tau^- \end{pmatrix}. \tag{16.37}$$

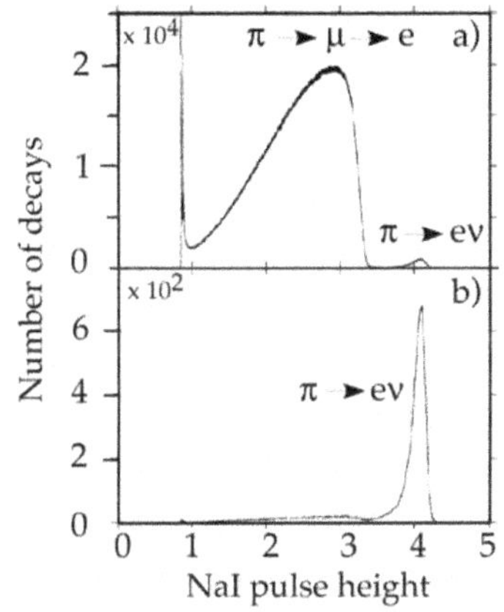

Figure 16.4. (*a*) Energy distribution of positrons from the decay of stopping positive pions. The positron is recorded within 30 ns of the pion stop signal. The fast increasing low energy enhancement stems from γ background. (*b*) The background from the Michel spectrum can be further suppressed by using the additional energy deposit from muons stopping in the target. After [7].

We have labelled the Dirac spinors with the corresponding lepton names. Formally, the weak charged current with W^- emission (e.g. $\mu^- \rightarrow \nu_\mu$) can be written as

$$j_C^\mu = \bar{u}\gamma^\mu\left(1-\gamma^5\right)\overbrace{\begin{bmatrix}0 & 1\\ 0 & 0\end{bmatrix}}^{\tau_+}u = \left(\bar{\nu}, \overline{\mu^-}\right)\gamma^\mu\left(1-\gamma^5\right)\overbrace{\begin{bmatrix}0 & 1\\ 0 & 0\end{bmatrix}\begin{pmatrix}\nu_\mu\\ \mu^-\end{pmatrix}}^{\begin{pmatrix}\mu^-\\ 0\end{pmatrix}}, \tag{16.38}$$

and the weak charged current with W^+ emission (e.g. $\nu_\mu \rightarrow \mu^-$) as

$$j_C^\mu = \bar{u}\gamma^\mu\left(1-\gamma^5\right)\overbrace{\begin{bmatrix}0 & 0\\ 1 & 0\end{bmatrix}}^{\tau_-}u = \left(\bar{\nu}, \overline{\mu^-}\right)\gamma^\mu\left(1-\gamma^5\right)\overbrace{\begin{bmatrix}0 & 0\\ 1 & 0\end{bmatrix}\begin{pmatrix}\nu_\mu\\ \mu^-\end{pmatrix}}^{\begin{pmatrix}0\\ \nu_\mu\end{pmatrix}}. \tag{16.39}$$

The 2×2 matrices are reminiscent of the ladder operators constructed from the Pauli matrices τ_1 and τ_2:

$$\tau^\pm \equiv \frac{1}{2}\left(\tau_1 \pm \mathrm{i}\tau_2\right). \tag{16.40}$$

The τ_+ (τ_-) operator transforms the lower (upper) component of the doublet (16.37) into its upper (lower) component.

We have seen that the *strong interaction* does not distinguish between upper and lower components of an isospinor (charge independence), and hence that $\vec{\tau}$ commutes with H(strong). Accordingly, the masses are degenerate within isospin multiplets. For *weak* charged currents the incoming and outgoing leptons belong to the same doublet (16.37). By analogy with strong interactions, it is natural to hypothesize the existence of an analogous symmetry in weak interactions: *weak isospin*. However, given the large difference between charged lepton and neutrino masses, this symmetry is badly broken. The doublets (16.37) are interpreted as $i = \frac{1}{2}$ isospinors of the weak isospin

$$\vec{I} = \frac{1}{2}\vec{\tau}. \tag{16.41}$$

The eigenvalues of the operator $I_3 = \frac{1}{2}\tau_3$ are

$$\begin{aligned} i_3 &= +\frac{1}{2} \quad \text{for } \nu_e,\ \nu_\mu,\ \nu_\tau, \\ i_3 &= -\frac{1}{2} \quad \text{for } e^-,\ \mu^-,\ \tau^-. \end{aligned} \tag{16.42}$$

For quarks the weak charged current may be written similarly. Consider for the time being only the three lightest quarks. The current $u \to d$ or $u \to s$ with W^+ emission reads

$$j_C^\mu = \cos\theta_C \cdot \bar{d}\gamma^\mu\left(1 - \gamma^5\right)u + \sin\theta_C \cdot \bar{s}\gamma^\mu\left(1 - \gamma^5\right)u, \tag{16.43}$$

where θ_C is the *Cabibbo* angle which takes the measured value $\sin\theta_C \simeq 0.23$.

For example, for pion β-decay, only the first term enters the transition matrix ($\pi^+ \to \pi^0 e^+ \nu_e$ with branching ratio $\propto G_F^2 \cos^2\theta_C$), while only the second term enters for kaon β-decay ($K^+ \to \pi^0 e^+ \nu_e$ with branching ratio $\propto G_F^2 \sin^2\theta_C$). Nonetheless, the branching ratio for the former is much smaller ($f = 10^{-8}$ for pion β-decay, compared to $f = 0.05$ for kaon β-decay) due to the strongly limited phase space. Let us define the doublet

$$u = \begin{pmatrix} u \\ d' \end{pmatrix} \quad \text{with} \quad \boxed{d' \equiv d \cdot \cos\theta_C + s \cdot \sin\theta_C}. \tag{16.44}$$

For ease of notation we have used the same symbol for the Dirac spinor of the u quark as for the isospin doublet. One can then formally write the quark current as

$$j_C^\mu = \bar{u}\gamma^\mu\left(1 - \gamma^5\right)\overbrace{\begin{bmatrix} 0 & 0 \\ 1 & 0 \end{bmatrix}}^{\tau_-}u. \tag{16.45}$$

The operator τ_- appears here also, and therefore the doublet (16.44) is interpreted as a weak isospinor with quantum numbers

$$\begin{aligned} i_3 &= +\frac{1}{2} \quad \text{for the } u \text{ quark}, \\ i_3 &= -\frac{1}{2} \quad \text{for the } d \text{ and } s \text{ quarks}. \end{aligned} \tag{16.46}$$

Note that the eigenstate of I_3 is the rotated state d' in (16.44) and not the flavour state d or s.

16.2.1 Weak neutral currents

Since the operators τ_1 and τ_2 appear in (16.38), (16.39) and (16.45), one would also expect currents involving τ_3, of the form

$$j_N^\mu = \bar{u}\gamma^\mu \underbrace{\tau_3}_{\begin{bmatrix}1 & 0\\ 0 & -1\end{bmatrix}} u, \tag{16.47}$$

thus consolidating our hypothesis of weak isospin symmetry. This equation describes a *weak neutral* current in which the quark or the lepton flavour do not change since τ_3 is diagonal. Hence a *neutral* heavy boson (Z^0) must be emitted. We have omitted here the $V - A$ factor since it is not yet clear whether neutral weak currents are also left-handed. Indeed, we shall see that both left-handed and right-handed weak neutral currents exist. Their relative contributions are predicted by the Glashow–Weinberg–Salam model (section 16.7).

The neutral currents were discovered in 1973 at CERN. The following deep inelastic reactions on nuclei were investigated (figure 16.5):

$$\nu_\mu N \to \mu \;\; + \text{hadrons}\left(\nu_\mu d \to \mu^- u, \;\; W^+ \text{exchange}\right), \tag{16.48}$$

$$\bar{\nu}_\mu N \to \mu^+ + \text{hadrons}\left(\bar{\nu}_\mu u \to \mu^+ d, \;\; W^- \text{ exchange}\right), \tag{16.49}$$

$$\nu_\mu N \to \nu_\mu + \text{hadrons}\left(\nu_\mu [u \text{ or } d] \to \nu_\mu [u \text{ or } d], \;\; Z^0 \text{exchange}\right), \tag{16.50}$$

$$\bar{\nu}_\mu N \to \bar{\nu}_\mu + \text{hadrons}\left(\bar{\nu}_\mu [u \text{ or } d] \to \bar{\nu}_\mu [u \text{ or } d], \;\; Z^0 \text{ exchange}\right). \tag{16.51}$$

The interaction between charged currents (W exchange) emits muons which are absent with neutral currents (Z^0 exchange).

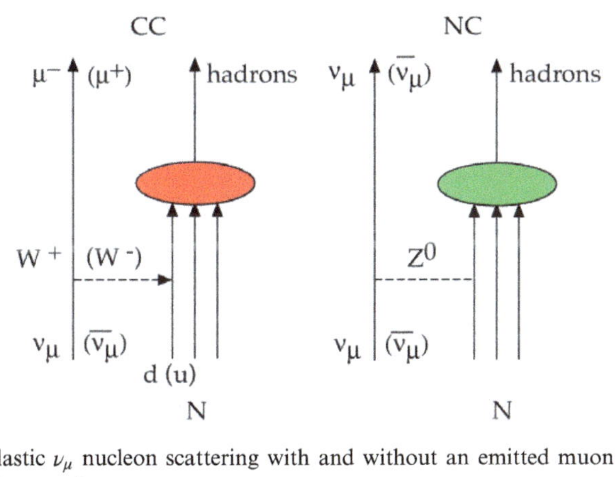

Figure 16.5. Deep inelastic ν_μ nucleon scattering with and without an emitted muon (CC refers to charged current, NC to neutral current).

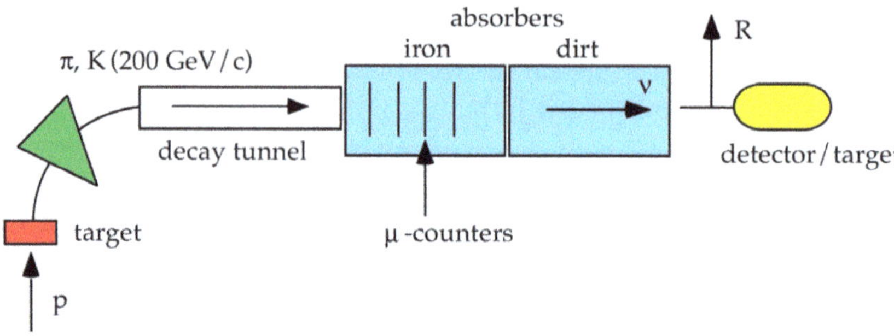

Figure 16.6. Principle of a broadband ν_μ beam.

Neutrinos are produced by π^+ and K^+ decay into $\mu^+\nu_\mu$ (and antineutrinos by π^- and K^- decay into $\mu^-\bar{\nu}_\mu$). The total cross section for neutrino-induced reactions grows linearly with s. Thus for neutrino interaction with protons,

$$\sigma \propto s \simeq 2m_p E_\nu, \tag{16.52}$$

for sufficiently high laboratory energies $E_\nu \gg m_p$, see figure 17.27(b) below and (8.64), hence neutrinos of the highest possible energies are advantageous. Furthermore, neutrinos are emitted tightly forward at high energy, so that neutrino beams can be produced.

Figure 16.6 shows the layout of a broadband neutrino beam produced by 200 GeV c^{-1} pions and kaons. Neutrinos or antineutrinos are selected by flipping the polarity of the magnet, which selects the charge of the meson. A parallel beam of pions and kaons is produced with a magnetic horn (figure 14.23). The pions and kaons decay while flying in a 300 m long evacuated pipe to avoid absorption. The surviving ones and the decay muons are stopped in a 200 m long iron absorber, followed by a 150 m thick dirt plug. The neutrino flux to the detector is monitored by measuring the muon intensity, for example with scintillation counters.

Figure 16.7(a) shows the correlation between neutrino momentum and distance R to the beam axis at the detector location. The contributions from $\pi \to \mu\nu_\mu$ and $K \to \mu\nu_\mu$ are clearly separated. Figure 16.7(b) and 16.7(c) show the corresponding $\bar{\nu}_\mu$ and ν_μ intensities. Note the characteristic box-like spectra due to the isotropic angular distribution in the rest frame of the decaying meson (section 8.1). The smearing is due to the solid angle acceptance which depends on the decay point in the tunnel.

The neutral currents were discovered in the Gargamelle bubble chamber [9] (figure 16.8 shows a photograph of the bubble chamber). The bubble chamber medium was freon ($BrCF_3$), chosen because of its high density (1.5 g cm^{-3}, compared to 0.07 g cm^{-3} for liquid hydrogen). The high density is also advantageous for identifying muons and distinguishing them from hadrons: hadrons produced in the liquid perform observable elastic and inelastic scatterings, while electromagnetically interacting muons from reactions (16.48) and (16.49) traverse and exit the chamber virtually undisturbed. Figure 16.9 (left) shows a typical neutral current event with an incoming (invisible) neutrino, leaving a hadron 'star' from reaction (16.50) and no muon in the final state.

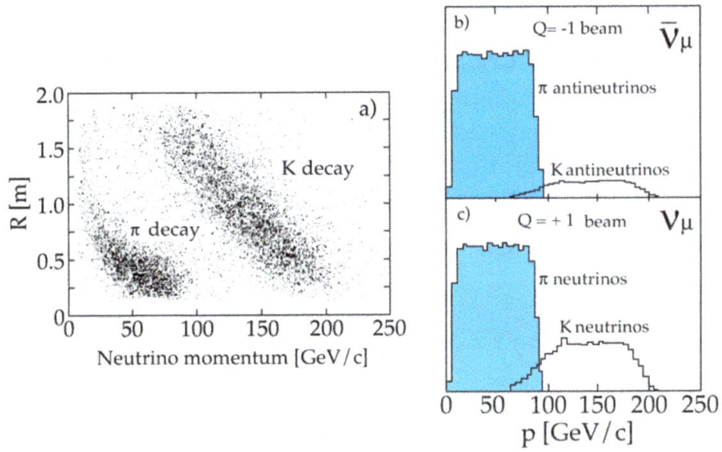

Figure 16.7. (*a*) Distance R versus neutrino momentum at the detector for 200 GeV c^{-1} mesons. (*b*) and (*c*) Intensity distributions at the detector. After [8].

Figure 16.8. Photograph of the Gargamelle bubble chamber. The bubble chamber was about 5 m long and contained 12 m^3 of liquid freon ($BrCF_3$). Image credit: CERN.

Since the neutrino–nucleus cross section is small one does not expect any appreciable decrease of the neutrino flux in the bubble chamber. Indeed, figure 16.9(*a*) and (*b*) show a constant distribution of the interaction point along the beam axis for $\bar{\nu}_\mu$ without (NC) and with (CC) an emitted muon, as well as their ratio $\overline{R}$. The distributions for ν_μ-induced events look similar [9].

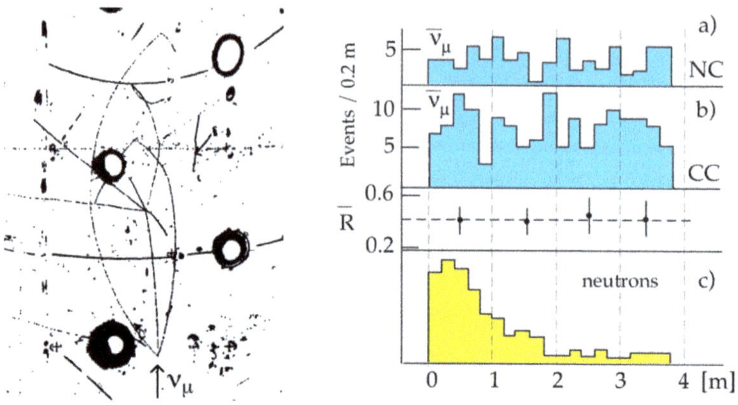

Figure 16.9. Left: a ν_μ-induced event in the Gargamelle bubble chamber without an emerging muon. After [10]. Right: longitudinal distribution of the interaction point for neutral (NC) and for charged current interactions (CC) with $\bar{\nu}_\mu$. $\overline{R}$ refers to the ratio of cross sections NC/CC for $\bar{\nu}_\mu$. The bottom figure shows the distribution for background neutrons. After [9].

A dangerous background stems from neutrons generated in the shielding in front of the detector. These neutrons may undergo nuclear reactions with target nucleons, but without muons, mimicking neutral current interactions. However, neutrons are strongly absorbed in the bubble chamber (strong interaction), so that the event distribution should decrease along the beam axis much more rapidly than for neutrino reactions. Neutron interactions can be observed in the bubble chamber by selecting events with outgoing protons. Indeed, such events tend to accumulate at the entrance of the bubble chamber (figure 16.9(*c*)). Hence the neutron background cannot account for the flat distribution in figure 16.9(*a*) and 16.9(*b*). The ratio of neutral to charged cross sections for neutrinos

$$R = \frac{\sigma\left(\nu_\mu N \to \nu_\mu \text{ hadrons}\right)}{\sigma\left(\nu_\mu N \to \mu^- \text{ hadrons}\right)} \sim 0.3, \tag{16.53}$$

and that for antineutrinos

$$\overline{R} = \frac{\sigma\left(\bar{\nu}_\mu N \to \bar{\nu}_\mu \text{ hadrons}\right)}{\sigma\left(\bar{\nu}_\mu N \to \mu^+ \text{ hadrons}\right)} \sim 0.4, \tag{16.54}$$

can be used to determine the Weinberg angle, as we shall see in section 16.7.

The observation of the reaction

$$\bar{\nu}_\mu e^- \to \bar{\nu}_\mu e^-, \tag{16.55}$$

in which only the recoil electron is observed, is the smoking gun evidence for neutral currents, and hence for the existence of a heavy neutral Z^0 boson (figure 16.10(*a*)). This process was also observed (figure 16.11), but the cross section $\sigma \propto 2m_e E_\nu$ is three orders of magnitude smaller than for neutrino–nucleon scattering.

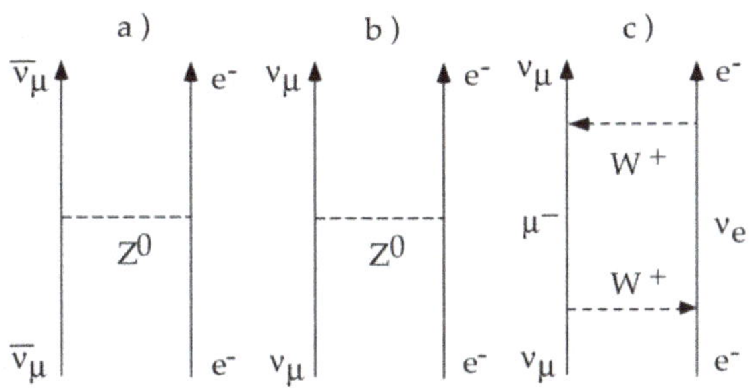

Figure 16.10. (*a*) $\bar{\nu}_\mu e^- \to \bar{\nu}_\mu e^-$, (*b*) $\nu_\mu e^- \to \nu_\mu e^-$, and (*c*) $\nu_\mu e^- \to \nu_\mu e^-$ in higher order with weak charged currents.

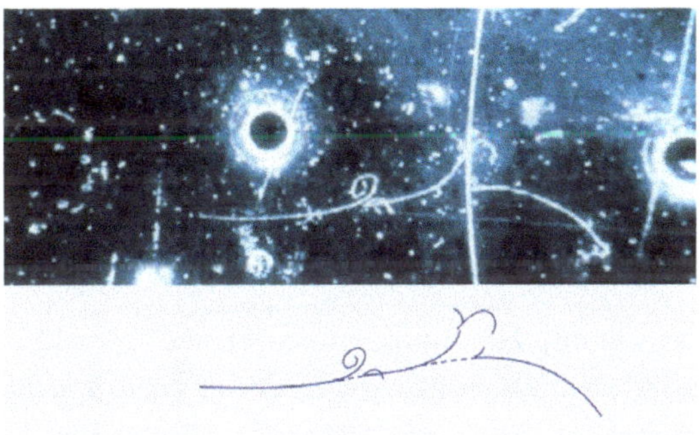

Figure 16.11. A purely leptonic neutral current event $\bar{\nu}_\mu e^- \to \bar{\nu}_\mu e^-$ observed in the Gargamelle bubble chamber. The $\bar{\nu}_\mu$ flying in from the left knocks an electron which induces an electromagnetic shower. Image credit: CERN.

Note that the analogous reaction with neutrinos

$$\nu_\mu e^- \to \nu_\mu e^-, \tag{16.56}$$

also proceeds through neutral current interactions, albeit only in first order (figure 16.10(*b*)). The process is possible in higher order without Z^0 exchange (figure 16.10(*c*)).

16.3 GIM mechanism

For quarks let us rewrite the neutral current (16.47) in terms of quark flavours:

$$\begin{aligned} j_N^\mu &= \left(\bar{u},\, \bar{d}\cos\theta_C + \bar{s}\sin\theta_C\right)\gamma^\mu \underbrace{\tau_3}_{\begin{bmatrix}1 & 0\\ 0 & -1\end{bmatrix}} \begin{pmatrix} u \\ d\cos\theta_C + s\sin\theta_C \end{pmatrix} \\ &= \bar{u}\gamma^\mu u - \bar{d}\gamma^\mu d\cdot\cos^2\theta_C - \bar{s}\gamma^\mu s\cdot\sin^2\theta_C \\ &\quad - \left(\bar{d}\gamma^\mu s + \bar{s}\gamma^\mu d\right)\cdot\sin\theta_C\cos\theta_C. \end{aligned} \tag{16.57}$$

The first three terms describe the flavour conserving currents with Z^0 emission while the final term describes the flavour changing transitions $s \to dZ^0$ and $d \to sZ^0$, with

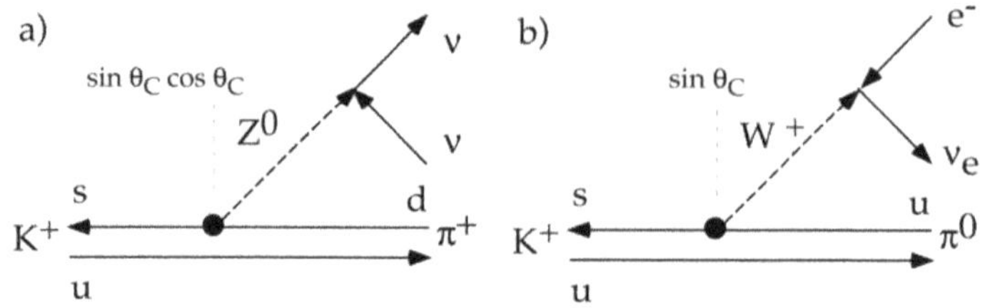

Figure 16.12. Comparison between $K^+ \to \pi^+\nu\bar{\nu}$ (*a*) and $K^+ \to \pi^0 e^+\nu_e$ (*b*). The GIM mechanism forbids flavour changing neutral currents (*a*) with Z^0 emission.

couplings proportional to $\sin\theta_C \cos\theta_C$. Consider for example the following two decays with branching ratios f (figure 16.12):

$$K^+ \overset{\bar{s}\to\bar{d}}{\overbrace{\longrightarrow}} \pi^+\nu\bar{\nu}, \qquad f \propto \left(\sin\theta_C \cos\theta_C\right)^2 \simeq \left(\sin\theta_C\right)^2 \tag{16.58}$$

and

$$K^+ \underset{\bar{s}\to\bar{u}}{\underbrace{\longrightarrow}} \pi^0 e^+\nu_e, \qquad f \propto \left(\sin\theta_C\right)^2. \tag{16.59}$$

The branching ratios should be comparable. However, one finds experimentally that the former decay is strongly suppressed, while the branching ratio for the latter is $f = 5\%$.

Figure 16.13 (left) shows as sketch of the equipment used to search for $K^+ \to \pi^+\nu\bar{\nu}$ at the Berkeley Bevatron [11]. Positive kaons selected by a water Čerenkov counter (1) are slowed down by the absorber (2) before being stopped in the scintillation counter (3). The emitted π^+ is detected by a set of spark chambers (4) interleaved with brass plates and scintillation counters (5), with which its range (and hence its energy) can be determined. A set of veto counters (V) is used to remove decay events with additional charged particles, while the lead glass detector (6) suppresses decays involving photons such as $K^+ \to \pi^+\gamma\gamma$, including one of the main decay modes of the kaon, $K^+ \to \pi^+\pi^0(\to \gamma\gamma)$ with $f = 21\%$.

The favoured decay mode of the kaon is $K^+ \to \mu^+\nu_\mu$ ($f = 64\%$). Figure 16.13 (right) shows the distribution of events as a function of range in copper. The 236 MeV c^{-1} muons (peak at 80 g cm^{-2}) can be suppressed by the water Čerenkov counter (7) and by adjusting the thickness of the brass, so that the decay pions stop while the faster muons traverse the final counter. Figure 16.13 also shows the distribution with and without coincidence from the γ counter (6). The experiment relies on the fact that the decay (16.58) leads to pions with momenta larger than pions from $K^+ \to \pi^+\pi^0$ (205 MeV c^{-1}, corresponding to a range of 45 g cm^{-2} in copper). There are no events above the $K^+ \to \pi^+\pi^0$ signal. The resulting upper limit for the decay (16.58) is $f = 1.2 \times 10^{-6}$. Hence, there are no strangeness changing neutral currents and, generalizing to all flavours, there are *no flavour changing* neutral currents.

In fact, the decay (16.58) violates the $\Delta S = \Delta Q$ rule, but occurs in higher order Feynman graphs involving charged currents (W exchanges) for which the standard

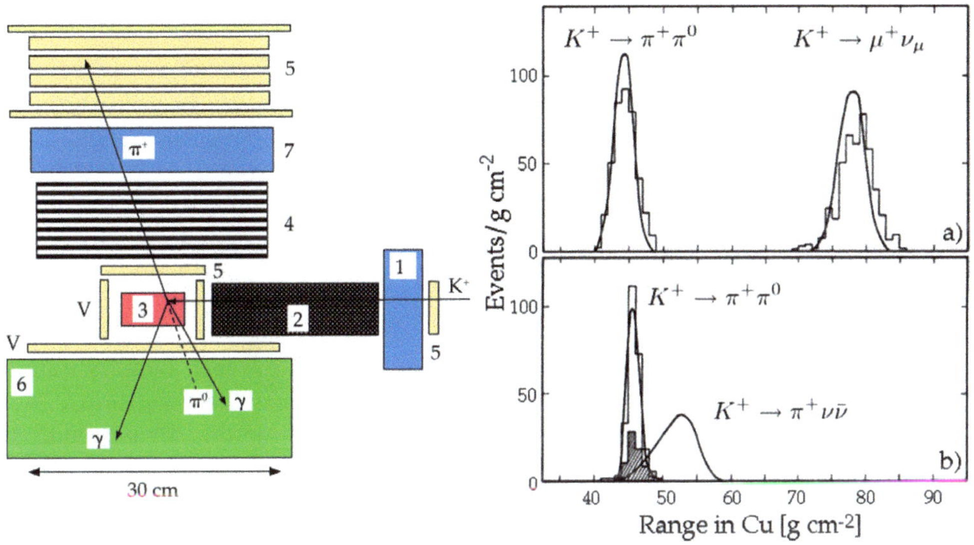

Figure 16.13. Left: the apparatus used to search for strangeness changing neutral currents. 1, 7: water Čerenkov; 2: absorber; 3: target; 4: spark chambers; 5: scintillators; 6: lead glass counter; V: veto counters. Right: range distribution of the π^+ and μ^+ in K^+ decay (*a*) and for suppressed muons without and with γ anticoincidence (hatched histogram) (*b*). The curve shows the signal expected from $K^+ \to \pi^+ \nu\bar{\nu}$ with an assumed branching ratio of 1.8×10^{-5}. No signal is observed. Adapted from [11].

model predicts a branching ratio $f \simeq 8 \times 10^{-11}$. The present experimental upper limit is $f = 1.7 \times 10^{-10}$.

The riddle was solved in 1970 by Sheldon Glashow, John Iliopoulos and Luciano Maiani (GIM), who introduced the second quark doublet

$$u = \begin{pmatrix} c \\ s' \end{pmatrix}, \quad \text{where} \boxed{s' \equiv -d \sin\theta_C + s \cos\theta_C}, \tag{16.60}$$

with a new hypothetical c quark. The neutral quark current then receives the additional term

$$\begin{aligned} j_N^{\prime\mu} &= \left(\bar{c}, -\bar{d} \sin\theta_C + \bar{s} \cos\theta_C\right)\gamma^\mu \begin{bmatrix} 1 & 0 \\ 0 & -1 \end{bmatrix} \begin{pmatrix} c \\ -d \sin\theta_C + s \cos\theta_C \end{pmatrix} \\ &= \bar{c}\gamma^\mu c - \bar{d}\gamma^\mu d \cdot \sin^2\theta_C - \bar{s}\gamma^\mu s \cdot \cos^2\theta_C \\ &\quad + \left(\bar{d}\gamma^\mu s + \bar{s}\gamma^\mu d\right) \cdot \sin\theta_C \cos\theta_C. \end{aligned} \tag{16.61}$$

The last strangeness changing terms in (16.57) and (16.61) now cancel: for figure 16.12(*a*) the term $s \to d$ in (16.57) is compensated by the corresponding one in (16.61). The total weak neutral current becomes:

$$\boxed{j_N^\mu = \bar{u}\gamma^\mu u + \bar{c}\gamma^\mu c - \bar{d}\gamma^\mu d - \bar{s}\gamma^\mu s}. \tag{16.62}$$

Note that the Cabibbo angle does not appear in weak neutral currents.

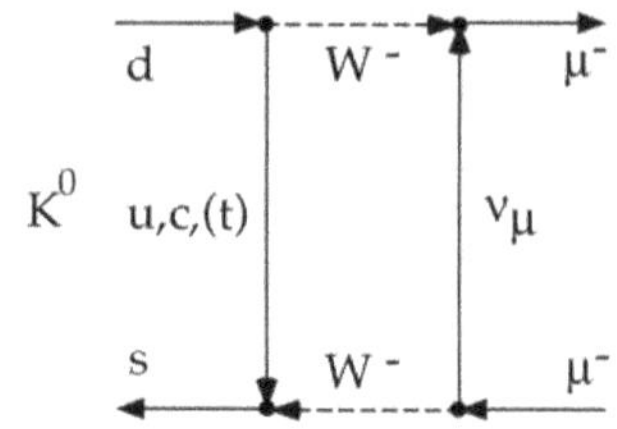

Figure 16.14. The decay $K^0 \to \mu^+\mu^-$ is forbidden through strangeness changing neutral currents ($d \to sZ^0$), but occurs through higher order diagrams such as the one shown here.

The hypothesis of the doublet (16.60) implies that the c quark couples mainly to the s quark (coupling $\propto \cos\theta_C$) via W^+ emission while the transition $c \to d$ (coupling $\propto \sin\theta_C$) is 'Cabibbo-suppressed'. Indeed, the D mesons decay mainly to kaons. For example,

$$\begin{aligned} D^+ &\overset{c\to s}{\longrightarrow} K^-\pi^+\pi^+, \qquad & f &= (9.1 \pm 0.2)\times 10^{-2}, \\ D^+ &\underset{c\to d}{\longrightarrow} \pi^-\pi^+\pi^+, \qquad & f &= (3.2 \pm 0.2)\times 10^{-3}, \end{aligned} \tag{16.63}$$

notwithstanding the larger phase space available to the latter decay.

The decay $K^0 \to \mu^+\mu^-$ would proceed in first order through strangeness changing neutral currents ($d\bar{s} \to Z^0 \to \mu^+\mu^-$) and is therefore forbidden. However, the measured branching ratio $f = 6.8 \times 10^{-9}$ shows that the decay occurs nonetheless, but is suppressed, proceeding in higher order through charged currents (figure 16.14). The two contributions with u and c exchanges interfere destructively, further reducing the branching ratio. The mass of the hypothetical c quark could be predicted in 1970 from the experimental branching ratio: $m_c \sim$ 1–3 GeV. The c quark was discovered four years later with the observation of the J/ψ (section 10.3.1).

A similar suppression is observed for the decays $B_s^0(\bar{b}s) \to \mu^+\mu^-$ and $B_d^0(\bar{b}d) \to \mu^+\mu^-$ which occur via higher order Feynman diagrams (figure 16.15, left). Figure 16.15 (right) shows the signals observed by the Compact Muon Solenoid (CMS) experiment at the Large Hadron Collider (LHC) (a description of CMS is given in section 17.3). A clear peak is observed for $B_s^0(\bar{b}s) \to \mu^+\mu^-$, corresponding to the branching ratio $f = 3 \times 10^{-9}$ [12]. We shall show below that transitions between the first and third quark families (e.g. $t \to d$) are suppressed, compared to transitions between the second and third families (e.g. $t \to s$). This explains why the branching ratio for $B_d^0 \to \mu^+\mu^-$ is expected to be even smaller. A suggestive signal from this decay is also shown in figure 16.15.

The d' and s' states which take part in the weak interaction are orthogonal linear combinations of the d and s quarks:

$$\begin{pmatrix} d' \\ s' \end{pmatrix} = \begin{pmatrix} \cos\theta_C & \sin\theta_C \\ -\sin\theta_C & \cos\theta_C \end{pmatrix} \begin{pmatrix} d \\ s \end{pmatrix} = \begin{pmatrix} V_{ud} & V_{us} \\ V_{cd} & V_{cs} \end{pmatrix} \begin{pmatrix} d \\ s \end{pmatrix}. \tag{16.64}$$

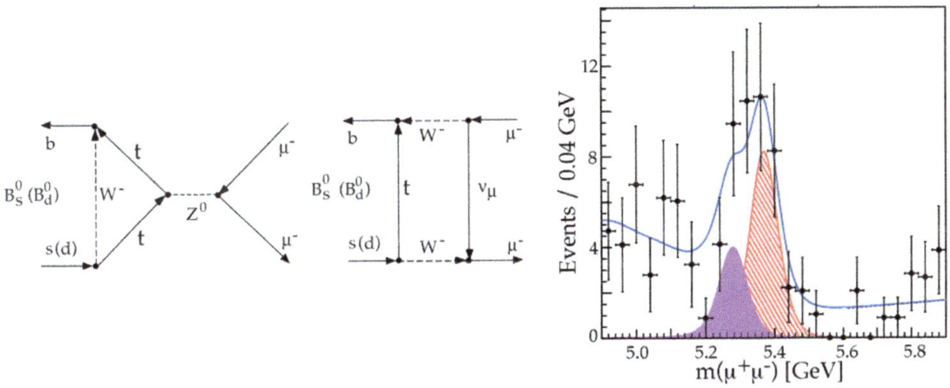

Figure 16.15. Left: $B_s^0 \to \mu^+\mu^-$ and $B_d^0 \to \mu^+\mu^-$ decays through higher order diagrams. Right: evidence for $B_s^0 \to \mu^+\mu^-$ (red hatched area). The pink area shows the possible contribution from $B_d^0 \to \mu^+\mu^-$. Adapted from [12].

With six quarks the Cabibbo mixing matrix is generalized to the 3×3 Cabibbo–Kobayashi–Maskawa (CKM) unitary matrix V_{CKM}:

$$\begin{pmatrix} d' \\ s' \\ b' \end{pmatrix} = V_{\text{CKM}} \begin{pmatrix} d \\ s \\ b \end{pmatrix}. \tag{16.65}$$

The matrix V_{CKM} describes the couplings between the (weak isospin) $i_3 = -\frac{1}{2}$ and $+\frac{1}{2}$ quarks ($d \to u$, $d \to c$, $d \to t$, $s \to u$, etc). As in the neutrino sector V_{CKM} contains three angles, θ_{12}, θ_{23}, θ_{13} and one phase δ, see (14.61). The recommended parametrization is[1]

$$V_{\text{CKM}} = \begin{pmatrix} V_{ud} & V_{us} & V_{ub} \\ V_{cd} & V_{cs} & V_{cb} \\ V_{td} & V_{ts} & V_{tb} \end{pmatrix} \tag{16.66}$$

$$= \begin{pmatrix} c_{12}c_{13} & s_{12}c_{13} & s_{13}\mathrm{e}^{-\mathrm{i}\delta} \\ -s_{12}c_{23} - c_{12}s_{23}s_{13}\mathrm{e}^{\mathrm{i}\delta} & c_{12}c_{23} - s_{12}s_{23}s_{13}\mathrm{e}^{\mathrm{i}\delta} & s_{23}c_{13} \\ s_{12}s_{23} - c_{12}c_{23}s_{13}\mathrm{e}^{\mathrm{i}\delta} & -c_{12}s_{23} - s_{12}c_{23}s_{13}\mathrm{e}^{\mathrm{i}\delta} & c_{23}c_{13} \end{pmatrix} \tag{16.67}$$

with $c_{ij} \equiv \cos\theta_{ij}$ and $s_{ij} \equiv \sin\theta_{ij}$. Unless $\delta = 0$ or π, the matrix is complex, leading to CP violation (appendix D). Chapters 19 and 20 describe CP violation in the $K^0 - \overline{K}^0$ system and in hadronic B decays. From the measurements in B decays one finds $\delta \simeq 70°$.

---

[1] V_{ud} ($\simeq\cos\theta_C$) is determined by comparing the muon lifetime to that for nuclear β-decay (section 13.2.4). V_{cs}/V_{cd} can be obtained from the branching ratios for $D^0 \to K^-e^+\nu_e$ and $D^0 \to \pi^-e^+\nu_e$, V_{cb}/V_{ub} from the branching ratios for $B^+ \to \overline{D}^0e^+\nu_e$ and $B^+ \to \pi^0e^+\nu_e$.

In contrast to the neutrino sector, the experiments show that the non-diagonal elements are small:

$$1 \gg \overbrace{\sin\theta_{12}}^{\sin\theta_C \simeq 0.23} \gg \overbrace{\sin\theta_{23}}^{\simeq 0.04} \gg \overbrace{\sin\theta_{13}}^{\simeq 0.004}. \tag{16.68}$$

The more convenient (Wolfenstein) parametrization is

$$V_{\mathrm{CKM}} = \begin{pmatrix} 1-\lambda^2/2 & \lambda & A\lambda^3(\rho - \mathrm{i}\eta) \\ -\lambda & 1-\lambda^2/2 & A\lambda^2 \\ A\lambda^3(1-\rho-\mathrm{i}\eta) & -A\lambda^2 & 1 \end{pmatrix} + \mathcal{O}(\lambda^4), \tag{16.69}$$

with $\lambda \equiv \sin\theta_C$, and where A, ρ and η are real. We shall return in more detail to the CKM matrix in chapter 20.

16.4 Gauge invariance in QED

Figure 16.16 shows the diffraction pattern created on a screen by an electron with wave function $\psi(\vec{r}, t)$ flying through two slits A and B. A different interference pattern would be obtained when applying the *local* phase transformation

$$\boxed{\psi \to \psi'(\vec{r}, t) = \mathrm{e}^{\mathrm{i}q\chi(\vec{r},t)}\psi(\vec{r}, t)} \tag{16.70}$$

with $q = -e$, and where $\chi(\vec{r}, t)$ is an arbitrary function. This can be seen easily by choosing two different phases at slits A and B, $\chi(\vec{A}, t) \neq \chi(\vec{B}, t)$. The phases are equivalent to different path lengths between A and B and the screen centre O, thus shifting the interference pattern to the left or to the right.

However, *gauge invariance* means that the interference pattern does not change when applied *simultaneously* with the phase transformation (16.70), the gauge transformation of the field (11.59, 11.60), that is, by replacing A^μ by A'^μ in the equation of motion of the electron:

$$A^\mu \to A'^\mu = A^\mu - \partial^\mu\chi, \tag{16.71}$$

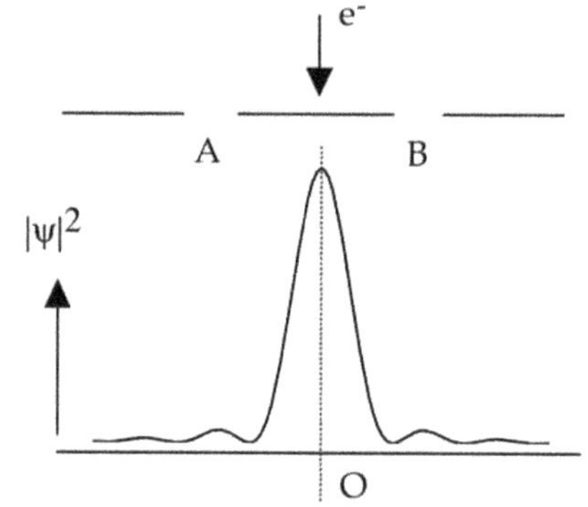

Figure 16.16. Diffraction pattern produced by an electron in a double slit experiment.

with $\partial^\mu = (\frac{\partial}{\partial t}, -\text{grad})$. In other words, the wave equation of the electron does not change. Let us prove this statement for the Dirac equation. First we formulate the equation of motion of a relativistic particle with mass m, charge q and spin $\frac{1}{2}$ in the field A^μ, by substituting p^μ by $p^\mu - qA^\mu$ into the free Dirac equation:

$$\left(\gamma_\mu p^\mu - m\right)\psi = 0 \Rightarrow \left(\gamma_\mu p^\mu - q\gamma_\mu A^\mu - m\right)\psi = 0. \tag{16.72}$$

Using the correspondence principle $p^\mu \to \mathrm{i}\partial^\mu = (\mathrm{i}\frac{\partial}{\partial t}, -\mathrm{i}\ \text{grad})$ one finds that

$$\left(\mathrm{i}\gamma_\mu \partial^\mu - q\gamma_\mu A^\mu - m\right)\psi = 0. \tag{16.73}$$

With the *covariant derivative*, defined as

$$\boxed{D^\mu \equiv \partial^\mu + \mathrm{i}qA^\mu}, \tag{16.74}$$

one obtains

$$\left(\mathrm{i}\gamma_\mu D^\mu - m\right)\psi = 0. \tag{16.75}$$

This equation has the same form as the free Dirac equation, except that ψ is no longer a free Dirac spinor. We are now in a position to prove the assertion of gauge invariance. We write the left-hand side of the Dirac equation (16.75) for the phase-transformed wave function (16.70) in the new fields (16.71):

$$\begin{aligned}
\left(\mathrm{i}\gamma_\mu D'^\mu - m\right)\psi' &= \left[\mathrm{i}\gamma_\mu\left(\partial^\mu + \mathrm{i}qA'^\mu\right) - m\right]\mathrm{e}^{\mathrm{i}q\chi}\psi \\
&= \mathrm{e}^{\mathrm{i}q\chi}\left(\mathrm{i}\gamma_\mu \partial^\mu \psi\right) + \left(\mathrm{i}\gamma_\mu \partial^\mu \mathrm{e}^{\mathrm{i}q\chi}\right)\psi + \Big[-\gamma_\mu q\overbrace{\left(A^\mu - \partial^\mu\chi\right)}^{A'^\mu} - m\Big]\mathrm{e}^{\mathrm{i}q\chi}\psi \\
&= \mathrm{e}^{\mathrm{i}q\chi}\left(\mathrm{i}\gamma_\mu \partial^\mu \psi\right) - \gamma_\mu q\left(\partial^\mu \chi\right)\mathrm{e}^{\mathrm{i}q\chi}\psi \\
&\quad + \left(-\gamma_\mu qA^\mu + \gamma_\mu q\partial^\mu\chi\right)\mathrm{e}^{\mathrm{i}q\chi}\psi - m\mathrm{e}^{\mathrm{i}q\chi}\psi \\
&= \mathrm{e}^{\mathrm{i}q\chi}\Big(\mathrm{i}\gamma_\mu \underbrace{\left[\partial^\mu + \mathrm{i}qA^\mu\right]}_{D^\mu} - m\Big)\psi.
\end{aligned} \tag{16.76}$$

The equivalence

$$\left(\mathrm{i}\gamma_\mu D^\mu - m\right)\psi = 0 \Longleftrightarrow \left(\mathrm{i}\gamma_\mu D'^\mu - m\right)\psi' = 0 \tag{16.77}$$

proves that the phase-transformed wave function also fulfils the Dirac equation in the new fields. The wave function ψ' qualifies equally well as ψ to describe the motion of the electron, the phase transformation (16.70) being *compensated* by the gauge transformation (16.71). However, this is only possible in the presence of the compensating field A^μ, that of the photon.

The Maxwell equations of the field A^μ read

$$\partial_\mu F^{\mu\nu} = j^\nu \tag{16.78}$$

where $F^{\mu\nu}$ is the field tensor

$$F^{\mu\nu} = \partial^\mu A^\nu - \partial^\nu A^\mu = \begin{pmatrix} 0 & -E_x & -E_y & -E_z \\ E_x & 0 & -B_z & B_y \\ E_y & B_z & 0 & -B_x \\ E_z & -B_y & B_x & 0 \end{pmatrix} \tag{16.79}$$

and

$$j^\nu = \left(\rho, \vec{j}\right), \qquad \vec{B} = \mathrm{rot}\,\vec{A}, \qquad \vec{E} = -\mathrm{grad}\,\phi - \frac{\partial \vec{A}}{\partial t}. \tag{16.80}$$

Hence (16.78) becomes

$$\partial_\mu \partial^\mu A^\nu - \partial_\mu \partial^\nu A^\mu = j^\nu \tag{16.81}$$

or with the d'Alembert operator $\Box \equiv \partial_\mu \partial^\mu$:

$$\partial_\mu F^{\mu\nu} = \Box A^\nu - \partial^\nu\left(\partial_\mu A^\mu\right) = j^\nu. \tag{16.82}$$

These equations are invariant under the gauge transformation (16.71). Indeed,

$$\begin{aligned} \Box A'^\nu - \partial^\nu\left(\partial_\mu A'^\mu\right) &= \Box\left(A^\nu - \partial^\nu \chi\right) - \partial^\nu\left(\partial_\mu\left[A^\mu - \partial^\mu \chi\right]\right) \\ &= \Box A^\nu - \partial^\nu\left(\partial_\mu A^\mu\right) - \Box \partial^\nu \chi + \partial^\nu(\underbrace{\partial_\mu \partial^\mu}_{\Box} \chi) \\ &= \Box A^\nu - \partial^\nu\left(\partial_\mu A^\mu\right) = j^\nu. \end{aligned} \tag{16.83}$$

Hence the transformed field A'^μ also qualifies as electromagnetic field, and the fields $\vec{E}$ and $\vec{B}$ defined by (16.80) are invariant. Gauge invariance is then abandoned by choosing a particular gauge. For example, the Lorentz gauge $\partial_\mu A^\mu = 0$ (or $\frac{\partial \phi}{\partial t} + \mathrm{div}\,\vec{A} = 0$) leads to the Maxwell equations

$$\Box A^\nu = j^\nu. \tag{16.84}$$

Summarizing:

In a *gauge theory* (such as QED) a local phase transformation can be compensated by a gauge transformation, leaving the equation of motion of a particle and the compensating fields invariant.

16.5 Spontaneous symmetry breaking

Maxwell's equations (16.82) receive an additional term with a massive photon of mass M:

$$\left(\Box + M^2\right)A^\nu - \partial^\nu\left(\partial_\mu A^\mu\right) = j^\nu . \tag{16.85}$$

This is the field equation of a heavy boson, as can be seen by assuming a free field ($j^\nu = 0$) and applying ∂_ν from the left:

$$\left(\Box + M^2\right)\partial_\nu A^\nu - \underbrace{\partial_\nu\partial^\nu}_{\Box}\partial_\mu A^\mu = 0, \tag{16.86}$$

hence

$$M^2\partial_\mu A^\mu = 0. \tag{16.87}$$

We are now forced to work in the Lorentz gauge $\partial_\mu A^\mu = 0$ in which (16.86) reduces to the Klein–Gordon equation of a heavy boson (compare to (7.30) for the pion field):

$$\left(\Box + M^2\right)A^\nu = 0. \tag{16.88}$$

For $M \neq 0$ the Maxwell equations (16.85) are not gauge invariant, as can be seen directly by substituting the gauge transformations (16.71) into (16.85):

$$\begin{aligned}\Box A'^\nu - \partial^\nu\left(\partial_\mu A'^\mu\right) + M^2 A'^\nu &= \Box A^\nu - \partial^\nu\left(\partial_\mu A^\mu\right) + M^2\left(A^\nu - \partial^\nu\chi\right)\\ &= j^\nu - M^2\partial^\nu\chi \neq j^\nu ,\end{aligned} \tag{16.89}$$

where we have used gauge invariance for the first two terms, see (16.83). QED would not be a gauge theory in a world with massive photons, and Maxwell's equations would read

$$\partial_\mu F^{\mu\nu} + M^2 A^\nu = \left(\Box + M^2\right)A^\nu - \partial^\nu\left(\partial_\mu A^\mu\right) = j^\nu . \tag{16.90}$$

With the definitions (16.80) one obtains the *Proca* equations (Alexandru Proca, 1897–1955):

$$\operatorname{div}\vec{E} = \rho - M^2\phi, \tag{16.91}$$

$$\operatorname{rot}\vec{B} = \vec{j} + \frac{\partial\vec{E}}{\partial t} - M^2\vec{A}, \tag{16.92}$$

with $\operatorname{div}\vec{B} = 0$ and $\operatorname{rot}\vec{E} = -\frac{\partial\vec{B}}{\partial t}$. These are Maxwell's equations (11.65) and (11.66) up to the M^2 terms in the two inhomogeneous equations (Gauss's law and Ampère's law, formulated in Heaviside–Lorentz units). From Gauss's law the M^2 term causes the E-field to decrease faster than $1/r^2$. However, the Cavendish experiment (figure 16.17) verifies that the electric charge always sits on the outer surface of an electrical conductor, as a consequence of the $1/r^2$ law. A modern version of the

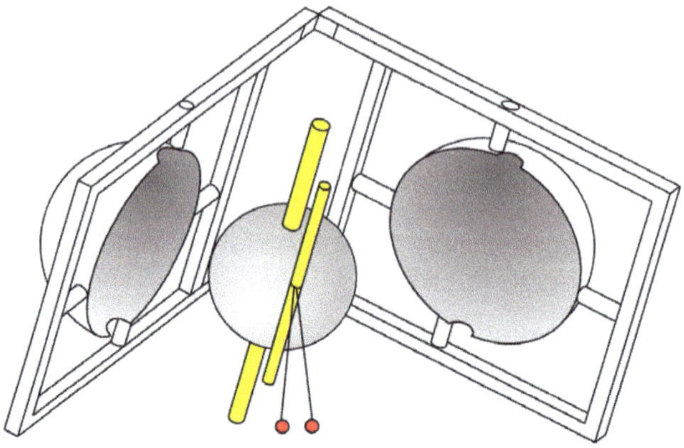

Figure 16.17. The Cavendish experiment (performed in 1773). The electrometer measures the electric charge which flows from the small charged sphere to the outer one, when the device is closed. After [13].

Cavendish experiment has established an upper limit for the photon mass, $M < 4.7 \times 10^{-12}$ eV [14]. According to Ampère's law, the B-field on the equator of a dipole magnet decreases faster than $1/r^3$. This fact was used during the 1974 Pioneer-10 fly-by of Jupiter to set an upper limit of $M = 6 \times 10^{-16}$ eV on the photon mass [15]. The current upper limit for the photon mass is 10^{-18} eV, based on the solar wind profile at the orbit of Pluto (see [16] for a bibliography).

We shall therefore assume that the mass of the photon is zero. Thus QED is a gauge theory and we investigate under which conditions a gauge invariance can be broken when introducing massive photons in QED. In the Lorentz gauge we expect that Maxwell's equations assume the form

$$\Box A^\mu + M^2 A^\mu = j^\mu \tag{16.93}$$

for a massive photon. These equations are not compatible with (16.84) unless the current on the right-hand side contains a term of the form $M^2 A^\mu$ which switches on *automatically* in the presence of A^μ. Such a situation is excluded in vacuum, but occurs on the surface of a superconductor. Denoting the *shielding current* $\vec{j}$ on the superconductor surface and assuming no further external currents, we obtain from Maxwell's equations (16.84) in the static case:

$$-\Delta \vec{A} = \vec{j}\,. \tag{16.94}$$

We assume that the shielding current is proportional to $\vec{A}$ and label the proportionality constant $-M^2$ (that is $\vec{j} = -M^2 \vec{A}$) and find (16.93) in the static case:

$$-\Delta \vec{A} + M^2 \vec{A} = 0 \Rightarrow \Delta \vec{B} - M^2 \vec{B} = 0, \tag{16.95}$$

with $\vec{B} = \text{rot}\,\vec{A}$. The one-dimensional solution $B(x) = B_0 \exp(-Mx)$ describes the behaviour of the $\vec{B}$-field inside the superconductor: the shielding current ensures that the magnetic field is expelled from the superconductor (Meissner–Ochsenfeld effect, figure 16.18). Therefore, if we were living inside a superconductor we would not

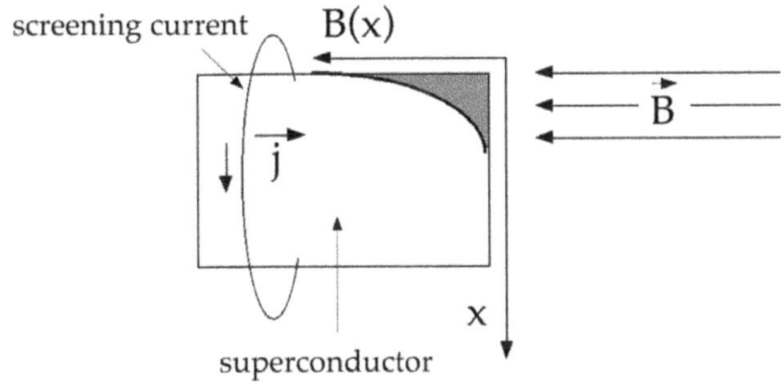

Figure 16.18. The Meissner–Ochsenfeld effect.

observe the shielding currents and would conclude that Maxwell's equations assume the form (16.93), the field equations of a photon with mass M. One speaks of *hidden gauge invariance*, or of *spontaneous symmetry breaking*.

In a superconductor the shielding current is produced by the field φ of the Cooper pairs (e^-e^- pairs that interact with the lattice). The field is scalar and the corresponding shielding current (Klein–Gordon current) fulfils the continuity equation (15.7). Up to the factor –i:

$$j^\mu = \mathrm{i}q\left(\varphi^*\left[\partial^\mu\varphi\right] - \left[\partial^\mu\varphi\right]^*\varphi\right). \tag{16.96}$$

In the standard model of weak interactions, the fields associated with the W and Z^0 bosons are massive. The shielding currents j^μ generated by the *Higgs* field φ are responsible for spontaneous symmetry breaking.

Before dealing with spontaneous symmetry breaking in weak interactions, let us examine more closely how the Higgs field operates when symmetry is spontaneously broken. The BEH mechanism was proposed in 1964 by Robert Brout, François Englert and Peter Higgs [17]. A real scalar field φ of mass m in vacuum obeys the Klein–Gordon equation

$$\partial_\nu\partial^\nu\varphi + m^2\varphi = 0. \tag{16.97}$$

This can easily be derived from the Lagrangian

$$\mathcal{L} = \frac{1}{2}(\partial_\nu\varphi)(\partial^\nu\varphi) - \frac{1}{2}m^2\varphi^2 = T - V, \tag{16.98}$$

by using Lagrange's equations of motion $\partial_\nu[\frac{\partial\mathcal{L}}{\partial(\partial_\nu\varphi)} - \frac{\partial\mathcal{L}}{\partial\varphi}] = 0$, where $\mathcal{L}$ depends on the generalized coordinates φ and $\partial_\nu\varphi$. For a stationary field, $\partial_\nu\partial^\nu$ is equal to the operator $-\Delta \propto \vec{p}^2$. Hence the first term in (16.98) is the kinetic energy T while the second term is the potential energy

$$V = \frac{1}{2}m^2\varphi^2. \tag{16.99}$$

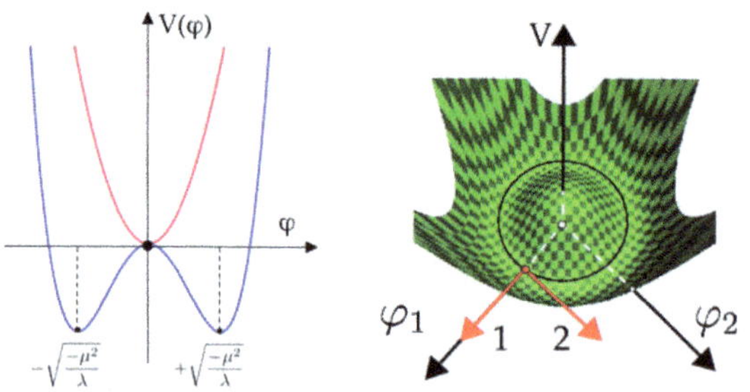

Figure 16.19. Left: $V(\varphi)$ as function of vacuum field φ for a potential energy of the type (16.99) (red curve) and of the symmetry breaking type (16.101) (blue curve with $\mu^2 < 0$). Right: two-dimensional contour of a potential energy of the type (16.105). Stability is achieved on the circle at the bottom of the valley. At $\varphi_2 = 0$ the curvature of $V(\varphi)$ is equal to $m_1^2 > 0$ along direction 1 and vanishes along direction 2 ($m_2^2 = 0$).

The field vanishes in vacuum since stability is reached at the minimum potential energy which occurs when $\varphi = 0$ (figure 16.19, left). Its mass is determined by the curvature of the potential energy at the minimum energy:

$$\frac{\partial^2 V}{\partial \varphi^2} = m^2. \tag{16.100}$$

We now add to V the next higher order term, excluding φ^3 since the field is scalar (i.e. $\mathcal{L}(-\varphi) = \mathcal{L}(\varphi)$):

$$V = \frac{1}{2}\mu^2\varphi^2 + \frac{\lambda}{4}\varphi^4 \tag{16.101}$$

with $\lambda > 0$. Again, the minimum is reached for $\varphi = 0$, unless the parameter μ^2 is negative, in which case the minimum is reached when $\frac{\partial V}{\partial \varphi} = 0$, for the *vacuum expectation value*

$$\langle\varphi\rangle = \pm\sqrt{\frac{-\mu^2}{\lambda}}. \tag{16.102}$$

The mass of the field is given by

$$m^2 = \left.\frac{\partial^2 V}{\partial \varphi^2}\right|_{\langle\varphi\rangle} = \mu^2 + 3\langle\varphi\rangle^2\lambda \Rightarrow m = \sqrt{-2\mu^2}. \tag{16.103}$$

Figure 16.19 (left) shows the behaviour of V in this case, as a function of φ. There are two stable solutions for the vacuum expectation value $\langle\varphi\rangle$. Whichever solution is chosen, reflection symmetry is spontaneously broken by requiring a non-zero field in vacuum. We now choose for φ the complex field

$$\varphi = \frac{1}{\sqrt{2}}(\varphi_1 + i\varphi_2), \tag{16.104}$$

for which the potential energy (16.101) reads

$$V = \frac{1}{2}\mu^2\varphi\varphi^* + \frac{\lambda}{4}(\varphi\varphi^*)^2 = \frac{1}{4}\mu^2\left(\varphi_1^2 + \varphi_2^2\right) + \frac{\lambda}{8}\left(\varphi_1^2 + \varphi_2^2\right)^2. \quad (16.105)$$

Stability is reached at the minimum, determined by the derivatives

$$\begin{aligned}\frac{\partial V}{\partial \varphi_1} &= \frac{1}{2}\mu^2\varphi_1 + \frac{\lambda}{2}\left(\varphi_1^2 + \varphi_2^2\right)\varphi_1 = 0,\\ \frac{\partial V}{\partial \varphi_2} &= \frac{1}{2}\mu^2\varphi_2 + \frac{\lambda}{2}\left(\varphi_1^2 + \varphi_2^2\right)\varphi_2 = 0,\end{aligned} \quad (16.106)$$

that is, on the circle defined by the vacuum expectation values $\langle\varphi_1\rangle$ and $\langle\varphi_2\rangle$ such that

$$\langle\varphi_1\rangle^2 + \langle\varphi_2\rangle^2 = -\frac{\mu^2}{\lambda}. \quad (16.107)$$

The contour of V resembles the bottom of a wine bottle (figure 16.19, right). We now choose the coordinate system so that $\langle\varphi_2\rangle = 0$ (φ becomes real). The vacuum expectation value of the field φ_1 is

$$\langle\varphi_1\rangle = \sqrt{\frac{-\mu^2}{\lambda}}. \quad (16.108)$$

This particular choice breaks the rotation symmetry. Furthermore, along direction 2 the curvature of V is zero, and hence one expects the mass m_2 of the field φ_2 to vanish, while along direction 1 the curvature is positive. Indeed, taking the second derivatives at $\langle\varphi_2\rangle = 0$ and for $\langle\varphi_1\rangle$ given by (16.108) leads to the masses

$$\begin{aligned}m_1^2 &= \frac{\partial^2 V}{\partial \varphi_1^2} = \frac{\mu^2}{2} + \frac{3\lambda}{2}\varphi_1^2 + \frac{\lambda}{2}\varphi_2^2 = -\mu^2 > 0,\\ m_2^2 &= \frac{\partial^2 V}{\partial \varphi_1^2} = \frac{\mu^2}{2} + \frac{3\lambda}{2}\varphi_2^2 + \frac{\lambda}{2}\varphi_1^2 = 0.\end{aligned} \quad (16.109)$$

Therefore, the choice of $\langle\varphi_2\rangle = 0$ reduces the complex scalar field to a real scalar field with mass

$$m_H = \sqrt{-\mu^2} \quad (16.110)$$

and vacuum expectation value

$$\langle\varphi\rangle = \frac{m_H}{\sqrt{\lambda}} = \frac{f}{\sqrt{2}}, \quad (16.111)$$

with $f \equiv \langle \varphi_1 \rangle$. To the field φ_2, which vanishes at our stability point, corresponds a massless (*Goldstone*) boson. The standard model that will be described in section 16.7 introduces a *doublet* of complex scalar fields

$$\varphi = \begin{pmatrix} \phi_1(\vec{r}, t) \\ \phi_2(\vec{r}, t) \end{pmatrix}. \tag{16.112}$$

According to (16.101) the potential energy is

$$V = \frac{1}{2}\mu^2 \varphi^\dagger \varphi + \frac{\lambda}{4}\left(\varphi^\dagger \varphi\right)^2. \tag{16.113}$$

The four real fields reduce to three fields with vanishing vacuum expectation values, associated with three massless Goldstone bosons. The fourth real *Higgs* field has the vacuum expectation value (16.111). The mass is of the Higgs boson is then

$$m_H = f\sqrt{\frac{\lambda}{2}}. \tag{16.114}$$

The constant f can be derived from the standard model (section 16.7) but m_H cannot be predicted since λ is not known.

16.6 Gauge invariance in the electroweak interaction

As we learned in chapter 11, a global phase invariance (i.e. $\chi = \text{constant}$) is related to a symmetry and a conservation law. For example:

- In the electromagnetic interaction, both wave functions $\psi' = U\psi \equiv e^{iq\chi}\psi$ and ψ describe the physical system. The (unitary U(1)) operation U commutes with H. In other words, the electric charge is conserved.
- In the strong interaction, both wave functions $\psi' = U\psi \equiv e^{i\vec{\chi}\cdot\vec{\tau}/2}\psi$ and ψ qualify to describe an isospin multiplet (see section 12.4 with $\vec{I} = \vec{\tau}/2$). The (unitary SU(2)) operation U, representing a rotation by the angle $|\vec{\chi}|$ around the direction $\vec{\chi}$ in isospin space, commutes with H. In other words i_3 is conserved (charge independence).

As we have seen, a weak isospin symmetry is also suspected for the weak interaction. Therefore, the same statement of global phase invariance can be made as for the strong interaction under SU(2). Hence we try to formulate a gauge theory for the weak interaction, and extend the statement of global phase invariance to local phase invariance: if ψ describes the physical system, then we *postulate* that the local phase transformation

$$\boxed{\psi' = e^{ig\vec{\chi}(\vec{r},t)\cdot\frac{\vec{\tau}}{2}}\psi} \tag{16.115}$$

qualifies equally well. Here ψ and ψ' are doublets of quarks or leptons and g is a coupling constant which plays the role of q in the electromagnetic interaction and describes the strength of the weak interaction. However, the phase transformation (16.115) must be compensated by a gauge transformation of certain fields. To formulate the Dirac equation in these fields we need the covariant derivative D^μ by comparing (16.115) with (16.70). The charge q in (16.70) has to be replaced by $g\vec{\tau}/2$. Since D^μ is a scalar while $\vec{\tau}$ is a vector we need three gauge fields, W_1^μ, W_2^μ and W_3^μ (Yang–Mills fields) which we write concisely as a three-dimensional vector $\vec{W}^\mu$. Hence the covariant derivative reads, following (16.74):

$$\boxed{D^\mu = \partial^\mu + \mathrm{i}g\frac{\vec{\tau}}{2} \cdot \vec{W}^\mu}. \tag{16.116}$$

The gauge fields are closely related to the $W^\pm$ and Z^0 bosons, as we show later. To establish a gauge theory we require that the gauge fields transform in such a way that the Dirac equation with the covariant derivatives (16.116) remains invariant when carrying out the phase transformation (16.115). Hence we require that

$$\left(\mathrm{i}\gamma_\mu D^\mu - m\right)\psi = 0 \Longleftrightarrow \left(\mathrm{i}\gamma_\mu D'^\mu - m\right)\psi' = 0. \tag{16.117}$$

The proof is given here for an infinitesimal phase transformation only, hence for $|\vec{\chi}|$ small enough so that

$$\psi' \simeq \left(1 + \mathrm{i}g\vec{\chi} \cdot \frac{\vec{\tau}}{2}\right)\psi \tag{16.118}$$

(see [18] for a finite transformation). The fields transform according to

$$\vec{W}^\mu \to \vec{W}'^\mu = \vec{W}^\mu - \partial^\mu\vec{\chi} - g\left[\vec{\chi} \times \vec{W}^\mu\right]. \tag{16.119}$$

To prove the last statement let us multiply the equation on the left-hand side of (16.117) by $(1 + \mathrm{i}g\vec{\chi} \cdot \vec{\tau}/2)$ to obtain

$$\begin{aligned} \mathrm{i}\gamma_\mu\left(1 + \mathrm{i}g\vec{\chi} \cdot \frac{\vec{\tau}}{2}\right)D^\mu\psi - m\left(1 + \mathrm{i}g\vec{\chi} \cdot \frac{\vec{\tau}}{2}\right)\psi &= 0 \\ \Rightarrow \mathrm{i}\gamma_\mu D'^\mu\psi' - m\psi' &= 0, \end{aligned} \tag{16.120}$$

thus the right-hand side of (16.117). We have used the lemma

$$D'^\mu\psi' = \left(1 + \mathrm{i}g\vec{\chi} \cdot \frac{\vec{\tau}}{2}\right)D^\mu\psi, \tag{16.121}$$

the proof of which follows from the commutation relations for Pauli matrices $\tau_i\tau_j = \delta_{ij} + \mathrm{i}\epsilon_{ijk}\tau_k$ (appendix E).

The field transformations (16.119) have the same shape as the QED ones, (16.71), except for the cross products. These terms come from the fact that the three *generators* of the SU(2) group, $(\tau_1, \tau_2, \tau_3) \equiv \vec{\tau}$, do not commute (see the proof of the lemma). Unlike QED, a U(1) theory, our gauge theory is *non-abelian*. The cross products in (16.119) transform the gauge fields into each other, leading to couplings between the $W^\pm$ and Z^0 bosons (figure 16.20(*a*)).

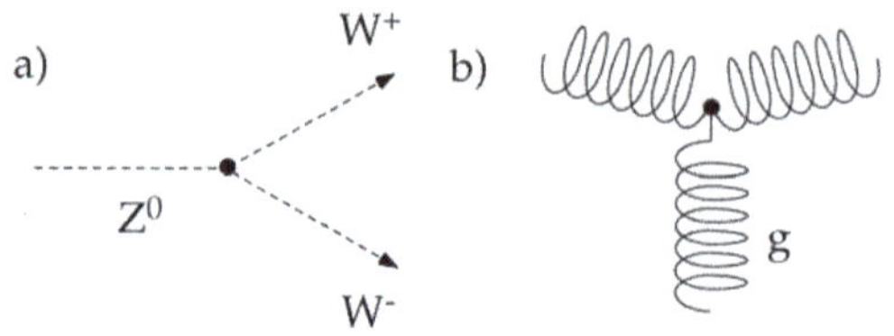

Figure 16.20. Coupling of the gauge fields in the weak interaction (*a*) and in QCD (*b*).

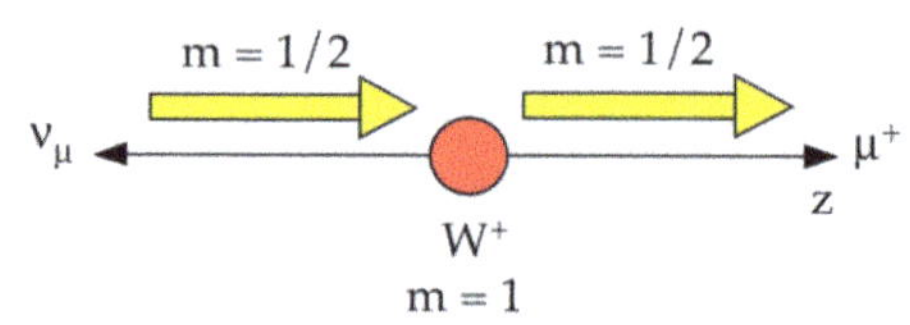

Figure 16.21. W^+ decay into $\mu^+\nu_\mu$.

The colour symmetry of the strong interaction is based on $SU(3)_c$, a unitary symmetry with eight generators $\vec{\lambda} = (\lambda_1,..., \lambda_8)$ [19]. Accordingly, the corresponding local phase transformation $e^{ig_s\vec{\lambda}\cdot\vec{\chi}(\vec{r},t)}$ requires eight gauge fields $\vec{G}^\mu$, the eight gluons introduced in section 10.4. This gauge theory (quantum chromodynamics (QCD)) is also non-abelian since among the eight generators only two commute. Therefore the field transformations contain cross terms responsible for interactions between gluons (figure 16.20(*b*)) which may lead to bound states (glueballs). These hadrons do not contain any quark but only radiation quanta (see figure 10.25(*b*) in section 10.4).

What are the equations of the fields $\vec{W}^\mu$? To this purpose let us consider the decay of the W^+ boson into $\mu^+\nu_\mu$ (figure 16.21). Due to the $V - A$ rule particles (antiparticles) are left-handed (right-handed), so that the projection of the total angular momentum on the z-axis is equal to +1. Conservation of angular momentum then requires the z-component of the W spin to be equal to +1. The spin of the W boson is thus $\geqslant$1. As the simplest possibility we assume a spin equal to 1, which makes the W boson behave like a massive photon, that is, satisfying (16.85).

It is therefore not possible to construct a gauge theory without spontaneous symmetry breaking. In analogy to superconductivity in QED, we shall formulate the gauge invariant Maxwell equations for three massless fields $\vec{W}^\mu$:

$$\Box\vec{W}^\mu - \partial^\mu\left(\partial_\nu\vec{W}^\nu\right) = \vec{j}^\mu, \tag{16.122}$$

and introduce a *model* for the Higgs field. The screening currents $(j_1^\mu, j_2^\mu, j_3^\mu) \equiv \vec{j}^\mu$ which switch on spontaneously with the fields $\vec{W}^\mu$, should be linearly proportional to the fields, with a proportionality constant $-M^2$ predicted by the model. These currents are then transferred to the left-hand side of (16.122), leading to field equations of the form (16.85) with massive gauge bosons. This is the subject of the following section.

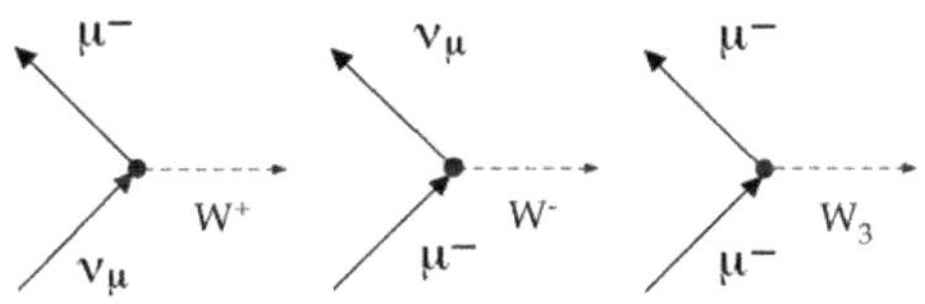

Figure 16.22. The three *W* fields.

16.7 The Glashow–Weinberg–Salam model

We have seen that both charged *and* neutral weak currents exist. Charged currents take the form

$$j_C^\mu = g\bar{u}\gamma^\mu\left(\frac{1-\gamma^5}{2}\right)\tau_\pm u, \tag{16.123}$$

where g is the coupling constant introduced in the previous section, and u stands for the doublets of leptons or quarks. We shall see shortly that the W^+ and W^- bosons are linear combinations of the W_1 and W_2 fields:

$$W^{\mu\pm} = \frac{1}{\sqrt{2}}\left(W_1^\mu \pm \mathrm{i}W_2^\mu\right). \tag{16.124}$$

For neutral currents one would expect for symmetry reasons that

$$j_N^\mu \overset{?}{\triangleq} g\bar{u}\gamma^\mu\left(\frac{1-\gamma^5}{2}\right)\tau_3 u. \tag{16.125}$$

Feynman graphs for the three fields $\vec{W}^\mu$ are drawn in figure 16.22. W_3 is responsible for the interaction of neutral currents and is electrically neutral. However, experiments show that charged weak currents are left-handed (factor $(1-\gamma^5)/2$ in (16.123)), while weak neutral currents (with the exception of neutrinos) have both left-handed $(V-A)$ and right-handed $(V+A)$ components (chapter 17). Consequently, the structure of the interaction with W_3 emission is not the same as the one with W^+ or W^-, and (16.125) must be extended to include $V+A$ currents.

How does this asymmetry between $W^\pm$ and W_3 come about? The solution was proposed in 1961 by Glashow: electromagnetic currents are 'neutral' (because they emit electrically neutral photons) and contain both left-handed and right-handed components in equal proportions. The *electromagnetic* current, with γ instead of W_3 emission in figure 16.22, contains the coupling

$$\gamma^\mu = \gamma^\mu\left[\left(\frac{1-\gamma^5}{2}\right) + \left(\frac{1+\gamma^5}{2}\right)\right]. \tag{16.126}$$

The contribution from right-handed neutral currents comes from the mixing of the electromagnetic and weak interactions (γ and W_3 emissions). This idea stems from the fact that neutrinos are purely left-handed: since the neutrino does not carry any electric charge, the contribution of the photon is absent. Neutrino currents are then left of the $V-A$ type.

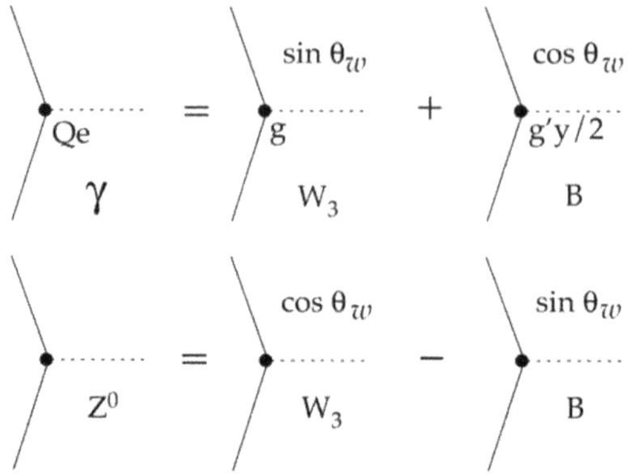

Figure 16.23. The fields W_3 and B mix to produce the physical fields of the γ and Z^0. Currents contributing to W_3 emission are purely left-handed, while B emission involves both left- and right-handed currents.

More precisely and as we shall see, the SU(2) field W_3^μ mixes (coupling g) with a U(1) field called B^μ (coupling $g'y/2$) to produce the observed Z^0 and γ (figure 16.23).

The mixing angle θ_w will be defined below. The parameter y depends on lepton or quark flavour, and is given by

$$\boxed{Q = i_3 + \frac{y}{2}}, \tag{16.127}$$

where the charge Q is in units of e. This formula is similar to the Gell–Mann–Nishijima formula (10.46) of strong interactions. However, i_3 is here the third component of the *weak* isospin and y is the *weak* hypercharge. In the Glashow–Weinberg–Salam model the following *left-handed* isospin *doublets* of Dirac spinors contribute to the charged and neutral currents:

$$u = \begin{pmatrix} \nu_e \\ e_L^- \end{pmatrix}, \begin{pmatrix} \nu_\mu \\ \mu_L^- \end{pmatrix}, \begin{pmatrix} \nu_\tau \\ \tau_L^- \end{pmatrix}, \begin{pmatrix} u_L \\ d_L' \end{pmatrix}, \begin{pmatrix} c_L \\ s_L' \end{pmatrix}, \begin{pmatrix} t_L \\ b_L' \end{pmatrix}. \tag{16.128}$$

The d_L', s_L' and b_L' are rotated d_L, s_L and b_L spinors, see (16.65). The isospin *singlets* do not contribute to charged currents, due to the presence of the ladder operators $\tau_\pm$. They contribute only to the U(1) part. They are *right-handed* and appear only in neutral currents:

$$u = e_R^-, \mu_R^-, \tau_R^-, u_R, d_R, c_R, s_R, t_R, b_R \tag{16.129}$$

(table 16.1). The d_R, s_R, b_R are the unrotated quark spinors. Neutrinos contribute only to left-handed charged and neutral currents. The corresponding antiparticles are not needed since for their Dirac spinors $v(p) = u(-p)$. The numerical values of y, computed from (16.127), are displayed in table 16.1.

Table 16.1. Weak isospin i_3 and weak hypercharge y for charged (CC) and neutral (NC) weak currents. Left-handed (LH) spinors contribute to both charged and neutral currents, right-handed (RH) spinors only to neutral currents.

LH i doublets	Q	i_3	y		RH i singlets	Q	i_3	y	
ν_e, ν_μ, ν_τ	0	$\frac{1}{2}$	-1	CC, NC					
e_L^-, μ_L^-, τ_L^-	-1	$-\frac{1}{2}$	-1	CC, NC	e_R^-, μ_R^-, τ_R^-	-1	0	-2	NC
u_L, c_L, t_L	$\frac{2}{3}$	$\frac{1}{2}$	$\frac{1}{3}$	CC, NC	u_R, c_R, t_R	$\frac{2}{3}$	0	$\frac{4}{3}$	NC
d_L', s_L', b_L'	$-\frac{1}{3}$	$-\frac{1}{2}$	$\frac{1}{3}$	CC, NC	d_R, s_R, b_R	$-\frac{1}{3}$	0	$-\frac{2}{3}$	NC

Since we have the additional field B^μ, we extend the symmetry to SU(2) × U(1) and postulate instead of (16.115) gauge invariance under the extended phase transformation

$$\psi' = e^{ig\vec{\alpha}(\vec{r},t)\cdot\frac{\vec{\tau}}{2} + ig'\frac{y}{2}\beta(\vec{r},t)}\psi. \tag{16.130}$$

The compensating gauge fields are the four fields $\vec{W}^\mu$ and B^μ. We refrain from giving the explicit gauge transformations here. The covariant derivative in the Dirac equation reads, extending (16.116):

$$D^\mu = \partial^\mu + ig\frac{\vec{\tau}}{2}\cdot\vec{W}^\mu + ig'\frac{y}{2}B^\mu. \tag{16.131}$$

We shall now write the transition amplitude for the interaction of weak currents with the gauge fields in the form $\mathcal{M}$ = currents × field, as explained in chapter 15. Here the couplings are $g\vec{\tau}/2$ and $g'y/2$, replacing q in the currents:

$$\mathcal{M} = g\left[\bar{u}\gamma_\mu\left(\frac{\vec{\tau}}{2}\cdot\vec{W}^\mu\right)u\right] + g'\frac{y}{2}\left[\bar{u}\gamma_\mu u\cdot B^\mu\right]. \tag{16.132}$$

On the other hand,

$$\begin{aligned}\frac{\vec{\tau}}{2}\cdot\vec{W}^\mu &= \frac{1}{2}\left(\tau_1 W_1^\mu + \tau_2 W_2^\mu + \tau_3 W_3^\mu\right)\\ &= \frac{1}{\sqrt{2}}\left(\tau_+ W^{-\mu} + \tau_- W^{+\mu}\right) + \frac{1}{2}\left(\tau_3 W_3^\mu\right),\end{aligned} \tag{16.133}$$

where we have used the fields (16.124) and the ladder operators $\tau_\pm = \frac{1}{2}(\tau_1 \pm i\tau_2)$. The transition amplitude becomes

$$\begin{aligned}\mathcal{M} = &\frac{g}{\sqrt{2}}\left[\bar{u}\gamma_\mu\tau_+ u\right]W^{-\mu} + \frac{g}{\sqrt{2}}\left[\bar{u}\gamma_\mu\tau_- u\right]W^{+\mu}\\ &+ g\left[\bar{u}\gamma_\mu\frac{\tau_3}{2}u\right]W_3^\mu + g'\frac{y}{2}\left[\bar{u}\gamma_\mu u\right]B^\mu.\end{aligned} \tag{16.134}$$

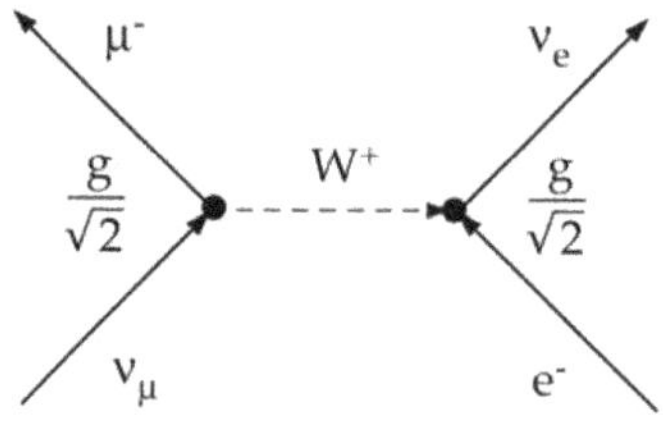

Figure 16.24. The interaction between two charged currents illustrated by the Feynman graph of the reaction $\nu_\mu e^- \to \mu^- \nu_e$.

The first two terms describe the interaction of charged weak currents with the fields (emission of a W^- or a W^+, see figure 16.22), up to the factor $(1 - \gamma^5)/2$ *inserted by hand* to guarantee the left-handedness of the doublets. The two last terms describe the weak interaction of neutral currents to which we shall return later.

We rewrite the fields for the charged currents in the form propagator × current of the quark or lepton doublet u'. For a heavy boson ($M_W^2 \gg q^2$) the propagator is simply $1/M_W^2$ (see (15.109)). Assuming a point-like interaction between two charged currents, such as the one shown in figure 16.24, one obtains the transition amplitude

$$\mathcal{M}_{\mathrm{CC}} = \frac{g^2}{2}\left[\bar{u}\gamma_\mu\left(\frac{1-\gamma^5}{2}\right)\tau_\pm u\right]\frac{1}{M_W^2}\left[\bar{u}'\gamma^\mu\left(\frac{1-\gamma^5}{2}\right)\tau_\mp u'\right]. \qquad (16.135)$$

Comparing this last expression with the Fermi point-like amplitude (16.12) one obtains the important relation

$$\boxed{\frac{G_F}{\sqrt{2}} = \frac{g^2}{8M_W^2}}. \qquad (16.136)$$

The coupling constant g can be determined from experiments with neutral currents, thus leading together with G_F to a prediction for the W mass. Before interpreting the two neutral current terms in (16.134) let us examine how the fields become massive. To this end we write, following (16.122) and assuming no *external* quark or lepton current,

$$\begin{aligned} \Box\vec{W}^\mu - \partial^\mu\left(\partial_\nu\vec{W}^\nu\right) &= \vec{j}^{\,\mu}(W) + \vec{j}^{\,\mu}(\varphi), \\ \Box B^\mu - \partial^\mu\left(\partial_\nu B^\nu\right) &= j_y^\mu(\varphi). \end{aligned} \qquad (16.137)$$

The currents $\vec{j}^{\,\mu}(\varphi)$ and $j_y^\mu(\varphi)$ are the screening currents that are automatically generated from the Higgs field φ once the gauge fields are switched on. The term $\vec{j}^{\,\mu}(W)$ comes from the non-abelian structure of the SU(2) group and appears even in the absence of external sources. Its exact form is of no interest here.

We now seek a matching *model* for φ such that (16.137) takes the form (16.85). Let us try a *scalar* (spin 0), *electrically neutral* field with weak isospin $i = \frac{1}{2}$, and furthermore assume that $|\varphi| = \text{constant}$. Then φ can be recast as follows (see (16.112)):

$$\varphi(\vec{r}, t) = \begin{pmatrix} \phi_1(\vec{r}, t) \\ \phi_2(\vec{r}, t) \end{pmatrix} = \mathrm{e}^{\mathrm{i}\frac{\vec{\tau}}{2}\cdot\vec{\theta}(\vec{r},t)} \begin{pmatrix} 0 \\ \frac{f}{\sqrt{2}} \end{pmatrix} \tag{16.138}$$

with $f = \text{constant}$ and real. The components ϕ_1 and ϕ_2 are complex functions. The operator on the right-hand side is a (local) rotation of an isospinor, corresponding to a constant field with $i_3 = -\frac{1}{2}$. An appropriate rotation direction $\vec{\theta}$ with rotation angle $|\vec{\theta}|$ can always be found. The field is constant with modulus $|\varphi| = f/\sqrt{2}$.

The screening currents are given by the Klein–Gordon currents (16.96). Due to the presence of gauge fields, we have to replace the derivatives by the covariant derivatives (16.131). One obtains since $\vec{\tau}$ is Hermitian:

$$\vec{j}^{\,\mu}(\varphi) = \mathrm{i}g\left(\varphi^{\dagger}\frac{\vec{\tau}}{2}\left[D^{\mu}\varphi\right] - \left[D^{\mu}\varphi\right]^{\dagger}\frac{\vec{\tau}}{2}\varphi\right) \tag{16.139}$$

and

$$j_y^{\mu}(\varphi) = \mathrm{i}g'\frac{y}{2}\left(\varphi^{\dagger}\left[D^{\mu}\varphi\right] - \left[D^{\mu}\varphi\right]^{\dagger}\varphi\right). \tag{16.140}$$

The field equations (16.137) are gauge invariant. We now choose the particular gauge $\vec{\alpha} \equiv -\frac{1}{g}\vec{\theta}$, $\beta \equiv 0$ thus abandoning gauge invariance, and perform the phase transformations (16.130). The Higgs field obtains the simple form

$$\varphi = \begin{pmatrix} 0 \\ \frac{f}{\sqrt{2}} \end{pmatrix}, \tag{16.141}$$

with vacuum expectation $\langle\varphi\rangle = |\varphi| = f/\sqrt{2} = \text{constant}$, see (16.111). The Higgs field becomes real and constant with isospin projection $i_3 = -\frac{1}{2}$ while symmetry is spontaneously broken.

We now derive explicitly the screening currents (16.139) and (16.140). Let us calculate the currents (16.139) component-wise with $\alpha = 1, 2, 3$. Since φ is constant, the derivative $\partial^{\mu}\varphi$ does not contribute. With $\varphi^{\dagger} = \varphi^{\top}$ one obtains

$$\begin{aligned} j_{\alpha}^{\mu}(\varphi) &= \mathrm{i}g\left[\varphi^{\top}\frac{\tau_{\alpha}}{2}\Bigg(\mathrm{i}g\overbrace{\frac{\vec{\tau}}{2}\cdot\vec{W}^{\mu}}^{\frac{1}{2}\tau_{\beta}W_{\beta}^{\mu}} + \mathrm{i}g'\frac{y}{2}B^{\mu}\Bigg)\varphi - \varphi^{\top}\Bigg(\mathrm{i}g\overbrace{\frac{\vec{\tau}}{2}\cdot\vec{W}^{\mu}}^{\frac{1}{2}\tau_{\beta}W_{\beta}^{\mu}} + \mathrm{i}g'\frac{y}{2}B^{\mu}\Bigg)^{\dagger}\frac{\tau_{\alpha}}{2}\varphi\right] \\ &= -\frac{g^2}{4}\varphi^{\top}2\delta_{\alpha\beta}W_{\beta}^{\mu}\varphi - gg'yB^{\mu}\left(\varphi^{\top}\frac{\tau_{\alpha}}{2}\varphi\right), \end{aligned} \tag{16.142}$$

where we have used the anti-commutation rule

$$\tau_\alpha \tau_\beta + \tau_\beta \tau_\alpha = 2\delta_{\alpha\beta}. \tag{16.143}$$

With the Pauli matrices

$$\tau_1 = \begin{bmatrix} 0 & 1 \\ 1 & 0 \end{bmatrix}, \qquad \tau_2 = \begin{bmatrix} 0 & -\mathrm{i} \\ \mathrm{i} & 0 \end{bmatrix}, \qquad \tau_3 = \begin{bmatrix} 1 & 0 \\ 0 & -1 \end{bmatrix} \tag{16.144}$$

one obtains

$$\varphi^{\mathrm{T}}\tau_1\varphi = 0, \qquad \varphi^{\mathrm{T}}\tau_2\varphi = 0, \qquad \varphi^{\mathrm{T}}\tau_3\varphi = -\frac{f^2}{2}, \qquad \varphi^{\mathrm{T}}\varphi = \frac{f^2}{2}. \tag{16.145}$$

The first two currents read

$$j_1^\mu(\varphi) = -\frac{g^2 f^2}{4} W_1^\mu, \qquad j_2^\mu(\varphi) = -\frac{g^2 f^2}{4} W_2^\mu \tag{16.146}$$

and display the desired proportionality. For the Higgs field $Q = 0$, $i_3 = -\frac{1}{2}$ and $y = 1$ (see equation (16.127)). Hence the third current is

$$j_3^\mu(\varphi) = -\frac{g^2 f^2}{4} W_3^\mu + \frac{g g' f^2}{4} B^\mu. \tag{16.147}$$

Similarly, the fourth current (16.140) reads

$$j_y^\mu(\varphi) = \frac{g g' f^2}{4} W_3^\mu - \frac{g'^2 f^2}{4} B^\mu. \tag{16.148}$$

Introducing the definitions

$$M \equiv \frac{fg}{2}, \qquad M' \equiv \frac{fg'}{2}, \tag{16.149}$$

leads with (16.137) to the four field equations

$$\boxed{\left(\Box + M^2\right) W_\alpha^\mu - \partial^\mu\left(\partial_\nu W_\alpha^\nu\right) = j_\alpha^\mu(W) \; (\alpha = 1, 2)} \tag{16.150}$$

and

$$\begin{aligned} \Box W_3^\mu - \partial^\mu\left(\partial_\nu W_3^\nu\right) &= j_3^\mu(W) - M^2 W_3^\mu + M M' B^\mu \quad (\alpha = 3), \\ \Box B^\mu - \partial^\mu\left(\partial_\nu B^\nu\right) &= M M' W_3^\mu - M'^2 B^\mu. \end{aligned} \tag{16.151}$$

The first two equations (16.150) describe massive fields W_1 and W_2 (or W^+ and W^- according to (16.124)) with masses $M_W \equiv M$. The two last equations (16.151) are coupled in the fields W_3^μ and B^μ. To decouple them let us multiply the first equation by M', the second one by M, and add:

$$\Box\left[M' W_3^\mu + M B^\mu\right] - \partial^\mu\left(\partial_\nu\left[M' W_3^\nu + M B^\nu\right]\right) = M' j_3^\mu(W). \tag{16.152}$$

Multiplying the first (second) equation by M (M') and subtracting gives

$$\begin{aligned}\Box[MW_3^\mu - M'B^\mu] - \partial^\mu(\partial_\nu[MW_3^\nu - M'B^\nu]) \\ = Mj_3^\mu(W) - [MW_3^\mu - M'B^\mu](M^2 + M'^2).\end{aligned} \tag{16.153}$$

The *Weinberg angle* θ_w is defined as

$$\boxed{\tan\theta_w \equiv \frac{g'}{g} = \frac{M'}{M}}. \tag{16.154}$$

Using the trigonometrical relations

$$\sin\theta_w = \frac{\tan\theta_w}{\sqrt{1+\tan^2\theta_w}} = \frac{M'}{\sqrt{M^2+M'^2}}, \tag{16.155}$$

$$\cos\theta_w = \frac{1}{\sqrt{1+\tan^2\theta_w}} = \frac{M}{\sqrt{M^2+M'^2}}, \tag{16.156}$$

and dividing (16.152) and (16.153) by $\sqrt{M^2+M'^2}$ leads to the field equations

$$\boxed{\Box A^\mu - \partial^\mu\partial_\nu A^\nu = \sin\theta_w \cdot j_3^\mu(W)} \qquad \text{and} \tag{16.157}$$

$$\boxed{\left(\Box + [M^2+M'^2]\right)Z^\mu - \partial^\mu\partial_\nu Z^\nu = \cos\theta_w \cdot j_3^\mu(W)}, \tag{16.158}$$

with the definitions of the fields A^μ and W_3^μ:

$$\boxed{A^\mu \equiv \cos\theta_w B^\mu + \sin\theta_w W_3^\mu}, \qquad \boxed{Z^\mu \equiv -\sin\theta_w B^\mu + \cos\theta_w W_3^\mu}. \tag{16.159}$$

Equation (16.157) describes a massless field, the photon, while (16.158) describes a massive field, the Z^0 with mass

$$M_{Z^0}^2 = M^2 + M'^2 = M^2\left(1+\frac{M'^2}{M^2}\right) = M^2(1+\tan^2\theta_w) = M_W^2/\cos^2\theta_w, \tag{16.160}$$

or

$$\boxed{M_{Z^0} = \frac{M_W}{\cos\theta_w}}. \tag{16.161}$$

We have reached our goal: (16.159) mean that the U(1) field mixes with the SU(2) field to produce the photon and the Z^0. A vanishing rotation angle θ_w would imply that $Z^0 = W_3$ and the weak interaction would be pure SU(2) with $g' = 0$ (equation (16.154)). In addition, (16.161) would require M_{Z^0} to be equal to M_W. However, the existence of right-handed neutral currents implies that $\sin\theta_w \neq 0$.

We are now in a position to interpret the neutral currents in (16.134). Inverting (16.159) and substituting B^μ and W_3^μ into (16.134) gives for the neutral current interaction

$$\mathcal{M}_{\rm NC} = g\left[\bar{u}\gamma_\mu \frac{\tau_3}{2} u\right]\overbrace{\left[\sin\theta_w A^\mu + \cos\theta_w Z^\mu\right]}^{W_3^\mu} + g'\frac{y}{2}\left[\bar{u}\gamma_\mu u\right]\overbrace{\left[\cos\theta_w A^\mu - \sin\theta_w Z^\mu\right]}^{B^\mu}, \tag{16.162}$$

or

$$\begin{aligned}\mathcal{M}_{\rm NC} = &\left[g\left(\bar{u}\gamma_\mu \frac{\tau_3}{2} u\right)\sin\theta_w + g'\frac{y}{2}\left(\bar{u}\gamma_\mu u\right)\cos\theta_w\right]A^\mu \\ &+\left[g\left(\bar{u}\gamma_\mu \frac{\tau_3}{2} u\right)\cos\theta_w - g'\frac{y}{2}\left(\bar{u}\gamma_\mu u\right)\sin\theta_w\right]Z^\mu.\end{aligned} \tag{16.163}$$

The term in A^μ describes the electromagnetic interaction of the form $eQ\bar{u}\gamma_\mu u A^\mu$. Comparing and using the relations $g'\cos\theta_w = g\sin\theta_w$ and $\frac{1}{2}\tau_3 u = i_3 u$ leads to

$$eQ = g\sin\theta_w\left(i_3 + \frac{y}{2}\right), \tag{16.164}$$

which must be valid for all quarks and leptons, that is all Q, i_3 and y values. Therefore,

$$\boxed{e = g\sin\theta_w} \tag{16.165}$$

and

$$\boxed{Q = i_3 + \frac{y}{2}}. \tag{16.166}$$

The first equation connects the electric charge e to the coupling constant g of the weak interaction through the Weinberg angle, a *fundamental constant* of nature which is usually expressed as $\sin^2\theta_w$. Our gauge theory is the unified *electroweak* theory.

The term Z^μ in (16.163) describes the interaction $j_{\mu N} Z^\mu$ between neutral currents, where

$$\begin{aligned} j_N^\mu &= g\bar{u}\gamma^\mu\left(\overbrace{\frac{\tau_3}{2}}^{i_3}\cos\theta_w - \overbrace{\frac{g'}{g}}^{\tan\theta_w}\frac{y}{2}\sin\theta_w\right)u \\ &= \frac{e}{\sin\theta_w\cos\theta_w}\left[\bar{u}\gamma^\mu\left(i_3\cos^2\theta_w - \underbrace{\frac{y}{2}}_{Q-i_3}\sin^2\theta_w\right)u\right] \\ &= \frac{e}{\sin\theta_w\cos\theta_w}\left[\bar{u}\gamma^\mu\left(i_3 - Q\sin^2\theta_w\right)u\right].\end{aligned} \tag{16.167}$$

Since there are left-handed ($i_3 = \pm 1/2$) and right-handed neutral currents ($i_3 = 0$) we introduce the corresponding operators $\left(\frac{1-\gamma^5}{2}\right)$ and $\left(\frac{1+\gamma^5}{2}\right)$ *by hand*:

$$j_N^\mu = \frac{e}{\sin\theta_w \cos\theta_w}\bar{u}\gamma^\mu\left[c_L\left(\frac{1-\gamma^5}{2}\right) + c_R\left(\frac{1+\gamma^5}{2}\right)\right]u, \tag{16.168}$$

where the *chiral couplings* are defined as

$$\boxed{c_L = i_3 - Q\sin^2\theta_w}, \qquad \boxed{c_R = -Q\sin^2\theta_w}. \tag{16.169}$$

The Weinberg angle determines the left-handed and right-handed contributions. For neutrino currents $c_L = \frac{1}{2}$ and $c_R = 0$.

Let us rewrite the transition amplitude for weak neutral currents (Z^μ term in (16.163)) in the form field = propagator × current with the current (16.168). We obtain, by assuming a point-like interaction,

$$\boxed{\mathcal{M}_{\mathrm{NC}} = \frac{e^2}{\sin^2\theta_w \cos^2\theta_w}\frac{1}{M_{Z^0}^2}J_N^\mu J_{N\mu}}, \tag{16.170}$$

where we have redefined the current as

$$J_N^\mu \equiv \bar{u}\gamma^\mu\left[c_L\left(\frac{1-\gamma^5}{2}\right) + c_R\left(\frac{1+\gamma^5}{2}\right)\right]u. \tag{16.171}$$

The prefactor in (16.170) can be rewritten with (16.136), (16.161) and (16.165) as

$$\frac{e^2}{\sin^2\theta_w \cos^2\theta_w}\frac{1}{M_{Z^0}^2} = \frac{g^2\sin^2\theta_w}{\sin^2\theta_w \cos^2\theta_w M_{Z^0}^2} = \frac{g^2}{M_W^2} = \frac{8G_F}{\sqrt{2}}, \tag{16.172}$$

leading to the compact expression

$$\boxed{\mathcal{M}_{\mathrm{NC}} = \frac{8G_F}{\sqrt{2}}J_N^\mu J_{N\mu}}. \tag{16.173}$$

Finally, we also rewrite the transition amplitude (16.135) for weak charged currents using (16.136):

$$\boxed{\mathcal{M}_{\mathrm{CC}} = \frac{4G_F}{\sqrt{2}}J_C^\mu J_{C\mu}}, \tag{16.174}$$

with the charged currents redefined as

$$\boxed{J_C^\mu \equiv \bar{u}\gamma^\mu\left(\frac{1-\gamma^5}{2}\right)u}. \tag{16.175}$$

Here and in (16.171) the spinors denote the components of the doublets or singlets. The spinor of the incoming (outgoing) particle is $u(\bar{u})$. The interaction depends on $\sin^2\theta_w$ and G_F being multiplied by the elements of the CKM matrix (16.67).

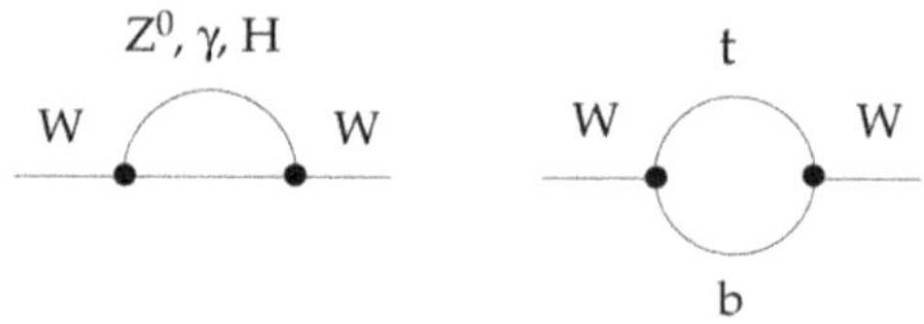

Figure 16.25. Radiative corrections to the W mass.

The Weinberg angle can be determined from the ratio (16.161) of W to Z^0 masses. Before the discovery of the gauge bosons in 1983, the Weinberg angle was obtained from deep inelastic $\nu_\mu(\bar{\nu}_\mu)$ scattering, namely from the ratios of cross sections $R = \sigma(\nu_\mu N \to \nu_\mu X)/\sigma(\nu_\mu N \to \mu^- X)$ and $\overline{R} = \sigma(\bar{\nu}_\mu N \to \bar{\nu}_\mu X)/\sigma(\bar{\nu}_\mu N \to \mu^+ X)$, see (16.53, 16.54) and also section 17.4.3. The Weinberg angle was also measured from $e^- d$ scattering data (section 17.1) and from the forward–backward asymmetry in $e^+e^- \to \mu^+\mu^-$ scattering (section 17.4.4). For example, with $\sin^2\theta_w = 0.23$ obtained from deep inelastic ν_μ scattering one obtains by using (16.136) and (16.165)

$$M_W = \sqrt{\frac{\sqrt{2}e^2}{8G_F \sin^2\theta_w}} = \sqrt{\frac{\alpha\pi}{\sqrt{2}G_F}}\frac{1}{\sin\theta_w} = 78 \text{ GeV}, \tag{16.176}$$

where we have used the Fermi coupling constant and $\alpha = 1/137$. Equation (16.161) then gives

$$M_{Z^0} = 89 \text{ GeV}. \tag{16.177}$$

The most precise value of the Weinberg angle is $\sin^2\theta_w = 0.23126 \pm 0.0005$ [16].

The measured masses for the $W^\pm$ and Z^0 gauge boson are 80.38 and 91.19 GeV, respectively. The relatively small discrepancies with our theoretical predictions stem from radiative corrections, such as the ones shown in figure 16.25. The corrected $W^\pm$ mass is

$$M_W(\text{corr}) = M_W\left(1 + \frac{\Delta r}{2}\right). \tag{16.178}$$

The main contribution to Δr stems from the running coupling constant α at the W mass:

$$\Delta r_0 = 1 - \frac{\alpha}{\alpha(W)} \simeq 1 - \frac{1/137}{1/129} = 0.062, \tag{16.179}$$

hence $M_W(\text{corr}) = 80.4$ GeV, using our prediction (16.176), which is now in excellent agreement with the data. Additional smaller contributions come from the mass m_t of the t quark and from the Higgs mass m_H[2] [20]. Although the Higgs mass cannot be predicted precisely in the standard model, the experimental values for

[2] $\Delta r \simeq \Delta r_0 - \frac{3}{8}\frac{G_F m_t^2}{\sqrt{2}\pi^2 \tan^2\theta_w} + \frac{11}{12}\frac{M_W^2 G_F}{\sqrt{2}\pi^2}\ln(\frac{m_H}{M_Z})$ with $\Delta r_0 = 0.062$. The second and third term contribute $\simeq 10^{-2}$ and $\simeq 10^{-3}\ln(\frac{m_H}{M_Z})$, respectively.

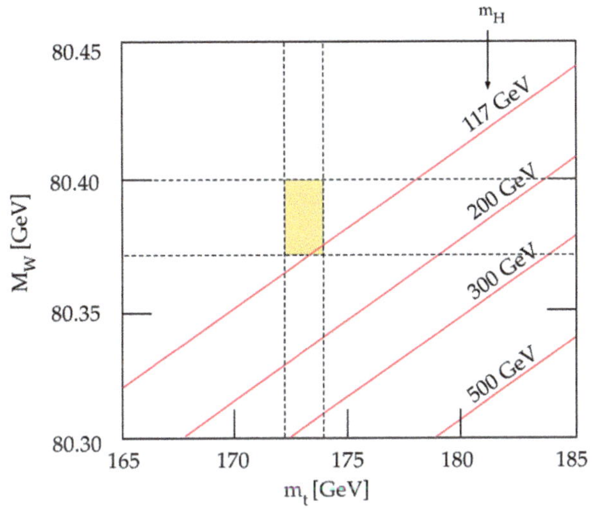

Figure 16.26. The Higgs mass can be estimated from radiative corrections to the W mass. The coloured area (± 1 standard deviation on M_W and m_t) shows the favoured region. Adapted from [21].

M_W and m_t provide an estimate of the range in which m_H should lie. Figure 16.26 shows the mass range allowed by radiative corrections. The Higgs with a mass of $\simeq$125 GeV was discovered in 2012 by the ATLAS and CMS experiments at the LHC (section 17.3).

The vacuum expectation value (16.141) of the Higgs field is, with (16.149) and (16.136):

$$\frac{f}{\sqrt{2}} = \frac{1}{\sqrt{2}}\left(\frac{2M_W}{g}\right) = \frac{1}{\sqrt{2}}\frac{1}{\left(\sqrt{2}\,G_F\right)^{\frac{1}{2}}}, \tag{16.180}$$

or

$$f = \frac{1}{\left(\sqrt{2}\,G_F\right)^{\frac{1}{2}}} = 246 \text{ GeV}. \tag{16.181}$$

Note that in the derivation of $\mathcal{M}_{\rm CC}$ we have used $M_{Z_0} = M_W/\cos\theta_w$. Assuming instead that the Higgs boson is a member of an isospin triplet ($i = 1$) would lead to $M_{Z^0} = M_W\sqrt{2}/\cos\theta_w$, in contradiction with the measured value of the Z^0 mass. In addition the transition amplitude (16.173) would have to be replaced by

$$\mathcal{M}_{\rm NC} = \rho\frac{8G_F}{\sqrt{2}}J_N^{\mu}J_{N\mu}, \tag{16.182}$$

with $\rho = \frac{1}{2}$ instead of $\rho = 1$. The parameter ρ has been determined in deep inelastic neutrino scattering [22]: $\rho = 1.027 \pm 0.023$, compatible with $\rho = 1$, that is, with an $i = \frac{1}{2}$ Higgs boson.

References

[1] Burkard H *et al* 1985 *Phys. Lett.* B **150** 242

[2] Corriveau F *et al* 1981 *Phys. Rev.* D **24** 2004

[3] Halzen F and Martin A D 1984 *Quarks and Leptons: An Introductory Course in Modern Particle Physics* (New York: Wiley)

[4] Beltrami I *et al* 1986 *Nucl. Phys.* A **451** 679

[5] Bardin G *et al* 1984 *Phys. Lett.* B **137** 135

[6] Webber D M *et al* 2011 *Phys. Rev. Lett.* **106** 041803

[7] Britton D I *et al* 1992 *Phys. Rev. Lett.* **68** 3000

[8] de Groot J G H *et al* 1979 *Z. Phys.* C **1** 143

[9] Hasert F J *et al* 1973 *Phys. Lett.* B **46** 138

[10] Hasert F J *et al* 1974 *Nucl. Phys.* B **73** 1

[11] Klems J H, Hildebrand R H and Stiening R 1970 *Phys. Rev. Lett.* **24** 1086

[12] Chatrchyan S *et al* (CMS Collaboration) 2013 *Phys. Rev. Lett.* **111** 101804

[13] Goldhaber A S and Nieto M M 1976 *Sci. Am.* **234** 86

[14] Chernikov M A *et al* 1992 *Phys. Rev. Lett.* **68** 3383

[15] Davis L, Goldhaber A and Nieto M 1975 *Phys. Rev. Lett.* **35** 1402

[16] Olive K *et al* (Particle Data Group) 2014 *Chin. Phys.* C **38** 090001

[17] Englert F and Brout R 1964 *Phys. Rev. Lett.* **13** 321
Higgs P W 1964 *Phys. Rev. Lett.* **13** 508

[18] Aitchison I and Hey A 1989 *Gauge Theories in Particle Physics* (Bristol: Hilger)

[19] Close F E 1979 *An Introduction to Quarks and Partons* (London: Academic)

[20] Marciano W J 2004 *SLAC Summer Institute on Particle Physics* (*Menlo Park, CA, 2–13 August 2004*) ed J Hewett *et al* (Stanford, CA: SLAC) L009

[21] Beringer J *et al* (Particle Data Group) 2012 *Phys. Rev.* D **86** 010001

[22] Jonker M *et al* 1981 *Phys. Lett.* B **99** 265

IOP Publishing

Nuclear and Particle Physics

Claude Amsler

Chapter 17

Applications of the standard model

17.1 Parity violation in electron–deuterium scattering

One way to measure the Weinberg angle and predict the mass of the gauge bosons is to scatter energetic polarized electrons off deuterium, that is to measure the cross section for *deep inleastic* scattering $e^-d \to e^-X$ with longitudinally polarized electrons (figure 17.1). Such an experiment was performed at SLAC with 20 GeV electrons [1]. At this energy the wavelength of the exchanged γ is short enough to break up the nucleon, and to interact with the constituent u and d quarks. The scattered quark then fragments into hadrons X. The 4-momentum transfer is, when neglecting the mass of the electron,

$$q^2 = (E - E')^2 - \left(\vec{k} - \vec{k}'\right)^2 \simeq -4EE' \sin^2 \frac{\theta}{2} \leqslant 0, \tag{17.1}$$

where E is the incident and E' the scattered electron energy, respectively (see (2.7)). We define by

$$y \equiv \frac{E - E'}{E} \tag{17.2}$$

the fraction of incident energy carried away by the *hadrons*. At fixed incident energy E the momentum transfer q^2 is fully determined once E' *and* the scattering angle θ of the electron are specified.

As a consequence of parity conservation the scattering cross section does not depend on the polarization of the electron for a pure electromagnetic interaction: consider the polarization of the electron in the mirrored reaction, in which the spin of the longitudinally polarized electron is flipped. Left-handed and right-handed electrons must be scattered with the same probability. Denoting by σ_L and σ_R the differential cross sections with left-handed and right-handed electrons, respectively, one expects the asymmetry

$$A = \frac{\sigma_R - \sigma_L}{\sigma_R + \sigma_L} = 0. \tag{17.3}$$

doi:10.1088/978-0-7503-1140-3ch17 © IOP Publishing Ltd 2015

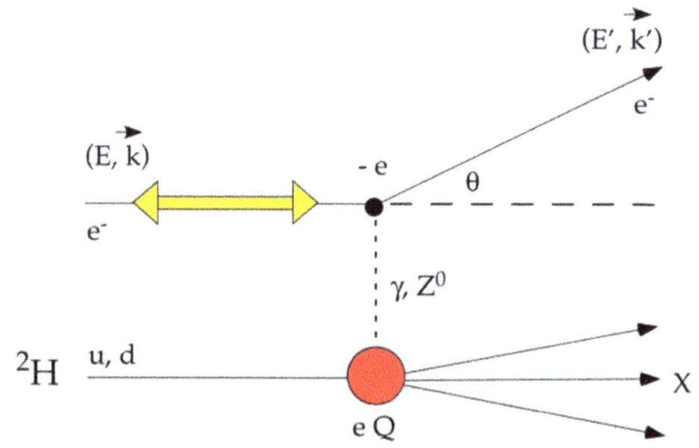

Figure 17.1. Deep inelastic electron–deuteron scattering with polarized electrons. The u or d quark interacts with the exchanged boson (γ and Z^0) and fragments into hadrons X.

However, since a Z^0 can also be exchanged the cross section contains a contribution from the parity violating weak interaction, leading to a small but finite asymmetry $A \neq 0$.

To calculate the asymmetry, consider the scattering of a left-handed or right-handed electron off a left-handed or right-handed quark. To avoid confusion with momentum transfer, we write $\tilde{q}$ for quarks and $\bar{q}$ for antiquarks. We start from the cross sections for deep inelastic neutrino scattering with an outgoing muon (charged current (CC)) derived in section 17.4.2. For ν_μ scattering the cross section is proportional to the square of the transition amplitude (16.174) while for $\bar{\nu}_\mu$ scattering the multiplicative factor $(1 - y)^2$ has to be appended:

$$\sigma\left(\nu_\mu \tilde{q} \to \mu^- X\right) \propto \left| \overbrace{\frac{1}{2}\left(\frac{8G_F}{\sqrt{2}}\right)}^{\frac{4G_F}{\sqrt{2}}} J_C^\mu(\ell) J_{C\mu}(\tilde{q}) \right|^2,$$
$$\sigma\left(\bar{\nu}_\mu \tilde{q} \to \mu^+ X\right) \propto \left| \frac{1}{2}\left(\frac{8G_F}{\sqrt{2}}\right) J_C^\mu(\bar{\ell}) J_{C\mu}(\tilde{q}) \right|^2 (1 - y)^2. \tag{17.4}$$

The lepton ℓ plays the role of the electron in ed scattering. Furthermore, since the transition amplitudes for charged and neutral interactions have the same form, it suffices to replace the factor $\frac{1}{2}$ in (17.4) by the chiral coupling constants $c_i(e)c_j(\tilde{q})$ to describe neutral current scattering (NC) between e_i and $\tilde{q}_j$ (where $i, j = L, R$). However, we still have to figure out in which case the factor $(1 - y)^2$ is required:

- $e_L \tilde{q}_L$ scattering (NC) is equivalent to $\nu \tilde{q}$ scattering (CC) since only left-handed fermions contribute to CC interactions.
- $e_R \tilde{q}_R$ scattering (NC) is equivalent to $\bar{\nu} \bar{q}$ scattering (CC) since only right-handed fermions contribute to CC interactions. However, $\bar{\nu} \bar{q}$ scattering is equivalent to $\nu \tilde{q}$ scattering due to CP invariance.

- $e_L\tilde{q}_R$ scattering (NC) is equivalent to $\nu\bar{q}$ scattering (CC), which in turn is equivalent to $\bar{\nu}\tilde{q}$ scattering due to *CP* invariance. Here the factor $(1-y)^2$ has to be appended to the cross section.
- $e_R\tilde{q}_L$ scattering (NC) is equivalent to $\bar{\nu}\tilde{q}$ scattering (CC). The cross section also obtains the factor $(1-y)^2$.

We now write the four contributions to the asymmetry A, and also take into account the propagator $-\frac{1}{q^2}$ of the photon as well as the electromagnetic couplings $(-e)(eQ)$, where Q is the charge of the quark:

$$\begin{aligned}
\sigma\left(e_R\tilde{q}_R\right) &\propto \left|-\frac{e^2Q}{q^2}+\frac{8G_F}{\sqrt{2}}c_R(e)c_R(\tilde{q})\right|^2,\\
\sigma\left(e_R\tilde{q}_L\right) &\propto \left|-\frac{e^2Q}{q^2}+\frac{8G_F}{\sqrt{2}}c_R(e)c_L(\tilde{q})\right|^2(1-y)^2,\\
\sigma\left(e_L\tilde{q}_R\right) &\propto \left|-\frac{e^2Q}{q^2}+\frac{8G_F}{\sqrt{2}}c_L(e)c_R(\tilde{q})\right|^2(1-y)^2,\\
\sigma\left(e_L\tilde{q}_L\right) &\propto \left|-\frac{e^2Q}{q^2}+\frac{8G_F}{\sqrt{2}}c_L(e)c_L(\tilde{q})\right|^2.
\end{aligned} \tag{17.5}$$

We have replaced q^2 by $-q^2$, hence $q^2=+4EE'\sin^2\frac{\theta}{2}\geqslant 0$. In the experiment to be described below q^2 is typically equal to $4\,\mathrm{GeV}^2\,c^{-2}$, therefore $e^2/q^2\approx 2\times 10^{-2}$ GeV^{-2}, while the weak contribution is much smaller, $8G_F/\sqrt{2}\approx 7\times 10^{-5}\ \mathrm{GeV}^{-2}$. Hence the Coulomb terms dominate and we can approximate the asymmetry by

$$A\simeq\frac{\text{interference terms }(e_R)-\text{interference terms }(e_L)}{\text{Coulomb terms}}. \tag{17.6}$$

We now calculate the asymmetry (17.6) by adding the contributions from the u and d quarks listed in table 17.1, since deuterium contains equal numbers of u and d quarks. One finds for the numerator of (17.6) with $e_R\equiv c_R(e)$, $u_L\equiv c_L(u)$, etc:

$$\begin{aligned}
&-\frac{16G_Fe^2}{\sqrt{2}q^2}\left[\frac{2}{3}e_Ru_R-\frac{1}{3}e_Rd_R-\left(\frac{2}{3}e_Lu_L-\frac{1}{3}e_Ld_L\right)\right]\\
&-\frac{16G_Fe^2}{\sqrt{2}q^2}\left[\frac{2}{3}e_Ru_L-\frac{1}{3}e_Rd_L-\left(\frac{2}{3}e_Lu_R-\frac{1}{3}e_Ld_R\right)\right](1-y)^2,
\end{aligned} \tag{17.7}$$

and for the denominator,

$$\frac{2e^4}{q^4}\left[\left(\frac{2}{3}\right)^2+\left(\frac{2}{3}\right)^2(1-y)^2+\left(\frac{1}{3}\right)^2+\left(\frac{1}{3}\right)^2(1-y)^2\right]. \tag{17.8}$$

Table 17.1. Products of chiral couplings $c_L = i_3 - Qa$, $c_R = -Qa$, where $a \equiv \sin^2\theta_w$.

$c_R(e)c_j(\tilde{q})$		$c_L(e)c_j(\tilde{q})$	
$e_R u_L$	$a\left(\frac{1}{2} - \frac{2}{3}a\right)$	$e_L u_L$	$\left(-\frac{1}{2} + a\right)\left(\frac{1}{2} - \frac{2}{3}a\right)$
$e_R u_R$	$-\frac{2}{3}a^2$	$e_L u_R$	$\left(-\frac{1}{2} + a\right)\left(-\frac{2}{3}a\right)$
$e_R d_L$	$a\left(-\frac{1}{2} + \frac{1}{3}a\right)$	$e_L d_L$	$\left(-\frac{1}{2} + a\right)\left(-\frac{1}{2} + \frac{1}{3}a\right)$
$e_R d_R$	$\frac{1}{3}a^2$	$e_L d_R$	$\left(-\frac{1}{2} + a\right)\left(\frac{1}{3}a\right)$

The asymmetry becomes, after substituting the values from table 17.1 and with $a \equiv \sin^2\theta_w$:

$$A = -\frac{q^2}{e^2}\frac{G_F}{\sqrt{2}}\frac{9}{5}\left[\frac{\frac{16}{9}a(1-y)^2 + 2 - \frac{56}{9}a}{1+(1-y)^2}\right] \tag{17.9}$$

or with $e^2/4\pi = \alpha$ and after several algebraic manipulations:

$$\boxed{A = -q^2\frac{9G_F}{20\sqrt{2}\pi\alpha}\left[1 - \frac{20}{9}\sin^2\theta_w + \frac{1-(1-y)^2}{1+(1-y)^2}\left(1 - 4\sin^2\theta_w\right)\right]}. \tag{17.10}$$

We now know that $\sin^2\theta_w \simeq 0.25$, thus A depends only weakly on y. Expected is a tiny asymmetry, $A \simeq -7.7\times 10^{-5}q^2$ [GeV^2 c^{-2}], of about 3×10^{-4} with $q^2 \simeq 4$ GeV^2 c^{-2}.

The SLAC experiment [1] is illustrated in figure 17.2. A laser beam (L) is linearly polarized with a Nicol prism (K), then circularly polarized by a quarter-wave plate. When the direction of the incident polarization coincides with the optical axis of the quarter-wave plate ($\phi_p = 0$), or is perpendicular to it ($\phi_p = 90°$), the outgoing light beam is still linearly polarized. When $\phi_p = \pm 45°$ the light becomes left-circularly or right-circularly polarized. The polarization is then transferred to a GaAs crystal which emits polarized electrons that are accelerated to $E = 20$ GeV. The (longitudinal) polarization of the electrons ($P_e \simeq \pm 37\%$) can be measured using Möller scattering (e^-e^- scattering) off a magnetized foil (M). The electron flux through the target is measured from the total charge deposit in a Faraday cage (F). The electrons that are scattered by the liquid deuterium target (LD_2) under the scattering angle $\theta = 4°$ are analysed using a magnetic spectrometer (S). Thus θ and the scattered energy E' determine the momentum transfer q^2. Electrons are discriminated from hadrons by a gaseous nitrogen Čerenkov counter (N) and a lead glass detector (P).

The momentum vector rotates during the transit time t between the GaAs crystal and the LD_2 target by the angle $\Theta_B = \frac{eB}{m\gamma}t$, where B is the average field in the beam transport system (figure 17.2). The laboratory angle between the momentum and

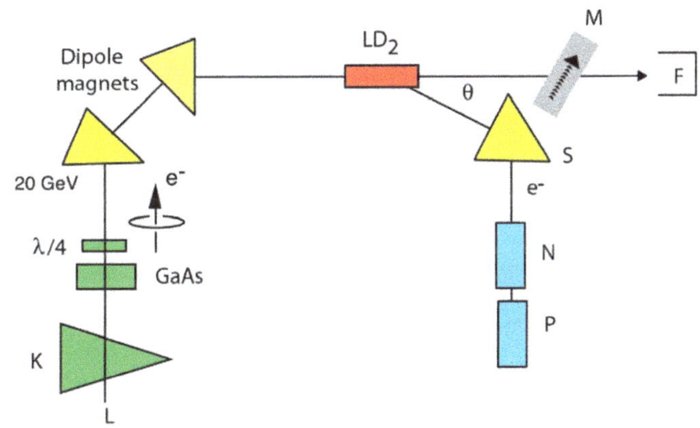

Figure 17.2. Diagram of the electron–deuterium scattering experiment [1] (see the text for details).

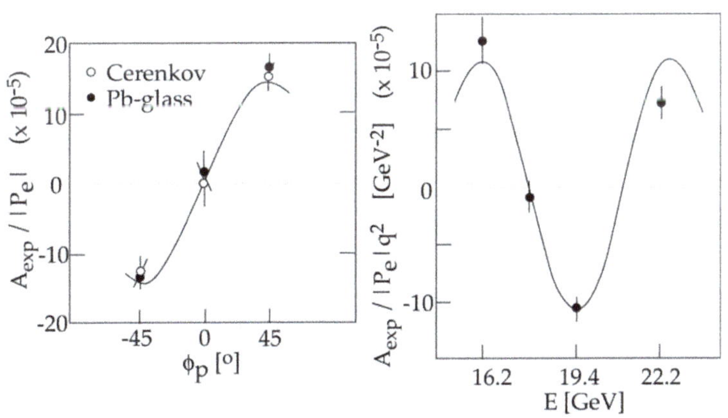

Figure 17.3. Measured asymmetry versus ϕ_p and versus E. After [1].

spin vectors at the LD_2 target is then $\omega_a t = \left(\frac{g-2}{2}\right)\frac{eB}{m}t = \gamma\Theta_B$, see (13.77). For a given setting of the spectrometer one measures the asymmetry

$$A_{\text{exp}} = A\left|P_e\right| \sin\left(2\phi_p\right) \cos\left(\frac{g-2}{2}\gamma\Theta_B\right). \tag{17.11}$$

Figure 17.3 shows the measured asymmetry A_{exp} as a function of prism angle ϕ_p, and its dependence on incident energy E. Figure 17.4 shows the final asymmetry A for various settings of the spectrometer (variable y) and different incident energies E.

Fitting (17.10) to the data leads to

$$\sin^2\theta_w = 0.224 \pm 0.020. \tag{17.12}$$

Note that the experiment is consistent with expectations from the standard model, but does not exclude other models also reproducing the asymmetry in figure 17.4. However, the experiment shows a clear violation of parity, induced by the weak

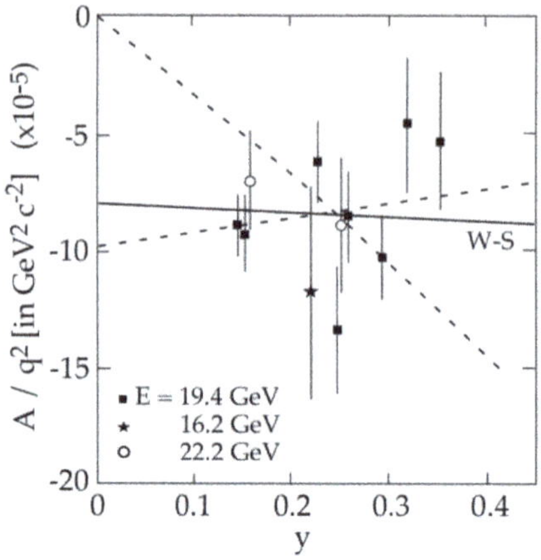

Figure 17.4. Asymmetry A/q^2 versus y for various incident energies. The lines show fits from various models. The line labelled W–S is the standard model fit. After [1].

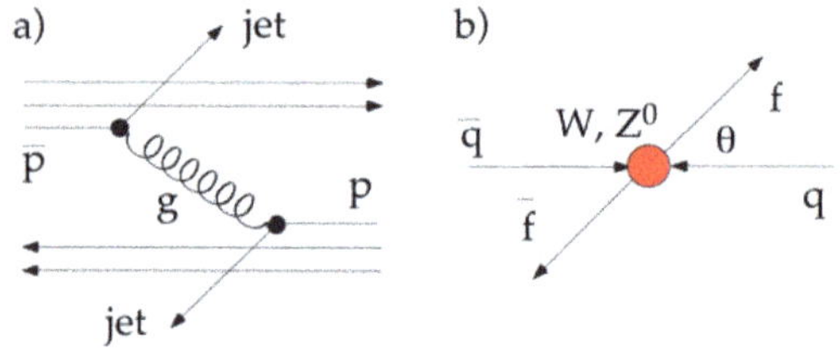

Figure 17.5. (*a*) Strong interaction processes such as gluon exchange dominate in $\bar{p}p$ collisions. (*b*) Z^0 and W production by $\bar{q}q$ pairs.

interaction. We have calculated the Weinberg angle from the standard model, with which the masses of the W and Z^0 bosons can be predicted. Their discovery at the expected masses is described in the following section.

17.2 The discovery of the $W^{\pm}$ and Z^0 bosons

The W and Z bosons were discovered in proton–antiproton collisions. The dominant underlying mechanism in high energy quark–antiquark scattering is gluon exchange (strong interaction) which leads to jets (figure 17.5(*a*)). Albeit with very low probability, a W^+ or a W^- (in a $u\bar{d}$ or $d\bar{u}$ collision, respectively) or a Z^0 (in a $u\bar{u}$ or $d\bar{d}$ collision) can also be produced (figure 17.5(*b*)). The heavy boson decays into fermion–antifermion pairs, for example the W into $e\nu$ and the Z^0 into e^+e^-, whereby leptons of about 40 GeV are emitted in the laboratory, since on average the bosons are created at rest. The remaining quarks interact through gluon exchange (figure 17.5(*a*)) generating dozens of hadrons and jets.

The Z^0 boson also appears as a resonance in e^+e^- scattering, while the W bosons are produced pairwise. Figure 17.6 (left) shows the Feynman diagrams for W^+W^- production in e^+e^- collisions and figure 17.6 (right) their contributions to the cross section. Neutrino exchange dominates but the contributions from γ and (virtual)

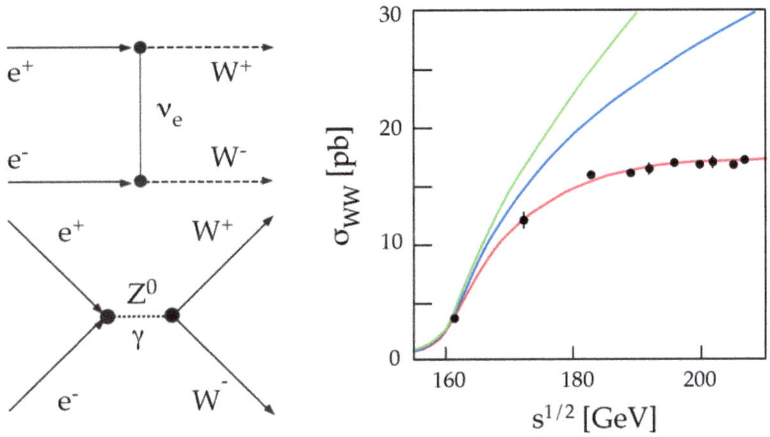

Figure 17.6. Left: pairwise W boson production in e^+e^- collisions by neutrino exchange or via virtual Z^0/γ production. Right: cross sections as a function of centre-of-mass energy. The green (top) curve shows the predicted contribution from neutrino exchange only, the blue (middle) curve the cross section when adding the intermediate γ. The red (bottom) curve also includes the Z^0 contribution. The points are measurements taken at LEP. Adapted from [2].

Z^0 exchange interfere destructively to reduce the cross section. Comprehensive tests of the standard model were performed with the Z^0 (and W) boson at the Large Electron–Positron storage ring (LEP) and the SLAC Linear Collider (SLC). Colliding beams of 45 GeV each (LEP-I at CERN and SLC) are required to excite the Z^0 resonance. Higher beam energies of at least 80 GeV are needed to produce W pairs. Due to the energy loss by synchrotron radiation, LEP-II was equipped with superconducting accelerating cavities. The LEP-II machine reached the centre-of-mass energy of 209 GeV in 2009.

The heavy bosons were discovered in 1983 at CERN by the UA1 [3, 4] and UA2 [5, 6] collaborations in high energy $\bar{p}p$ collisions at the SPS, operating as a collider (see figure 9.8). Let us estimate how much energy was required. One needs 45 GeV/quark to produce 90 GeV bosons in quark–antiquark collisions. However, since 50% of the nucleon energy is carried by gluons (section 17.4.2), one needs to collide $6 \times 45 = 270$ GeV protons against 270 GeV antiprotons (ignoring the momentum distribution of quarks in the nucleon).

Reactions without heavy boson production (figure 17.5(*a*)) are much more frequent since strong interaction dominates the interaction. One expects at $\sqrt{s}$ = 540 GeV the ratio of cross sections

$$\frac{\sigma(\bar{p}p \to WX \to e\nu X)}{\sigma_T(\bar{p}p)} \approx 10^{-8}, \tag{17.13}$$

where X stands for hadrons. This unfavourable ratio sets enormous requirements on the detector, because only one event in 10^8 collisions contains a detectable heavy boson. Figure 17.7 shows the UA1 detector and figure 17.8 an event containing a decay of the type $W \to e\nu_e$ (with an undetected neutrino).

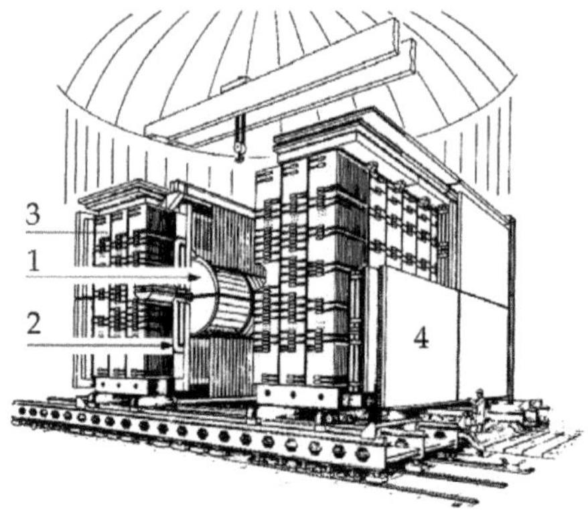

Figure 17.7. The UA1 detector at the SPS $\bar{p}p$ collider. The magnetic field (0.7 T) is horizontal. 1: inner drift chamber and electromagnetic calorimeter; 2: coil; 3: hadronic calorimeter; 4: muon chamber. After [7].

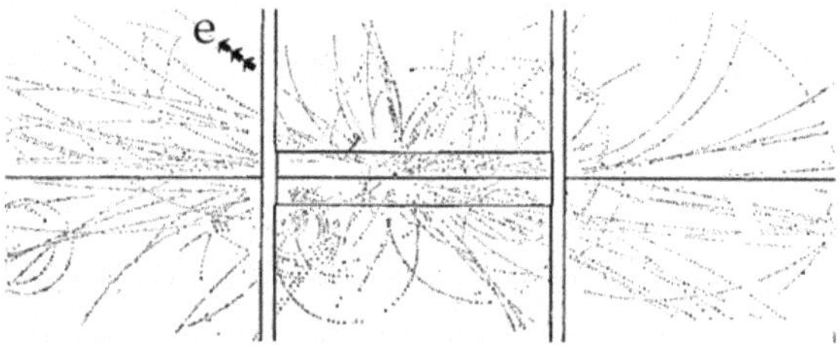

Figure 17.8. A typical event in which an energetic electron from the decay $W \to e\nu$ is detected in the electromagnetic calorimeter of the UA1 detector. The points are reconstructed coordinates of charged tracks in the drift chamber (figure 17.9 below). After [3].

Searching for 40 GeV electrons or positrons is not sufficient to identify W events, because other processes (such as semileptonic decays of D and B mesons) also emit high energy leptons. One therefore computes the total momentum $\vec{p}(X)$ of the hadrons by measuring the energies and angles of all hadrons and photons (e.g. from π^0 decays), adding the momenta *vectorially*. Genuine W events have a 'missing' momentum $\vec{p}(\nu)$, that of the unobserved neutrino. From momentum conservation

$$\vec{p}(\nu) = -\vec{p}(X) - \vec{p}(e). \tag{17.14}$$

The missing neutrino momentum must be equal to that of the electron since the W bosons are on average produced in rest. In addition, $\vec{p}(\nu)$ must be antiparallel to $\vec{p}(e)$. However, particles emitted in the very forward directions are not detected due to the holes left in the detector for the incident beams. Equation (17.13) is therefore not useful. One applies instead momentum conservation to the momenta $\vec{p}_\perp(\nu)$, $\vec{p}_\perp(X)$, $\vec{p}_\perp(e)$ *transverse* to the beam, hence

$$\vec{p}_\perp(\nu) = -\vec{p}_\perp(X) - \vec{p}_\perp(e). \tag{17.15}$$

The missing transverse momenta of forward emitted particles are small and can be neglected.

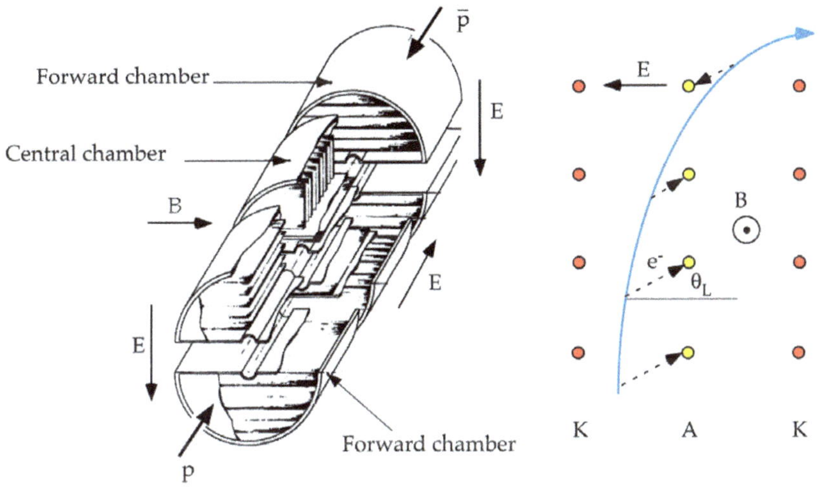

Figure 17.9. Diagram of the UA1 drift chambers. As for silicon (section 9.9), the electrons do not drift in the directions of the applied electric field $\vec{E}$, but under the Lorentz angle θ_L, due to the orthogonal magnetic field $\vec{B}$ (see figure 9.33). K: cathode wires; A: anode wires. After [8].

The sophisticated UA1 drift chambers that detect charged particles are shown in figure 17.9. The drift time of the electron cloud to the anode plane is measured on every wire. The starting time is determined by the collision time of the beam bunches. The coordinates along the anode wires are determined by comparing the pulse heights from both ends of the wires. The electric field is parallel to the beam axis in the central drift chamber and perpendicular in the two forward chambers, to maintain large angles between the tracks and drift directions. For high energy particles the curvature in the magnetic field is very small and hence the momentum resolution is modest (±8 GeV at 40 GeV). However, the lepton charge can still be determined, and therefore also the charge of the $W^{\pm}$. This is important to verify that equal numbers of W^+ as W^- are produced, as expected in $\bar{p}p$ interactions.

The energies of the hadrons are measured using hadronic calorimeters and the energies of the photons, electrons and positrons using electromagnetic calorimeters (figure 17.10). An iron-scintillator sandwich is used for hadrons. The iron also serves as magnetic flux return. To maximize the tightness of the detector, the usual curved plexiglas light guides between the scintillator and photomultiplier are avoided and replaced by wavelength shifters which absorb and re-emit the light produced by the scintillators. The electromagnetic calorimeter (figure 17.10, right) consists of a lead–scintillator sandwich (with a resolution of ±500 MeV for 40 GeV electrons). The impact point (angle of the electron or photon) is obtained by comparing the signal amplitudes from both ends of the light guide.

Figure 17.11 shows the projection of $\vec{p}_\perp(\nu)$ on $\vec{p}_\perp(e)$ as a function of $p_\perp(e)$. For the 43 W candidates, the calculated neutrino momenta are comparable and antiparallel to the momenta of the electrons or positrons. Because of the finite resolution of the detector, and since the W bosons are not always at rest, the measured momenta are not exactly antiparallel.

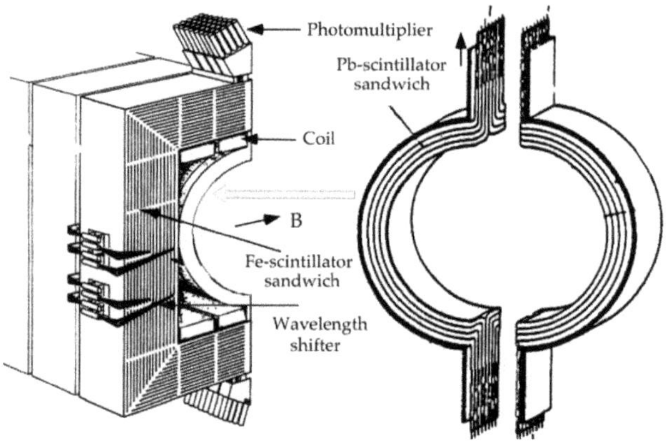

Figure 17.10. Left: hadronic calorimeter. After [9]. Right: electromagnetic calorimeter. After [7].

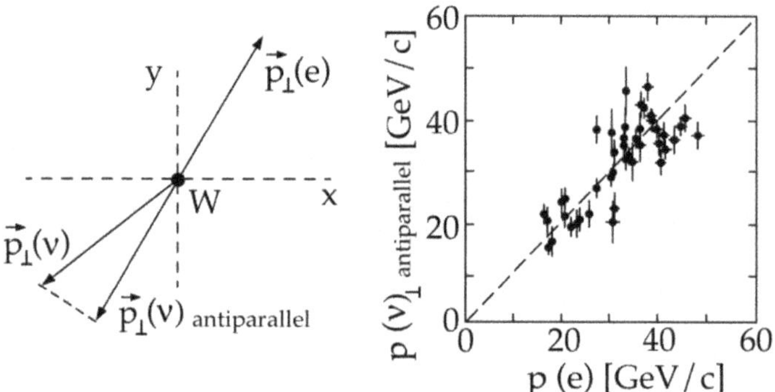

Figure 17.11. Left: lepton momenta transverse to the beam. Right: transverse neutrino momentum antiparallel to the electron transverse momentum, as a function of electron transverse momentum (43 events). After [10].

One still needs to be convinced that the 43 events stem from $W \rightarrow e\nu$, and not from the three-body decay $e\nu\bar{\nu}$ of a new heavy lepton. Furthermore, the W mass should be consistent with expectation. Let us calculate the mass under the simplifying assumption that the angular distribution of the electron is isotropic (this is not quite the case, as explained in section 17.2.1). The distribution of transverse momentum for the electron or positron, emitted under the angle θ with respect to the beam axis, is given by

$$\frac{\mathrm{d}N}{\mathrm{d}p_\perp} = \frac{\mathrm{d}N}{d\cos\theta} \cdot \frac{d\cos\theta}{\mathrm{d}p_\perp} \propto \frac{d\cos\theta}{\mathrm{d}p_\perp}. \tag{17.16}$$

Using

$$p_\perp = \frac{M_W}{2}\sin\theta = \frac{M_W}{2}\sqrt{1-\cos^2\theta} \tag{17.17}$$

one obtains

$$\frac{dp_\perp}{d\cos\theta} = \left(\frac{M_W}{2}\right)\frac{\cos\theta}{\sin\theta} = \left(\frac{M_W}{2}\right)\frac{\sqrt{1-\sin^2\theta}}{\sin\theta} = \left(\frac{M_W}{2}\right)^2\frac{\sqrt{1-\frac{4p_\perp^2}{M_W^2}}}{p_\perp}, \qquad (17.18)$$

hence with (17.16),

$$\frac{dN}{dp_\perp}(p_\perp) \propto \frac{p_\perp}{\sqrt{M_W^2 - 4p_\perp^2}}. \qquad (17.19)$$

A pole arises in the $p_\perp$ distribution at $p_\perp = M_W/2$. The distribution shown in figure 17.12 (left) is compatible with (17.19), leading to the mass $M_W \sim 80$ GeV. The smearing of the distribution around the pole is due to the fact that (i) the W bosons are only on average at rest; (ii) they have a finite width; and (iii) the detector resolution is finite. More accurate values of the mass and width can be obtained with a Monte Carlo simulation.

The $Z^0 \to e^+e^-$ was also discovered by UA1 (four events) and by UA2 (eight events). In this case a 45 GeV electron recoils against a 45 GeV positron. The two leptons are emitted back-to-back and are directly detected by their charged tracks pointing to large energy deposits in the electromagnetic calorimeter (figure 17.12, right).

It is instructive to compare these early results with the 'routine' distributions currently obtained at the Large Hadron Collider (LHC). Figure 17.13 (left) shows the transverse mass distribution for 12 257 W decays into $\mu\nu$ obtained by the Compact Muon Solenoid (CMS) at 7 TeV (the apparatus will be described in section 17.3 below). The transverse mass is defined as $M_\perp \simeq 2p_\perp$. Its distribution is given by

$$\frac{dN}{dM_\perp}(M_\perp) \propto \frac{M_\perp}{\sqrt{M_W^2 - M_\perp^2}}, \qquad (17.20)$$

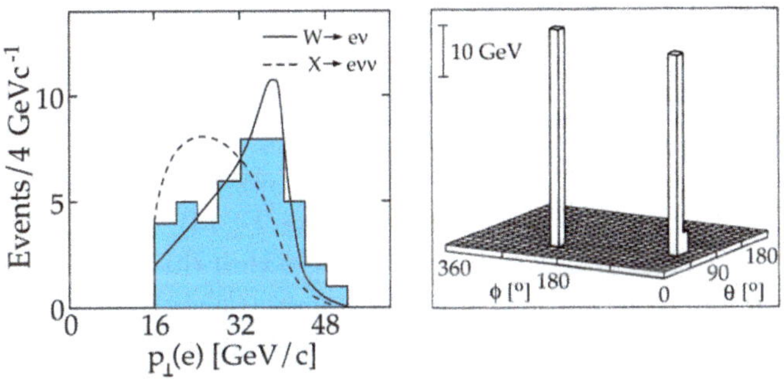

Figure 17.12. Left: transverse momentum distribution of the UA1 electrons or positrons (43 events). The solid curve shows the prediction for $W \to e\nu$ (Monte Carlo simulation). The dashed curve shows the expectation for a hypothetical heavy lepton X decaying into $e\nu\bar{\nu}$. After [10]. Right: a $Z^0 \to e^+e^-$ event observed by the UA2 detector. Plotted is the energy deposit in the electromagnetic calorimeter versus θ and ϕ (azimuthal angle). The leptons are emitted back-to-back. After [6].

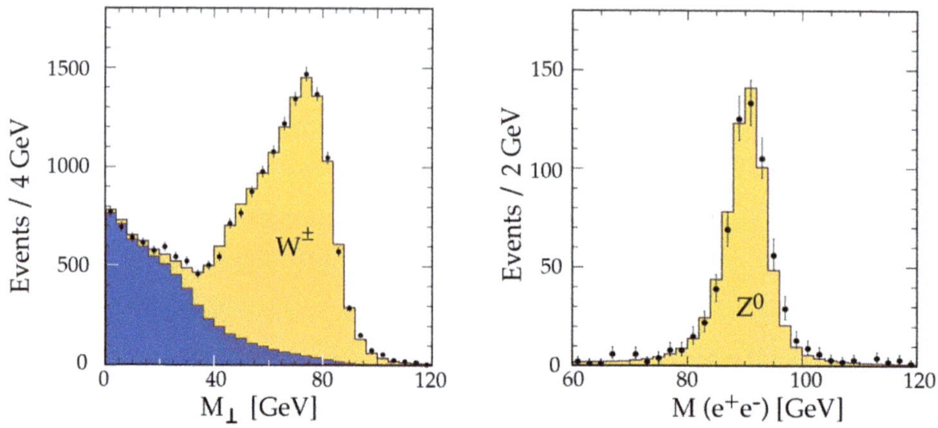

Figure 17.13. Left: transverse mass distribution for $W \to \mu\nu$. The points are data and the histograms simulations of the signal (yellow) and of the background (blue). Right: e^+e^- invariant mass distribution for $Z^0 \to e^+e^-$ (data from CMS). After [11].

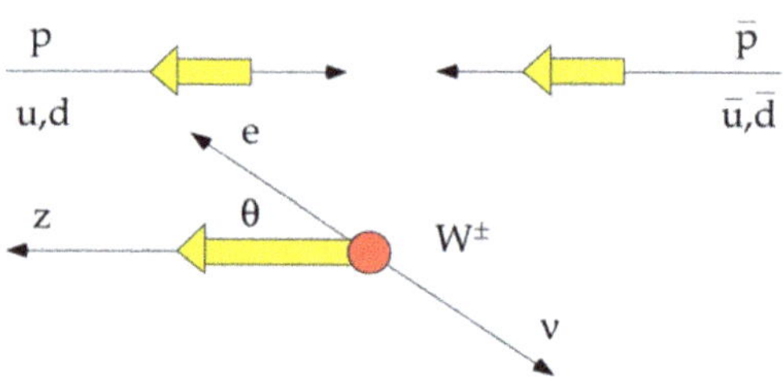

Figure 17.14. The polarization of the W boson produced in $\bar{p}p$ collisions.

using (17.19). The distribution has a pole at $M_\perp = M_W \simeq 80$ GeV. Figure 17.13 (right) shows the e^+e^- invariant mass distribution for 677 Z^0 decays into e^+e^-. The masses of the gauge bosons are now known with very good accuracy: $M_W = 80.385 \pm 0.015$ GeV and $M_{Z^0} = 91.1876 \pm 0.0021$ GeV [12].

17.2.1 *W* width

Due to $V - A$ and the conservation of angular momentum the spin of the W boson points in the direction of the antiproton (figure 17.14): the W bosons are 100% polarized in the direction of the antiproton. (Note that this does not apply to the Z^0 boson since both left-handed and right-handed quark currents are involved.)

Let us compute the transition amplitude for the decay $W^- \to e^-\bar{\nu}_e$. We have seen that the interaction field between charged currents could be written in the form propagator × current, (16.134) and (16.135). The decay amplitude for the charged current interaction is therefore

$$\mathcal{M}_{\mathrm{CC}} = \frac{4G_F}{\sqrt{2}} J_C^\mu J_{C\mu} = \left(\frac{g}{\sqrt{2}} J_C^\mu\right) \frac{1}{M_W^2} \left(\frac{g}{\sqrt{2}} J_{C\mu}\right) = \frac{g}{\sqrt{2}} J_C^\mu W_\mu^-. \tag{17.21}$$

For a 100% polarized W^- in the z-direction, the field is described by a right-circularly polarized wave propagating along z (see (11.81) on the helicity of the photon):

$$W_\mu^- = \frac{1}{\sqrt{2}}(0,\ 1,\ \mathrm{i},\ 0). \tag{17.22}$$

Therefore with $F \equiv \mathcal{M}_{\mathrm{CC}}$:

$$|F|^2 = \frac{g^2}{2} \underbrace{\left[\bar{u}(e)\gamma^\mu\left(\frac{1-\gamma^5}{2}\right)\overbrace{v(\nu)}^{u(-k)}\right]\left[\bar{u}(e)\gamma^\nu\left(\frac{1-\gamma^5}{2}\right)v(\nu)\right]^\dagger}_{\equiv L^{\mu\nu}/4} W_\mu^- W_\nu^{-*}. \tag{17.23}$$

The product between currents can be calculated using the trace techniques explained in chapter 15. Denoting the 4-momentum of the electron by p and that of the antineutrino by k, and neglecting the mass of both leptons, one obtains

$$\begin{aligned}
L^{\mu\nu} &= \sum_{ss'} \bar{u}_\alpha\left[\gamma^\mu\left(1-\gamma^5\right)\right]_{\alpha\beta} v_\beta \bar{v}_\gamma \left[\gamma^\nu\left(1-\gamma^5\right)\right]_{\gamma\delta} u_\delta \\
&= \sum_s u_\delta \bar{u}_\alpha\left[\gamma^\mu\left(1-\gamma^5\right)\right]_{\alpha\beta} \sum_{s'} v_\beta \bar{v}_\gamma \left[\gamma^\nu\left(1-\gamma^5\right)\right]_{\gamma\delta} \\
&= \mathrm{Tr}\left[\not{p}\gamma^\mu\left(1-\gamma^5\right)\not{k}\gamma^\nu\left(1-\gamma^5\right)\right] \\
&= 2\,\mathrm{Tr}\left[\not{p}\gamma^\mu\left(1-\gamma^5\right)\not{k}\gamma^\nu\right] \\
&= 2\,\mathrm{Tr}\left(\not{p}\gamma^\mu\not{k}\gamma^\nu\right) - 2\,\mathrm{Tr}\left(\not{p}\gamma^\mu\gamma^5\not{k}\gamma^\nu\right) \\
&= 2\,\mathrm{Tr}\left(\gamma^\alpha\gamma^\mu\gamma^\beta\gamma^\nu\right)p_\alpha k_\beta - 2\,\mathrm{Tr}\left(\gamma^\alpha\gamma^\mu\gamma^5\gamma^\beta\gamma^\nu\right)p_\alpha k_\beta.
\end{aligned} \tag{17.24}$$

Using the following properties of γ matrices from table 15.1,

$$\begin{aligned}
\mathrm{Tr}\left(\gamma^\alpha\gamma^\mu\gamma^\beta\gamma^\nu\right) &= 4\left(g^{\alpha\mu}g^{\beta\nu} + g^{\alpha\nu}g^{\mu\beta} - g^{\alpha\beta}g^{\mu\nu}\right), \\
\mathrm{Tr}\left(\gamma^\alpha\gamma^\mu\gamma^5\gamma^\beta\gamma^\nu\right) &= 4\mathrm{i}\epsilon^{\alpha\mu\beta\nu} = -4\mathrm{i}\epsilon^{\mu\nu\alpha\beta},
\end{aligned} \tag{17.25}$$

one obtains

$$\begin{aligned}
L^{\mu\nu} &= 8\left[g^{\alpha\mu}g^{\beta\nu} + g^{\alpha\nu}g^{\mu\beta} - g^{\alpha\beta}g^{\mu\nu} + \mathrm{i}\epsilon^{\mu\nu\alpha\beta}\right]p_\alpha k_\beta \\
&= 8\left(p^\mu k^\nu + p^\nu k^\mu - g^{\mu\nu}pk + \mathrm{i}\epsilon^{\mu\nu\alpha\beta}p_\alpha k_\beta\right).
\end{aligned} \tag{17.26}$$

Therefore

$$|F|^2 = g^2\left[p^\mu k^\nu + p^\nu k^\mu - g^{\mu\nu}pk + \mathrm{i}\epsilon^{\mu\nu\alpha\beta}p_\alpha k_\beta\right]W_\mu^- W_\nu^{-*}. \tag{17.27}$$

Introducing the components of the 4-momenta

$$\begin{aligned}
p &= \frac{M_W}{2}(1,\ \sin\theta\cos\phi,\ \sin\theta\sin\phi,\ \cos\theta), \\
k &= \frac{M_W}{2}(1,\ -\sin\theta\cos\phi,\ -\sin\theta\sin\phi,\ -\cos\theta),
\end{aligned} \tag{17.28}$$

and the field (17.22) one obtains the angular distribution of the electron,

$$\boxed{|F|^2 = \frac{g^2 M_W^2}{4}(1 - \cos\theta)^2}, \tag{17.29}$$

where θ is the angle between the electron and $\bar{p}$-directions (the proof is easy when choosing the x- and y-axes such that $\phi = 0$).

The angular distribution of the electron is anisotropic and for $\theta = 0$ the amplitude vanishes. The reason becomes obvious when looking at figure 17.14: a left-handed electron flying in the direction of the W spin would violate angular momentum conservation. Hence electrons fly preferably in the direction of the incident proton. On the other hand, in the decay $W^+ \to e^+\nu_e$ the positron is right-handed and therefore prefers the direction of the antiproton with the angular distribution $\propto(1 + \cos\theta)^2$, where θ is the angle between the positron and antiproton directions.

At UA1 electrons can be distinguished from positrons by their curvatures in the magnetic field. Figure 17.15 shows the measured angular distribution, where θ is now the angle between the electron and proton directions, or between the positron and antiproton directions. Figure 17.15 is the smoking gun confirmation of the $V - A$ interaction in W decay.

Let us compute the partial width for the decay $W^- \to e^-\bar{\nu}_e$, and then the total width (inversely proportional to the lifetime) of the W boson. The differential partial decay width is, using (12.22) for 100% polarization,

$$\frac{d\Gamma}{d\Omega} = \frac{1}{64\pi^2 M_W}|F|^2, \tag{17.30}$$

where we have set $k = M_W/2$. Therefore,

$$d\Gamma = \frac{|F|^2}{64\pi^2 M_W}\overbrace{d(2\pi\cos\theta)}^{d\Omega}. \tag{17.31}$$

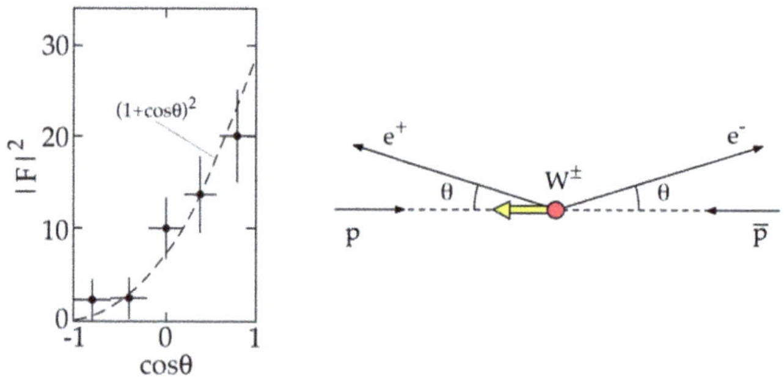

Figure 17.15. Angular distribution $\propto (1 + \cos\theta)^2$ of the electrons or positrons in the decay of polarized W bosons (left). The curve shows the $V - A$ prediction from (17.29). After [10]. The angle θ is between the p- and e^--directions or between the $\bar{p}$- and e^+-directions (right).

Substituting $|F|^2$ from (17.29) and integrating over $\cos\theta$ leads to the prediction for the partial width

$$\Gamma(W \to e\nu) = \left(\frac{g^2 M_W^2}{4}\right)\frac{2\pi}{64\pi^2 M_W}\overbrace{\int_{-1}^{1}(1-\cos\theta)^2 d\cos\theta}^{\frac{8}{3}}$$

$$= \frac{g^2 M_W}{48\pi} = \left(\frac{G_F}{\sqrt{2}}\right)\frac{M_W^3}{6\pi} = 224\ \text{MeV}, \qquad (17.32)$$

with $M_W = 80$ GeV, and where we have used the relation $G_F/\sqrt{2} = g^2/8M_W^2$.

To compute the *total* width we employ the universality of the couplings in W decay which states that

- in a very good approximation, since the lepton masses are negligible compared to the W mass, the partial widths for $W^- \to e^-\bar{\nu}$, $\mu^-\bar{\nu}$ and $\tau^-\bar{\nu}$ are equal;
- the partial widths for $W^- \to \bar{u}d$ and $\bar{c}s$ are equal to the partial width for $e^-\bar{\nu}$, up to the factor $\cos^2\theta_C \sim 1$. The decay into $\bar{t}b$ is hindered by the very large t mass;
- the partial widths for $W^- \to \bar{u}s$, $\bar{c}d$, $\bar{u}b$ and $\bar{c}b$ are much smaller, since transitions between quark generations are suppressed by the CKM matrix (16.69).

The quark currents exist in three colours which are conserved. Since the W is colour-blind, one expects with three lepton currents and the quark currents $\bar{u}d$ and $\bar{c}s$ in three colours, the total width

$$\Gamma_T(W) \simeq 9 \times \Gamma(W \to e\nu) = 2.02\ \text{GeV}, \qquad (17.33)$$

compared to the measured value $\Gamma_T(W) = 2.085 \pm 0.042$ GeV [12]. This confirms (i) the standard model, (ii) the existence of three quark colours, and (iii) that there are no further unknown light fermions with the same couplings. A more precise statement for the latter is obtained from the Z^0 width.

17.2.2 Z^0 width

Let us first calculate the partial width for the (unobservable) decay $Z^0 \to \nu\bar{\nu}$. For the interaction between neutral currents we express again the field as propagator × current:

$$\mathcal{M}_{\text{NC}} = \frac{8G_F}{\sqrt{2}} J_N^\mu J_{N\mu} = gJ_N^\mu \frac{1}{M_W^2} gJ_{N\mu} = \left(\frac{g}{\cos\theta_w} J_N^\mu\right)\frac{1}{M_{Z^0}^2}\left(\frac{g}{\cos\theta_w} J_{N\mu}\right)$$

$$= \frac{g}{\cos\theta_w} J_N^\mu Z_\mu. \qquad (17.34)$$

The neutral current J_N^μ of the neutrino, being pure $V - A$, has the same form as J_C^μ, apart from a factor $c_L(\nu) = \frac{1}{2}$. Hence by comparing (17.34) with (17.21) we perform the substitution

$$g \to \frac{g\sqrt{2}}{\cos\theta_w} c_L(\nu) \qquad \text{and} \qquad M_W \to M_{Z^0} \qquad (17.35)$$

in (17.32), and obtain

$$\Gamma(Z^0 \to \nu\bar{\nu}) = \frac{g^2 M_{Z^0}}{96\pi \cos^2\theta_w}, \tag{17.36}$$

or with $\frac{G_F}{\sqrt{2}} = \frac{g^2}{8M_W^2}$,

$$\Gamma(Z^0 \to \nu\bar{\nu}) = \left(\frac{G_F}{\sqrt{2}}\right)\frac{M_{Z^0}^3}{12\pi} = 165 \text{ MeV} \tag{17.37}$$

for $M_{Z^0} = 91$ GeV. To compute the total width, let us list all possible decays into fermion–antifermion pairs (table 17.2). Dividing (17.37) by $c_L(\nu)^2 = 1/4$ and multiplying by $c_L(f)^2 + c_R(f)^2$ (factor K) one obtains with n generations

$$Kn = 3 \times 1 + 3 \times 0.5 + 6 \times 0.57 + 9 \times 0.74 = 14.6 \tag{17.38}$$

and the total width

$$\Gamma_T(Z^0) = Kn\Gamma(Z^0 \to \nu\bar{\nu}) = 2.41 \text{ GeV}, \tag{17.39}$$

to be compared to the measured 2.4952 ± 0.0023 GeV [12].

The width can be measured directly in e^+e^- collisions, where the Z^0 appears as a resonance. The total cross section for $e^+e^- \to Z^0 \to f\bar{f}$ is given by the Breit–Wigner distribution

$$\sigma = \frac{4\pi}{s}\left[\frac{2J+1}{(2S+1)^2}\right]\frac{\Gamma(e^+e^-)\Gamma(f\bar{f})}{\left(\sqrt{s} - M_{Z^0}\right)^2 + \Gamma_T^2(Z^0)/4}, \tag{17.40}$$

where $J = 1$ is the spin of the Z^0, $S = \frac{1}{2}$ is the spin of the electron (positron), and $\sqrt{s}$ is the total energy. $\Gamma(e^+e^-)$ and $\Gamma(f\bar{f})$ are the partial widths for the decays $Z^0 \to e^+e^-$ and $Z^0 \to f\bar{f}$, respectively. Figure 17.16 shows the cross section σ versus $\sqrt{s}$

Table 17.2. Partial width $\Gamma(Z^0 \to f\bar{f})$ for n generations of fermions; $K = 4[c_L(f)^2 + c_R(f)^2]$ and $a \equiv \sin^2\theta_w = 0.23$.

$f\bar{f}$	K	n
$\nu\bar{\nu}$	1	3
e^+e^-, $\mu^+\mu^-$, $\tau^+\tau^-$	$4\left[\left(-\frac{1}{2}+a\right)^2 + a^2\right] = 0.50$	3
$u\bar{u}$, $c\bar{c}$	$4\left[\left(\frac{1}{2}-\frac{2}{3}a\right)^2 + \frac{4}{9}a^2\right] = 0.57$	2 × 3 colours
$d\bar{d}$, $s\bar{s}$, $b\bar{b}$	$4\left[\left(-\frac{1}{2}+\frac{1}{3}a\right)^2 + \frac{1}{9}a^2\right] = 0.74$	3 × 3 colours

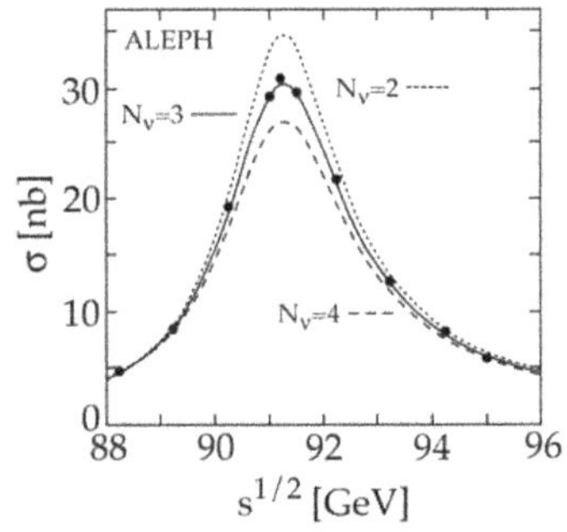

Figure 17.16. The Z^0 resonance in e^+e^- scattering into hadrons. The curves show the predictions assuming the existence of 2, 3 and 4 light neutrinos. After [13].

measured by the ALEPH experiment at LEP, where quark–antiquark pairs in the final state fragment into hadrons. The $1/s$ dependence of the electromagnetic contribution to the cross section (γ exchange) has been divided out. To measure the cross section, the incident particle flux (luminosity integrated over data taking time) must be known. This is achieved by simultaneously recording Bhabha events (elastic e^+e^- scattering) at *small* scattering angles. The cross section for Bhabha scattering grows with decreasing angle θ as $\sin^{-4}\theta/2$ (in analogy to Mott scattering) and is dominated by γ exchange at small scattering angles.

The Z^0 width can be calculated precisely in the standard model including radiative corrections. The three curves in figure 17.16 show the predictions of the standard model with $N_\nu = 2$, 3 or 4 neutrino families. The best fit confirms the existence of three families, excluding a fourth one. From the experiments at LEP and at the SLC [14] one finds

$$\boxed{N_\nu = 2.984 \pm 0.008} \tag{17.41}$$

light neutrinos. However, the existence of additional quarks (apart from the t quark) and neutrinos with masses larger than half the Z^0 mass ($\gtrsim$45 GeV) is not excluded, nor is the existence of 'sterile' neutrinos that do not interact weakly.

Significant deviations from the standard model have not been observed so far. As our final examples, let us estimate with table 17.2 the branching ratios for Z^0 decay into leptons and hadrons. In first order the branching ratios

$$f(\mu^+\mu^-) = \frac{\Gamma(\mu^+\mu^-)}{\Gamma_T} = \frac{0.5}{14.6} = 3.4\%, \tag{17.42}$$

and

$$f(\text{hadrons}) = \frac{\Gamma(\text{hadrons})}{\Gamma_T} = \frac{6 \times 0.57 + 9 \times 0.74}{14.6} = 69.0\% \tag{17.43}$$

are in very good agreement with the experimental values of $(3.366 \pm 0.007)\%$ and $(69.91 \pm 0.06)\%$, respectively [12].

17.3 The discovery of the Higgs boson

The Higgs boson was sought without success at LEP in e^+e^- collisions and at the Fermilab $\bar{p}p$ collider. At LEP the lower mass limit was 114 GeV while the CDF and D0 experiments at Fermilab excluded a Higgs boson around 165 GeV. The main production modes in e^+e^-, $\bar{p}p$ and pp colliders are depicted in figure 17.17 (left). In $\bar{p}p$ colliders the dominant contributions are from gluon–gluon fusion (a) and Higgs emission (or 'higgsstrahlung') by virtual $W^\pm$ or Z^0 bosons (b). In pp colliders gluon–gluon fusion dominates (a) followed by vector boson fusion (c) which played an important role in the Higgs discovery (see below). At LEP higgsstrahlung dominates (b) followed by vector boson fusion (c). Figure 17.17 (right) shows the corresponding Higgs production cross sections in pp collisions at a centre-of-mass energy of 14 TeV. The total pp cross section is around 110 mb. A 126 GeV Higgs is produced with a tiny probability of a few 10^{-10}, associated with a recoiling system X involving typically 100 charged particles.

The decay branching ratios are shown in figure 17.18, as a function of Higgs mass. A heavy Higgs would mainly decay into pairs of gauge bosons while a light Higgs couples mainly to a pair of heavy fermions. The decay mode $H \rightarrow b\bar{b}$ is very hard to observe due to the huge QCD background from direct $b\bar{b}$ production. Therefore it is mostly sought in associated production with the W or Z^0 (higgs-strahlung in figure 17.17(*b*)). The decay $H \rightarrow \gamma\gamma$ proceeds via virtual charged pairs such as W^+W^- and $t\bar{t}$ loops, and the decay $H \rightarrow gg$ over $t\bar{t}$ loops. The table in figure 17.18 lists the predicted main branching ratios for a 126 GeV Higgs boson in the standard model.

The Higgs boson was discovered in 2012 at the LHC running at 7–8 TeV by the CMS and ATLAS Collaborations [17, 18]. The CMS (figure 17.19) consists of a superconducting (niobium–titanium) solenoid of 6 m internal diameter, providing a

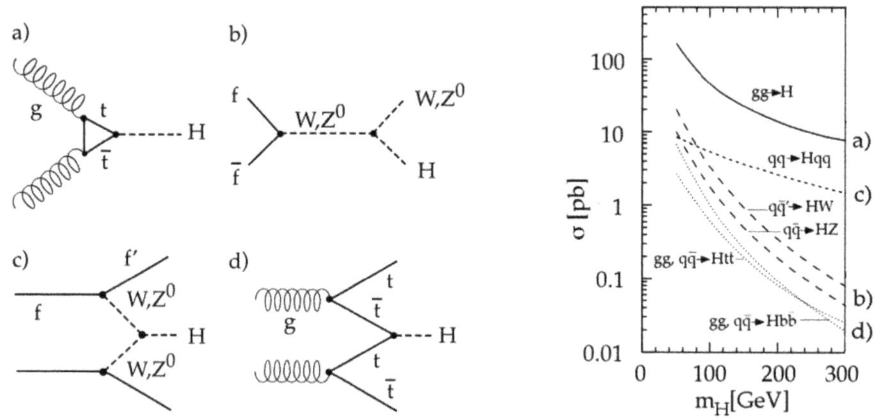

Figure 17.17. Left: Feynman diagrams of the main Higgs production processes at an e^+e^- collider, the Tevatron ($\bar{p}p$) and the LHC (pp); (*a*) gluon–gluon fusion; (*b*) higgsstrahlung (f stands for an electron or quark); (*c*) vector boson fusion; (*d*) $t\bar{t}$ fusion. Right: corresponding production cross sections in pp collisions at $\sqrt{s}$ = 14 TeV, as a function of Higgs mass. Adapted from [15].

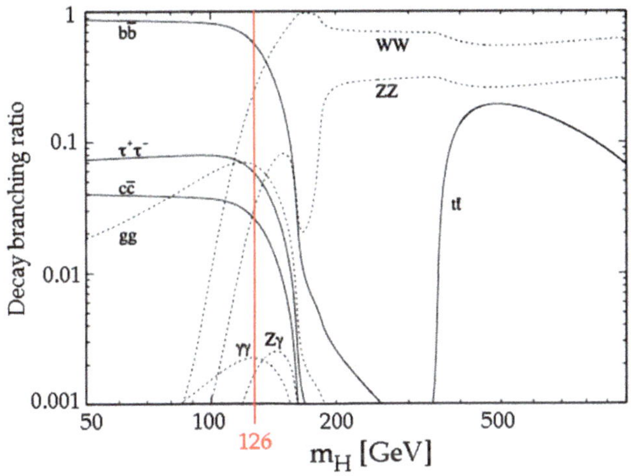

Mode	B.r.
$b\bar{b}$	5.8×10^{-1}
W^+W^-	2.2×10^{-1}
$\tau^+\tau^-$	6.3×10^{-2}
Z^0Z^0	2.6×10^{-2}
$c\bar{c}$	2.4×10^{-2}
$\gamma\gamma$	2.3×10^{-3}

Figure 17.18. Left: main Higgs decay modes in the standard model as a function of mass. After [16]. Right: branching ratios (B.r.) for a Higgs boson at 126 GeV.

magnetic field of 3.8 T. The solenoid contains a silicon pixel detector (66 million pixels of dimensions $100 \times 150\ \mu m^2$, in three cylindrical layers and two endcap disks) surrounding the beam pipe, followed by a microstrip tracker (9.3 million strips). A crystal calorimeter made of 75 848 lead tungstate ($PbWO_3$) crystals detects photons, electrons and positrons. Electromagnetic showers are narrow in lead tungstate due to the small Molière radius (9.81) of 21 mm. The γ impact point is determined by a lead radiator (preshower) which initiates the electromagnetic shower, followed by microstrip sensors. The rms energy resolution on a 126 GeV Higgs decaying into two photons is 1.1 GeV. A hadronic calorimeter (brass–scintillator sandwich), the thickness of which varies from 7 to 11 interaction lengths, completes the internal detector. Muons are measured in drift tubes, cathode strip chambers and resistive plate chambers located in the field-return steel yoke. Track coordinates in the transverse plane can be measured with a precision of typically 100 μm. Endcap calorimeters cover the forward detector regions. The CMS detector is 29 m long, 15 m in diameter and weighs 14 000 t. Details on the apparatus can be found in [19].

The inner detector of ATLAS consists of a silicon pixel device, a silicon microstrip detector, and a transition radiation tracker (figure 17.19, bottom). The inner detector is surrounded by a thin superconducting solenoid providing a magnetic field of 2 T, and by a high-granularity liquid argon electromagnetic calorimeter. The electromagnetic calorimeter is divided into a central barrel and two endcaps. The central barrel is radially segmented into three layers. The rms energy resolution on a 126 GeV Higgs decaying into two photons is 1.6 GeV. An iron–scintillator sandwich calorimeter detects the hadrons. The forward regions are instrumented with liquid argon calorimeters for both electromagnetic and hadronic measurements. A muon spectrometer consisting of air-core superconducting coils instrumented with tracking chambers provides a toroidal magnetic field. The ATLAS detector is 44 m long, 22 m in diameter and weighs 7 000 t. Details on the apparatus can be found in [20].

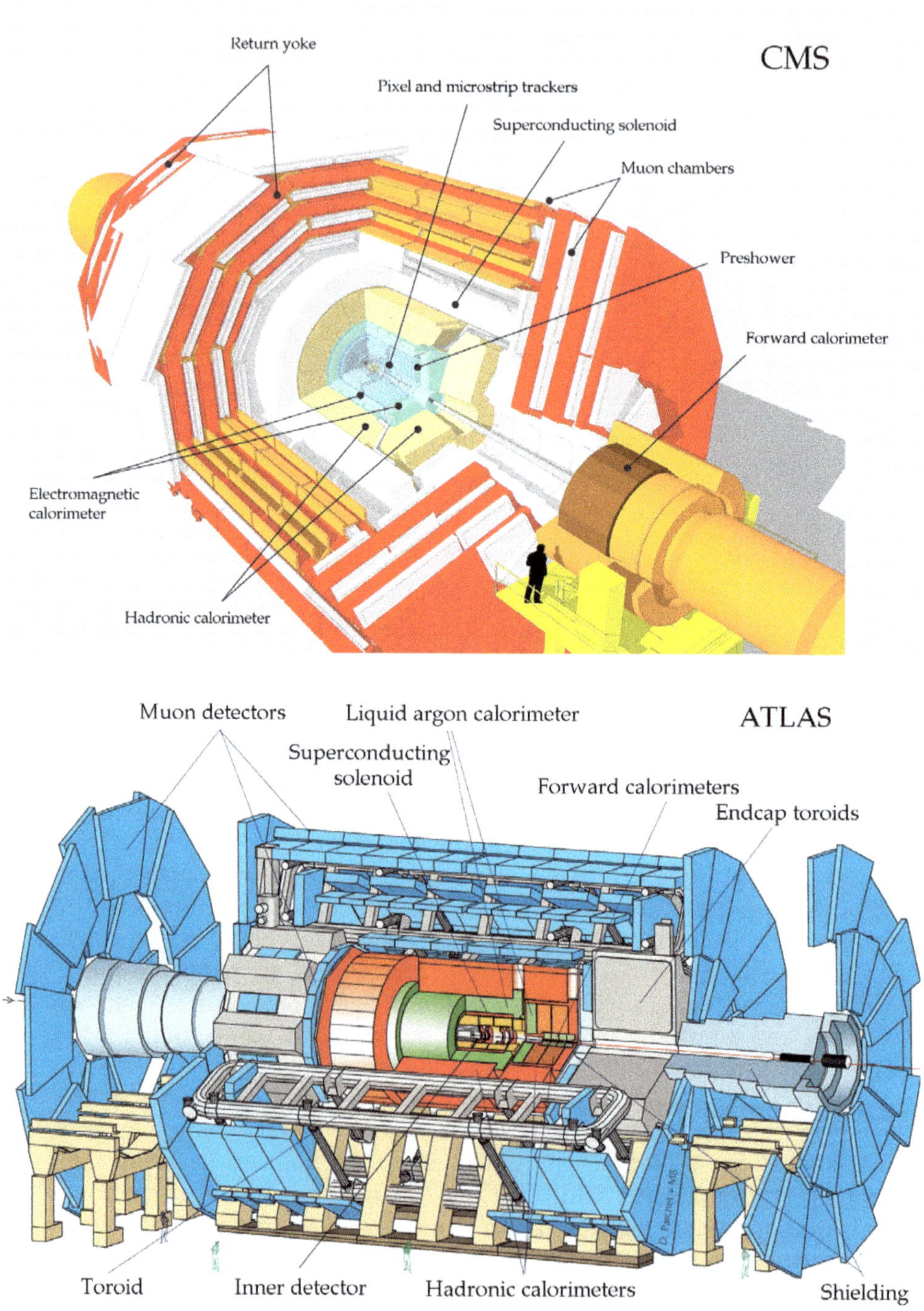

Figure 17.19. The CMS and ATLAS detectors (see the text for details). Figure 17.20 shows a photograph of the CMS detector. Image credit: CERN.

Figure 17.20. A front view of the Compact Muon Solenoid (CMS) detector at the LHC. Image credit: CERN.

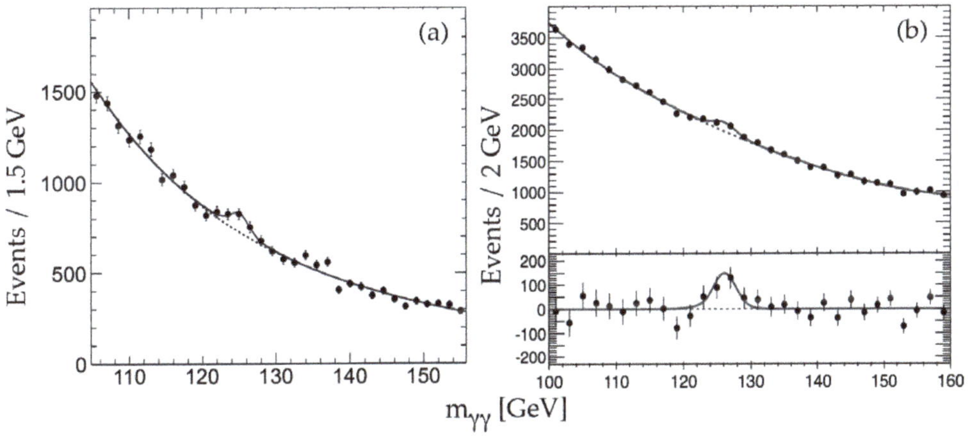

Figure 17.21. Observation of the 126 GeV Higgs signal in $\gamma\gamma$ with pp collisions at 7 and 8 TeV, (*a*) by CMS [17] and (*b*) by ATLAS [18]. For the latter the background subtracted data are also plotted. The full lines show the fits and the dotted polynomial curves show the background contributions.

The most sensitive decay channel is $H \to \gamma\gamma$, which has a relatively small branching ratio, but which can be identified by the presence of a narrow peak in the 2γ invariant mass, consistent with the detector resolution. The best production channel was vector boson fusion (figure 17.17(*c*)) for which the background can be minimized by requiring the presence of two associated quark jets. Figure 17.21(*a*) shows the 2γ invariant mass distribution measured by CMS. A clear signal is

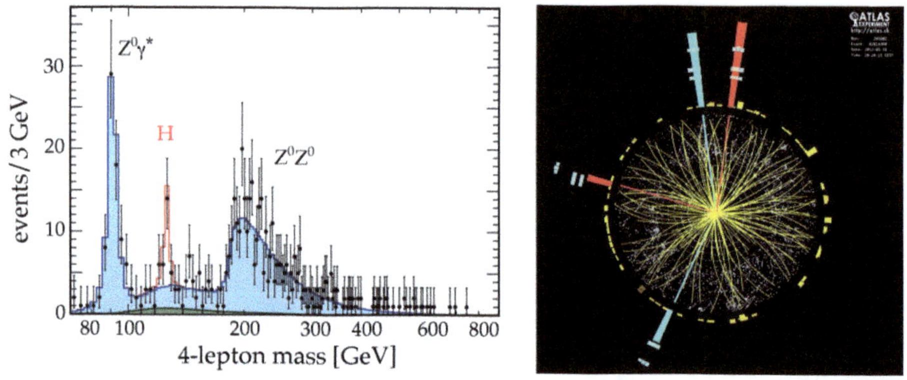

Figure 17.22. Left: $4e$, $2e2\mu$, and 4μ invariant mass distribution above 70 GeV for events consistent with the production of a Z^0Z^0 (virtual) pair. The shaded blue histogram is the background and the red histogram is the expected signal for $m_H = 126$ GeV. There are 17.3 ± 1.3 events in the 126 GeV peak (CMS data). After [21]. Right: a $H \to Z^0Z^0 \to (e^+e^-)(e^+e^-)$ event observed by ATLAS. The towers show the energy deposits in the electromagnetic calorimeters. Image credit: CERN.

observed at 126 GeV. The signal strength (measured cross section multiplied by decay branching ratio, divided by the expected value) is $\mu = 1.6 \pm 0.4$, in agreement with the standard model prediction. Figure 17.21(*b*) shows the similar result from ATLAS.

At the time of discovery signals were also found for decays into two gauge bosons, albeit with lower significance. The analysis of further data taken at 7 and 8 TeV led to stronger evidence for these decay modes, as well as for $H \to \tau^+\tau^-$. Figure 17.22 (left) shows the 4-lepton invariant mass distribution recorded by CMS with a clear signal for a state at 126 GeV. The peak at 90 GeV stems from background pairs associated with the production of single Z^0 bosons. Figure 17.22 (right) shows a typical ATLAS event involving the production of a Higgs boson decaying into $e^+e^-e^+e^-$ via a pair of Z^0 bosons.

Figure 17.23 shows the signal strengths for the four decay modes observed by ATLAS and CMS. All measurements are in good agreement with the theoretical predictions for a standard model Higgs.

The current best value for the Higgs mass is 125.7 ± 0.4 GeV [12]. There is strong evidence that the observed particle is indeed a $J^P = 0^+$ boson, as expected from the standard model. The observation of the mode $H \to \gamma\gamma$ excludes a spin $J = 1$ by virtue of the Landau–Yang theorem (section 11.6). For spin 0 one would expect the two charged leptons from $H \to W^+W^- \to 2\ell^\pm + 2\nu$ to be emitted close to one another. This is due to the asymmetric decay distribution (17.29) of the W boson, a consequence of $V - A$, as illustrated in figure 17.24.

The spin can also be determined by studying the angular correlation between the decay planes spanned by the two lepton pairs in $H \to Z^0Z^0 \to 4\ell^\pm$ [22]. For spin 0 one expects a distribution of the form $1 + \chi \cos 2\phi$, where ϕ is the angle between the planes. This is similar to the decay $\pi^0 \to e^+e^-e^+e^-$ from which we have determined the spin and parity of the π^0 (section 11.6). However, π^0 decay involves transverse photons while massive gauge bosons can be transverse (helicity ± 1) or longitudinal (helicity 0).

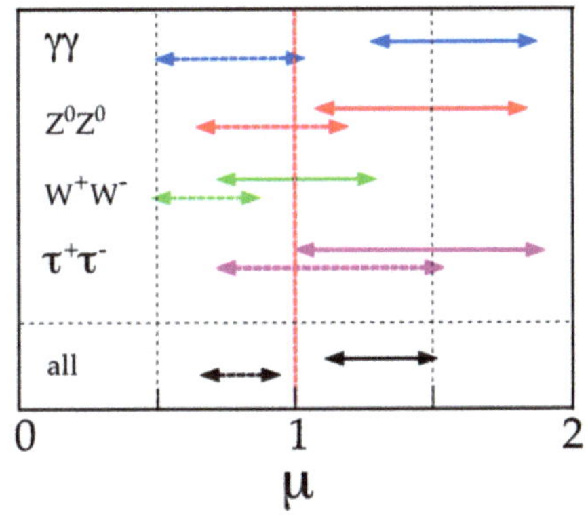

Figure 17.23. Signal strength μ (measured cross section × decay branching ratio divided by the expected standard model value) for the CMS (bottom dashed arrows) and ATLAS experiments. The $\pm 1\sigma$ ranges are plotted. After Carena *et al*, p 161 of [12].

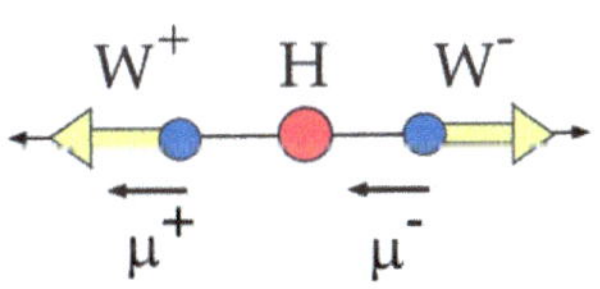

Figure 17.24. A spin-0 Higgs boson decays into two spin-1 W bosons with antiparallel polarizations. The positive leptons are preferably emitted into the forward hemisphere with respect to the W spin while the negative leptons tend to fly backwards. Thus the two leptons prefer to fly into the same hemisphere, close to one another.

Angular correlations of this type were studied at the LHC. The decays $H \to W^+W^- \to 2\ell^\pm + 2\nu$ and $H \to Z^0Z^0 \to 4\ell^\pm$ have been used to test alternative J^P assignments such as 0^- or 2^+. The quantum numbers $J^P = 0^+$ are preferred. For example, the $J^P = 0^-$ hypothesis is rejected with a probability of 97.8% by $H \to 4\ell^\pm$ alone [23]. Larger statistical samples are needed to reach the final certitude that the signal at 126 GeV is indeed due to a scalar boson.

17.3.1 Width of the Higgs boson

The coupling of the Higgs boson to two fermions is proportional to the fermion mass m, while the coupling to two vector bosons is much larger and proportional to the square of the boson mass M (see Carena *et al*, p 161 of [12])

$$g_{Hf\bar{f}} = \frac{m}{f}, \qquad g_{HVV} = \frac{2M^2}{f}, \tag{17.44}$$

where $f = 246$ GeV as given by (16.181). However, the branching ratio into two vector bosons is strongly reduced by the mass of the Higgs boson which lies below threshold. Therefore H decays to *virtual* W^+W^- or Z^0Z^0 pairs. The decay into $t\bar{t}$ pairs is also suppressed by kinematics. The Higgs boson width is hence relatively narrow, dominated by the $b\bar{b}$ width and followed by the W^+W^- one (table in figure 17.18). Let us estimate the width of the Higgs boson by calculating the dominating partial width into $b\bar{b}$ pairs. The Higgs field interacts with a scalar current. The transition amplitude is therefore given by

$$\mathcal{M} = \frac{m}{f}\bar{u}(p)v(p'), \tag{17.45}$$

where m is the b quark mass, and $p(p')$ denote the 4-momenta of the b quark ($\bar{b}$ antiquark). Squaring gives

$$|\mathcal{M}|^2 = \left(\frac{m}{f}\right)^2 \bar{u}(p)v(p')\overbrace{\left[\bar{u}(p)v(p')\right]^\dagger}^{v^\dagger\gamma^0 u} = \left(\frac{m}{f}\right)^2 \bar{u}(p)v(p')\bar{v}(p')u(p). \tag{17.46}$$

Using (15.55c) and (15.55d) and the trace theorems in table 15.1 one finds, summing over spins,

$$\begin{aligned}|\mathcal{M}|^2 &= \left(\frac{m}{f}\right)^2 \mathrm{Tr}\left(\not{p}' - m\right)(\not{p} + m) = \left(\frac{m}{f}\right)^2 \mathrm{Tr}\left(\not{p}'\not{p} - m^2\right)\\ &= \left(\frac{m}{f}\right)^2 4\left(p'_\alpha p^\alpha - m^2\right) = \left(\frac{m}{f}\right)^2 8EE' = \left(\frac{m}{f}\right)^2 2m_H^2,\end{aligned} \tag{17.47}$$

where $E = E' = \frac{m_H}{2}$ in the rest frame of the Higgs, and where we have neglected the mass m of the b quark. The partial width is obtained by integrating the differential partial width (12.22) over the solid angle (factor 4π):

$$\Gamma = \frac{4\pi k}{32\pi^2 m_H^2}|\mathcal{M}|^2 = \frac{m_H}{8\pi}\left(\frac{m}{f}\right)^2, \tag{17.48}$$

where $k = \frac{m_H}{2}$. The partial width has to be multiplied by 3 to take the three quark colours into account, hence

$$\Gamma_H \simeq \Gamma_H(b\bar{b}) = \frac{3m_H}{8\pi}\left(\frac{m}{f}\right)^2 = 4.3\ \mathrm{MeV}, \tag{17.49}$$

with $m = 4.18$ GeV and $f = 246$ GeV. It is interesting to exhibit the dependence on the standard model parameters. Recalling that

$$\frac{1}{f^2} = \sqrt{2}\,G_F = \frac{g^2}{4M_W^2} \qquad \text{and} \qquad g^2 = \frac{4\pi\alpha}{\sin^2\theta_w}, \tag{17.50}$$

one finds that

$$\frac{\Gamma_H}{m_H} \simeq \frac{3\alpha m^2}{8M_W^2\sin^2\theta_w} \sim 3\times 10^{-5}. \tag{17.51}$$

Although the mass of the Higgs is not predicted by the standard model, the relative width of the Higgs can be estimated. We emphasize that the result (17.49) is valid only for a relatively light Higgs, neglecting higher order corrections. The current experimental *upper limit* for the width ($\Gamma_H = 3.4$ GeV [21]) is limited by detector resolution. A much lower upper limit has been determined indirectly from on-shell and off-shell production and decay rates into Z^0Z^0: $\Gamma_H < 22$ MeV [24].

17.4 Neutrino scattering

We now deal with high energy neutrinos scattering off spin-$\frac{1}{2}$ fermions. We first derive the cross sections for charged current scattering off the electron, then for deep inelastic scattering in the nucleon. Finally, we extend the cross sections to include neutral currents and determine the Weinberg angle from experiments in deep inelastic νN and νe scattering.

17.4.1 Charged currents

Table 17.3 lists the possible two-body ν_e and ν_μ reactions on the electron. Let us first derive the differential cross sections for the two reactions $\nu_\mu e^- \to \nu_e \mu^-$ and $\bar{\nu}_e e^- \to \bar{\nu}_\mu \mu^-$, to which only charged currents contribute. We denote the 4-momenta of the incident neutrino (or antineutrino) by q_2, that of the target electron by k, that of the outgoing neutrino (or antineutrino) by q_1, and that of the muon by p. In the centre-of-mass system the 4-momentum of the muon will be denoted by $p^* = (E^*, \vec{p}^*)$. The masses are negligible at high energies and the energies and momenta in the centre of mass are all equal. Hence

$$\left|\vec{q}_2^*\right| \simeq \left|\vec{k}^*\right| \simeq \left|\vec{q}_1^*\right| \simeq \left|\vec{p}^*\right| \simeq E^*. \tag{17.52}$$

For $\nu_\mu e^- \to \nu_e \mu^-$ (figure 17.25(*a*)) the transition amplitude is

$$\mathcal{M}_{\mathrm{CC}} = \frac{4G_F}{\sqrt{2}}\left(\bar{u}(\mu)\gamma^\alpha\left[\frac{1-\gamma^5}{2}\right]u(\nu_\mu)\right) \times \left(\bar{u}(\nu_e)\gamma_\alpha\left[\frac{1-\gamma^5}{2}\right]u(e)\right). \tag{17.53}$$

Table 17.3. Charged currents (CC) and neutral currents (NC) contributing to neutrino–electron scattering in first order; $Z = Z^0$ exchange; $W^a = W$ exchange; W^b = intermediate W boson.

Initial state	Final state	CC	NC
$\nu_\mu e^-$	$\nu_\mu e^-$	×	Z
	$\nu_e \mu^-$	W^a	×
$\bar{\nu}_\mu e^-$	$\bar{\nu}_\mu e^-$	×	Z
$\nu_e e^-$	$\nu_e e^-$	W^a	Z
$\bar{\nu}_e e^-$	$\bar{\nu}_e e^-$	W^b	Z
	$\bar{\nu}_\mu \mu^-$	W^b	×

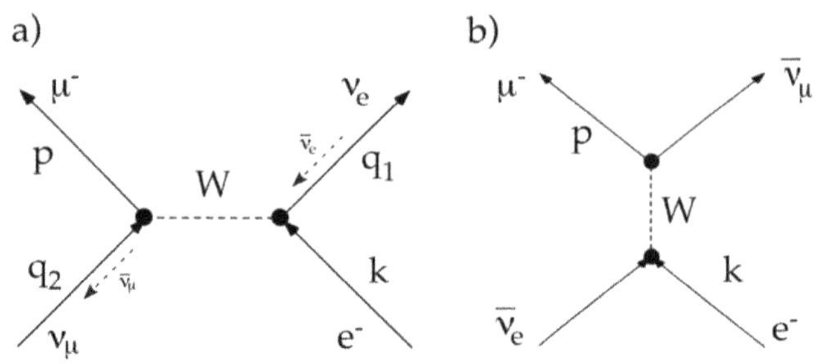

Figure 17.25. (*a*) $\nu_\mu e^- \to \nu_e \mu^-$. (*b*) $\bar{\nu}_e e^- \to \bar{\nu}_\mu \mu^-$.

One can show that

$$|\mathcal{M}_{\mathrm{CC}}|^2 = 128\ G_F^2 (q_2 \cdot k)(p \cdot q_1) \tag{17.54}$$

by applying the trace techniques described in chapter 15 (see [25]). According to (12.21) the differential cross section in the centre-of-mass system is given by

$$\frac{\mathrm{d}\sigma}{\mathrm{d}\Omega^*} = \left(\frac{1}{2}\right)\frac{1}{64\pi^2\, s}\,|\mathcal{M}_{\mathrm{CC}}|^2, \tag{17.55}$$

with the multiplicity factor $K = 1$ since only left-handed fermions contribute. Neutrinos are left-handed but half of the target electrons are right-handed, hence the factor $\frac{1}{2}$. From 4-momentum conservation $q_2 + k = p + q_1$ it follows with

$$(q_2 + k)^2 = (p + q_1)^2 \tag{17.56}$$

that

$$q_2 \cdot k \simeq p \cdot q_1 = E^{*2} - \vec{p}^*(-\vec{p}^*) = 2E^{*2} = \frac{s}{2}, \tag{17.57}$$

at high energies and in the centre-of-mass system (we have neglected the squared masses q_2^2, k^2, p^2 and q_1^2). With (17.54) the differential cross section becomes

$$\boxed{\frac{\mathrm{d}\sigma}{\mathrm{d}\Omega^*}(\nu_\mu e^- \to \nu_e \mu^-) = \frac{G_F^2}{4\pi^2} s}. \tag{17.58}$$

The angular distribution is therefore *isotropic* in the centre-of-mass system. Furthermore, since $s \simeq 2m_e E_\nu$ the cross section grows *linearly* with laboratory energy E_ν. However, at extremely high energies (~10^5 TeV) the cross section violates unitarity. The assumption of point-like interaction collapses and the full W propagator (15.109) has to be taken into account.

For $\bar{\nu}_e e^- \to \bar{\nu}_\mu \mu^-$ the outgoing ν_e in the previous reaction can be replaced by an incoming $\bar{\nu}_e$ (momentum $-q_1$) and the incoming ν_μ by an outgoing $\bar{\nu}_\mu$ (momentum $-q_2$), as shown by the dashed lines in figure 17.25(*a*). Replacing q_1 by $-q_2$ and q_2 by $-q_1$

in (17.54) yields the transition amplitude for the scattering process $\bar{\nu}_e e^- \to \bar{\nu}_\mu \mu^-$ (figure 17.25(*b*)):

$$|\mathcal{M}_{CC}|^2 = 128\, G_F^2 (q_1 \cdot k)(p \cdot q_2). \tag{17.59}$$

Since now $p - q_2 = k - q_1$,

$$p \cdot q_2 \simeq q_1 \cdot k = E^{*2} - \vec{p}^* \cdot \vec{q}_2^* = E^{*2}(1 - \cos\theta^*) = \frac{s}{2}\left[\frac{1 - \cos\theta^*}{2}\right], \tag{17.60}$$

where θ^* denotes the angle between the incoming $\bar{\nu}_e$ and the outgoing muon in the centre-of-mass system. With (17.55) the differential cross section is

$$\boxed{\frac{d\sigma}{d\Omega^*}(\bar{\nu}_e e^- \to \bar{\nu}_\mu \mu^-) = \frac{G_F^2}{4\pi^2} s \left[\frac{1 - \cos\theta^*}{2}\right]^2}\,. \tag{17.61}$$

We are already familiar with the term $(1 - \cos\theta^*)^2$, encountered when dealing with W^- decay, see (17.29). This term follows from the $V - A$ interaction, implying that the antineutrino cannot be scattered backwards and the muon emitted forwards ($\theta^* = 0$) without violating angular momentum conservation (figure 17.26(*a*)).

Let us calculate the energy distribution of the outgoing neutrino in the laboratory. For $\nu_\mu e^- \to \nu_e \mu^-$ one expects a constant (box) distribution, in analogy to π^0 decay (section 8.1) since the angular distribution (17.58) is isotropic. The electron energy in the centre of mass

$$E^* = \gamma m_e, \tag{17.62}$$

is equal to that of the incident massless neutrino,

$$E^* = \gamma E_\nu - \beta\gamma E_\nu. \tag{17.63}$$

The velocity of the centre of mass is therefore

$$\beta = \frac{E_\nu - m_e}{E_\nu} \simeq 1. \tag{17.64}$$

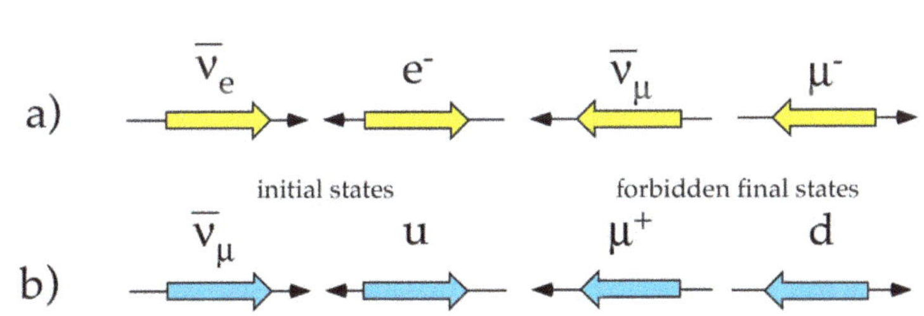

Figure 17.26. (*a*) The muon cannot be emitted forwards in $\bar{\nu}_e e^- \to \bar{\nu}_\mu \mu^-$ scattering, similarly, (*b*) the μ^+ in $\bar{\nu}_\mu u \to \mu^+ d$ cannot be emitted backwards.

The laboratory energy E_ν' of the outgoing ν_e is

$$E_\nu' = \gamma E^* + \beta\gamma q_1^* \cos\theta_\nu^* \simeq \gamma E^*\left(1 + \cos\theta_\nu^*\right), \tag{17.65}$$

where θ_ν^* is the angle of the neutrino in the centre-of-mass system. The energy distribution of the outgoing neutrino is

$$\frac{\mathrm{d}\sigma}{\mathrm{d}E_\nu'} = \frac{\mathrm{d}\sigma}{\mathrm{d}\Omega^*}\cdot\frac{\mathrm{d}\Omega^*}{\mathrm{d}E_\nu'} = \frac{\mathrm{d}\sigma}{\mathrm{d}\Omega^*}2\pi\overbrace{\frac{d\cos\theta_\nu^*}{\mathrm{d}E_\nu'}}^{\frac{1}{\gamma E^*}} = \frac{\mathrm{d}\sigma}{\mathrm{d}\Omega^*}\frac{2\pi}{\gamma E^*} = \frac{G_F^2\overbrace{s}^{4E^{*2}}}{4\pi^2}\frac{2\pi}{\gamma E^*} = \frac{2G_F^2E^*}{\gamma\pi}, \tag{17.66}$$

or with $\gamma = E^*/m_e$,

$$\boxed{\frac{\mathrm{d}\sigma}{\mathrm{d}E_\nu'}\left(\nu_\mu e^- \to \nu_e\mu^-\right) = \frac{2G_F^2m_e}{\pi}}, \tag{17.67}$$

which is indeed independent of E_ν'.

For $\bar{\nu}_e e^- \to \bar{\nu}_\mu\mu^-$ the angle of the outgoing $\bar{\nu}_\mu$ is $\theta_\nu^* = \pi - \theta^*$. The term in square brackets in (17.61) becomes

$$\frac{1}{4}\left(1 + \cos\theta_\nu^*\right)^2 = \frac{E_\nu'^2}{4\gamma^2E^{*2}} = \frac{E_\nu'^2m_e^2}{4E^{*4}} = \left(\frac{E_\nu'}{E_\nu}\right)^2, \tag{17.68}$$

since

$$4E^{*2} = s \;\Rightarrow\; E^{*4} = \frac{s^2}{16} = \frac{m_e^2E_\nu^2}{4}. \tag{17.69}$$

As expected the energy distribution

$$\boxed{\frac{\mathrm{d}\sigma}{\mathrm{d}E_\nu'}\left(\bar{\nu}_e e^- \to \bar{\nu}_\mu\mu^-\right) = \frac{2G_F^2m_e}{\pi}\left(\frac{E_\nu'}{E_\nu}\right)^2} \tag{17.70}$$

depends on the energy E_ν' of the outgoing antineutrino. Integrating (17.67) and (17.70) leads to the ratio of total cross sections

$$\frac{\sigma\left(\nu_\mu e^- \to \nu_e\mu^-\right)}{\sigma\left(\bar{\nu}_e e^- \to \bar{\nu}_\mu\mu^-\right)} = 3. \tag{17.71}$$

17.4.2 Neutrino–quark scattering

We now deal with *deep inelastic* neutrino–nucleon scattering $\nu_\mu N \to \mu^- X$ and $\bar{\nu}_\mu N \to \mu^+ X$, where X stands for hadrons produced by the fragmentation of the struck quark (see figure 16.5). For $\nu_\mu N$ the neutrino scatters off a d quark and turns into a μ^- while for $\bar{\nu}_\mu N$ it scatters off a u quark and turns into a μ^+. Let us first assume that quarks are free in the nucleon with effective mass m, and derive from (17.67) and (17.70) the energy distribution of the muon. For $\bar{\nu}_\mu u \to \mu^+ d$, the μ^+ cannot be emitted backwards (figure 17.26(b)), hence the muon plays here the role of the outgoing $\bar{\nu}_\mu$ in $\bar{\nu}_e e^- \to \bar{\nu}_\mu \mu^-$ of the previous section. Let E be the laboratory energy of the incoming neutrino (or antineutrino) and E' that of the muon. One obtains from (17.67) and (17.70) the cross sections for charged current scattering

$$\frac{\mathrm{d}\sigma_C}{\mathrm{d}E'} = \frac{2G^2 m}{\pi} \quad \text{and} \quad \frac{\mathrm{d}\bar{\sigma}_C}{\mathrm{d}E'} = \frac{2G^2 m}{\pi}\left(\frac{E'}{E}\right)^2, \tag{17.72}$$

describing deep inelastic $\nu_\mu N$ and $\bar{\nu}_\mu N$ scattering, respectively, where

$$G \equiv G_F \cos\theta_C. \tag{17.73}$$

Let us denote by y the fraction of the incident energy carried away by the outgoing quark (hence by the emitted hadrons):

$$\boxed{y \equiv \frac{E - E'}{E}}, \quad \text{where } 0 \leqslant y \leqslant 1. \tag{17.74}$$

The cross sections (17.72) can then be rewritten as a function of y:

$$\frac{\mathrm{d}\sigma_C}{\mathrm{d}y} = \frac{\mathrm{d}\sigma_C}{\mathrm{d}E'} \cdot \frac{\mathrm{d}E'}{\mathrm{d}y} = -\frac{2G^2 mE}{\pi} \quad \text{and} \quad \frac{\mathrm{d}\bar{\sigma}_C}{\mathrm{d}y} = -\frac{2G^2 mE}{\pi}(1-y)^2. \tag{17.75}$$

In reality the quarks are bound in the nucleon. We define the quark *effective* mass as $m = xM_N$, where x is the Bjorken scaling variable (chapter 18). Let us denote by $q(x)\mathrm{d}x$ the number of quarks carrying a fraction of the nucleon mass between x and $x + \mathrm{d}x$. For the time being, we consider only valence quarks and neglect sea quarks. The *structure function* $q(x)$ is then normalized to three quarks:

$$\int_0^1 q(x)\mathrm{d}x = 3. \tag{17.76}$$

The differential cross sections for scattering off the nucleon are, averaged between proton and neutron and using (17.75):

$$\begin{aligned} -\frac{\mathrm{d}^2\sigma_C}{\mathrm{d}x\mathrm{d}y} &= \frac{1}{2}\cdot\frac{2G^2\overbrace{xM_N}^{m}E}{\pi}q(x) \quad \text{and} \\ -\frac{\mathrm{d}^2\bar{\sigma}_C}{\mathrm{d}x\mathrm{d}y} &= \frac{1}{2}\cdot\frac{2G^2\overbrace{xM_N}^{m}E}{\pi}q(x)(1-y)^2. \end{aligned} \tag{17.77}$$

The factor $\frac{1}{2}$ arises from the fact that ν_μ and $\bar{\nu}_\mu$ scatter only off d and u quarks, respectively. These formulas apply to so-called *isoscalar* targets that contain equal numbers of protons and neutrons (equal numbers of u and d quarks), such as deuterium or ^{12}C. With $s \simeq 2M_N E$ one obtains

$$\frac{d^2\sigma_C}{dx dy} = \frac{G^2}{2\pi} sxq(x) \qquad \text{and} \qquad \frac{d^2\bar{\sigma}_C}{dx dy} = \frac{G^2}{2\pi} sxq(x)(1-y)^2. \tag{17.78}$$

Integrating over x leads to

$$\frac{d\sigma_C}{dy} = \frac{G^2}{2\pi} s \int_0^1 xq(x) dx = \frac{G^2 \bar{x}}{2\pi} s \qquad \text{and} \qquad \frac{d\bar{\sigma}_C}{dy} = \frac{G^2 \bar{x}}{2\pi} s(1-y)^2, \tag{17.79}$$

where

$$\bar{x} \equiv \int_0^1 xq(x) dx. \tag{17.80}$$

Assuming that the three quarks carry the whole nucleon mass one expects $\bar{x} = 1$, which is the total contribution from quarks to the nucleon mass, since the average fraction of the mass carried by one quark is

$$\frac{\int_0^1 xq(x) dx}{\int_0^1 q(x) dx} = \frac{1}{3}. \tag{17.81}$$

Integrating (17.79) from $y = 1 (E' = 0)$ to $y = 0 (E' = E)$ gives with the total cross sections for deep inelastic neutrino–nucleon scattering $\int_1^0 (1-y)^2 dy = -\frac{1}{3}$

$$\boxed{\sigma_C = \frac{G^2 \bar{x}}{2\pi} s} \qquad \text{and} \qquad \boxed{\bar{\sigma}_C = \frac{G^2 \bar{x}}{2\pi} \frac{s}{3}}, \tag{17.82}$$

as well as the energy-independent ratio

$$\boxed{\frac{\sigma_C}{\bar{\sigma}_C} = 3}. \tag{17.83}$$

By measuring the energies of the hadrons and of the muon (E') one can determine the incident energy E, and thus s and y. Figure 17.27(a) shows the y dependence of the cross sections $d\sigma_C/dy$ and $d\bar{\sigma}_C/dy$ for $\nu_\mu N \to \mu^- X$ and $\bar{\nu}_\mu N \to \mu^+ X$. For neutrinos the cross section is not quite constant but decreases slightly as a function of y. This is due to transient virtual quark–antiquark pairs in the nucleon (*sea* quarks and sea antiquarks). The cross section for neutrino–antiquark scattering is equal to that for antineutrino–quark scattering (*CP* invariance) which contributes a small admixture of the term $(1-y)^2$.

The measurements agree with (17.79) and are consistent with the existence of spin-$\frac{1}{2}$ scattering centres in the nucleon, the *partons*. The parton model was introduced in 1969 by Richard Feynman to describe deep inelastic scattering. When direct evidence for quarks as real entities was obtained—such as the observation of jets—partons were identified as the quarks and gluons of the quark model.

Figure 17.27(b) shows that the total cross sections σ_C and $\bar{\sigma}_C$ indeed grow proportionally to the laboratory incident energy E by at least two orders of magnitude.

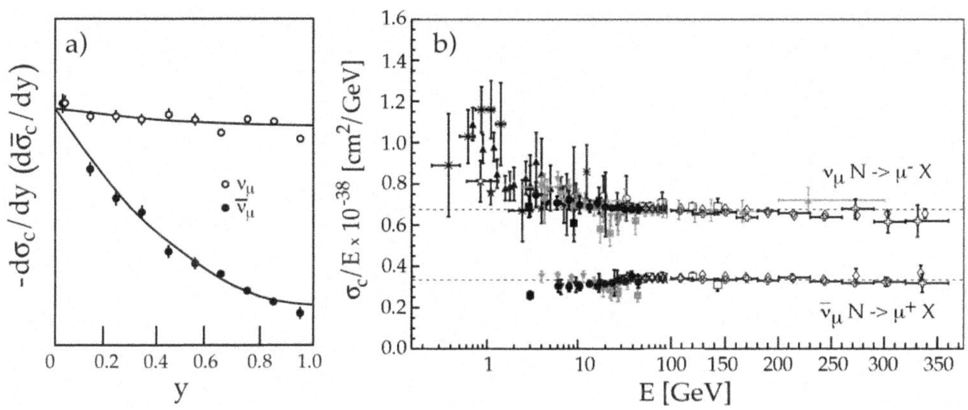

Figure 17.27. (*a*) Cross sections for $\nu_\mu N \to \mu^- X$ and $\bar{\nu}_\mu N \to \mu^+ X$ (inclusive scattering) between 30 and 200 GeV as a function of y (in arbitrary units). The curves are fits of the form $a + b(1-y)^2$. After [26]. (*b*) Total cross sections divided by E from various experiments (for details see [12]) as a function of laboratory energy E. Note the transition between logarithmic and linear horizontal scales at 100 GeV.

However, the measured ratio $\sigma_C/\bar{\sigma}_C$ differs from the expected value of 3. This is again due to the contribution of sea quarks and antiquarks. The fraction ϵ of sea antiquarks in the nucleon can be calculated from the measured ratio

$$\frac{\sigma_C}{\bar{\sigma}_C} = 3\left(\frac{1 + \frac{1}{3}\epsilon}{1 + 3\epsilon}\right) \simeq 2 \;\Rightarrow\; \epsilon \simeq 20\%. \tag{17.84}$$

Furthermore, one can estimate from the measured total cross sections (for instance $\sigma_C/E = 0.67 \times 10^{-38}$ cm^2 GeV^{-1} for $\nu_\mu N$) the total contribution $\bar{x}$ of quarks to the nucleon mass. One finds with (17.82)

$$\frac{\sigma_C}{s} = \frac{G^2}{2\pi}\left(1 + \frac{\epsilon}{3}\right)\bar{x} \;\Rightarrow\; \bar{x} \simeq 0.43, \tag{17.85}$$

where $s = 2M_N E$. Roughly half of the nucleon mass is carried by quarks, the other half by gluons. Structure functions will be discussed in more detail in chapter 18.

17.4.3 Neutral currents

We now treat deep inelastic neutrino scattering off the nucleon without a muon in the final state (NC, see figure 16.5). The transition amplitude for νq scattering reads, according to (16.182):

$$\mathcal{M}_{\rm NC} = \frac{8\rho G_F}{\sqrt{2}}\left(\bar{u}(\nu)\gamma^\alpha c_L(\nu)\left[\frac{1-\gamma^5}{2}\right]u(\nu)\right)$$
$$\times\left(\bar{u}(q)\gamma_\alpha c_L(q)\left[\frac{1-\gamma^5}{2}\right]u(q) + \bar{u}(q)\gamma_\alpha c_R(q)\left[\frac{1+\gamma^5}{2}\right]u(q)\right). \tag{17.86}$$

The two terms in the quark current describe neutrino scattering off left-handed and right-handed quarks, respectively. Let us compare this expression with (17.53) for charged currents. For scattering off left-handed quarks we replace G^2 by $4\rho^2 G_F^2 c_L(\nu)^2 c_L(q)^2$ in the cross section (17.78). Neutrino scattering off right-handed quarks is equivalent to antineutrino scattering off left-handed antiquarks, thus requiring the multiplicative term $(1-y)^2$. Therefore we replace G^2 by $4\rho^2 G_F^2 c_L(\nu)^2 c_R(q)^2 (1-y)^2$ for scattering off right-handed quarks. Hence with $c_L(\nu)^2 = \frac{1}{4}$:

$$-\frac{\mathrm{d}^2\sigma_N}{\mathrm{d}x\mathrm{d}y} = \frac{\rho^2 G_F^2}{\pi} sxq(x)\left[c_L(q)^2 + c_R(q)^2(1-y)^2\right], \tag{17.87}$$

valid for $\nu_\mu N$ scattering. We have removed the factor $\frac{1}{2}$ in (17.78) since neutrinos scatter on both u and d quarks with Z^0 exchange. The interference term between c_L and c_R is proportional to M_N/E and is negligible at high energies. Similarly, for $\bar{\nu}_\mu N$ scattering:

$$-\frac{\mathrm{d}^2\bar{\sigma}_N}{\mathrm{d}x\mathrm{d}y} = \frac{\rho^2 G_F^2}{\pi} sxq(x)\left[c_L(q)^2(1-y)^2 + c_R(q)^2\right]. \tag{17.88}$$

Integrating over x and y leads to the total cross sections

$$\begin{aligned}\sigma_N &= \frac{\rho^2 G_F^2}{\pi} s\bar{x}\left[c_L(q)^2 + \frac{1}{3}c_R(q)^2\right],\\ \bar{\sigma}_N &= \frac{\rho^2 G_F^2}{\pi} s\bar{x}\left[\frac{1}{3}c_L(q)^2 + c_R(q)^2\right].\end{aligned} \tag{17.89}$$

One measures in ν_μ and $\bar{\nu}_\mu$ beams the ratios R (16.53) and $\overline{R}$ (16.54) without and with muon emission. The ratios of neutral to charged current cross sections are with (17.82):

$$\begin{aligned}R &\equiv \frac{\sigma_N}{\sigma_C} = 2\rho^2\left[c_L(q)^2 + \frac{1}{3}c_R(q)^2\right],\\ \overline{R} &\equiv \frac{\bar{\sigma}_N}{\bar{\sigma}_C} = 6\rho^2\left[\frac{1}{3}c_L(q)^2 + c_R(q)^2\right].\end{aligned} \tag{17.90}$$

An isoscalar target contains as many u as d quarks, therefore we take the average:

$$\begin{aligned}c_L(q)^2 &= \frac{1}{2}\left[c_L(u)^2 + c_L(d)^2\right],\\ c_R(q)^2 &= \frac{1}{2}\left[c_R(u)^2 + c_R(d)^2\right].\end{aligned} \tag{17.91}$$

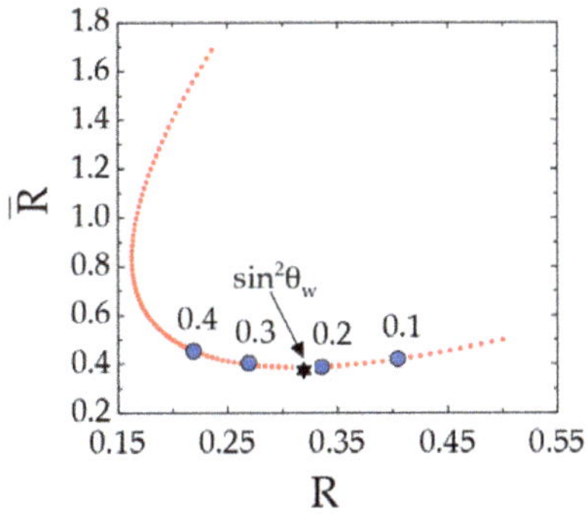

Figure 17.28. $\overline{R}$ as a function of R ('Weinberg nose') for different values of $a = \sin^2\theta_w$ (computed from (17.92) with $\rho = 1$). The triangle shows the result from the CHARM-I experiment [27].

Neglecting the contribution from sea quarks and substituting the chiral couplings (16.169) leads to

$$R = \rho^2\left(\frac{1}{2} - a + \frac{20}{27}a^2\right),$$
$$\overline{R} = \rho^2\left(\frac{1}{2} - a + \frac{20}{9}a^2\right), \tag{17.92}$$

with $a \equiv \sin^2\theta_w$. These ratios have been measured by the CHARM-I experiment at CERN [27]: $R = 0.320 \pm 0.010$ and $\overline{R} = 0.377 \pm 0.020$, from which the Weinberg angle and the parameter ρ could be extracted: $a = 0.247 \pm 0.038$ and $\rho = 1.027 \pm 0.023$, consistent with the expected $\rho = 1$ in the standard model. Figure 17.28 shows the relation between R and $\overline{R}$ for $\rho = 1$. Fixing $\rho = 1$ leads to the more precise value $a = 0.230 \pm 0.023$.

17.4.4 Purely leptonic reactions

Let us finally calculate the cross sections for the remaining leptonic reactions in table 17.3 and from there determine $\sin^2\theta_w$. A major advantage of pure leptonic reactions is the absence of QCD corrections, though the cross sections are three orders of magnitude smaller than for νN scattering since $s = 2m_e E \ll 2m_N E$. Leptonic cross sections were nonetheless measured before the discovery of the gauge bosons to test the consistency of the standard model.

The total cross sections for $\nu_\mu e^- \to \nu_\mu e^-$ and $\bar{\nu}_\mu e^- \to \bar{\nu}_\mu e^-$ read, according to (17.89) with $\rho = 1$ and $\bar{x} = 1$:

$$\sigma\left(\nu_\mu e^- \to \nu_\mu e^-\right) = \frac{G_F^2}{\pi}s\left[c_L(e)^2 + \frac{1}{3}c_R(e)^2\right] = \frac{G_F^2}{4\pi}s\left[(1 - 2a)^2 + \frac{4}{3}a^2\right],$$
$$\bar{\sigma}\left(\bar{\nu}_\mu e^- \to \bar{\nu}_\mu e^-\right) = \frac{G_F^2}{\pi}s\left[\frac{1}{3}c_L(e)^2 + c_R(e)^2\right] = \frac{G_F^2}{4\pi}s\left[\frac{1}{3}(1 - 2a)^2 + 4a^2\right], \tag{17.93}$$

with $c_L(e) = -\frac{1}{2} + a$ and $c_R(e) = a$. The contribution from charged currents has to be added for the reactions $\nu_e e^- \to \nu_e e^-$ and $\bar{\nu}_e e^- \to \bar{\nu}_e e^-$. Substituting c_L by $c_L + 1$ in (17.93) leads to the cross sections

$$\begin{aligned} \sigma(\nu_e e^- \to \nu_e e^-) &= \frac{G_F^2}{4\pi} s \left[(1 + 2a)^2 + \frac{4}{3} a^2 \right], \\ \bar{\sigma}(\bar{\nu}_e e^- \to \bar{\nu}_e e^-) &= \frac{G_F^2}{4\pi} s \left[\frac{1}{3} (1 + 2a)^2 + 4a^2 \right]. \end{aligned} \tag{17.94}$$

It is convenient to introduce two coupling constants, g_V and g_A defined as

$$g_V \equiv c_L(e) + c_R(e) \qquad \text{and} \qquad g_A \equiv c_L(e) - c_R(e). \tag{17.95}$$

In the standard model they take the simple values

$$g_V = -\frac{1}{2} + 2\,a \simeq 0 \qquad \text{and} \qquad g_A = -\frac{1}{2}. \tag{17.96}$$

In a g_A versus g_V diagram the cross sections (17.93) and (17.94) define ellipses. Figure 17.29 shows the allowed regions delimited by the data.

The twofold ambiguity (dark zones) can be solved in favour of the standard model by measuring the differential cross section for $e^+e^- \to \mu^+\mu^-$ scattering. Muon pair production (figure 17.30, left) proceeds via the electromagnetic interaction (intermediate γ) and the weak interaction (intermediate Z^0). The interference between the two contributions leads to an angular distribution of the form

$$\frac{d\sigma}{d\Omega} = \frac{\alpha^2}{4s} \left[A_0 \left(1 + \cos^2\theta \right) + A_1 \cos\theta) \right] \tag{17.97}$$

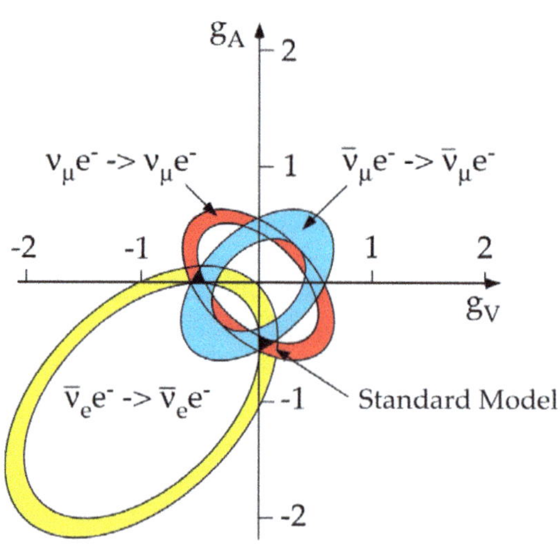

Figure 17.29. The cross sections for pure leptonic scattering define ellipses in a g_A versus g_V diagram with widths determined by the experimental uncertainties. The ellipses cross at $(g_V, g_A) \simeq (-\frac{1}{2}, 0)$ and around the standard model value of $(0, -\frac{1}{2})$. After [28].

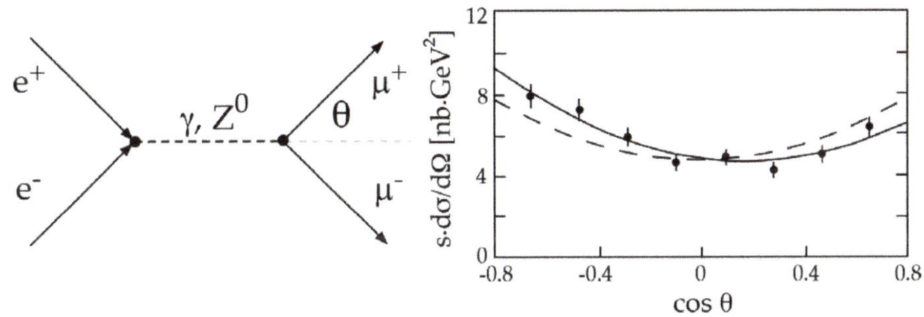

Figure 17.30. Left: Feynman graph of muon pair production in $e^+e^- \to \mu^+\mu^-$. Right: differential cross section measured by the PLUTO experiment at $\sqrt{s} = 34.7\,\mathrm{GeV}$. The full curve shows the standard model fit, the dashed curve the symmetric contribution from the intermediate γ only. After [30].

in the centre-of-mass system [25], where θ is the angle between the μ^+ and the e^+. The term linear in $\cos\theta$ induces a forward–backward asymmetry given by [29]

$$A_{FB} = \frac{3A_1}{8A_0} \simeq -\frac{3g_A^2}{\sqrt{2}}\left(\frac{G_F}{4\pi\alpha}\right)s \tag{17.98}$$

for energies well below the Z^0 resonance ($s \ll M_{Z^0}^2$). Figure 17.30 (right) shows the angular distribution at $\sqrt{s} = 34.7\,\mathrm{GeV}$ measured in the PETRA e^+e^- storage ring at DESY [30]. A clear asymmetry $A_{FB} = -(13.4 \pm 3.1)\%$ is observed, to be compared to the -8.1% predicted from (17.98) when assuming the standard model value $g_A = -\frac{1}{2}$. The value $g_A = 0$ would lead to a vanishingly small asymmetry.

The CHARM-II experiment at CERN has studied ν_μ and $\bar{\nu}_\mu$ scattering off atomic shell electrons [31]. Figure 17.31 shows the apparatus. As discussed below the recoil electrons are emitted forward under very small angles. Accurate measurements of the electron angles require a target with small atomic number Z to reduce multiple scattering and shower development in the vicinity of the interaction point. Glass (SiO_2) is used as the target material. The energy of the electron-induced shower and its direction are measured by streamer tubes (gaseous detectors operated at high

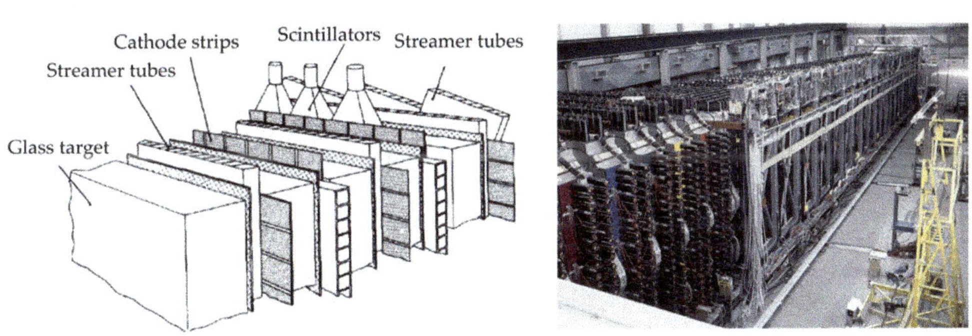

Figure 17.31. Left: diagram of the CHARM-II apparatus (five out of 420 modules) used to measure the $\nu_\mu e$ and $\bar{\nu}_\mu e$ scattering cross sections. After [31]. Right: a photograph of the detector. Image credit: CERN.

electric fields near the voltage breakdown point). The 35 m long detector consists of 420 glass/streamer tube modules. A magnetic spectrometer located at the rear rejects muons (or is used to measure the charged current cross sections). The average neutrino energy is 24 GeV (broadband neutrino beam, see figure 16.6) and the energies of the recoil electrons lie between 3 and 24 GeV.

Let us denote the 4-momentum of the incident neutrino by $k = (E, \vec{k})$, that of the scattered neutrino by $k' = (E', \vec{k}')$, the 4-momentum of the target electron by $p = \left(m_e, 0\right)$ and that of the recoil electron by $p' = \left(E_e', \vec{p}\,'\right)$. By using energy and momentum conservation $k + p = k' + p'$ one obtains by squaring the momentum transfer $q = k - p' = k' - p$,

$$q^2 \simeq -2\, kp' \simeq -2\left(EE_e' - EE_e' \cos\theta\right) = -4EE_e' \sin^2\!\left(\frac{\theta}{2}\right), \tag{17.99}$$

where all masses have been neglected and where E, E_e' and θ (the recoil angle of the electron) refer to the laboratory system. Furthermore,

$$q^2 \simeq -2k'p = -2m_e E', \tag{17.100}$$

whence

$$E_e' \sin^2\!\left(\frac{\theta}{2}\right) = -\frac{q^2}{4E} = \frac{m_e E'}{2E} = \frac{m_e}{2}\left(1 - \frac{E_e'}{E}\right). \tag{17.101}$$

For example, with 12 GeV electrons and 25 GeV neutrinos the recoil angle is quite small: $\theta \simeq 6$ mrad. A good approximation for the previous equation is therefore

$$E_e'\theta^2 \simeq 2m_e\left(1 - \frac{E_e'}{E}\right). \tag{17.102}$$

Electrons are emitted forward under tiny angles, in contrast to hadrons in deep inelastic scattering. The maximum laboratory angle, under which an electron of energy E_e' can be emitted in a broadband neutrino beam, is

$$\theta_m \simeq \sqrt{\frac{2m_e}{E_e'}} \tag{17.103}$$

for $E \to \infty$. Figure 17.32 shows the measured distributions $E_e'\theta^2$. The peaks at small values stem from the leptonic reactions, the wide distributions from deep inelastic (ν quark) scattering off nucleons. The dominant contribution to the background arises from π^0 production and decay with subsequent conversion of the two photons into electron–positron showers. The Weinberg angle can be calculated from the ratio of cross sections (17.93):

$$r = \frac{\sigma\left(\nu_\mu e^-\right)}{\bar{\sigma}\left(\bar{\nu}_\mu e^-\right)} = 3\left(\frac{1 - 4a + \frac{16}{3}a^2}{1 - 4a + 16a^2}\right) \tag{17.104}$$

plotted in figure 17.33. The CHARM-II result is $\sin^2\theta_w = 0.232 \pm 0.014$ [31] for which the cross sections are roughly equal.

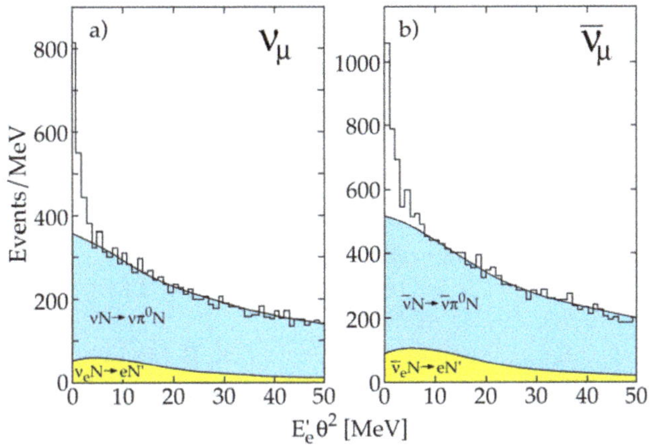

Figure 17.32. $E'_e\theta^2$ distribution for $\nu_\mu e^- \to \nu_\mu e^-$ (*a*) and $\bar{\nu}_\mu e^- \to \bar{\nu}_\mu e^-$ (*b*). The energy of the electron recoiling under the angle θ is E'_e. Note the small contribution from ν_e and $\bar{\nu}_e$, contaminating the muon neutrino beam. After [31].

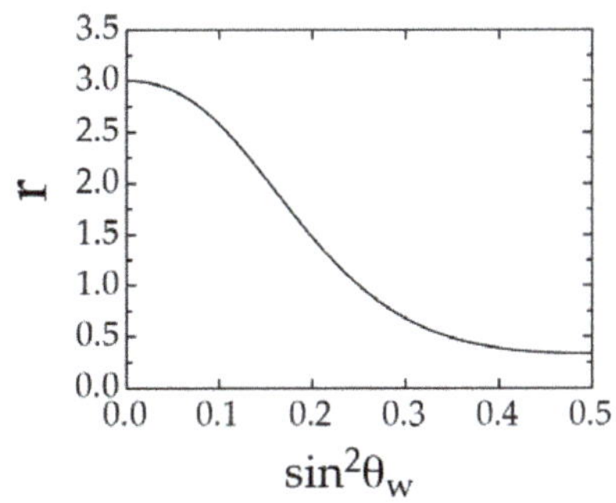

Figure 17.33. Ratio of $\nu_\mu e^- \to \nu_\mu e^-$ to $\bar{\nu}_\mu e^- \to \bar{\nu}_\mu e^-$ cross sections as a function of $a \equiv \sin^2\theta_W$.

References

[1] Prescott C Y *et al* 1978 *Phys. Lett.* B **77** 347
Prescott C Y *et al* 1979 *Phys. Lett.* B **84** 524

[2] LEP Electroweak Working Group 2013 *Phys. Rep.* **532** 119

[3] Arnison G *et al* 1983 *Phys. Lett.* B **122** 103

[4] Arnison G *et al* 1983 *Phys. Lett.* B **126** 398

[5] Banner G *et al* 1983 *Phys. Lett.* B **122** 476

[6] Bagnaia G *et al* 1983 *Phys. Lett.* B **129** 130

[7] Cochet C *et al* 1986 *Nucl. Instrum. Methods Phys. Res.* A **243** 45

[8] Barranco Luque M *et al* 1980 *Nucl. Instrum. Methods Phys. Res.* A **176** 175

[9] Corden M J *et al* 1985 *Nucl. Instrum. Methods Phys. Res.* A **238** 273

[10] Arnison G *et al* 1983 *Phys. Lett.* B **129** 273

[11] Khachatryan V *et al* (CMS Collaboration) 2011 *J. High Energy Phys.* JHEP01(2011)080

[12] Olive K *et al* (Particle Data Group) 2014 *Chin. Phys.* C **38** 090001

[13] For a review see Steinberger J 1991 *Phys. Rep.* **203** 346
[14] LEP Electroweak Working Group 2006 *Phys. Rep.* **427** 257
[15] Spira M and Zerwas P M 1998 *Lecture Notes in Physics* vol 512 (Berlin: Springer) p 161
[16] Djouadi A, Kalinowski J and Spira M 1998 *Comput. Phys. Commun.* **108** 56
[17] Chatrchyan S *et al* (CMS Collaboration) 2012 *Phys. Lett.* B **716** 30
[18] Aad G *et al* (ATLAS Collaboration) 2012 *Phys. Lett.* B **716** 1
[19] Chatrchyan S *et al* (CMS Collaboration) 2008 *J. Instrum.* **3** S08004
[20] Aad G *et al* (ATLAS Collaboration) 2008 *J. Instrum.* **3** S08003
[21] Chatrchyan S *et al* (CMS Collaboration) 2014 *Phys. Rev.* D **89** 092007
[22] Gao Y *et al* 2010 *Phys. Rev.* D **81** 075022
[23] Aad A *et al* (ATLAS Collaboration) 2013 *Phys. Lett.* B **726** 120
[24] Khachatryan V *et al* (CMS Collaboration) 2014 *Phys. Lett.* B **736** 64
[25] Halzen F and Martin A D 1984 *Quarks and Leptons: An Introductory Course in Modern Particle Physics* (New York: Wiley)
[26] de Groot J G H *et al* 1979 *Z. Phys.* C **1** 143
[27] Jonker M *et al* 1981 *Phys. Lett.* B **99** 265
[28] Hung P Q and Sakurai J J 1981 *Ann. Rev. Nucl. Part. Sci.* **31** 375
[29] Wu S L 1984 *Phys. Rep.* **107** 59
[30] Berger Ch *et al* 1983 *Z. Phys.* C **21** 53
[31] Geiregat D *et al* 1989 *Phys. Lett.* B **232** 539

IOP Publishing

Nuclear and Particle Physics

Claude Amsler

Chapter 18

Deep inelastic electron–proton scattering

The experimental evidence for spin-$\frac{1}{2}$ constituents in the nucleon has already been discussed for the *weak* interaction in neutrino–quark scattering (section 17.4.2). We now examine in more detail the quark structure of the nucleon from another point of view, that of inelastic electron–proton scattering involving the *electromagnetic* interaction between electrons and quarks. We calculate the differential cross section for the *elastic* scattering of a spin-$\frac{1}{2}$ electron (or muon) off a point-like spin-$\frac{1}{2}$ charge +1 target, such as an idealized point-like proton. We derive the Rosenbluth formula (2.42) apart from the electromagnetic form factors, that is for $G_E = 1$ and $G_M = 1$. The result is then applied to elastic electron–quark scattering and compared to inelastic electron–proton scattering in which the quarks fragment into hadrons.

According to (15.105)–(15.107) the first order transition amplitude for *elastic* electron–proton scattering is

$$\mathcal{M} = \left[\bar{u}(k')\gamma^{\mu}u(k)\right]\left(-\frac{e^2}{q^2}\right)\left[\bar{u}(p')\gamma_{\mu}u(p)\right]. \tag{18.1}$$

Figure 18.1 (left) shows the one-photon exchange diagram. The 4-momenta of the electron and proton are k, k' and p, p', respectively, and q is the momentum transfer (4-momentum of the exchanged photon). Squaring $\mathcal{M}$ gives

$$|\mathcal{M}|^2 = \left(\frac{e}{q}\right)^4 \left[\bar{u}(k')\gamma^{\mu}u(k)\right]\overbrace{\left[\bar{u}(k')\gamma^{\nu}u(k)\right]^{\dagger}}^{=\left[\bar{u}(k)\gamma^{\nu}u(k')\right]}\left[\bar{u}(p')\gamma_{\mu}u(p)\right]\overbrace{\left[\bar{u}(p')\gamma_{\nu}u(p)\right]^{\dagger}}^{=\left[\bar{u}(p)\gamma_{\nu}u(p')\right]}. \tag{18.2}$$

doi:10.1088/978-0-7503-1140-3ch18

© IOP Publishing Ltd 2015

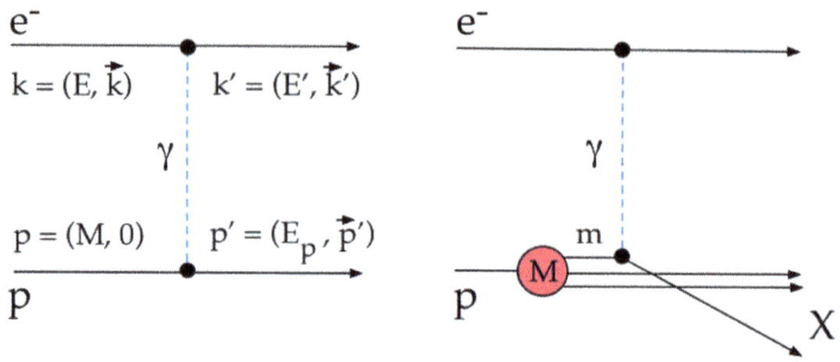

Figure 18.1. Elastic (left) and deep inelastic (right) electron–proton scattering.

The terms in brackets written in components are

$$\sum_{ss'} \bar{u}(k')_\alpha \gamma^\mu_{\alpha\beta} u(k)_\beta \bar{u}(k)_\delta \gamma^\nu_{\delta\gamma} u(k')_\gamma \cdot \sum_{ss'} \bar{u}(p')_\alpha \gamma_{\mu_{\alpha\beta}} u(p)_\beta \bar{u}(p)_\delta \gamma_{\nu_{\delta\gamma}} u(p')_\gamma$$
$$= \underbrace{\sum_{ss'} u(k')_\gamma \bar{u}(k')_\alpha \gamma^\mu_{\alpha\beta} u(k)_\beta \bar{u}(k)_\delta \gamma^\nu_{\delta\gamma}}_{\mathrm{Tr}\left[\not{k}'\gamma^\mu \not{k}\gamma^\nu\right]} \cdot \underbrace{\sum_{ss'} u(p')_\gamma \bar{u}(p')_\alpha \gamma_{\mu_{\alpha\beta}} u(p)_\beta \bar{u}(p)_\delta \gamma_{\nu_{\delta\gamma}}}_{\mathrm{Tr}\left[(p' + M)\gamma_\mu (p + M)\gamma_\nu\right]}, \tag{18.3}$$

where M is the proton mass and where we have neglected the electron mass. The trace for the electron gives

$$\begin{aligned} \mathrm{Tr}\left[\not{k}'\gamma^\mu \not{k}\gamma^\nu\right] &= \mathrm{Tr}\left(k'_\alpha \gamma^\alpha \gamma^\mu k_\beta \gamma^\beta \gamma^\nu\right) = \mathrm{Tr}\left(\gamma^\alpha \gamma^\mu \gamma^\beta \gamma^\nu\right) k'_\alpha k_\beta \\ &= 4g^{\alpha\mu} g^{\beta\nu} k'_\alpha k_\beta + 4g^{\alpha\nu} g^{\mu\beta} k'_\alpha k_\beta - 4g^{\alpha\beta} g^{\mu\nu} k'_\alpha k_\beta \\ &= 4k'^\mu k^\nu + 4k'^\nu k^\mu - 4g^{\mu\nu}(k' \cdot k). \end{aligned} \tag{18.4}$$

Since the trace of an odd number of γ matrices vanishes one finds for the proton

$$\begin{aligned} \mathrm{Tr}\left[(\not{p}' + M)\gamma_\mu (\not{p} + M)\gamma_\nu\right] &= \mathrm{Tr}\left[\not{p}'\gamma_\mu \not{p}\gamma_\nu + M^2 \gamma_\mu \gamma_\nu\right] \\ &= \mathrm{Tr}\left(p'^\alpha \gamma_\alpha \gamma_\mu p^\beta \gamma_\beta \gamma_\nu\right) + M^2\, \mathrm{Tr}\left(\gamma_\mu \gamma_\nu\right) \\ &= \mathrm{Tr}\left(\gamma_\alpha \gamma_\mu \gamma_\beta \gamma_\nu\right) p'^\alpha p^\beta + 4M^2 g_{\mu\nu} \\ &= \left(4g_{\alpha\mu} g_{\beta\nu} + 4g_{\alpha\nu} g_{\mu\beta} - 4g_{\alpha\beta} g_{\mu\nu}\right) p'^\alpha p^\beta + 4M^2 g_{\mu\nu} \\ &= 4p'_\mu p_\nu + 4p'_\nu p_\mu - 4g_{\mu\nu}(p' \cdot p) + 4M^2 g_{\mu\nu}. \end{aligned} \tag{18.5}$$

We have used the properties of γ matrices listed in table 15.1. Multiplying (18.4) by (18.5) gives

$$\begin{aligned} |\mathcal{M}|^2 &= 32\left(\frac{e}{q}\right)^4 \left[k'^\mu p'_\mu k^\nu p_\nu + k'^\mu p_\mu k^\nu p'_\nu - M^2(k' \cdot k)\right] \\ &= 32\left(\frac{e}{q}\right)^4 \left[(k' \cdot p')(k \cdot p) + (k' \cdot p)(k \cdot p') - M^2(k' \cdot k)\right]. \end{aligned} \tag{18.6}$$

Let us express the *laboratory* 4-momenta in components and neglect the electron mass:

$$
\begin{aligned}
k &= (E, E, 0, 0), \\
k' &= (E', E' \cos\theta, E' \sin\theta, 0), \\
p &= (M, 0, 0, 0), \\
p' &= \left(E_p, \vec{p}\,'\right).
\end{aligned}
\tag{18.7}
$$

Substituting $p' = k - k' + p$ (4-momentum conservation) into (18.6) and ignoring the two negligible terms $(k \cdot k) = (k' \cdot k') = m_e^2$ gives

$$
\begin{aligned}
|\mathcal{M}|^2 &= 32\left(\frac{e}{q}\right)^4 \left[(k' \cdot k)(k \cdot p) - (k' \cdot k)(k' \cdot p) + 2(k \cdot p)(k' \cdot p) - M^2(k' \cdot k)\right] \\
&= 32\left(\frac{e}{q}\right)^4 \left[(k' \cdot k)\{(k \cdot p) - (k' \cdot p)\} + 2(k \cdot p)(k' \cdot p) - M^2(k' \cdot k)\right] \\
&= 32\left(\frac{e}{q}\right)^4 \left[EE'(1 - \cos\theta)M(E - E') + 2EE'M^2 - M^2EE'(1 - \cos\theta)\right] \\
&= 32\left(\frac{e}{q}\right)^4 \left[2EE'M(E - E')\sin^2\frac{\theta}{2} + 2EE'M^2 - 2M^2EE'\sin^2\frac{\theta}{2}\right] \\
&= 64\left(\frac{e}{q}\right)^4 EE'M^2\left[\frac{E - E'}{M}\sin^2\frac{\theta}{2} + \cos^2\frac{\theta}{2}\right].
\end{aligned}
\tag{18.8}
$$

Since $p + q = p'$ squaring gives with $p^2 = p'^2$ $(= M^2)$:

$$
2p \cdot q = 2M(E - E') = -q^2 = 4EE'\sin^2\frac{\theta}{2} \equiv Q^2. \tag{18.9}
$$

Substituting into (18.8) leads to the result

$$
|\mathcal{M}|^2 = 64\left(\frac{e}{Q}\right)^4 EE'M^2\left[\cos^2\frac{\theta}{2} + \frac{Q^2}{2M^2}\sin^2\frac{\theta}{2}\right]. \tag{18.10}
$$

The cosine-term describes Mott scattering and the sine-term the interaction of the magnetic moments in the limit of point-like particles. We now derive the differential cross section by evaluating (12.2) *in the laboratory*. The flux factor F is equal to $4 \times \sqrt{(k \cdot p)^2 - m_e^2 M^2} \simeq 4EM$. Hence

$$
d\sigma = \frac{1}{16\pi^2 EMK} \int \frac{d^3\vec{k}'}{2E'} \frac{d^3\vec{p}\,'}{2E_p} |\mathcal{M}|^2 \delta\left(\vec{k}' + \vec{p}\,' - \vec{k}\right) \delta\left(E' + E_p - M - E\right). \tag{18.11}
$$

Integrating over $\vec{p}\,'$ and writing $d^3\vec{k}\,'$ as $E'^2 dE' d\Omega$ gives the differential cross section

$$\frac{d\sigma}{d\Omega} = \frac{1}{64\pi^2 EMK} \int \frac{E'^2 dE'}{E' E_p} |\mathcal{M}|^2 \delta\underbrace{\left(E' + E_p(E') - M - E\right)}_{g(E')}. \qquad (18.12)$$

Since

$$E_p^2 = M^2 + \left(\vec{k} - \vec{k}'\right)^2 = M^2 + E^2 + E'^2 - 2EE' \cos\theta, \qquad (18.13)$$

the derivative $\frac{dg}{dE'}$ is

$$\begin{aligned} \frac{dg}{dE'} &= 1 + \frac{2E' - 2\cos\theta}{2E_p} = \frac{E_p + E' - E\cos\theta}{E_p} \\ &= \frac{E + M - E\cos\theta}{E_p} = \frac{M + 2E\sin^2\frac{\theta}{2}}{E_p}. \end{aligned} \qquad (18.14)$$

The integration over E' gives with the property (12.13) of the δ-function,

$$\frac{d\sigma}{d\Omega} = \frac{1}{64\pi^2 EMK}\left(\frac{E'}{M + 2E\sin^2\frac{\theta}{2}}\right)|\mathcal{M}|^2 = \frac{1}{64\pi^2 M^2 K}\left(\frac{E'}{E}\right)^2 |\mathcal{M}|^2, \qquad (18.15)$$

where we have used (2.6):

$$M + 2E\sin^2\frac{\theta}{2} = M\frac{E}{E'}, \qquad (18.16)$$

which is easily verified from energy conservation $E_p = E + M - E'$ and by comparing with (18.13).

The spin multiplicity factor is $K = 4$. Substituting (18.10) into (18.15) leads to the final result

$$\frac{d\sigma}{d\Omega} = \frac{4\alpha^2 E'^2}{Q^4} \cdot \frac{E'}{E}\left[\cos^2\frac{\theta}{2} + \frac{Q^2}{2M^2}\sin^2\frac{\theta}{2}\right]. \qquad (18.17)$$

This is the Rosenbluth formula (2.42) for a point-like proton. The generalization to a finite-size proton is described in [1]. With the definition

$$\boxed{\nu \equiv E - E'} \qquad (18.18)$$

one finds together with (18.9) an important relation between Q^2 and ν, valid for *elastic* scattering:

$$\boxed{\nu = \frac{Q^2}{2M}}. \qquad (18.19)$$

In deep *inelastic* electron–proton scattering the proton breaks up into hadrons (figure 18.1, right) so that two variables can be chosen independently, such as ν and Q^2 or θ and E'. Hence we first express the differential cross section (18.17) for elastic scattering in terms of the two variables ν and Q^2:

$$\frac{\mathrm{d}^2\sigma}{\mathrm{d}\Omega\mathrm{d}E'} = \frac{4\alpha^2E'^2}{Q^4}\left[\cos^2\frac{\theta}{2} + \frac{Q^2}{2M^2}\sin^2\frac{\theta}{2}\right]\delta\left(\frac{Q^2}{2M} - \nu\right). \tag{18.20}$$

This can be proven by integrating over E':

$$\begin{aligned}\frac{\mathrm{d}\sigma}{\mathrm{d}\Omega} &= \int\frac{4\alpha^2E'^2}{Q^4}\left[\cos^2\frac{\theta}{2} + \frac{Q^2}{2M^2}\sin^2\frac{\theta}{2}\right]\delta\left(\frac{Q^2}{2M} - \nu\right)\mathrm{d}E' \\ &= \frac{4\alpha^2E'^2}{Q^4}\left[\cos^2\frac{\theta}{2} + \frac{Q^2}{2M^2}\sin^2\frac{\theta}{2}\right]\frac{1}{\left|\frac{\mathrm{d}g}{\mathrm{d}E'}\right|},\end{aligned} \tag{18.21}$$

where

$$g(E') \equiv \frac{Q^2}{2M} - \nu. \tag{18.22}$$

With

$$\frac{\mathrm{d}g}{\mathrm{d}E'} = \frac{1}{2M}\frac{\mathrm{d}Q^2}{\mathrm{d}E'} - \frac{\mathrm{d}\nu}{\mathrm{d}E'} = \frac{4E\sin^2\frac{\theta}{2}}{2M} + 1 = \frac{Q^2}{2ME'} + 1 = \frac{E}{E'} \tag{18.23}$$

one retrieves (18.17).

Partons in the proton can be resolved, provided that the momentum transfer squared Q^2 carried by the photon is large enough (*deep* inelastic scattering). Let us assume that the electron scatters *elastically* off a spin-$\frac{1}{2}$ quark with effective mass $m = xM$. The differential cross section is given by, following (18.20)

$$\frac{\mathrm{d}^2\sigma}{\mathrm{d}\Omega\mathrm{d}E'} = \frac{4\alpha^2E'^2}{Q^4}\left[\cos^2\frac{\theta}{2} + \frac{Q^2}{2m^2}\sin^2\frac{\theta}{2}\right]\delta\left(\frac{Q^2}{2m} - \nu\right). \tag{18.24}$$

For reasons that will become clear below, we express the previous equation as a function of x. This requires rewriting the δ distribution as an explicit function of x. The cosine-term is to be multiplied by the function

$$\delta\left(\frac{Q^2}{2xM} - \nu\right) = \delta\left(x - \frac{Q^2}{2M\nu}\right)\frac{1}{\left|\frac{\mathrm{d}g}{\mathrm{d}x}\right|} = \delta\left(x - \frac{Q^2}{2M\nu}\right)\frac{x}{\nu}, \tag{18.25}$$

where $g(x) \equiv Q^2/2xM$, and the sine-term multiplied by

$$\frac{Q^2}{2x^2M^2}\delta\left(\frac{Q^2}{2xM} - \nu\right) = \frac{Q^2}{2x^2M^2}\delta\left(x - \frac{Q^2}{2M\nu}\right)\frac{x}{\nu} = \delta\left(x - \frac{Q^2}{2M\nu}\right)\frac{1}{M}. \tag{18.26}$$

The differential cross section (18.20) can be rewritten as

$$\frac{\mathrm{d}^2\sigma}{\mathrm{d}\Omega\mathrm{d}E'} = \frac{4\alpha^2 E'^2}{Q^4}\left[\underbrace{x\delta\left(x - \frac{Q^2}{2M\nu}\right)\frac{1}{\nu}}_{\equiv F_2(x)}\cos^2\frac{\theta}{2} + \underbrace{\delta\left(x - \frac{Q^2}{2M\nu}\right)\frac{1}{M}}_{\equiv 2F_1(x)}\sin^2\frac{\theta}{2}\right], \tag{18.27}$$

where we have introduced the two *structure functions* $F_1(x)$ and $F_2(x)$. They are related through the Callan–Gross relation:

$$\boxed{F_2(x) = 2xF_1(x)}. \tag{18.28}$$

On the other hand the differential cross section (18.20) for deep inelastic scattering off the proton should incorporate two form factors W_1 and W_2 to take into account the finite size of the proton:

$$\frac{\mathrm{d}^2\sigma}{\mathrm{d}\Omega\mathrm{d}E'} = \frac{4\alpha^2 E'^2}{Q^4}\left[W_2\left(Q^2, \nu\right)\cos^2\frac{\theta}{2} + 2W_1\left(Q^2, \nu\right)\sin^2\frac{\theta}{2}\right]. \tag{18.29}$$

We have removed the δ-function since q^2 and ν are now independent. Comparing (18.27) and (18.29) one expects that at sufficiently high momentum transfer

$$W_2\left(Q^2, \nu\right) \to \frac{F_2(x)}{\nu} \qquad \text{and} \qquad W_1\left(Q^2, \nu\right) \to \frac{F_1(x)}{M}. \tag{18.30}$$

We have so far assumed that the struck quark carries a fraction x of the proton mass. The structure function $F_2(x)$ therefore needs to be folded with an effective mass distribution of the type $q(x)$ introduced in section 17.4.2. Let $q_i(x)$ describe the x distribution of the quark with flavour i and charge eQ_i ($\alpha \to \alpha Q_i$). We sum over all flavours and fold with the δ-function, thus replacing $F_2(x)$ by

$$F_2(x) = \int_0^1 \sum_i q_i(x')x'\delta\left(x' - \frac{Q^2}{2M\nu}\right)Q_i^2\mathrm{d}x' = \sum_i q_i(x)xQ_i^2. \tag{18.31}$$

The differential cross section (18.27) becomes

$$\begin{aligned}\frac{\mathrm{d}^2\sigma}{\mathrm{d}\Omega\mathrm{d}E'} &= \frac{4\alpha^2 E'^2}{Q^4}F_2(x)\left[\frac{1}{\nu}\cos^2\frac{\theta}{2} + \frac{1}{xM}\sin^2\frac{\theta}{2}\right] \\ &= \frac{4\alpha^2 E'^2}{Q^4}F_2(x)\left[\frac{1}{\nu} + \sin^2\frac{\theta}{2}\left(\frac{1}{xM} - \frac{1}{\nu}\right)\right] \\ &= \frac{4\alpha^2 E'^2}{Q^4}F_2(x)\left[\frac{1}{yE} + \frac{Q^2}{4EE'}\left(\frac{1}{xM} - \frac{1}{yE}\right)\right].\end{aligned} \tag{18.32}$$

We now perform a variable transformation and express the cross section as a function of the fraction x of the proton mass carried by the quark, and the fraction y

of the incident energy removed by the outgoing hadrons. The variables x and y are given by

$$x = \frac{Q^2}{2M\nu} = \frac{EE'(1 - \cos\theta)}{M(E - E')} \qquad \text{and} \qquad y \equiv \frac{\nu}{E} = \frac{E - E'}{E}, \tag{18.33}$$

where $0 \leqslant x \leqslant 1$ and $0 \leqslant y \leqslant 1$. The differential cross section becomes

$$\begin{aligned} \frac{d^2\sigma}{d\Omega dE'} &= \frac{1}{2\pi}\left(\frac{d^2\sigma}{dx dy}\right)\frac{\partial(x, y)}{\partial(\cos\theta, E')} = \frac{1}{2\pi}\left(\frac{d^2\sigma}{dx dy}\right)\frac{\partial x}{\partial \cos\theta}\cdot\frac{\partial y}{\partial E'} \\ &= \frac{1}{2\pi}\left(\frac{d^2\sigma}{dx dy}\right)\frac{EE'}{M(E - E')}\cdot\frac{1}{E} = \frac{1}{2\pi}\left(\frac{d^2\sigma}{dx dy}\right)\frac{E'}{MEy} \end{aligned} \tag{18.34}$$

or in other words

$$\frac{d^2\sigma}{dx dy} = \left(\frac{d^2\sigma}{d\Omega dE'}\right)\overbrace{2ME}^{\simeq s}\frac{\pi y}{E'}. \tag{18.35}$$

Introducing (18.32) one obtains

$$\begin{aligned} \frac{d^2\sigma}{dx dy} &= \frac{2\pi\alpha^2 s}{Q^4}F_2(x)\left[\frac{2E'}{E} + \frac{Q^2 y}{2E}\left(\frac{1}{xM} - \frac{1}{yE}\right)\right] \\ &\simeq \frac{2\pi\alpha^2 s}{Q^4}F_2(x)\left[2(1 - y) + \underbrace{\frac{Q^2 y}{2ExM}}_{=\frac{\nu y}{E}}\right], \end{aligned} \tag{18.36}$$

where we neglect the term $\frac{1}{yE} \ll \frac{1}{xM}$ at high energies. The final result is

$$\boxed{\frac{d^2\sigma}{dx dy} = \frac{2\pi\alpha^2 s}{Q^4}\left[2(1 - y) + y^2\right]F_2(x)}. \tag{18.37}$$

Let us summarize: elastic electron–proton scattering is described by two electromagnetic form factors $G_E(Q^2)$ and $G_M(Q^2)$ (Rosenbluth formula). For inelastic scattering the form factors are replaced by $W_2(Q^2, \nu)$ and $2W_1(Q^2, \nu)$ which are functions of two independent variables, Q^2 and ν. In deep inelastic scattering the wavelength of the exchanged photon becomes small enough to resolve the proton constituents which we have assumed to be spin-$\frac{1}{2}$ particles. The differential cross section is described by only one structure function $F_2(x)$ which depends on one dimensionless variable only, the (Bjorken) *scaling* variable x. Scaling means that the cross section at fixed x does not depend on momentum transfer. In other words the size of the constituents cannot be measured by reducing the photon wavelength since the partons are *point-like*.

Figure 18.2 (left) illustrates how the differential cross section (18.29) varies with invariant hadron mass M'. The strong peak on the left stems from elastic

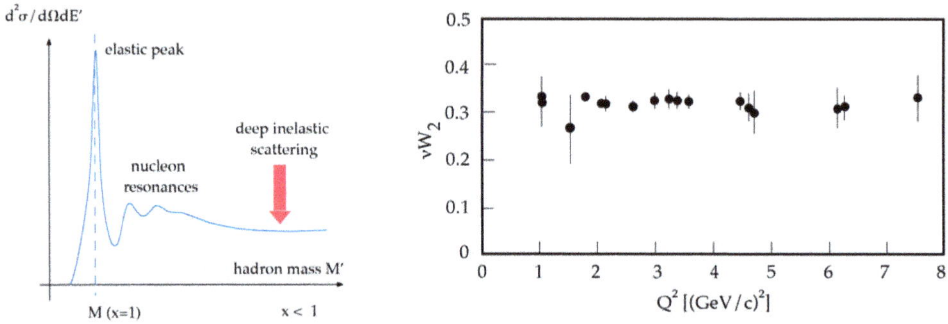

Figure 18.2. Left: typical behaviour of the electron–proton differential cross section as a function of the invariant hadron mass. The peak at $M' = M$ is due to elastic scattering ($x = 1$), the bumps at higher M' are due to nucleon resonances. Right: $F_2 = \nu W_2$ as a function of Q^2 for $x = 0.25$ and $M' > 2\,\text{GeV}$ (deep inelastic scattering). Adapted from [2].

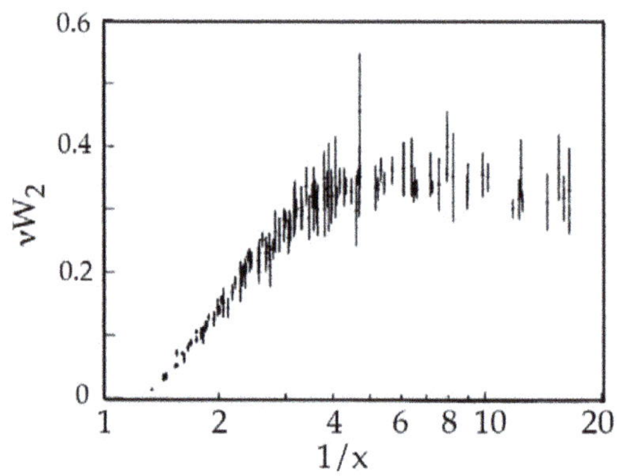

Figure 18.3. Structure function F_2 of the proton, plotted as a function of $\frac{1}{x}$ (for details see [2]).

electron–proton scattering ($M' = M$) and the bumps further to the right are nucleon resonances such as $\Delta(1232)$, $N(1520)$ and $\Delta(1920)$. Figure 18.2 (right) shows a measurement of $F_2 = \nu W_2$ in electron–proton scattering at fixed $x = 0.25$, obtained with electrons up to 21 GeV from the SLAC linear accelerator. The energy E' of the scattered electron is measured by a magnetic spectrometer for various scattering angles θ (see figure 2.8). As expected, the structure function does not depend on Q^2. Figure 18.3 shows measurements of F_2 plotted as a function of $\frac{1}{x}$. The structure function saturates when $x \to 0$, which implies that some of the functions $q_i(x)$ grow to infinity. The low-x values are dominated by the contributions from sea quarks and sea antiquarks which increase with decreasing x (see figure 18.5 below).

According to (18.31) the structure function of the proton is given by

$$\frac{F_2^p(x)}{x} = \left(\frac{2}{3}\right)^2 u(x) + \left(\frac{1}{3}\right)^2 d(x) + \frac{4}{3} S(x), \tag{18.38}$$

and that of the neutron by

$$\frac{F_2^n(x)}{x} = \left(\frac{1}{3}\right)^2 u(x) + \left(\frac{2}{3}\right)^2 d(x) + \frac{4}{3}S(x), \tag{18.39}$$

where $u(x) \equiv q_u(x)$ and $d(x) \equiv q_d(x)$. The normalization is such that

$$\int_0^1 \left[u(x) - \bar{u}(x)\right]\mathrm{d}x = 2, \int_0^1 \left[d(x) - \bar{d}(x)\right]\mathrm{d}x = 1. \tag{18.40}$$

We have assumed that the x distribution of the u quark in the proton is identical to that of the d quark in the neutron. The term $S(x)$ stands for sea quarks and antiquarks, and the factor $\frac{4}{3}$ for the sum of squared charges of the sea quarks u, $\bar{u}$, d, $\bar{d}$, s and $\bar{s}$. To simplify we assume that their structure functions are identical and neglect contributions from the heavier c and b quarks. The ratio of the proton to neutron structure functions is

$$\frac{F_2^p(x)}{F_2^n(x)} = \frac{\frac{4}{9}u(x) + \frac{1}{9}d(x) + \frac{4}{3}S(x)}{\frac{1}{9}u(x) + \frac{4}{9}d(x) + \frac{4}{3}S(x)}. \tag{18.41}$$

The structure function of the neutron is measured with a deuterium target by subtracting the contribution from the proton. For small x the sea dominates while becoming negligible for large x. From this ratio one therefore expects that

$$\begin{aligned} F_2^p(x) &= F_2^n(x) \quad \text{for } x \to 0, \\ F_2^p(x) &= 4F_2^n(x) \quad \text{for } x \to 1. \end{aligned} \tag{18.42}$$

The data show that this is indeed the case (figure 18.4).

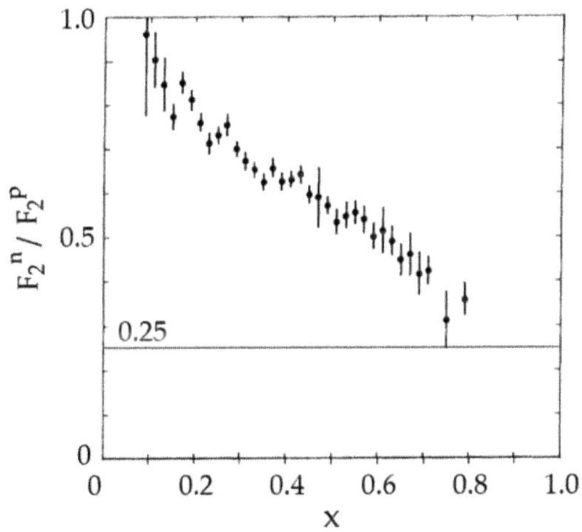

Figure 18.4. Ratio of electron–neutron to electron–proton deep inelastic cross sections as a function of x. After [2].

Let us estimate the fraction $\bar{x}$ of the proton mass carried by quarks. For 3 valence quarks (18.31) leads to

$$\int_0^1 F_2(x)\mathrm{d}x = \frac{\bar{x}}{3}\sum_i Q_i^2 = \bar{x} \times \begin{bmatrix} \frac{1}{3} \text{ for the proton} \\ \frac{2}{9} \text{ for the neutron} \end{bmatrix}. \tag{18.43}$$

Fitting F_2 to the cross sections gives for the integral $\simeq$0.18 for the proton and $\simeq$0.12 for the neutron, hence $\bar{x} = 54\%$. As for deep inelastic neutrino scattering (section 17.4.2), one finds that about half of the proton mass is carried by gluons. Figure 18.5 shows the individual contributions of the valence quarks, the sea quarks and the gluons to the proton.

It is instructive to compare deep inelastic neutrino scattering with deep inelastic electron scattering. Figure 18.6 shows the structure function $F_2(x)$ from neutrino

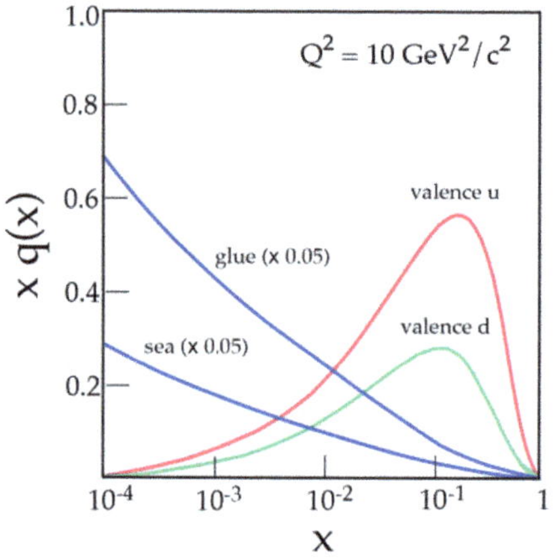

Figure 18.5. Structure functions of the valence quarks, sea quarks and gluons as a function of x. Adapted from [3].

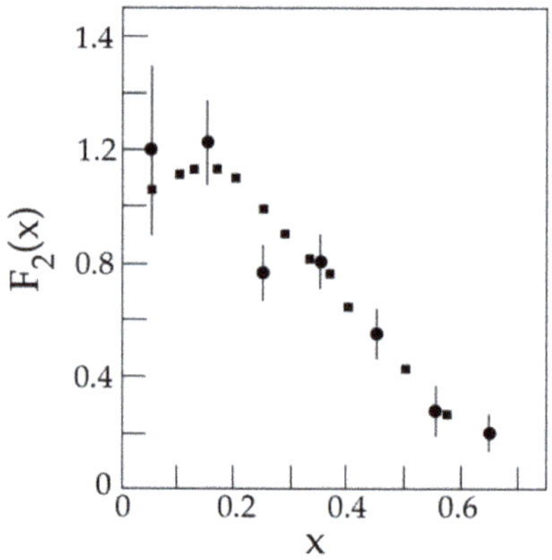

Figure 18.6. Comparison between deep inelastic neutrino–nucleon (circles) and deep inelastic electron–nucleon scattering (squares). The electron data have been divided by $\frac{5}{18}$. After [4].

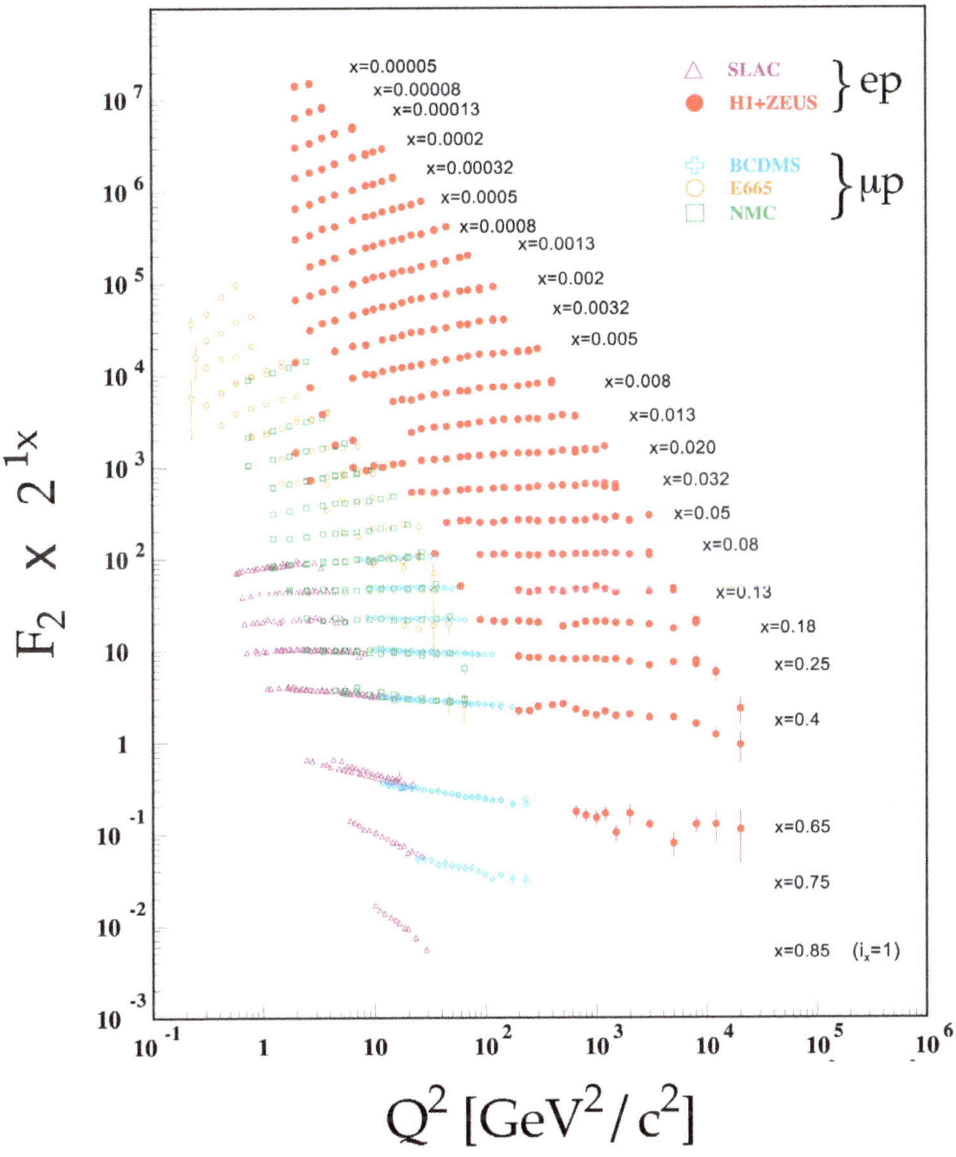

Figure 18.7. Proton structure function F_2^p measured by various experiments in *ep* and *μp* scattering (details of the data can be found in [5]). The data are offset for clarity by the factor 2^{i_x}, where i_x is the *x*-bin number ($i_x = 1 \ldots 24$). After [5].

data at CERN, and from $e^- p$ and $e^- n$ scattering at SLAC. The average squared electric charge encountered by the electron striking a quark is $\frac{1}{3}$ in the proton and $\frac{2}{9}$ in the neutron. The electron data have therefore been divided by $\frac{5}{18}$ (the average between the proton and neutron) for comparison with neutrino data. The agreement

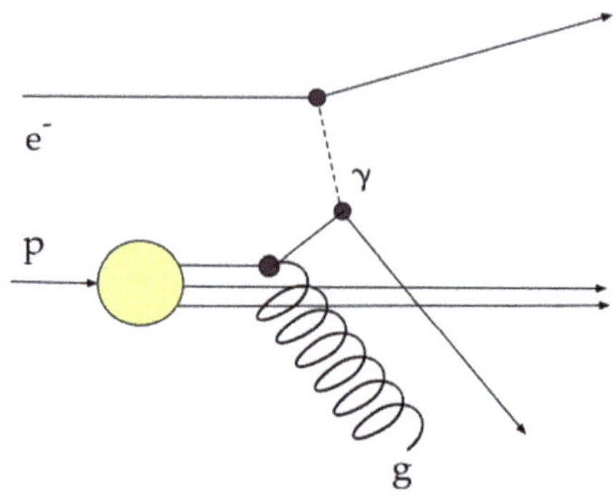

Figure 18.8. Scaling violation is due to higher order processes such as initial gluon emission.

between the electron and neutrino scattering is excellent, thus showing that both probes see the same point-like objects in the nucleon.

Figure 18.7 shows the structure function F_2 of the proton as a function of Q^2 and x, measured over the years by many experiments with electrons, positrons and muons. Most of the low-x data points have been collected at the HERA electron–proton collider. For low and high values of x one observes a strong dependence on Q^2. This violation of the scaling law is due to higher order processes, such as the one shown in figure 18.8. Scaling—and the evidence for the existence of quarks—was therefore discovered by a stroke of luck. This is reminiscent of the accidental discovery of the atomic nucleus made with α projectiles that were not energetic enough to penetrate the nuclear core (chapter 1).

References

[1] Halzen F and Martin A D 1984 *Quarks and Leptons: An Introductory Course in Modern Particle Physics* (New York: Wiley)

[2] Friedman J L and Kendall H W 1972 *Ann. Rev. Nucl. Sci.* **22** 203

[3] HERA Structure Functions Working Group 2009 *Int. Conf. on Deep Inelastic Scattering (DIS2009) (Madrid, Spain, April 2009)*

[4] Perkins D H 2000 *Introduction to High Energy Physics* (Cambridge: Cambridge University Press)

[5] Olive K *et al* (Particle Data Group) 2014 *Chin. Phys.* C **38** 090001

IOP Publishing

Nuclear and Particle Physics

Claude Amsler

Chapter 19

The $K^0 - \overline{K}^0$ system

The kaons form two doublets of the strong isospin. The isospin component i_3 can be easily derived from the Gell–Mann–Nishijima formula $Q = i_3 + \frac{S}{2}$. The two isospin doublets are [1]

$$\begin{pmatrix} K^+[u\bar{s}] \\ K^0[d\bar{s}] \end{pmatrix} \quad \text{and} \quad \begin{pmatrix} \overline{K}^0[\bar{d}s] \\ -K^-[\bar{u}s] \end{pmatrix}. \tag{19.1}$$

with $S = +1$ for the K^0 and $S = -1$ for the $(\overline{K}^0)$. The discovery of the K^0 meson was described in section 10.1. Due to strangeness conservation neutral kaons are produced by the *strong* interaction, together with other strange particles. The K^0 is produced with a hyperon, for example via the reaction $\pi^- p \to \Lambda K^0$. The production of the $\overline{K}^0$ is somewhat more difficult because of its negative strangeness. Strangeness conservation requires the production of additional K^+ (K^0) or antihyperons (and thus further baryons) in reactions such as $\pi^+ p \to K^+ \overline{K}^0 p$ or $\pi^- p \to \bar{\Lambda}\overline{K}^0 nn$. More K^0 than $\overline{K}^0$ mesons are therefore produced at low energies.

In contrast to the production mechanism, neutral kaons decay through the *weak* interaction with $\Delta S = \pm 1$, since no lighter strange hadrons exist. They decay mainly into 2π, 3π and semileptonically, the K^0 into $\pi^- e^+ \nu_e$ and $\pi^- \mu^+ \nu_\mu$, the $\overline{K}^0$ into $\pi^+ e^- \bar{\nu}_e$ and $\pi^+ \mu^- \bar{\nu}_\mu$, following the $\Delta S = \Delta Q$ rule (chapter 13).

19.1 The K_1 and K_2 mesons

The K^0 and $\overline{K}^0$ mesons have spin 0 and parity -1 (section 11.4):

$$P\left|K^0\left(\overline{K}^0\right)\right\rangle = -\left|K^0\left(\overline{K}^0\right)\right\rangle. \tag{19.2}$$

doi:10.1088/978-0-7503-1140-3ch19 © IOP Publishing Ltd 2015

Since

$$C^2\left|K^0\left(\overline{K}^0\right)\right\rangle = \left|K^0\left(\overline{K}^0\right)\right\rangle \tag{19.3}$$

charge conjugation gives

$$C\left|K^0\left(\overline{K}^0\right)\right\rangle = +\left|\overline{K}^0\left(K^0\right)\right\rangle \quad \text{or} \quad C\left|K^0\left(\overline{K}^0\right)\right\rangle = -\left|\overline{K}^0\left(K^0\right)\right\rangle. \tag{19.4}$$

We choose the minus sign:

$$\begin{aligned} C\left|K^0\right\rangle &\equiv -\left|\overline{K}^0\right\rangle, \\ C\left|\overline{K}^0\right\rangle &\equiv -\left|K^0\right\rangle, \end{aligned} \tag{19.5}$$

and therefore

$$\boxed{CP\left|K^0\right\rangle = \left|\overline{K}^0\right\rangle \quad \text{and} \quad CP\left|\overline{K}^0\right\rangle = \left|K^0\right\rangle}. \tag{19.6}$$

Hence the K^0 and $\overline{K}^0$ mesons are not eigenstates of the combined CP operation. Let us write the two states as basis vectors:

$$\left|K^0\right\rangle \equiv \begin{pmatrix} 1 \\ 0 \end{pmatrix} \quad \text{and} \quad \left|\overline{K}^0\right\rangle \equiv \begin{pmatrix} 0 \\ 1 \end{pmatrix}. \tag{19.7}$$

Following (19.6) we can express the CP operator as

$$CP = \begin{pmatrix} 0 & 1 \\ 1 & 0 \end{pmatrix}, \tag{19.8}$$

which is identical to the Pauli matrix σ_x. CP violation is a weak effect, so let us first assume that CP is conserved in weak interactions. The physical states (eigenstates of the weak Hamiltonian H_w) are eigenstates of CP:

$$\left[H_w, CP\right] = 0. \tag{19.9}$$

The most general 2×2 complex matrix is

$$H_w = A \cdot I + \vec{B} \cdot \vec{\sigma} \tag{19.10}$$

with *complex* parameters A, B_x, B_y and B_z. The Pauli matrices σ_y and σ_z do not commute with σ_x. Hence H_w commutes with $CP = \sigma_x$ if $B_y = B_z = 0$, in which case the time-independent Schrödinger equation reads

$$H_w|\psi\rangle = \begin{pmatrix} A & B_x \\ B_x & A \end{pmatrix}|\psi\rangle = E|\psi\rangle. \tag{19.11}$$

The secular equation

$$(A - E)^2 - B_x^2 = 0 \tag{19.12}$$

has two solutions for the eigenvalues:

$$E_1 = A + B_x \qquad \text{and} \qquad E_2 = A - B_x. \tag{19.13}$$

The corresponding eigenstates are

$$\begin{pmatrix} -B_x & B_x \\ B_x & -B_x \end{pmatrix} |\psi_1\rangle = 0 \Rightarrow |\psi_1\rangle = \frac{1}{\sqrt{2}}\begin{pmatrix} 1 \\ 1 \end{pmatrix}$$

$$= \boxed{\frac{1}{\sqrt{2}}\left[|K^0\rangle + |\overline{K}^0\rangle\right] \equiv |K_1\rangle}, \tag{19.14}$$

$$\begin{pmatrix} B_x & B_x \\ B_x & B_x \end{pmatrix} |\psi_2\rangle = 0 \Rightarrow |\psi_2\rangle = \frac{1}{\sqrt{2}}\begin{pmatrix} 1 \\ -1 \end{pmatrix}$$

$$= \boxed{\frac{1}{\sqrt{2}}\left[|K^0\rangle - |\overline{K}^0\rangle\right] \equiv |K_2\rangle}. \tag{19.15}$$

Obviously, K_1 and K_2 are eigenstates of CP with eigenvalues 1 and −1, respectively. Indeed, using (19.6),

$$\boxed{CP|K_1\rangle = +|K_1\rangle, \qquad CP|K_2\rangle = -|K_2\rangle}. \tag{19.16}$$

Conversely, the K^0 and $\overline{K}^0$ mesons are quantum superpositions of K_1 and K_2:

$$|K^0\rangle = \frac{1}{\sqrt{2}}\left[|K_1\rangle + |K_2\rangle\right] \qquad \text{and} \qquad |\overline{K}^0\rangle = \frac{1}{\sqrt{2}}\left[|K_1\rangle - |K_2\rangle\right]. \tag{19.17}$$

In the absence of weak interaction the K^0 and $\overline{K}^0$ mesons would be produced with well-defined degenerate masses, as a consequence of CPT invariance:

$$m(K^0) = m(\overline{K}^0) \equiv m_0. \tag{19.18}$$

The weak interaction H_w acts as a perturbation that mixes the K^0 and $\overline{K}^0$ states, leading the system to oscillate between K^0 and $\overline{K}^0$. The degeneracy (19.18) is lifted and there are two states, K_1 and K_2 with distinct masses. We shall see that the lifetime of the K_1 is much shorter than that of the K_2, which is why we call their decay constants λ_S ('S' for 'short') and λ_L ('L' for 'long'). Accordingly, we also label the masses with the symbols m_S and m_L, respectively. The masses of the K^0 and $\overline{K}^0$ mesons are actually not defined. An ideal measurement with infinitely good resolution (for example by using constrained kinematics in the reaction $\pi^- p \rightarrow K^0\Lambda$) would deliver two possible values for the K^0 mass, m_S and m_L. Figure 19.1 shows the $\Delta S = 2$ processes that are responsible for the oscillation between K^0 and $\overline{K}^0$.

Since $\Delta S = 2$ the mixing mechanism is sometimes referred to as 'superweak'. Note the analogy with the NH_3 maser in which the nitrogen atom swings through the

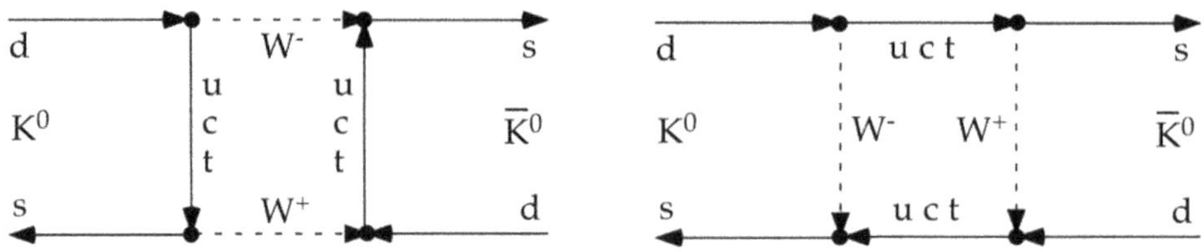

Figure 19.1. $\Delta S = 2$ contributions to $K^0 - \overline{K}^0$ oscillations.

plane spanned by the three hydrogen atoms. The maser emits or absorbs microwaves between the two levels 1 and 2.

The time evolution of an unstable system ψ with mass m and decay constant λ is given by $\psi\,\exp(-\mathrm{i}Et)$, where E is complex and equal to $m - \mathrm{i}\frac{\lambda}{2}$. The decay probability is then proportional to $|\psi|^2\exp(-\lambda t)$. The time evolution of the K_1 or K_2 states is given by

$$\left|K_{1,2}(t)\right\rangle = \left|K_{1,2}\right\rangle \mathrm{e}^{-\mathrm{i}E_{1,2}t} \tag{19.19}$$

in their rest frames, where E_1 and E_2 are complex. Thus, we rewrite E_1 and E_2 as

$$\begin{aligned} E_1 &= A + B_x \equiv m_S - \mathrm{i}\frac{\lambda_S}{2}, \\ E_2 &= A - B_x \equiv m_L - \mathrm{i}\frac{\lambda_L}{2}, \end{aligned} \tag{19.20}$$

where $m_{S,L}$ and $\lambda_{S,L}$ are real, and obtain

$$\left|K_{1,2}(t)\right\rangle = \left|K_{1,2}\right\rangle \mathrm{e}^{-\mathrm{i}m_{S,L}t - \frac{\lambda_{S,L}}{2}t}. \tag{19.21}$$

The states K_1 and K_2 are orthonormal, therefore

$$\left\langle K_{1,2}(t)\middle|K_{1,2}(t)\right\rangle = \left\langle K_{1,2}\middle|K_{1,2}\right\rangle \mathrm{e}^{-\lambda_{S,L}t} = \mathrm{e}^{-\lambda_{S,L}t}. \tag{19.22}$$

Let us now examine the CP-eigenvalues of the decay final states. According to (11.108) the charge conjugation of a pion pair gives

$$C\left|\pi^+\pi^-\right\rangle = (-1)^{\ell+s}\left|\pi^+\pi^-\right\rangle \tag{19.23}$$

with spin $s = 0$, while parity gives with (11.33):

$$P\left|\pi^+\pi^-\right\rangle = (-1)^{\ell}\left|\pi^+\pi^-\right\rangle. \tag{19.24}$$

For the combined CP operation one obtains

$$CP\left|\pi^+\pi^-\right\rangle = \left|\pi^+\pi^-\right\rangle. \tag{19.25}$$

The pion pair from neutral kaon decay is a CP-eigenstate with eigenvalue +1. This is also true for a pair of neutral pions:

$$CP\left|\pi^0\pi^0\right\rangle = \left|\pi^0\pi^0\right\rangle. \tag{19.26}$$

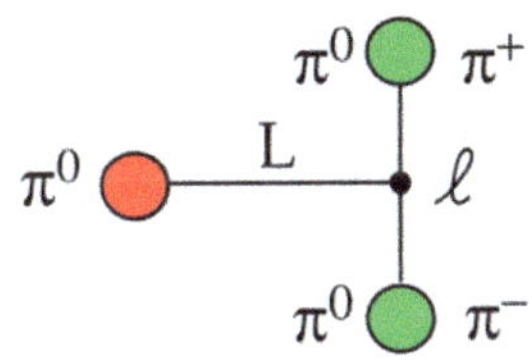

Figure 19.2. Decomposition of the three-pion system into a neutral pion and a dipion ($\pi^0\pi^0$ or $\pi^+\pi^-$).

Let us now examine the three-pion decay of the neutral kaon which is slightly more complicated. We group the three pions into a neutral dipion with angular momentum ℓ (figure 19.2). The relative angular momentum between the dipion and the third pion is $L = \ell$ since the spin of the kaon is zero. One obtains with $C(\pi^0) = 1$, $P(\pi^0) = -1$:

$$\begin{aligned} C|\pi^0\pi\pi\rangle &= C\left(\pi^0\right)(-1)^\ell\left|\pi^0\pi\pi\right\rangle = (-1)^\ell\left|\pi^0\pi\pi\right\rangle, \\ P|\pi^0\pi\pi\rangle &= P\left(\pi^0\right)P(\pi)^2\underbrace{(-1)^\ell}_{P(2\pi)}(-1)^L\left|\pi^0\pi\pi\right\rangle = -\left|\pi^0\pi\pi\right\rangle. \end{aligned} \tag{19.27}$$

The $3\pi^0$ system is a C-eigenstate with eigenvalue $C(\pi^0)^3 = +1$, hence ℓ is even (this also follows from the symmetry of the $2\pi^0$ wave function for identical bosons). The $3\pi^0$ system is then a CP-eigenstate with eigenvalue -1:

$$CP\left|\pi^0\pi^0\pi^0\right\rangle = -\left|\pi^0\pi^0\pi^0\right\rangle. \tag{19.28}$$

The CP-eigenvalue of the $\pi^+\pi^-\pi^0$ system depends on the angular momentum ℓ. Equation (19.27) gives for the combined operation

$$CP\left|\pi^+\pi^-\pi^0\right\rangle = (-1)^{\ell+1}\left|\pi^+\pi^-\pi^0\right\rangle. \tag{19.29}$$

The phase space is limited by the small Q-value (83 MeV), and therefore the angular momentum $\ell = L = 0$ prevails. Thus the CP-eigenvalue is *dominantly* equal to -1. Since CP is conserved in the decay one expects from (19.25), (19.26) and (19.28), (19.29) that

$$\boxed{K_1 \to 2\pi, \qquad K_2 \to 3\pi}. \tag{19.30}$$

Thus CP conservation forbids the 2π-decay of the K_2 meson. The decay $K_1 \to \pi^+\pi^-\pi^0$ is allowed, but suppressed with a branching ratio $f \simeq 3 \times 10^{-7}$.

19.1.1 Lifetimes of the K_1 and K_2 mesons

The small phase space for the three-body decay leads to a much larger mean life of the K_2 state, that is $\tau_L = 1/\lambda_L \gg \tau_S = 1/\lambda_S$. Figure 19.3 shows a measurement of the K_1 lifetime. The K^0 mesons are produced in a hydrogen bubble chamber with a $1.5\,\mathrm{GeV}\,c^{-1}$ K^+ beam, inducing the reaction $K^+p \to K^0p\pi^+$. The momenta of the

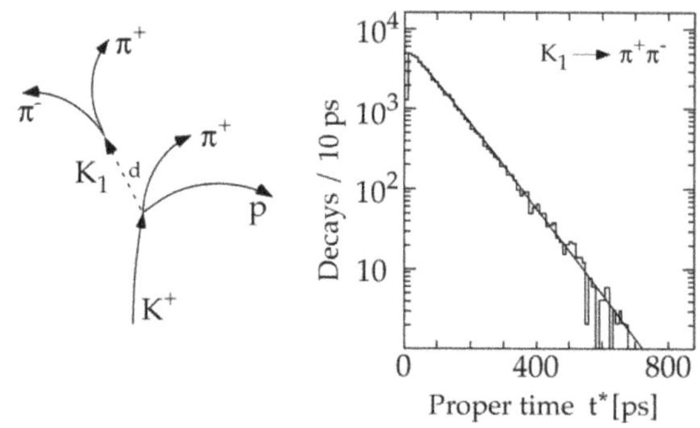

Figure 19.3. Decay of the K_1 component. Left: a typical event. Right: proper time distribution with fit. After [2].

pions are measured from the track curvatures, and thus also the velocity βc and the γ-factor of the invisible neutral K_1. The decay time is calculated from the distance d between the production and decay points. Figure 19.3 shows the distribution of the proper time $t^* = d/\gamma\beta c$ in the K_1 rest frame. The mean life of the K_1 is $\tau_S = 89.54 \pm 0.04$ [3]. This corresponds to a mean decay length $\bar{d} = \beta\gamma c\tau_S = \beta\gamma \times 2.7$ cm in the laboratory.

The K_2 meson was discovered in 1956 in a cloud chamber at the Brookhaven Cosmotron [4]. Its lifetime was measured at the Princeton-Pennsylvania Accelerator (figure 19.4(a)). Neutral kaons produced by 3 GeV protons are detected at the end of an evacuated tube by a detector mounted on a trolley (figure 19.4(b)). Photons are converted by a lead foil and charged particles are removed by a clearing magnet.

The detector (figure 19.4(c)) is more elaborate than required to simply measure the lifetime since the branching ratios for the 3π and semileptonic decays are measured at the same time. It consists of arrays of plastic scintillators and lead sheets to detect charged pions, photons from π^0 decays, and electrons or positrons from semileptonic decays. The background from cosmic rays is suppressed by veto counters. The velocity of the K_2 (and thus its γ-factor) is measured by using the time of flight t between the production point and the detector located at the distance d. The start time is determined by the high frequency signal of the accelerator. The counting rate at the detector is measured by the detector, as a function of time t. This rate is proportional to the K_2 flux at the detector position:

$$I = I_0\left[\frac{\Omega(d)}{4\pi}\right]\exp\underbrace{\left(-t/\gamma\tau_L\right)}_{\left(-t^*/\tau_L\right)}, \tag{19.31}$$

where the solid angle $\Omega(d)$ is equal to the surface of the collimator hole divided by d^2. Figure 19.5 shows the temporal distribution of I/I_0. A mean life $\tau_L = 51.54 \pm 0.44$ ns is measured which is more than two orders of magnitude larger than τ_S.

A more precise measurement of the K_2 lifetime is notoriously difficult due to background from cosmic rays or neutral particles, such as photons and neutrons which are produced more abundantly. An elegant method is to use the decay

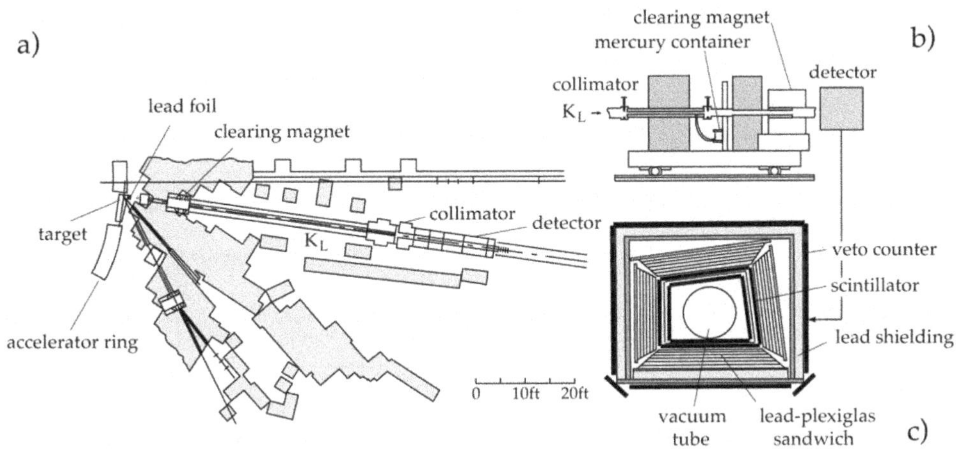

Figure 19.4. The equipment used to measure the lifetime of the long-lived K_2 meson. After [5].

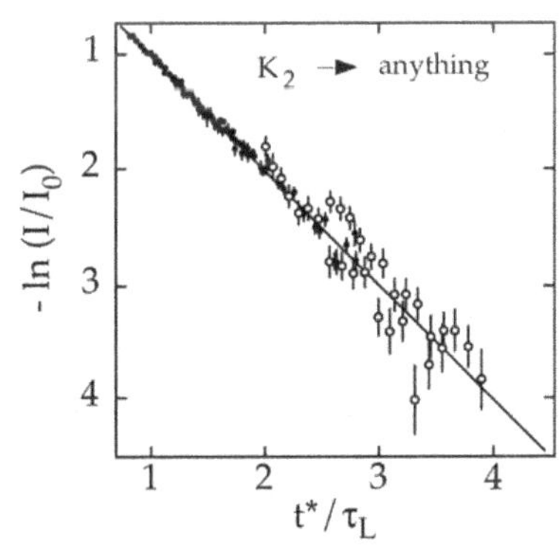

Figure 19.5. Attenuation of the K_2 intensity versus proper time (in units of τ_L). After [5].

$\phi(1020) \rightarrow K_1K_2$ and to *tag* the K_1, in other words to collect only events associated with the observation of the decay $K_1 \rightarrow \pi^+\pi^-$. This was performed at the DAϕNE collider at Frascati by exciting the ϕ resonance in e^+e^- collisions. The ϕ is produced at rest and decays dominantly into K^+K^- ($f = 49\%$) and K_1K_2 ($f = 34\%$).

The quantum numbers of the ϕ are $J^{PC} = 1^{--}$, hence $CP(\phi) = +1$. The ϕ decays into K_1K_2, but not into K_1K_1 due to CP conservation:

$$CP\left(K_1K_2\right) = CP\left(K_1\right) \cdot CP\left(K_2\right)(-1)^{\ell} = +1 \tag{19.32}$$

with $\ell = 1$ in ϕ decay (since kaons are spinless), while $CP(K_1K_1) = -1$ with $\ell = 1$. The K_1 and K_2 are produced back-to-back with momenta of 110 MeV c^{-1} ($\beta = 0.21$). The observation of a low energy K_1 (mean free path of 0.6 cm) signals the emission of a 110 MeV c^{-1} K_2 into the opposite direction, along which its decay point is to be sought. The KLOE detector, a solenoidal magnet equipped with drift chambers and calorimeters, obtained $\tau_L = 50.72 \pm 0.37$ ns and measured various decay branching ratios of the K_2 with enhanced precision [6].

Table 19.1. Main decay branching ratios of the K_1 and K_2 mesons.

		f [%]
$K_1 \to$	$\pi^+\pi^-$	69
	$\pi^0\pi^0$	31
$K_2 \to$	$\pi^0\pi^0\pi^0$	20
	$\pi^+\pi^-\pi^0$	13
	$\pi^\pm e^\mp \nu_e$	40
	$\pi^\pm \mu^\mp \nu_\mu$	27

Table 19.1 lists the largest decay branching ratios of the K_1 and K_2 mesons. For K_1 the ratio of $\pi^+\pi^-$ to $\pi^0\pi^0$ decay widths ($\simeq$2) is in good agreement with the prediction from the $|\Delta i| = \frac{1}{2}$ rule (13.10).

The semileptonic branching ratios contribute essentially only to K_2 decay since

$$\begin{aligned} f\left(K_1 \to \pi e \nu\right) &= \frac{\Gamma\left(K^0\left(\overline{K}^0\right) \to \pi e \nu\right)}{\Gamma_S} = \frac{\Gamma\left(K^0\left(\overline{K}^0\right) \to \pi e \nu\right)}{\Gamma_L} \cdot \frac{\Gamma_L}{\Gamma_S} \\ &= f\left(K_2 \to \pi e \nu\right)\frac{\tau_S}{\tau_L} \simeq 7.0 \times 10^{-4} \end{aligned} \tag{19.33}$$

with $\Gamma_{S,L} = \lambda_{S,L}$. Similarly, $f\left(K_1 \to \pi \mu \nu\right) \simeq 4.7 \times 10^{-4}$.

19.1.2 $K^0 - \overline{K}^0$ oscillations

Let us assume that only K^0 mesons are produced, for instance by the reaction $K^+p \to K^0 p \pi^+$ with incident K^+ mesons. Because of the $\Delta S = 2$ perturbation the K^0 oscillates between K^0 and $\overline{K}^0$ (figure 19.1). According to (19.17) the K^0 is a superposition of K_1 and K_2. At time $t = 0$,

$$\left|K^0\right\rangle = \frac{1}{\sqrt{2}}\left[\left|K_1\right\rangle + \left|K_2\right\rangle\right] \equiv |K(t=0)\rangle. \tag{19.34}$$

The K^0 propagates with proper time as

$$\begin{aligned} |K(t)\rangle &= \frac{1}{\sqrt{2}}\left[\left|K_1\right\rangle e^{-im_S t - \frac{\lambda_S t}{2}} + \left|K_2\right\rangle e^{-im_L t - \frac{\lambda_L t}{2}}\right], \\ &= \frac{1}{2}\left|K^0\right\rangle\left(e^{-im_S t - \frac{\lambda_S t}{2}} + e^{-im_L t - \frac{\lambda_L t}{2}}\right) \\ &\quad + \frac{1}{2}\left|\overline{K}^0\right\rangle\left(e^{-im_S t - \frac{\lambda_S t}{2}} - e^{-im_L t - \frac{\lambda_L t}{2}}\right). \end{aligned} \tag{19.35}$$

The probability of finding a K^0 at time t is

$$w(t) = \left|\left\langle K^0 \middle| K(t)\right\rangle\right|^2 = \frac{1}{4}\left[e^{-\lambda_S t} + e^{-\lambda_L t} + 2\, e^{-\frac{(\lambda_S+\lambda_L)t}{2}} \cos\underbrace{(m_L - m_S)}_{\equiv \Delta m} t\right], \tag{19.36}$$

since $\langle K^0|\overline{K}^0\rangle = 0$. Similarly, the probability of finding a $\overline{K}^0$ at time t is

$$\overline{w}(t) = \left|\left\langle \overline{K}^0 \middle| K(t)\right\rangle\right|^2 = \frac{1}{4}\left[e^{-\lambda_S t} + e^{-\lambda_L t} - 2\, e^{-\frac{(\lambda_S+\lambda_L)t}{2}} \cos \Delta m t\right], \tag{19.37}$$

and the probability of finding either a K^0 or a $\overline{K}^0$ is

$$w(t) + \overline{w}(t) = \frac{1}{2}\left[e^{-\lambda_S t} + e^{-\lambda_L t}\right]. \tag{19.38}$$

Figure 19.6 shows the time dependence of the probabilities $w(t)$ and $\overline{w}(t)$. At the beginning, the decay rate is determined by the mean life $\tau_S = 1/\lambda_S$ of the short-lived K_1 component. The *oscillation frequency* between K^0 and $\overline{K}^0$ is Δm. The K^0 mesons at time t can be distinguished from $\overline{K}^0$ by detecting the semileptonic decays $K^0 \to \pi^- e^+ \nu_e$ or $\overline{K}^0 \to \pi^+ e^- \overline{\nu}_e$ or by inserting a proton target and observing a reaction such as $\overline{K}^0 p \to \Lambda \pi^+$. The mass difference Δm can then be determined directly from the measured oscillation curve $w(t)$. The recommended current value of Δm is (see also section 19.2.1) [3]:

$$\begin{aligned} \Delta m &= m_L - m_S = (0.5293 \pm 0.0009) \times 10^{10}\ \mathrm{s}^{-1} \\ &\simeq 3.5 \times 10^{-6}\ \mathrm{eV} \ll m_0 = 498\ \mathrm{MeV}. \end{aligned} \tag{19.39}$$

Note that the sign of Δm cannot be obtained with this method. The corresponding oscillation period $2\pi/\Delta m$ is around 1 ns, compared to the lifetime of the

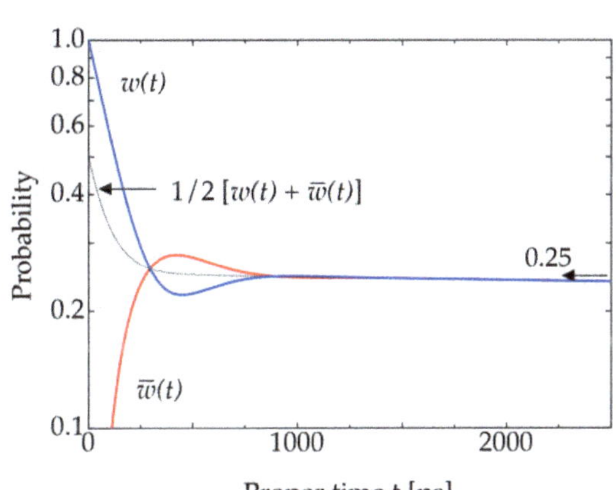

Figure 19.6. Probability $w(t)$ of observing a K^0 when a K^0 was produced at $t = 0$ (logarithmic scale). The probability $\overline{w}(t)$ of observing a $\overline{K}^0$ is also shown. The long tail beyond 1000 ps is due to the slowly decaying K_2 component.

K_1 component of about 0.1 ns. Flavour oscillations could be observed in other systems such as $B^0 - \overline{B}^0$, $B_s^0 - \overline{B}_s^0$ or $D^0 - \overline{D}^0$, provided that the lifetime of the short-lived component is not too short compared to $2\pi/\Delta m$. The condition for *complete mixing* is defined as $\Delta m \cdot \tau_S \gtrsim 1$. For the $K^0 - \overline{K}^0$ system this condition is only marginally fulfilled:

$$\Delta m \cdot \tau_S \simeq 0.5. \tag{19.40}$$

Therefore only the onset of the oscillation is observed as figure 19.6 shows. Complete mixing is observed in the $B^0 - \overline{B}^0$ system ($\Delta m \cdot \tau_S \simeq 0.8$) and in the $B_s^0 - \overline{B}_s^0$ system ($\Delta m \cdot \tau_S \simeq 27$) see chapter 20, but not in the $D^0 - \overline{D}^0$ system ($\Delta m \cdot \tau_S \simeq 0.01$).

19.1.3 Regeneration

Consider a sufficiently long beam of neutral kaons in which the K_1 component has decayed. We will show that K_1 mesons can be restored (*regenerated*) by inserting a thin absorber into the beam line. In front of the absorber the kaon component is the K_2:

$$|K_2\rangle = \frac{1}{\sqrt{2}}\left[|K^0\rangle - |\overline{K}^0\rangle\right]. \tag{19.41}$$

The wave function behind the absorber is given by the superposition

$$|\psi\rangle = \frac{1}{\sqrt{2}}\left[f(0)\,|K^0\rangle - \bar{f}(0)\,|\overline{K}^0\rangle\right], \tag{19.42}$$

where $f(0)$ and $\bar{f}(0)$ are the forward elastic scattering amplitudes off an absorber nucleus N, for K^0N and $\overline{K}^0N$ scattering, respectively. The forward elastic scattering amplitude (scattering angle $\theta = 0$) is related to the total cross section through the optical theorem $\sigma_T = (4\pi/k)\,\mathrm{Im}\,f(0)$ where k is the kaon momentum. Since the total cross sections for K^0N and $\overline{K}^0N$ scattering are different, $f(0)$ is not equal to $\bar{f}(0)$: due to the negative strangeness of the $\overline{K}^0$, hyperons are produced at low energy (e.g. via $\overline{K}^0p \to \Lambda\pi^+$) or intermediate resonances are excited (e.g. $\overline{K}^0p \to |uus\rangle \to \overline{K}^0p$), while the K^0 with positive strangeness scatters elastically without resonance excitation ($uu\bar{s}$ hadrons do not exist). One can think intuitively that the $\overline{K}^0$ component is more efficiently removed from the K_2 beam. Substituting the superpositions (19.17) one obtains

$$|\psi\rangle = \frac{1}{2}\left[f(0) + \bar{f}(0)\right]\cdot|K_2\rangle + \frac{1}{2}\left[f(0) - \bar{f}(0)\right]\cdot|K_1\rangle. \tag{19.43}$$

Hence K_1 mesons reappear behind the absorber with the *regeneration amplitude*

$$f_R = \frac{1}{2}\left[f(0) - \bar{f}(0)\right]. \tag{19.44}$$

Let us consider an absorber of thickness L (figure 19.7(*a*)) and compute the ratio of K_1 to K_2 intensities after the absorber. The contribution to the K_1 amplitude

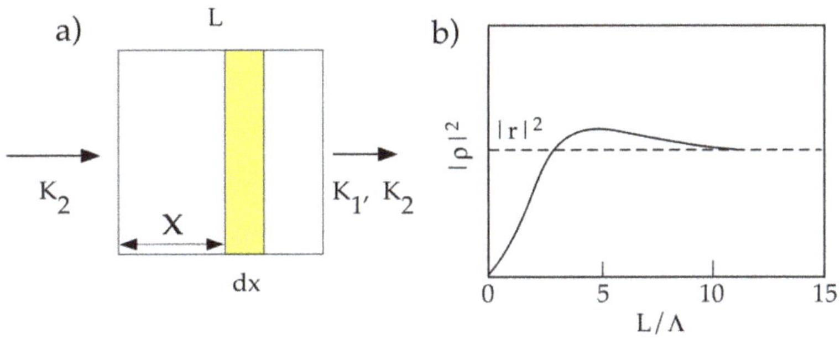

Figure 19.7. (*a*) Principle of regeneration. (*b*) Relative K_1 to K_2 intensity after the absorber of thickness L.

behind the absorber by an infinitesimally thin target of thickness $\mathrm{d}x$ is, see figure 19.7(*a*):

$$\mathrm{d}a_1 = \underbrace{a\mathrm{e}^{\mathrm{i}k_2x}}_{K_2\text{ wave}} \overbrace{\mathrm{e}^{\mathrm{i}k_1(L-x)}}^{K_1\text{ wave}} \mathrm{e}^{-\frac{L-x}{2\Lambda}} \mathrm{e}^{-\frac{L}{2\mu}} f_R n \mathrm{d}x. \tag{19.45}$$

The parameter Λ is the mean decay length of the K_1, μ is the mean absorption length, n is the density of the target nuclei, and k_1, k_2 are the momenta in the laboratory. The mean absorption lengths for K_1 and K_2 are equal since they contain equal numbers of K^0 and $\overline{K}^0$ mesons. We have also assumed that the mean decay length Λ of the K_2 is much larger than the absorber thickness L. Integrating over x gives the regenerated amplitude

$$a_1 = \int_0^L \mathrm{d}a_1 = \frac{af_R n}{\mathrm{i}(k_2 - k_1) + \frac{1}{2\Lambda}} \left[\mathrm{e}^{\mathrm{i}k_2L} - \mathrm{e}^{\mathrm{i}k_1L}\mathrm{e}^{-\frac{L}{2\Lambda}}\right]\mathrm{e}^{-\frac{L}{2\mu}}$$

$$= \frac{af_R n}{\frac{1}{2\Lambda} - \mathrm{i}\frac{\delta}{\Lambda}} \mathrm{e}^{\mathrm{i}k_1L}\left[\mathrm{e}^{-\mathrm{i}\frac{L\delta}{\Lambda}} - \mathrm{e}^{-\frac{L}{2\Lambda}}\right]\mathrm{e}^{-\frac{L}{2\mu}}, \tag{19.46}$$

where $\delta \equiv (k_1 - k_2)\Lambda$. The K_2 amplitude behind the absorber is

$$a_2 = a\mathrm{e}^{\mathrm{i}k_2L}\mathrm{e}^{-\frac{L}{2\mu}}. \tag{19.47}$$

The ratio of the amplitudes is then

$$\rho \equiv \frac{a_1}{a_2} = \underbrace{\frac{f_R n\Lambda}{\frac{1}{2} - \mathrm{i}\delta}}_{\equiv r}\left[1 - \mathrm{e}^{\mathrm{i}\frac{L\delta}{\Lambda} - \frac{L}{2\Lambda}}\right], \tag{19.48}$$

and the ratio of intensities behind the target is

$$|\rho|^2 = |r|^2\left[1 + \mathrm{e}^{-\frac{L}{\Lambda}} - 2\mathrm{e}^{-\frac{L}{2\Lambda}}\cos\left(\frac{L\delta}{\Lambda}\right)\right]. \tag{19.49}$$

The parameter δ can be expressed as a function of the mass difference Δm. From the approximation

$$k_1^2 + m_S^2 \simeq k_2^2 + m_L^2 \tag{19.50}$$

it follows that

$$\underbrace{(k_1 - k_2)}\underbrace{(k_1 + k_2)}_{\simeq 2k_1} \simeq \underbrace{(m_L - m_S)}_{\Delta m}\overbrace{(m_L + m_S)}^{\simeq 2m_S} \Rightarrow k_1 - k_2 = m_S\frac{\Delta m}{k_1}. \tag{19.51}$$

Since

$$\Lambda = \gamma\beta c\tau_S = \frac{k_1}{m_S}\tau_S \tag{19.52}$$

one obtains

$$\delta = (k_1 - k_2)\Lambda = m_S\frac{\Delta m}{k_1}\Lambda = \Delta m \cdot \tau_S. \tag{19.53}$$

From (19.49) the ratio of K_1/K_2 intensities behind the absorber is finally

$$\boxed{|\rho|^2 = |r|^2\left[1 + e^{-\frac{L}{\Lambda}} - 2e^{-\frac{L}{2\Lambda}}\cos\left(\frac{L}{\Lambda}\cdot\Delta m\cdot\tau_S\right)\right]}. \tag{19.54}$$

Figure 19.7(*b*) shows how the regenerated intensity $|\rho|^2$ depends on absorber thickness. The mass difference $|\Delta m|$ can be determined by measuring the $K_1 \to \pi^+\pi^-$ decay rate as a function of absorber thickness [7]. For this purpose the regeneration method is experimentally superior to that of the $K^0 - \overline{K}^0$ oscillation (section 19.1.2) because the measurement is carried out far away from the high background source at the kaon production target.

19.2 *CP* violation

The *CP*-violating decay $K_2 \to \pi^+\pi^-$ was discovered in 1964 in Brookhaven [8]. The experiment (figure 19.8) consists of two magnetic spectrometers equipped with spark chambers to measure the π^+ and π^- momenta. The K^0 mesons are produced by 30 GeV protons from the Alternating Gradient Synchrotron (AGS) impinging on a beryllium target. The collimator is located about 20 m from the kaon production target. To reduce multiple scattering and K_1 regeneration the decay volume is filled with helium gas (the density of helium gas is 0.179 kg m^{-3}, compared to 1.29 kg m^{-3} for air at 0°C and 1 atm).

The kinematics of the K_2 decay is shown in the inset of figure 19.8. The total momentum vector of the two charged pions is not parallel to the beam direction ($\theta \neq 0$) since the momentum of the undetected π^0 in the three-body decay has not

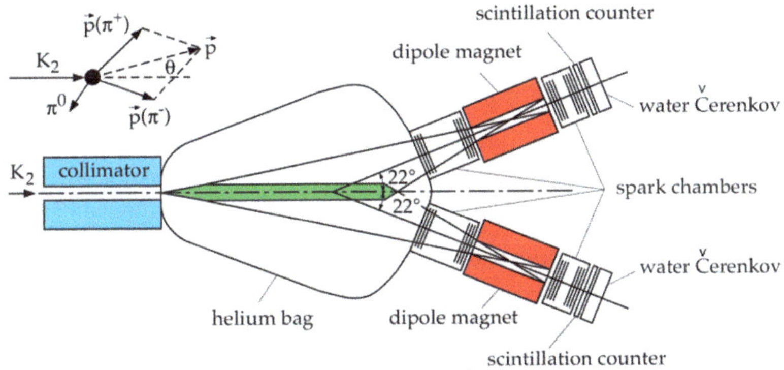

Figure 19.8. Diagram of the experiment that discovered *CP* violation in the decay $K_2 \to \pi^+\pi^-$. After [8].

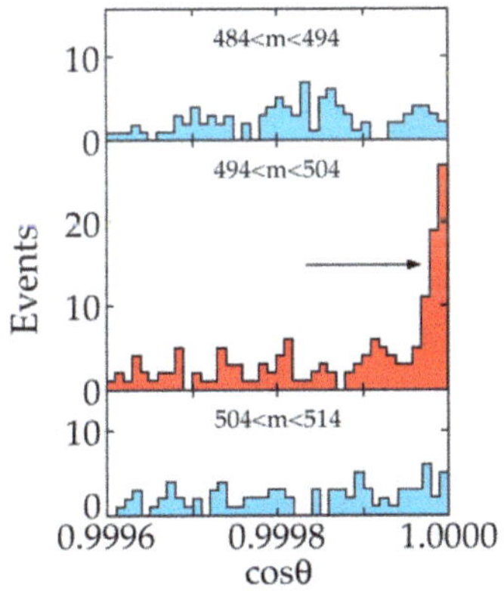

Figure 19.9. Angular distribution of the $\pi^+\pi^-$ momentum around the K_2 beam axis for three invariant mass intervals (in MeV). The arrow shows the signal from $K_2 \to \pi^+\pi^-$. After [8].

been taken into account. In contrast, $\theta = 0$ for $K_2 \to \pi^+\pi^-$ and the two-body invariant mass must coincide with the kaon mass:

$$m^2 = \left[E\left(\pi^+\right) + E\left(\pi^-\right)\right]^2 - \left[\vec{p}\left(\pi^+\right) + \vec{p}\left(\pi^-\right)\right]^2 = m_0^2 = (498\ \text{MeV})^2. \quad (19.55)$$

Figure 19.9 shows the cos θ distribution for three intervals of the $\pi^+\pi^-$ invariant mass. The isotropic distribution stems from the three-body decay of the K_2 meson. The forward peak, due to $K_2 \to \pi^+\pi^-$, appears at 499 MeV (central plot). This can be proven by inserting a regenerator to enhance the signal and also to measure the mean momentum of the kaon beam (1100 MeV c^{-1}). (In fact, *CP* violation had been observed earlier but had been mistakenly interpreted as 'anomalous' regeneration of K_1 mesons from K_2 mesons [9].) The ratio of partial widths

$$\frac{\Gamma\left(K_2 \to \pi^+\pi^-\right)}{\Gamma_L} = (2.3 \pm 0.4) \times 10^{-3} \quad (19.56)$$

shows that the *CP*-violating effect is rather weak. *CP* violation can occur in the $K^0 - \overline{K}^0$ oscillation mechanism and in the decay of the K_2 state. The first case is

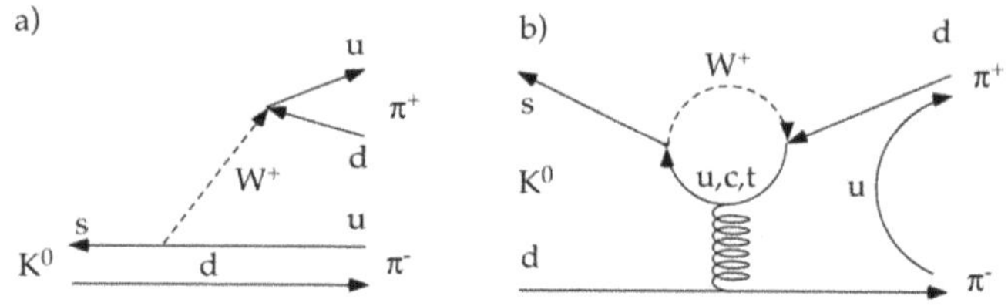

Figure 19.10. Contributions to the decay of the neutral kaon: 'tree' (*a*) and 'penguin' (*b*) diagrams. The main contributions stem from the matrix element V_{td}.

referred to as *indirect CP* violation, the second case as *direct CP* violation. Figure 19.10 shows the contributions to K^0 decay. In the standard model, the imaginary coupling constants in the CKM matrix (16.67) are responsible for violating the *CP* symmetry. These couplings appear in the currents for indirect *CP* violation (figure 19.1) and in those for direct *CP* violation (figure 19.10(*b*)). We will show experimentally that indirect *CP* violation dominates in the $K^0 - \overline{K}^0$ system. Because of *CPT* invariance also *T* invariance must be violated.

19.2.1 Indirect *CP* violation

Let us return to the general expression (19.10) for H_w but now allow *CP* violation in weak interactions, that is $[H_w, CP] \neq 0$. The condition on $B_y = 0$ can be relaxed. However, *CPT* invariance still requires B_z to vanish since

$$\begin{aligned} \left\langle K^0 \middle| H_w \middle| K^0 \right\rangle &= A + B_z = m\left(K^0\right), \\ \left\langle \overline{K}^0 \middle| H_w \middle| \overline{K}^0 \right\rangle &= A - B_z = m\left(\overline{K}^0\right) = m\left(K^0\right) \Rightarrow B_z = 0, \end{aligned} \tag{19.57}$$

because σ_z is diagonal. Hence the weak Hamiltonian becomes

$$H_w = \begin{pmatrix} A & B_x - \mathrm{i}B_y \\ B_x + \mathrm{i}B_y & A \end{pmatrix} = \begin{pmatrix} \alpha & \dfrac{\beta}{r} \\ \beta r & \alpha \end{pmatrix}, \tag{19.58}$$

where the complex numbers α, β, and r are defined as

$$\begin{aligned} \alpha &\equiv A, \\ \beta &\equiv \sqrt{B_x^2 + B_y^2}, \\ r &\equiv \frac{B_x + \mathrm{i}B_y}{\sqrt{B_x^2 + B_y^2}}. \end{aligned} \tag{19.59}$$

The secular equation of the eigenvalue problem $H_w|\psi\rangle = E|\psi\rangle$

$$(\alpha - E)^2 - \beta^2 = 0 \tag{19.60}$$

has the two solutions

$$E_1 = \alpha + \beta \qquad \text{and} \qquad E_2 = \alpha - \beta. \tag{19.61}$$

The eigenstates are easily found:

$$\begin{pmatrix} \alpha & \frac{\beta}{r} \\ \beta r & \alpha \end{pmatrix} |\psi_1\rangle = (\alpha + \beta)|\psi_1\rangle \Rightarrow |\psi_1\rangle \propto \begin{pmatrix} 1 \\ r \end{pmatrix},$$
$$\begin{pmatrix} \alpha & \frac{\beta}{r} \\ \beta r & \alpha \end{pmatrix} |\psi_2\rangle = (\alpha - \beta)|\psi_2\rangle \Rightarrow |\psi_2\rangle \propto \begin{pmatrix} 1 \\ -r \end{pmatrix}. \tag{19.62}$$

CP conservation would lead to $r = 1$, see (19.14) and (19.15). Since CP violation is a small effect one expects that $|r| \simeq 1$ and introduces the complex number ϵ:

$$r \equiv \frac{1 - \epsilon}{1 + \epsilon} \tag{19.63}$$

with $|\epsilon| \ll 1$. The eigenstates become

$$|\psi_1\rangle \propto \begin{pmatrix} 1 \\ \frac{1-\epsilon}{1+\epsilon} \end{pmatrix} \propto \begin{pmatrix} 1+\epsilon \\ 1-\epsilon \end{pmatrix} \propto \frac{1}{\sqrt{2}}\left[\begin{pmatrix} 1 \\ 1 \end{pmatrix} + \epsilon \begin{pmatrix} 1 \\ -1 \end{pmatrix}\right]$$
$$= |K_1\rangle + \epsilon|K_2\rangle, \tag{19.64}$$

$$|\psi_2\rangle \propto \begin{pmatrix} 1 \\ \frac{-1+\epsilon}{1+\epsilon} \end{pmatrix} \propto \begin{pmatrix} 1+\epsilon \\ -1+\epsilon \end{pmatrix} \propto \frac{1}{\sqrt{2}}\left[\begin{pmatrix} 1 \\ -1 \end{pmatrix} + \epsilon \begin{pmatrix} 1 \\ 1 \end{pmatrix}\right]$$
$$= |K_2\rangle + \epsilon|K_1\rangle. \tag{19.65}$$

After normalization the physical states are

$$\boxed{|K_S\rangle \equiv \frac{1}{\sqrt{1+|\epsilon|^2}}\left[|K_1\rangle + \epsilon|K_2\rangle\right] \quad \text{K 'short'}}, \tag{19.66}$$

$$\boxed{|K_L\rangle \equiv \frac{1}{\sqrt{1+|\epsilon|^2}}\left[\epsilon|K_1\rangle + |K_2\rangle\right] \quad \text{K 'long'}}. \tag{19.67}$$

The observed states are therefore not exactly K_1 and K_2 but rather K_S and K_L with energy eigenvalues

$$E_S \equiv E_1 = \alpha + \beta = m_S - \mathrm{i}\frac{\lambda_S}{2},$$
$$E_L \equiv E_2 = \alpha - \beta = m_L - \mathrm{i}\frac{\lambda_L}{2}. \tag{19.68}$$

In the limit of CP conservation ($\epsilon = 0$) one recovers $|K_S\rangle = |K_1\rangle$ and $|K_L\rangle = |K_2\rangle$. Indirect CP violation therefore *mixes* the K_1 and K_2 states. The 2π decay of the long-lived component stems from the small admixture of K_1 with the K_2-component. Incidentally, the states K_S and K_L are not orthogonal:

$$\left\langle K_S \middle| K_L \right\rangle = \frac{\epsilon\left\langle K_1 \middle| K_1 \right\rangle + \epsilon^*\left\langle K_2 \middle| K_2 \right\rangle}{1 + |\epsilon|^2} = \frac{2 \text{ Re } \epsilon}{1 + |\epsilon|^2} \simeq 2 \text{ Re } \epsilon. \tag{19.69}$$

Let us now describe a method to measure ϵ and the two ratios of amplitudes

$$\begin{aligned} \eta_{+-} &\equiv \frac{A\left(K_L \to \pi^+\pi^-\right)}{A\left(K_S \to \pi^+\pi^-\right)} \equiv \left|\eta_{+-}\right| \mathrm{e}^{\mathrm{i}\phi_{+-}}, \\ \eta_{00} &\equiv \frac{A\left(K_L \to \pi^0\pi^0\right)}{A\left(K_S \to \pi^0\pi^0\right)} \equiv \left|\eta_{00}\right| \mathrm{e}^{\mathrm{i}\phi_{00}}. \end{aligned} \tag{19.70}$$

The amplitudes for K_L and K_S decay into $\pi^+\pi^-$ are

$$A\left(K_S \to \pi^+\pi^-\right) = \left\langle \pi^+\pi^- \middle| V \middle| K_S \right\rangle \propto \left\langle \pi^+\pi^- \middle| V \middle| K_1 \right\rangle + \epsilon\left\langle \pi^+\pi^- \middle| V \middle| K_2 \right\rangle, \tag{19.71}$$

$$A\left(K_L \to \pi^+\pi^-\right) = \left\langle \pi^+\pi^- \middle| V \middle| K_L \right\rangle \propto \epsilon\left\langle \pi^+\pi^- \middle| V \middle| K_1 \right\rangle + \left\langle \pi^+\pi^- \middle| V \middle| K_2 \right\rangle. \tag{19.72}$$

We have labelled the transition operator as V. Similarly, for the $2\pi^0$ decay

$$A\left(K_S \to \pi^0\pi^0\right) = \left\langle \pi^0\pi^0 \middle| V \middle| K_S \right\rangle \propto \left\langle \pi^0\pi^0 \middle| V \middle| K_1 \right\rangle + \epsilon\left\langle \pi^0\pi^0 \middle| V \middle| K_2 \right\rangle, \tag{19.73}$$

$$A\left(K_L \to \pi^0\pi^0\right) = \left\langle \pi^0\pi^0 \middle| V \middle| K_L \right\rangle \propto \epsilon\left\langle \pi^0\pi^0 \middle| V \middle| K_1 \right\rangle + \left\langle \pi^0\pi^0 \middle| V \middle| K_2 \right\rangle. \tag{19.74}$$

Since we have assumed indirect *CP* violation, *CP* is conserved in the decay:

$$[V, CP] = 0. \tag{19.75}$$

Then $V\,|K_2\rangle$ is a *CP*-eigenstate with eigenvalue -1. Since $CP(2\pi) = +1$ the second terms on the right hand sides of (19.71)–(19.74) vanish:

$$\left\langle \pi^+\pi^- \middle| V \middle| K_2 \right\rangle = 0 \qquad \text{and} \qquad \left\langle \pi^0\pi^0 \middle| V \middle| K_2 \right\rangle = 0. \tag{19.76}$$

One then obtains for indirect *CP* violation:

$$\boxed{\eta_{+-} = \eta_{00} = \epsilon} \Rightarrow \left|\eta_{+-}\right| = \left|\eta_{00}\right| = |\epsilon|, \qquad \phi_{+-} = \phi_{00} = \phi, \tag{19.77}$$

with

$$\epsilon \equiv |\epsilon|\, \mathrm{e}^{\mathrm{i}\phi}. \tag{19.78}$$

The following relation also holds for indirect *CP* violation:

$$\boxed{\tan\phi = \frac{2\Delta m}{\lambda_S} \Rightarrow \phi \simeq 43.5^\circ}. \tag{19.79}$$

The proof follows from CPT invariance and from unitarity in the decay: the initial state density decreases with the same rate as the final state density increases, see [10].

The complex parameters η_{+-} and η_{00} can be measured. Let us assume that only K^0 mesons are produced at time $t = 0$. The time evolution of the K^0 beam is given by

$$|K(t)\rangle = a\mathrm{e}^{-\mathrm{i}E_S t}\left|K_S\right\rangle + b\mathrm{e}^{-\mathrm{i}E_L t}\left|K_L\right\rangle. \tag{19.80}$$

The K_S and K_L states defined by (19.66), (19.67) can be rewritten as functions of K^0 and $\overline{K}^0$ using (19.14), (19.15):

$$\begin{aligned} \left|K_S\right\rangle &= \frac{1}{\sqrt{1+|\epsilon|^2}}\left(\frac{1+\epsilon}{\sqrt{2}}\left|K^0\right\rangle + \frac{1-\epsilon}{\sqrt{2}}\left|\overline{K}^0\right\rangle\right), \\ \left|K_L\right\rangle &= \frac{1}{\sqrt{1+|\epsilon|^2}}\left(\frac{1+\epsilon}{\sqrt{2}}\left|K^0\right\rangle - \frac{1-\epsilon}{\sqrt{2}}\left|\overline{K}^0\right\rangle\right). \end{aligned} \tag{19.81}$$

We have assumed that $|K(0)\rangle = |K^0\rangle$, therefore

$$a = b \equiv N = \frac{\sqrt{1+|\epsilon|^2}}{\sqrt{2}(1+\epsilon)} \Rightarrow |N|^2 \simeq \frac{1}{2}. \tag{19.82}$$

With the definitions (19.70) the decay amplitude into $\pi^+\pi^-$ becomes

$$\left\langle \pi^+\pi^-\middle|V\middle|K(t)\right\rangle = N\left\langle \pi^+\pi^-\middle|V\middle|K_S\right\rangle\left[\mathrm{e}^{-\mathrm{i}E_S t} + \eta_{+-}\mathrm{e}^{-\mathrm{i}E_L t}\right]. \tag{19.83}$$

The decay probability is

$$\begin{aligned} w = \left|\left\langle \pi^+\pi^-\middle|V\middle|K(t)\right\rangle\right|^2 = |N|^2\left|\left\langle \pi^+\pi^-\middle|V\middle|K_S\right\rangle\right|^2 \Big[\mathrm{e}^{-\lambda_S t} + \left|\eta_{+-}\right|^2 \mathrm{e}^{-\lambda_L t} \\ + 2\,\mathrm{Re}\left(\eta_{+-}^*\mathrm{e}^{-\mathrm{i}\left(E_S - E_L^*\right)t}\right)\Big]. \end{aligned} \tag{19.84}$$

The interference term can be rewritten as

$$\begin{aligned} 2\,\mathrm{Re}\left(\eta_{+-}^*\mathrm{e}^{-\mathrm{i}\left(E_S-E_L^*\right)t}\right) &= 2\,\mathrm{Re}\left|\eta_{+-}\right|\mathrm{e}^{\mathrm{i}\left[\left(E_L^*-E_S\right)t-\phi_{+-}\right]} \\ &= 2\left|\eta_{+-}\right|\mathrm{e}^{-\frac{\lambda_L}{2}t}\mathrm{e}^{-\frac{\lambda_S}{2}t}\,\mathrm{Re}\,\mathrm{e}^{\mathrm{i}\left[(m_L-m_S)t-\phi_{+-}\right]} \\ &= 2\left|\eta_{+-}\right|\mathrm{e}^{-\left(\frac{\lambda_L+\lambda_S}{2}\right)t}\cos\left(\Delta m t - \phi_{+-}\right). \end{aligned} \tag{19.85}$$

Assuming instead that only $\overline{K}^0$ mesons are produced one obtains for the decay rate $\overline{w}$ the same result, but with a *minus* sign for the interference term in (19.84) since now

$a = -b$. At the energy of the experiment described below more K^0 than $\overline{K}^0$ mesons emerge from the production target. Denoting by I_0 and $\overline{I}_0$ the intensities of the produced K^0 and $\overline{K}^0$ mesons, respectively, one obtains with the *dilution factor*

$$D \equiv \frac{I_0 - \overline{I}_0}{I_0 + \overline{I}_0} \tag{19.86}$$

the decay rate $R(t) \propto I_0 w + \overline{I}_0 \overline{w}$ into $\pi^+\pi^-$ pairs:

$$R(t) = R_0\left[e^{-\lambda_S t} + \left|\eta_{+-}\right|^2 e^{-\lambda_L t} + 2D\left|\eta_{+-}\right| e^{-\left(\frac{\lambda_L+\lambda_S}{2}\right)t} \cos\left(\Delta m t - \phi_{+-}\right)\right], \tag{19.87}$$

where

$$R_0 \propto \left(I_0 + \overline{I}_0\right) |N|^2 \simeq \frac{I_0 + \overline{I}_0}{2}. \tag{19.88}$$

The derivation of the decay rate into $\pi^0\pi^0$ is obtained simply by replacing η_{+-} by η_{00} and ϕ_{+-} by ϕ_{00}. Close to the production target one observes the decay of the K_S-component into $\pi^+\pi^-$ or $\pi^0\pi^0$ (first term in (19.87)) and far from the target the CP-violating decay of the K_L-component (second term). In the intermediate region both amplitudes are of comparable strengths and interfere (third term). The interference term disappears with equal numbers of generated K^0 and $\overline{K}^0$ mesons ($D = 0$).

Figure 19.11 shows a sketch of the apparatus used at the CERN Proton Synchrotron to measure η_{+-} in a K_L beam [11]. The mean K_L momentum is 10 GeV c^{-1}. The 9 m long decay tube is filled with helium gas. The momentum vectors $\vec{p}\left(\pi^+\right)$ and $\vec{p}\left(\pi^-\right)$ (and hence the K_L momentum), the decay point and the decay distance d are determined using the dipole magnet and the proportional chambers. The semi-leptonic decays $K_L \to \pi e \nu_e$ and $K_L \to \pi \mu \nu_\mu$ are also studied in this experiment (section 19.3). The hydrogen Čerenkov counter is used to detect electrons or positrons from the decay $K_L \to \pi e \nu_e$. The concrete absorber stops the pions, but not

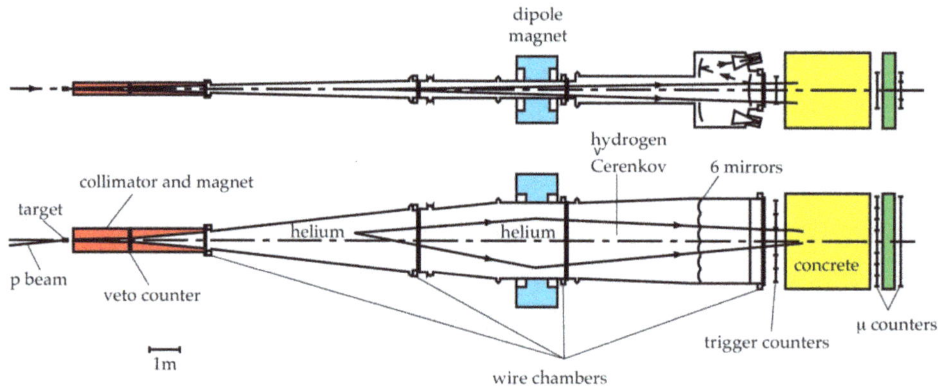

Figure 19.11. The equipment used to measure the $K_L \to \pi^+\pi^-$ and semileptonic K_L decay rates. After [11].

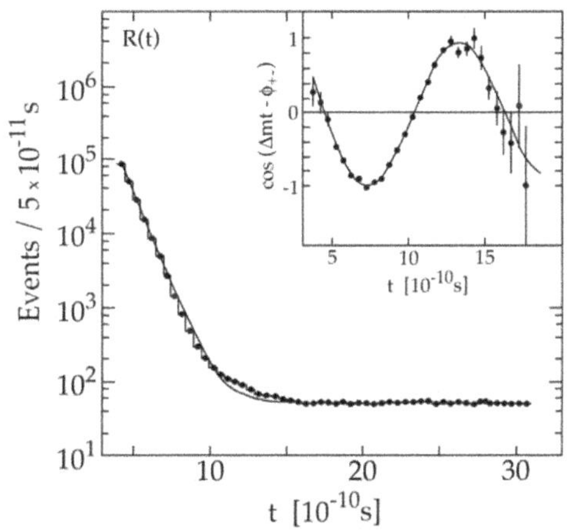

Figure 19.12. Rate R as a function of proper time t for K_L decay into $\pi^+\pi^-$. The data are corrected for acceptance. The curve shows the fit without cosine oscillation. The fitted oscillation is shown in the inset. After [12].

the muons from $K_L \to \pi\mu\nu_\mu$ which are detected by the rear scintillation counters. The acceptance corrected rate $R(t)$ is shown in figure 19.12 as a function of proper time $t = d/\gamma\beta c$. The results for η_{+-} and Δm are [12, 13]:

$$\left|\eta_{+-}\right| = (2.30 \pm 0.04) \times 10^{-3}, \qquad \phi_{+-} = (43.3 \pm 1.1)^\circ,$$
$$\Delta m = m_L - m_S = (5.34 \pm 0.03) \times 10^9 \text{ s}^{-1}. \tag{19.89}$$

For a measurement of the sign of Δm ($m_L > m_S$) see [13, 14]. The phase ϕ_{+-} agrees with the prediction (19.79). We will show in the following section that the contribution from direct CP violation is very small in the $K^0 - \overline{K}^0$ system and that η_{00} and η_{+-} are almost equal.

19.2.2 Direct *CP* violation

Let us now assume that CP invariance is also violated in the decay of the K_2 meson, therefore

$$\left\langle 2\pi \middle| V \middle| K_2 \right\rangle \neq 0. \tag{19.90}$$

Let us decompose the dipion into its isospin $i = 0$, 1, and 2 components. Since the relative angular momentum ℓ is zero the $i = 1$ contribution to the dipion wave function vanishes by virtue of Bose–Einstein symmetry. Decomposing the isospin wave function one obtains with the help of the table on Clebsch–Gordan coefficients (figure 11.6)

$$\left\langle \pi^+\pi^- \right| \propto \sqrt{\frac{1}{3}}\langle 0| + \sqrt{\frac{1}{6}}\langle 2| \Rightarrow \left\langle \pi^+\pi^- \right| = \sqrt{\frac{2}{3}}\langle 0| + \sqrt{\frac{1}{3}}\langle 2|,$$
$$\left\langle \pi^0\pi^0 \right| = -\sqrt{\frac{1}{3}}\langle 0| + \sqrt{\frac{2}{3}}\langle 2|, \tag{19.91}$$

where we have renormalized the wave functions to dipions from K^0 decay. The η_{+-} and η_{00} ratios become

$$\eta_{+-} = \frac{\langle \pi^+\pi^- | V | K_L \rangle}{\langle \pi^+\pi^- | V | K_S \rangle} = \frac{\sqrt{2}\langle 0 | V | K_L \rangle + \langle 2 | V | K_L \rangle}{\sqrt{2}\langle 0 | V | K_S \rangle + \langle 2 | V | K_S \rangle} \tag{19.92}$$

and

$$\eta_{00} = \frac{\langle \pi^0\pi^0 | V | K_L \rangle}{\langle \pi^0\pi^0 | V | K_S \rangle} = \frac{-\langle 0 | V | K_L \rangle + \sqrt{2}\langle 2 | V | K_L \rangle}{-\langle 0 | V | K_S \rangle + \sqrt{2}\langle 2 | V | K_S \rangle}. \tag{19.93}$$

The parameters ϵ_0, ϵ' and ω are defined as

$$\begin{aligned}\epsilon_0 &\equiv \frac{\langle 0 | V | K_L \rangle}{\langle 0 | V | K_S \rangle} = \frac{\epsilon\langle 0 | V | K_1 \rangle + \langle 0 | V | K_2 \rangle}{\langle 0 | V | K_1 \rangle + \epsilon\langle 0 | V | K_2 \rangle} \\ &= \frac{(1+\epsilon)\langle 0 | V | K^0 \rangle - (1-\epsilon)\langle 0 | V | \overline{K}^0 \rangle}{(1+\epsilon)\langle 0 | V | K^0 \rangle + (1-\epsilon)\langle 0 | V | \overline{K}^0 \rangle},\end{aligned} \tag{19.94}$$

$$\epsilon' \equiv \frac{1}{\sqrt{2}} \cdot \frac{\langle 2 | V | K_L \rangle}{\langle 0 | V | K_S \rangle} = \frac{1}{\sqrt{2}} \cdot \frac{\epsilon\langle 2 | V | K_1 \rangle + \langle 2 | V | K_2 \rangle}{\langle 0 | V | K_1 \rangle + \epsilon\langle 0 | V | K_2 \rangle}, \tag{19.95}$$

$$\omega \equiv \frac{\langle 2 | V | K_S \rangle}{\langle 0 | V | K_S \rangle} = \frac{\langle 2 | V | K_1 \rangle + \epsilon\langle 2 | V | K_2 \rangle}{\langle 0 | V | K_1 \rangle + \epsilon\langle 0 | V | K_2 \rangle}. \tag{19.96}$$

Dividing the numerator and denominator in (19.92) by $\langle 0 | V | K_S \rangle$ leads to

$$\eta_{+-} = \frac{\sqrt{2}(\epsilon_0 + \epsilon')}{\sqrt{2} + \omega}. \tag{19.97}$$

Similarly for (19.93):

$$\eta_{00} = \frac{\epsilon_0 - 2\epsilon'}{1 - \sqrt{2}\omega}. \tag{19.98}$$

Let us abbreviate the two K^0 decay amplitudes as A_0 and A_2,

$$A_{0,2} \equiv \langle 0, 2 | V | K^0 \rangle, \tag{19.99}$$

and show[1] that $\epsilon_0 = \epsilon$. From CPT invariance it follows that

$$\begin{aligned}\langle 0 | V | \overline{K}^0 \rangle &= \langle 0 | (CPT)^{-1} V \; CPT | \overline{K}^0 \rangle = \langle 0 | T^{-1} V T | K^0 \rangle = \langle K^0 | V | 0 \rangle \\ &= \langle 0 | V | K^0 \rangle^* = A_0^*.\end{aligned} \tag{19.100}$$

[1] A_0 and A_0^* should be multiplied by the factor $e^{i\delta_0}$, where the phase δ_0 describes the strong interaction in the dipion final state. However, this factor cancels in the expression (19.101). Similarly, the amplitudes A_2 and A_2^* should be multiplied by the strong phase factor $e^{i\delta_2}$.

One of the amplitudes, for example A_0, can be chosen to be real (Wu–Yang convention). Then (19.94) can be rewritten as

$$\epsilon_0 = \frac{(1+\epsilon)A_0 - (1-\epsilon)A_0^*}{(1+\epsilon)A_0 + (1-\epsilon)A_0^*} = \frac{(1+\epsilon)-(1-\epsilon)}{(1+\epsilon)+(1-\epsilon)} = \epsilon. \tag{19.101}$$

Since the isospin of the K^0 is $i = \frac{1}{2}$ the magnitude of ω should be small by virtue of the $|\Delta i| = \frac{1}{2}$ rule (chapter 13): $|\omega| \ll 1$. Thus the ratios (19.97) and (19.98) become, in a good approximation,

$$\boxed{\eta_{+-} \simeq \epsilon + \epsilon'} \qquad \text{and} \qquad \boxed{\eta_{00} \simeq \epsilon - 2\epsilon'}. \tag{19.102}$$

The parameter ϵ' is the contribution from direct CP violation. In the extreme case of pure direct CP violation (19.95) would give with $\epsilon = 0$:

$$\eta_{+-} \simeq \epsilon' = \frac{1}{\sqrt{2}} \cdot \frac{\langle 2|V|K_2\rangle}{\langle 0|V|K_1\rangle}. \tag{19.103}$$

Hence direct CP violation originates from the $|\Delta i| = \frac{3}{2}$ admixture to neutral kaon decay. The parameter ϵ' can be determined from the double ratio

$$R_{2\pi} \equiv \frac{\Gamma\left(K_L \to \pi^0\pi^0\right)}{\Gamma\left(K_S \to \pi^0\pi^0\right)} : \frac{\Gamma\left(K_L \to \pi^+\pi^-\right)}{\Gamma\left(K_S \to \pi^+\pi^-\right)} = \left|\frac{\eta_{00}}{\eta_{+-}}\right|^2. \tag{19.104}$$

Since η_{+-} and η_{00} are almost equal and the angle ϕ_{+-} agrees with the prediction (19.79), one expects indirect CP violation to dominate. With (19.102) and $|\epsilon'| \ll |\epsilon|$ the double ratio becomes

$$\begin{aligned} R_{2\pi} &\simeq \left|\frac{\epsilon - 2\epsilon'}{\epsilon + \epsilon'}\right|^2 = \left|\frac{1 - 2\epsilon'/\epsilon}{1 + \epsilon'/\epsilon}\right|^2 \simeq |(1 - 2\epsilon'/\epsilon)(1 - \epsilon'/\epsilon)|^2 \\ &\simeq \left|1 - 3\left(\frac{\epsilon'}{\epsilon}\right)\right|^2 \simeq 1 - 6\,\mathrm{Re}\left(\frac{\epsilon'}{\epsilon}\right). \end{aligned} \tag{19.105}$$

One can show that the phases of ϵ and ϵ' are nearly equal [3], hence

$$\mathrm{Re}\left(\frac{\epsilon'}{\epsilon}\right) \simeq \frac{1}{6}\left(1 - R_{2\pi}\right) \simeq \left|\frac{\epsilon'}{\epsilon}\right|. \tag{19.106}$$

The double ratio $R_{2\pi}$ (19.104) was measured at the CERN SPS in a 100 GeV neutral kaon beam [15]. Figure 19.13 (left) shows the principle of the experiment. At 100 GeV the decay products $\pi^+\pi^-$ and $\pi^0\pi^0 \to 4\gamma$ fly into the very forward direction. The four photons are detected in a liquid argon calorimeter (a sandwich

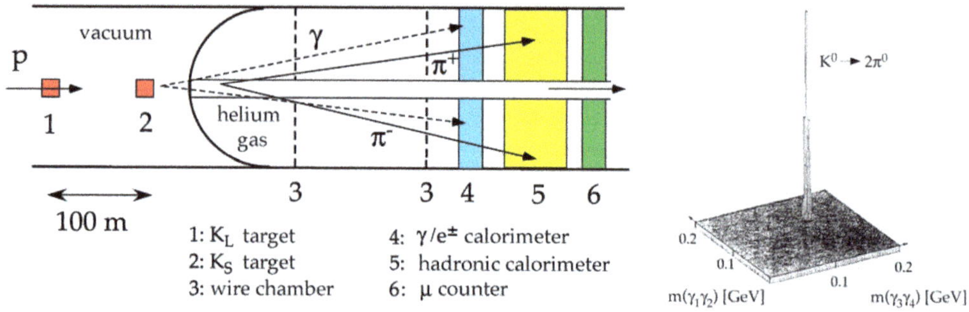

Figure 19.13. Left: principle of the NA31 experiment at CERN. The 450 GeV protons from the SPS enter from the left. Right: mass distribution of the 2γ pairs. After [16].

of lead and liquid argon with readout strips) and the two neutral pions are reconstructed. Figure 19.13 (right) shows the 2γ invariant mass distribution for 4γ events. The prominent peak stems from the $2\pi^0$ decay of the neutral kaon. The energies and angles of the two charged pions are measured in the hadronic calorimeter (an iron–scintillator sandwich).

For the K_L beam the production target is located about 100 m upstream of the detector. The target can be moved in the vacuum enclosure towards the sensitive decay volume, to collect the K_S data. Thus K_S and K_L decays can be investigated without modifying the apparatus layout. This reduces systematic errors from apparatus inefficiencies which cancel out in the ratios (19.104). The result $|\epsilon'/\epsilon| = (2.3 \pm 0.7) \times 10^{-3}$ was the first experimental evidence for direct CP violation.

Indirect CP violation is therefore the dominant contribution to CP violation in the $K^0 - \overline{K}^0$ system. The current values of the CP-violating parameters are [3]:

$$
\begin{aligned}
|\epsilon'/\epsilon| \simeq \mathrm{Re}\left(\frac{\epsilon'}{\epsilon}\right) \simeq \left(\frac{\epsilon'}{\epsilon}\right) &= (1.66 \pm 0.23) \times 10^{-3}, \\
|\epsilon| &= (2.228 \pm 0.011) \times 10^{-3}, \\
\phi_{+-} &= (43.4 \pm 0.5)^\circ, \\
\phi_{00} - \phi_{+-} &= (0.34 \pm 0.32)^\circ.
\end{aligned}
\tag{19.107}
$$

19.3 Semileptonic K_L decays

As a consequence of the $\Delta S = \Delta Q$ rule the K^0 decays semileptonically into $\pi^- e^+ \nu_e$ or $\pi^- \mu^+ \nu_\mu$, and the $\overline{K}^0$ into $\pi^+ e^- \bar{\nu}_e$ or $\pi^+ \mu^- \bar{\nu}_\mu$. Figure 19.14 shows a bubble chamber picture of the decay $K^0 \to \pi^- e^+ \nu_e$. The K^0 was produced by the reaction $K^+ p \to K^0 p \pi^+$. Note the characteristic spiral-shaped track of stopping positrons (or electrons) in light materials such as hydrogen (critical energy $E_c = 370$ MeV): in spite of its low momentum which strongly curves its path in the magnetic field, the range

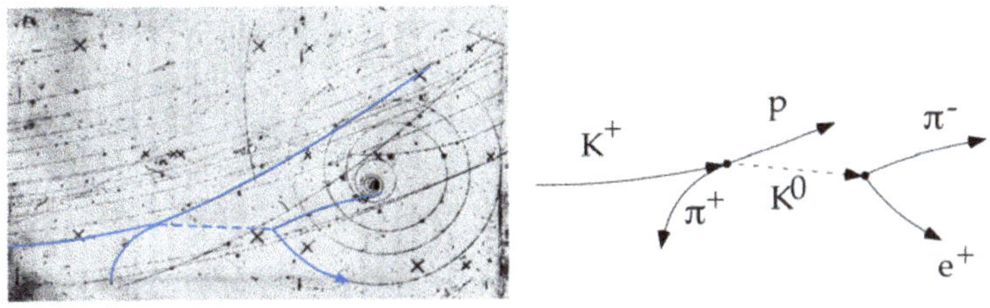

Figure 19.14. Semileptonic decay of the neutral kaon into $\pi^- e^+ \nu_e$. From [10]. Copyright 1976 Springer Science and Business Media.

of the positron in figure 19.14 is large due to its small mass. It loses energy by ionization of the medium and by bremsstrahlung and thus spirals.

Let us write the decay amplitudes as

$$\begin{aligned} A_+ &\equiv \left\langle \pi^- e^+ \nu_e \middle| V \middle| K^0 \right\rangle, \\ A_- &\equiv \left\langle \pi^+ e^- \bar{\nu}_e \middle| V \middle| \overline{K}^0 \right\rangle. \end{aligned} \tag{19.108}$$

The K_L state is given by the superposition (19.81):

$$\left| K_L \right\rangle = \frac{1}{\sqrt{1 + |\epsilon|^2}} \left(\frac{1+\epsilon}{\sqrt{2}} \left| K^0 \right\rangle - \frac{1-\epsilon}{\sqrt{2}} \left| \overline{K}^0 \right\rangle \right). \tag{19.109}$$

The partial width of the decay $K_L \to \pi^- e^+ \nu_e$ is proportional to the factor $|1 + \epsilon|^2/2$, that of the decay $K_L \to \pi^+ e^- \bar{\nu}_e$ to $|1 - \epsilon|^2/2$. Because of CPT invariance the semileptonic decay rates of the K^0 and $\overline{K}^0$ are equal: $|A_+|^2 = |A_-|^2 \equiv |A|^2$. The partial widths for the K_L meson are then given by

$$\begin{aligned} \Gamma_+ &\equiv \Gamma\left(K_L \to \pi^- e^+ \nu_e\right) \propto |A|^2 |1 + \epsilon|^2, \\ \Gamma_- &\equiv \Gamma\left(K_L \to \pi^+ e^- \bar{\nu}_e\right) \propto |A|^2 |1 - \epsilon|^2. \end{aligned} \tag{19.110}$$

Thus, CP violation introduces an asymmetry δ in the semileptonic decay rates of the K_L into e^+ and e^-:

$$\delta \equiv \frac{\Gamma_+ - \Gamma_-}{\Gamma_+ + \Gamma_-} = \frac{|A|^2|1+\epsilon|^2 - |A|^2|1-\epsilon|^2}{|A|^2|1+\epsilon|^2 + |A|^2|1-\epsilon|^2} \simeq \frac{4\,\mathrm{Re}\,\epsilon}{2} = 2\,\mathrm{Re}\,\epsilon. \tag{19.111}$$

Similar arguments apply to $K_L \to \pi\mu\nu_\mu$. The asymmetry δ was measured by the experiment shown in figure 19.11. The results for the $e^\pm$ and $\mu^\pm$ final states are compatible and lead to the average asymmetry [11]

$$\mathrm{Re}\,\epsilon = \frac{\delta}{2} = (1.67 \pm 0.08) \times 10^{-3} > 0, \tag{19.112}$$

which is in excellent agreement with the result $\mathrm{Re}\,\eta_{+-} = 1.66 \times 10^{-3}$ obtained from the 2π decay (19.89).

Hence, semileptonic final states with positrons or positively charged muons are slightly preferred. *CP* violation provides a way to distinguish matter from antimatter in the Universe: a resident of a distant galaxy could be asked whether the long-lived component of the $K^0 - \overline{K}^0$ system prefers to decay into a lepton with the same charge as the one circling the nucleus of the hydrogen atom. Shaking hands with the alien would not be recommended in this case!

19.3.1 *T* violation

Under the assumption of *CPT* invariance, *CP* violation implies that time reversal invariance *T* is also violated. Nonetheless, let us describe an experiment measuring *T* violation in the $K^0 - \overline{K}^0$ system without any assumptions on *CP* violation. Time reversal invariance requires that the probability P of finding a $\overline{K}^0$ when at $t=0$ a K^0 was produced (see (19.37)) be equal to the probability $\overline{P}$ of finding a K^0 when at $t=0$ a $\overline{K}^0$ was generated. The asymmetry

$$A_T \equiv \frac{\overline{P}\left(\overline{K}^0(0) \to K^0(t) \to \pi^- e^+ \nu_e\right) - P\left(K^0(0) \to \overline{K}^0(t) \to \pi^+ e^- \overline{\nu}_e\right)}{\overline{P}\left(\overline{K}^0(0) \to K^0(t) \to \pi^- e^+ \nu_e\right) + P\left(K^0(0) \to \overline{K}^0(t) \to \pi^+ e^- \overline{\nu}_e\right)} \tag{19.113}$$

has been measured at CERN. The CPLEAR experiment (figure 19.15) studied the annihilation channels $\bar{p}p \to K^+\overline{K}^0\pi^-$ and $K^-K^0\pi^+$. Antiprotons from the Low Energy Antiproton Ring (LEAR) with momenta of 200 MeV c^{-1} are stopped in a liquid hydrogen target. The momenta of the charged particles and the charge signs are determined with wire chambers (PC) and drift chambers (DC) in the solenoidal magnetic field. The $K^\pm$, $\pi^\pm$ and $e^\pm$ are identified by the Čerenkov counters (C), by the time of flight, and by dE/dx measurements in the plastic scintillators (S_1, S_2). Showers from $e^\pm$ are detected in the electromagnetic calorimeter (ECAL). A photograph of the detector is shown in figure 19.17.

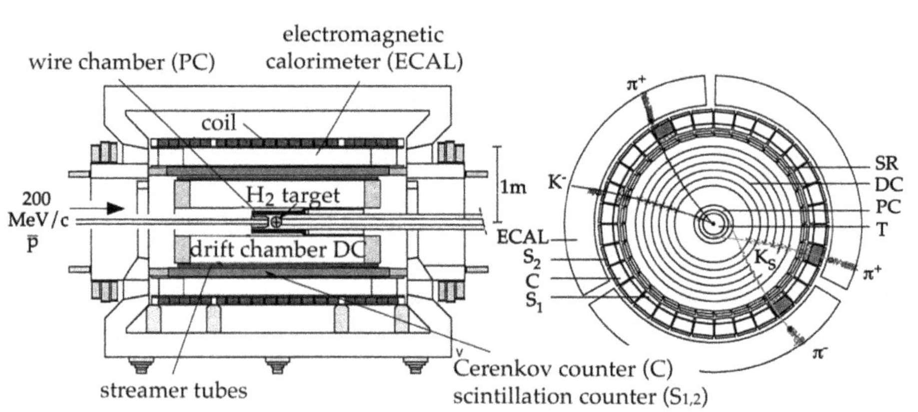

Figure 19.15. Left: diagram of the CPLEAR experiment at CERN (see the text for details). Right: a typical annihilation $\bar{p}p \to K^-K^0\pi^+$ with $K_S \to \pi^+\pi^-$. After [17].

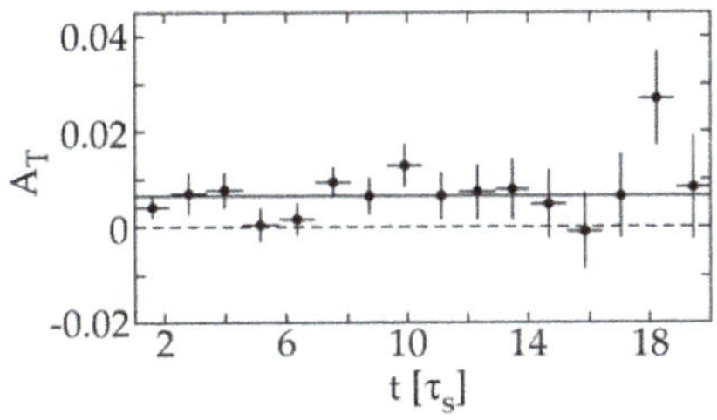

Figure 19.16. Asymmetry A_T as a function of proper time. After [18].

Figure 19.17. The CPLEAR detector at CERN's LEAR studied *CP* and *T* violation in the neutral kaon system. Image credit: CERN.

The strangeness of the neutral kaon at time $t = 0$ is determined by the sign of the charged kaon while the strangeness of the decaying kaon is obtained from the charge of the decay lepton. Figure 19.15 (right) shows a $K^-K^0\pi^+$ event (in this illustration the K^0 decays as a K_S into $\pi^+\pi^-$).

Figure 19.16 shows the measurement of the asymmetry A_T as a function of proper time. The asymmetry averaged over t does not vanish but gives the finite value

$$A_T = (6.6 \pm 1.6) \times 10^{-3}, \tag{19.114}$$

proving that T is violated in the neutral kaon system. One can also show [18] that the asymmetry A_T does not depend on t for $t > \tau_S$ and that $A_T \simeq 4 \text{ Re } \epsilon = 6.8 \times 10^{-3}$, in excellent agreement with the measured value (19.114).

References

[1] Close F E 1979 *An Introduction to Quarks and Partons* (London: Academic)
[2] Skjeggestad O *et al* 1972 *Nucl. Phys.* B **48** 343
[3] Olive K *et al* (Particle Data Group) 2014 *Chin. Phys.* C **38** 090001
[4] Lande K, Booth E T, Impeduglia J and Lederman L M 1956 *Phys. Rev.* **103** 1901
[5] Vosburgh K *et al* 1972 *Phys. Rev.* D **6** 1834
[6] Ambrosino F *et al* 2006 *Phys. Lett.* B **632** 43
[7] Good R H *et al* 1961 *Phys. Rev.* **124** 1223
[8] Christenson J H *et al* 1964 *Phys. Rev. Lett.* **13** 138
[9] Leipuner L B *et al* 1963 *Phys. Rev.* **132** 2285
[10] Paul E 1976 *Springer Tracts in Modern Physics* vol 79 (Berlin: Springer) p 53
[11] Geweniger C *et al* 1974 *Phys. Lett.* B **48** 483
[12] Geweniger C *et al* 1974 *Phys. Lett.* B **48** 487
[13] Geweniger C *et al* 1974 *Phys. Lett.* B **52** 108
[14] Hill D G *et al* 1971 *Phys. Rev.* D **4** 7
[15] Burkhardt H *et al* 1988 *Phys. Lett.* B **206** 169
[16] Burkhardt H *et al* 1988 *Nucl. Instrum. Methods Phys. Res.* A **268** 116
[17] Angelopoulos A *et al* 2003 *Phys. Rep.* **374** 165
[18] Angelopoulos A *et al* 1998 *Phys. Lett.* B **444** 43

IOP Publishing

Nuclear and Particle Physics

Claude Amsler

Chapter 20

The $B^0 - \overline{B}^0$ system

20.1 The CKM matrix

The Cabibbo–Kobayashi–Maskawa matrix (CKM matrix) has been introduced in section 16.3. The d', s' and b' quark states that are involved in the interaction of weak charged currents are linear combinations of the d, s and b quarks:

$$\begin{pmatrix} d' \\ s' \\ b' \end{pmatrix} = V_{\rm CKM}\begin{pmatrix} d \\ s \\ b \end{pmatrix} = \begin{pmatrix} V_{ud} & V_{us} & V_{ub} \\ V_{cd} & V_{cs} & V_{cb} \\ V_{td} & V_{ts} & V_{tb} \end{pmatrix}\begin{pmatrix} d \\ s \\ b \end{pmatrix}. \tag{20.1}$$

Explicit expressions for the matrix elements V_{ij} are given by (16.67) and (16.69). The couplings V_{ij} appear in the weak charged currents. For example, the current associated with the emission of a W^- boson in $d \to uW^-$ is

$$J^\mu(d \to u) = \bar{u}\,\gamma^\mu\left[\frac{1-\gamma^5}{2}\right]V_{ud}\;d. \tag{20.2}$$

The current associated with $b \to uW^-$ is given by the same expression, with V_{ud} replaced by V_{ub}. The time-reversed current is associated with the emission of a W^+ boson in $u \to bW^+$:

$$J^\mu(u \to b) = b^\dagger V_{ub}^*\left[\frac{1-\gamma^5}{2}\right]^\dagger \gamma^{\mu\dagger}\gamma^0 u = \bar{b}\;V_{ub}^*\,\gamma^\mu\left[\frac{1-\gamma^5}{2}\right]u. \tag{20.3}$$

The transition amplitude for $du \to ub$ scattering

$$\mathcal{M} = \frac{4G_F}{\sqrt{2}}\left(\bar{u}\gamma^\mu\left[\frac{1-\gamma^5}{2}\right]V_{ud}\;d\right)\left(\bar{b}\;V_{ub}^*\,\gamma_\mu\left[\frac{1-\gamma^5}{2}\right]u\right) \tag{20.4}$$

is driven by the coupling $V_{ud}V_{ub}^*$.

doi:10.1088/978-0-7503-1140-3ch20

© IOP Publishing Ltd 2015

The CKM matrix is unitary and contains $n_f = (N-1)^2 = 4$ real parameters with $N = 3$ generations (see (14.61)). Unitarity implies that the rows (or the columns) are orthogonal to one another. For example, the first and the third columns fulfil the orthogonality condition

$$V_{ud}V_{ub}^* + V_{cd}V_{cb}^* + V_{td}V_{tb}^* = 0 \tag{20.5}$$

which defines a triangle in the complex plane. Let us divide (20.5) by $V_{cd}V_{cb}^*$:

$$\frac{V_{ud}V_{ub}^*}{V_{cd}V_{cb}^*} + 1 + \frac{V_{td}V_{tb}^*}{V_{cd}V_{cb}^*} = 0. \tag{20.6}$$

The corresponding triangle is drawn in figure 20.1. Introducing the explicit expressions (16.69) for the matrix elements, one obtains

$$\frac{V_{ud}V_{ub}^*}{V_{cd}V_{cb}^*} = \frac{\lambda^3 A[\rho + i\eta]\left(1 - \frac{\lambda^2}{2}\right)}{-\lambda^3 A} = -[\rho + i\eta]\left(1 - \frac{\lambda^2}{2}\right). \tag{20.7}$$

The coordinates of the top vertex of the triangle are given by the tip of the vector defined by $-V_{ud}V_{ub}^*/V_{cd}V_{cb}^*$:

$$\bar{\rho} \equiv \rho\left(1 - \frac{\lambda^2}{2}\right), \qquad \bar{\eta} \equiv \eta\left(1 - \frac{\lambda^2}{2}\right). \tag{20.8}$$

The angles α, β and γ (figure 20.1(*a*)) are given by

$$\alpha = \arg\left(-\frac{V_{td}V_{tb}^*}{V_{ud}V_{ub}^*}\right), \qquad \beta = \arg\left(-\frac{V_{cd}V_{cb}^*}{V_{td}V_{tb}^*}\right), \qquad \gamma = \arg\left(-\frac{V_{ud}V_{ub}^*}{V_{cd}V_{cb}^*}\right). \tag{20.9}$$

The geometrical proof is given for the angle α in figure 20.1(*b*).

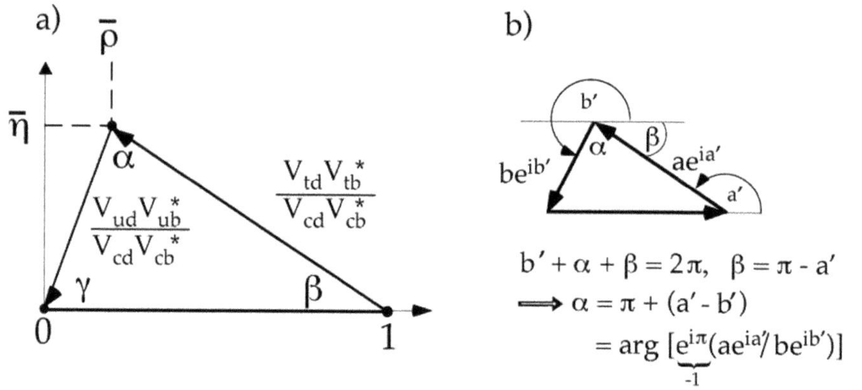

Figure 20.1. (*a*) Definitions of the angles α, β and γ associated with (20.6). (*b*) Diagram proving that $\alpha = \arg\left(-\frac{V_{td}V_{tb}^*}{V_{ud}V_{ub}^*}\right)$.

The data yield $\lambda \simeq 0.225$, $A \simeq 0.81$, $\bar{\rho} \simeq 0.14$ and $\bar{\eta} \simeq 0.34$ for the elements of the matrix (16.69), and the values $\alpha \simeq 89°$ and $\beta \simeq 21°$ for the angles of the triangle in figure 20.1 [1]. The area of the triangle is finite and therefore CP is violated. The measurement of γ is still subject to a large uncertainty: $\gamma \simeq (72 \pm 15)°$ [2]. The values of the three angles add up to $(182 \pm 15)°$.

20.2 $B^0 - \overline{B}^0$ oscillation

As in the $K^0 - \overline{K}^0$ system the neutral B^0 mesons oscillate, the $B^0(d\bar{b})$ into $\overline{B}^0(\bar{d}b)$ and the $B_s^0(s\bar{b})$ into $\overline{B}_s^0(\bar{s}b)$. The B^0 meson ($M = 5280$ MeV) decays preferably into $\overline{D}^0(\bar{c}u)X$ or $D^-(\bar{c}d)X$ and the B_s^0 meson ($M = 5367$ MeV) decays dominantly into $D_s^-(\bar{c}s)X$.

We treat the $B^0 - \overline{B}^0$ system and denote the eigenstates of the weak interaction by B_L ('L' for 'light') with mass M_L and width Γ_L, and B_H ('H' for 'heavy') with mass M_H and width Γ_H. The eigenstates are linear combinations of B^0 and $\overline{B}^0$:

$$|B_L\rangle = p|B^0\rangle + q|\overline{B}^0\rangle, \tag{20.10}$$

$$|B_H\rangle = p|B^0\rangle - q|\overline{B}^0\rangle. \tag{20.11}$$

The B_L meson corresponds to the K_S meson of chapter 19 and the B_H to the K_L. For $p = q = 1/\sqrt{2}$ the CP-eigenstates are B_L and B_H with $CP = +1$ and $CP = -1$ respectively. As for the neutral kaon system we expect that $|p/q| \simeq 1$. The time evolution of the $B^0 - \overline{B}^0$ system is given by

$$|B(t)\rangle = a|B_H\rangle e^{-i\left(m_H - i\frac{\Gamma_H}{2}\right)t} + b|B_L\rangle e^{-i\left(m_L - i\frac{\Gamma_L}{2}\right)t}, \tag{20.12}$$

with $\Gamma_L \equiv \lambda_L$ and $\Gamma_H \equiv \lambda_H$. If a B^0 is produced at time $t = 0$ then

$$\begin{aligned} |B(0)\rangle &= a|B_H\rangle + b|B_L\rangle \\ &= a\left(p|B^0\rangle - q|\overline{B}^0\rangle\right) + b\left(p|B^0\rangle + q|\overline{B}^0\rangle\right) = |B^0\rangle, \end{aligned} \tag{20.13}$$

and hence $a = b = \frac{1}{2p}$. Therefore

$$\begin{aligned} |B(t)\rangle &= \frac{1}{2p}\left(p|B^0\rangle - q|\overline{B}^0\rangle\right)e^{-i\left(m_H - i\frac{\Gamma_H}{2}\right)t} + \frac{1}{2p}\left(p|B^0\rangle + q|\overline{B}^0\rangle\right)e^{-i\left(m_L - i\frac{\Gamma_L}{2}\right)t} \\ &= |B^0\rangle g_+(t) + \frac{q}{p}|\overline{B}^0\rangle g_-(t), \end{aligned} \tag{20.14}$$

with

$$g_\pm(t) = \frac{1}{2}\left[e^{-i\left(m_L - i\frac{\Gamma_L}{2}\right)t} \pm e^{-i\left(m_H - i\frac{\Gamma_H}{2}\right)t}\right]. \tag{20.15}$$

Let us define the masses of the heavy and light states

$$m_H \equiv m + \frac{\Delta m_d}{2} \qquad \text{and} \qquad m_L \equiv m - \frac{\Delta m_d}{2} \tag{20.16}$$

with

$$m \equiv \frac{m_H + m_L}{2} \qquad \text{and} \qquad \boxed{\Delta m_d \equiv m_H - m_L}. \tag{20.17}$$

The subscript d refers to the d quark in the B^0 meson to distinguish Δm_d from Δm_s in the $B_s^0 - \overline{B}_s^0$ system. The lifetimes of the eigenstates ($\tau \simeq 1.5$ ps) are comparable due to the high mass of the B meson and its numerous decay channels. Let us therefore assume that $\Gamma_H \sim \Gamma_L \equiv \Gamma$. This is justified for both $B^0 - \overline{B}^0$ and $B_s^0 - \overline{B}_s^0$ systems ($\Delta\Gamma/\Gamma < 0.01$ and $\Delta\Gamma_s/\Gamma_s \simeq 0.14$, respectively). Hence

$$\begin{aligned} g_\pm(t) &= \frac{1}{2} e^{-\frac{\Gamma}{2}t}\left(e^{-im_L t} \pm e^{-im_H t}\right) \\ &= e^{-\frac{\Gamma}{2}t} e^{-imt} \frac{\left(e^{i\frac{\Delta m_d t}{2}} \pm e^{-i\frac{\Delta m_d t}{2}}\right)}{2} = e^{-\frac{\Gamma}{2}t} e^{-imt} \begin{pmatrix} \cos\frac{\Delta m_d t}{2} \\ i \sin\frac{\Delta m_d t}{2} \end{pmatrix}. \end{aligned} \tag{20.18}$$

The time evolution (20.14) becomes

$$|B(t)\rangle = e^{-\frac{\Gamma}{2}t} e^{-imt}\left[|B^0\rangle \cos\frac{\Delta m_d}{2}t + \left(\frac{iq}{p}\right)|\overline{B}^0\rangle \sin\frac{\Delta m_d}{2}t\right]. \tag{20.19}$$

The probability of finding a B^0 at time t is

$$w(t) = |\langle B^0|B(t)\rangle|^2 = e^{-\Gamma t}\cos^2\frac{\Delta m_d}{2}t = \frac{1}{2}e^{-\Gamma t}\left(1 + \cos\Delta m_d t\right), \tag{20.20}$$

while the probability of finding a $\overline{B}^0$ is

$$\overline{w}(t) = |\langle \overline{B}^0|B(t)\rangle|^2 = \left|\frac{q}{p}\right|^2 e^{-\Gamma t}\sin^2\frac{\Delta m_d}{2}t = \frac{1}{2}e^{-\Gamma t}\left(1 - \cos\Delta m_d t\right), \tag{20.21}$$

with $|q/p| \simeq 1$ since (indirect) CP violation is a small effect. On the other hand, if a $\overline{B}^0$ is produced at time $t=0$ then with $a = -b = -\frac{1}{2q}$ in (20.13):

$$\begin{aligned} |\overline{B}(t)\rangle &= -\frac{1}{2q}\left(p|B^0\rangle - q|\overline{B}^0\rangle\right)e^{-i\left(m_H - i\frac{\Gamma_H}{2}\right)t} + \frac{1}{2q}\left(p|B^0\rangle + q|\overline{B}^0\rangle\right)e^{-i\left(m_L - i\frac{\Gamma_L}{2}\right)t} \\ &= \frac{p}{q}|B^0\rangle g_-(t) + |\overline{B}^0\rangle g_+(t) \end{aligned} \tag{20.22}$$

$$= e^{-\frac{\Gamma}{2}t} e^{-imt}\left[\left(\frac{ip}{q}\right)|B^0\rangle \sin\frac{\Delta m_d}{2}t + |\overline{B}^0\rangle \cos\frac{\Delta m_d}{2}t\right]. \tag{20.23}$$

The probability of finding a $\overline{B}^0$ at time t is

$$w(t) = |\langle \overline{B}^0 | \overline{B}(t) \rangle|^2 = \mathrm{e}^{-\Gamma t} \cos^2 \frac{\Delta m_d}{2} t = \frac{1}{2} \mathrm{e}^{-\Gamma t} \big(1 + \cos \Delta m_d t\big), \tag{20.24}$$

while the probability of finding a B^0 is

$$\overline{w}(t) = |\langle B^0 | \overline{B}(t) \rangle|^2 = \underbrace{\left| \frac{p}{q} \right|^2}_{\simeq 1} \mathrm{e}^{-\Gamma t} \sin^2 \frac{\Delta m_d}{2} t = \frac{1}{2} \mathrm{e}^{-\Gamma t} \big(1 - \cos \Delta m_d t\big). \tag{20.25}$$

Thus $w(t)$ is the probability of finding the original flavour at time t, while $\overline{w}(t)$ is the probability to find the opposite flavour.

The b quark in the $\overline{B}^0$ meson decays preferably into a c quark which in turn decays into an s quark, for example $\overline{B}^0 \to D^0 X \to K^- \pi^+ X$, and correspondingly $B^0 \to \overline{D}^0 X \to K^+ \pi^- X$. The charge of the kaon determines the flavour or antiflavour of the B meson. The flavour can also be obtained from the charge of the lepton ℓ in semileptonic decay modes such as $b \to c\ell^- \bar{\nu}_\ell$ and $\bar{b} \to \bar{c}\ell^+ \nu_\ell$. Figure 20.2 shows the oscillation of a B^0 meson into a $\overline{B}^0$. The B^0 is produced by exciting the $\Upsilon(4S)$ resonance at the DORIS e^+e^- storage ring at DESY ($\sqrt{s} = 10.58$ GeV, see section 10.3.2). The $\Upsilon(4S)$ decays into $B^0\overline{B}^0$ and B^+B^- with branching ratio $f(B^0\overline{B}^0 + B^+B^-) > 96\%$. The 'event display' from the ARGUS detector (figure 20.2) shows the production of a $B^0\overline{B}^0$ pair and the subsequent transformation of the B^0 into a $\overline{B}^0$.

We now describe the BABAR experiment which determined the oscillation frequency Δm_d of the $B^0 - \overline{B}^0$ system. Electrons are collided with positrons at the

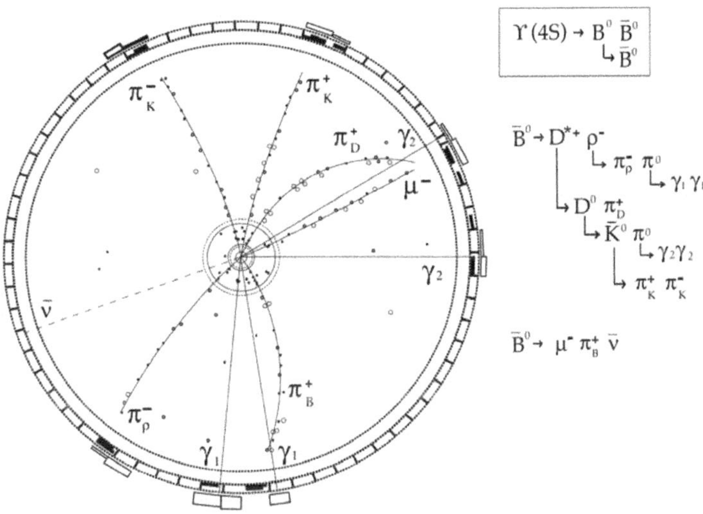

Figure 20.2. $B^0\overline{B}^0$ production in $e^+e^- \to \Upsilon(4S) \to B^0\overline{B}^0$ followed by the oscillation of the B^0 into $\overline{B}^0$, as seen by the ARGUS detector. One of the $\overline{B}^0$ mesons decays semileptonically into $\pi^+\mu^-\bar{\nu}_\mu$ and the other hadronically into $D^{*+}\rho^-$. After [3].

PEP-II storage ring at SLAC and excite the $\Upsilon(4S)$ resonance. The electron and positron bunches have finite dimensions, so that the interaction point varies from event to event. In a symmetric collider, such as DORIS colliding 5.3 GeV c^{-1} electrons and positrons, the B^0 and $\overline{B}^0$ mesons are emitted in opposite directions, so that the production point cannot be determined. In contrast, at the PEP-II asymmetric collider the beam momenta are different: $p_- = 9.0$ GeV c^{-1} for the electrons and $p_+ = 3.1$ GeV c^{-1} for the positrons. The B mesons are then emitted under small angles in the direction of the electron beam with a maximum relative angle θ of about 6° (figure 20.3). Thus the production point of the two B mesons, and therefore their individual decay times t_1 and t_2, are reconstructed from the momentum vectors of the decay products.

Since the $\Upsilon(4S)$ is not produced at rest the $B^0 - \overline{B}^0$ system moves along the electron direction with momentum $\Delta p = p_- - p_+$ and velocity $\beta = \Delta p/\sqrt{M[\Upsilon(4S)]^2 + \Delta p^2} \simeq 0.5$ ($\gamma \simeq 1.15$). The mesons oscillate until they decay at time t_1 and t_2 in their respective rest frames. We will show below that the time difference $\Delta t = t_2 - t_1$ only is relevant for measuring the oscillation frequency. Since the velocities of the B mesons in the rest frame of the $\Upsilon(4S)$ resonance are very low ($\beta = 0.06$), Δt is in a good approximation equal to the time difference in the rest frame of the $\Upsilon(4S)$. In the laboratory system the time difference is then $\gamma\Delta t$. The mean life of the B meson is 1.5 ps and its typical decay length 260 μm in the laboratory. One measures the distance $\Delta z \simeq c\beta\gamma\Delta t$ between the decay points (figure 20.3).

The BABAR detector (figure 20.4) measures the momenta of charged particles with a drift chamber (DCH) in a solenoidal magnetic field of 1.5 T. Photons from π^0 and η decays are detected by an array of 6580 CsI (Tl) crystals (EMC). Charged pions are identified and separated from kaons through their Čerenkov light. The light is produced in a barrel of 144 quartz rods (DIRC) and transmitted to photomultipliers by total internal reflection. The silicon strip detector (SVT) is able to determine the decay points of the B mesons with an rms resolution of 70 μm.

If one of the B mesons decays, say as a $\overline{B}^0$, then *at the same moment* the other meson must be a B^0: the states B^0B^0 and $\overline{B}^0\overline{B}^0$ are forbidden by Bose symmetry since the wave function would have to be symmetric under the permutation of the two bosons, and thus the relative angular momentum ℓ would be even. However, here $\ell = 1$ since B mesons are spinless and $J(\Upsilon) = 1$. Let us assume that one of the mesons decays as a $\overline{B}^0$ at time t_1, and that therefore the other meson is a B^0 at that time. The probability that the latter decays later at time t_2 as a $\overline{B}^0$ is given by (20.21)

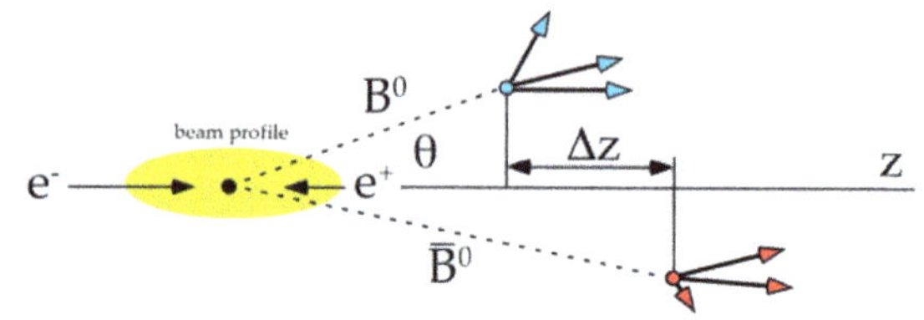

Figure 20.3. Decay of the B^0 and $\overline{B}^0$ mesons produced at the $\Upsilon(4S)$ in an asymmetric e^+e^- collider.

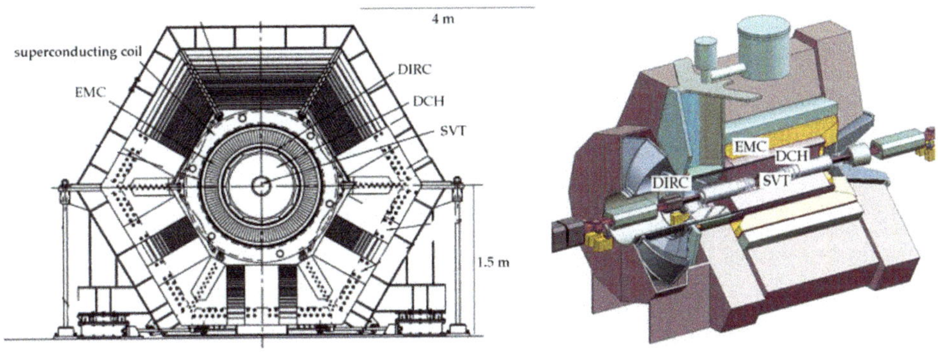

Figure 20.4. The BABAR detector at the PEP-II storage ring (see the text for details). After [4].

with $\Delta t = t_2 - t_1$, and the probability that it decays as a B^0 is given by (20.20). Hence, the probability of detecting the same flavours (B^0B^0 or $\overline{B}^0\overline{B}^0$) is

$$w_{=}(\Delta t) \propto \frac{1}{2}\mathrm{e}^{-\Gamma\Delta t}(1 - \cos \Delta m_d \Delta t), \tag{20.26}$$

while the probability of observing opposite flavours ($B^0\overline{B}^0$) is

$$w_{\neq}(\Delta t) \propto \frac{1}{2}\mathrm{e}^{-\Gamma\Delta t}(1 + \cos \Delta m_d \Delta t). \tag{20.27}$$

One obtains the time-dependent asymmetry

$$\boxed{\mathcal{A}(\Delta t) = \frac{w_{\neq}(\Delta t) - w_{=}(\Delta t)}{w_{\neq}(\Delta t) + w_{=}(\Delta t)} = \cos\left(\Delta m_d \Delta t\right)}. \tag{20.28}$$

In the BABAR experiment [5] the flavour of one of the B mesons is 'tagged' by the charge of the kaon in hadronic $B \to D \to K$ decays, and the flavour of the second B by the charge of the lepton (μ or e) in semileptonic decays. Figure 20.5 shows the measured asymmetry. The amplitude of the oscillation is attenuated by the imperfect identification of the flavour and by the decreasing experimental resolution towards large Δt. The mass difference between the eigenstates B_H and B_L is calculated from the oscillation period $T = 2\pi/\Delta m_d$: $\Delta m_d = (0.516 \pm 0.019)\ \mathrm{ps}^{-1}$ [5].

20.3 $B_s^0 - \overline{B}_s^0$ oscillation

In e^+e^- collisions the B_s^0 mesons can be pair-produced by exciting the $\Upsilon(10860)$ resonance. However, most of the data on the $B_s^0 - \overline{B}_s^0$ system stem from high energy $\bar{p}p$ and pp collisions at Fermilab and at the Large Hadron Collider (LHC), respectively. The oscillation frequency in the $B_s^0 - \overline{B}_s^0$ system was first measured by the CDF experiment at the Tevatron [6]. The B_s^0 and $\overline{B}_s^0$ mesons are produced in pairs by the reaction $\bar{p}p \to B_s^0\overline{B}_s^0X$ at 2 TeV. The production point is determined from the tracks of the outgoing hadrons X. The flavours of the B_s mesons are tagged

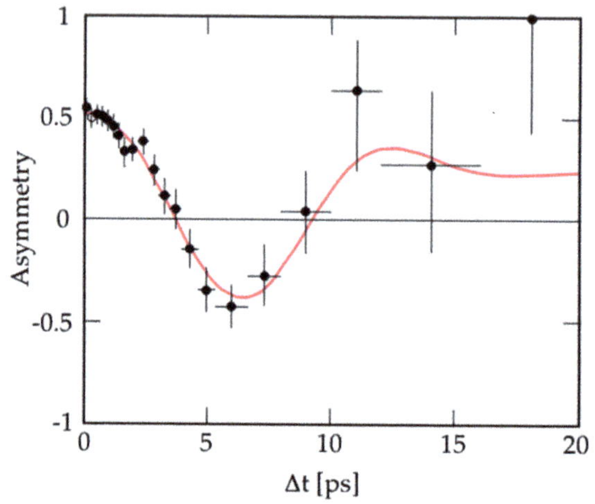

Figure 20.5. Asymmetry $\mathcal{A}(\Delta t)$ in the $B^0 - \overline{B}^0$ oscillation. The curve shows the fit from (20.28) when detector efficiency and resolution are taken into account. After [5].

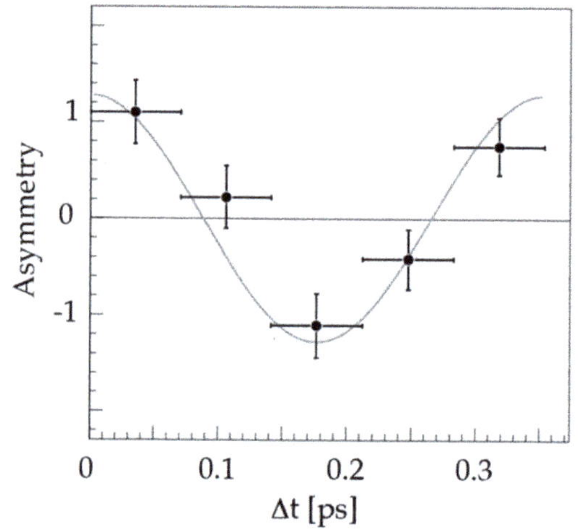

Figure 20.6. Asymmetry $\mathcal{A}(\Delta t)$ in the $B_s^0 - \overline{B}_s^0$ oscillation measured at the Tevatron (modulo $2\pi/\Delta m_s$). The curve shows the fit. After [6].

by their D_s^-, D_s^+ or semileptonic decays. An outstanding position resolution on the decay points is essential in view of the high oscillation frequency.

Figure 20.6 shows the measured asymmetry $\mathcal{A}(\Delta t)$. The period T of the oscillation is much smaller than the lifetime of the B_s meson: $T = 0.35$ ps $\ll \tau \simeq 1.5$ ps from which the mass difference Δm_s between the eigenstates B_{sH} and B_{sL} of the $s\bar{b} - \bar{s}b$ system is derived: $\Delta m_s = (17.77 \pm 0.12)$ ps^{-1} [6].

Figure 20.7 shows the measurement performed recently by LHCb at 7 TeV [7]. The detector is a single-arm forward spectrometer installed at one of the intersection points in the LHC ring. The B_s^0 and $\overline{B}_s^0$ mesons are reconstructed from their decay

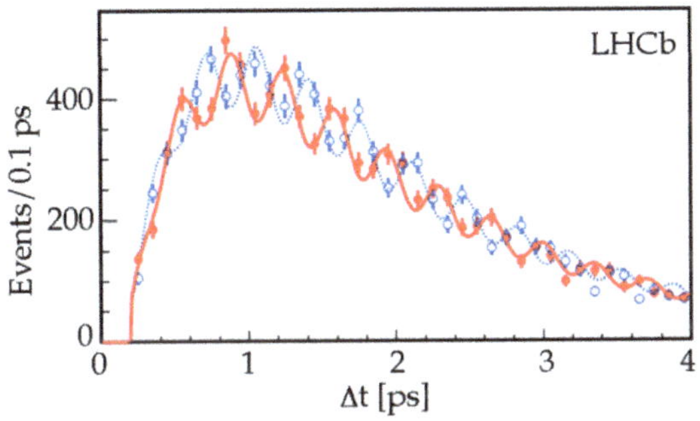

Figure 20.7. $B_s^0 - \overline{B}_s^0$ oscillation measured at the LHC. The dotted (blue) curve shows the oscillation for equal flavours ($B_s^0 B_s^0$ or $\overline{B}_s^0 \overline{B}_s^0$) and the full (red) curve the oscillation for opposite flavours ($B_s^0 \overline{B}_s^0$). After [7].

modes into D_s^- or D_s^+, where the D_s decays into various channels such as $D_s^+ \to \overline{K}^{*0}(\to K^-\pi^+)K^+$ and $D_s^- \to K^{*0}(\to K^+\pi^-)K^-$. A silicon strip detector surrounding the pp interaction region provides the time resolution required to observe the $B_s^0 - \overline{B}_s^0$ oscillation. The distance between the sensitive region of the vertex detector and the beam is only 8 mm. A resolution of 20 μm for tracks with high transverse momenta is achieved. The measured oscillation frequency is $\Delta m_s = (17.768 \pm 0.024)\ \text{ps}^{-1}$ [7].

20.4 CP violation in the $B^0 - \overline{B}^0$ system

Consider now the decay of the B^0 and $\overline{B}^0$ mesons into a *common* final state f with amplitudes

$$A_f \equiv \langle f | H_w | B^0 \rangle, \tag{20.29}$$

$$\overline{A}_f \equiv \langle f | H_w | \overline{B}^0 \rangle. \tag{20.30}$$

If a B^0 is produced at time $t = 0$ the time evolution of the decay amplitude is, see (20.14),

$$\langle f | H_w | B(t) \rangle = A_f g_+(t) + \overline{A}_f \frac{q}{p} g_-(t) = A_f \left[g_+(t) + \lambda_f g_-(t) \right], \tag{20.31}$$

where we have defined

$$\boxed{\lambda_f \equiv \frac{q}{p} \frac{\overline{A}_f}{A_f}}. \tag{20.32}$$

The time dependence of the decay intensity into the final state f is

$$I(t) \propto \left| \langle f | H_w | B(t) \rangle \right|^2 = \left| A_f \right|^2 \left[\left| g_+ \right|^2 + \left| \lambda_f \right|^2 \left| g_- \right|^2 + 2\,\mathrm{Re}\left(g_+^* g_- \lambda_f \right) \right]. \tag{20.33}$$

The three terms can be written as

$$\left|g_+\right|^2 = \frac{1}{4}\left[e^{-\Gamma_L t} + e^{-\Gamma_H t} + 2e^{-\left(\frac{\Gamma_L+\Gamma_H}{2}\right)t}\cos\Delta m_d t\right],$$

$$\left|g_-\right|^2 = \frac{1}{4}\left[e^{-\Gamma_L t} + e^{-\Gamma_H t} - 2e^{-\left(\frac{\Gamma_L+\Gamma_H}{2}\right)t}\cos\Delta m_d t\right],$$

$$2\,\mathrm{Re}\left(g_+^* g_- \lambda_f\right) = \frac{1}{2}\mathrm{Re}\,\lambda_f\left[e^{im_L t - \frac{\Gamma_L}{2}t} + e^{im_H t - \frac{\Gamma_H}{2}t}\right]\left[e^{-im_L t - \frac{\Gamma_L}{2}t} - e^{-im_H t - \frac{\Gamma_H}{2}t}\right]$$

$$= \frac{1}{2}\mathrm{Re}\,\lambda_f\left[e^{-\Gamma_L t} - e^{-\Gamma_H t} + \underbrace{e^{i\Delta m_d t}e^{-\left(\frac{\Gamma_L+\Gamma_H}{2}\right)t} - e^{-i\Delta m_d t}e^{-\left(\frac{\Gamma_L+\Gamma_H}{2}\right)t}}_{2ie^{-\left(\frac{\Gamma_L+\Gamma_H}{2}\right)t}\sin\Delta m_d t}\right]. \tag{20.34}$$

One then obtains with $\Gamma_H \sim \Gamma_L \equiv \Gamma$

$$\left|g_+\right|^2 = \frac{1}{2}e^{-\Gamma t}\left[1 + \cos\Delta m_d t\right],$$

$$\left|g_-\right|^2 = \frac{1}{2}e^{-\Gamma t}\left[1 - \cos\Delta m_d t\right],$$

$$2\,\mathrm{Re}\left(g_+^* g_- \lambda_f\right) = e^{-\Gamma t}\,\mathrm{Re}\left(i\lambda_f\right)\sin\Delta m_d t = -e^{-\Gamma t}\,\mathrm{Im}\,\lambda_f\cdot\sin\Delta m_d t. \tag{20.35}$$

Introducing into (20.33) gives the decay intensity into the final state f

$$I(t) \propto |\langle f\,|H_w|B(t)\rangle|^2 = \left|A_f\right|^2 e^{-\Gamma t}$$

$$\times\left[\frac{1}{2}\left(1 + \cos\Delta m_d t\right) + \frac{\left|\lambda_f\right|^2}{2}(1 - \cos\Delta m_d t) - \mathrm{Im}\,\lambda_f\cdot\sin\Delta m_d t\right], \tag{20.36}$$

hence

$$I(t) = \left|A_f\right|^2 e^{-\Gamma t}\left[\frac{1 + \left|\lambda_f\right|^2}{2} + \frac{1 - \left|\lambda_f\right|^2}{2}\cos\Delta m_d t - \mathrm{Im}\,\lambda_f\cdot\sin\Delta m_d t\right]. \tag{20.37}$$

If a $\overline{B}^0$ is produced at time $t = 0$ the time evolution of the decay amplitude is

$$\langle f\,|H_w|\overline{B}\,(t)\rangle = A_f\frac{p}{q}g_-(t) + \overline{A}_f g_+(t) = \overline{A}_f\left[g_+(t) + \underbrace{\frac{A_f}{\overline{A}_f}\frac{p}{q}}_{=\frac{1}{\lambda_f}}g_-(t)\right] \tag{20.38}$$

see (20.22). By comparing with (20.31) and replacing A_f by $\overline{A}_f$ and λ_f by $\frac{1}{\lambda_f}$ in (20.37) one obtains the time dependence of the decay intensity into the final state *f*:

$$\bar{I}(t) \propto |\langle f|H_w|\overline{B}(t)\rangle|^2 = \left|\overline{A}_f\right|^2 \mathrm{e}^{-\Gamma t} \times \left[\frac{1 + 1/\left|\lambda_f\right|^2}{2} + \frac{1 - 1/\left|\lambda_f\right|^2}{2} \cos \Delta m_d t - \mathrm{Im}(1/\lambda_f) \cdot \sin \Delta m_d t\right]. \quad (20.39)$$

With

$$\mathrm{Im}\left(1/\lambda_f\right) = \mathrm{Im}\, \frac{\lambda_f^*}{\left|\lambda_f\right|^2} = -\frac{\mathrm{Im}\, \lambda_f}{\left|\lambda_f\right|^2} \quad (20.40)$$

the intensity becomes

$$\bar{I}(t) = \frac{\left|\overline{A}_f\right|^2}{\left|\lambda_f\right|^2} \mathrm{e}^{-\Gamma t}\left[\frac{1 + \left|\lambda_f\right|^2}{2} - \frac{1 - \left|\lambda_f\right|^2}{2} \cos \Delta m_d t + \mathrm{Im}\, \lambda_f \cdot \sin \Delta m_d t\right]. \quad (20.41)$$

Since $|q/p| \simeq 1$ (20.32) gives

$$\left|\lambda_f\right|^2 = \left|\frac{q}{p}\right|^2 \left|\frac{\overline{A}_f}{A_f}\right|^2 \simeq \left|\frac{\overline{A}_f}{A_f}\right|^2. \quad (20.42)$$

Direct *CP* violation is signalled by $|\overline{A}_f/A_f| \neq 1$. The decay intensity is finally given by

$$\bar{I}(t) = \left|A_f\right|^2 \mathrm{e}^{-\Gamma t}\left[\frac{1 + \left|\lambda_f\right|^2}{2} - \frac{1 - \left|\lambda_f\right|^2}{2} \cos \Delta m_d t + \mathrm{Im}\, \lambda_f \cdot \sin \Delta m_d t\right]. \quad (20.43)$$

In the BABAR experiment [5] the time difference $\Delta t = t_2 - t_1$ is measured from Δz (figure 20.3). The *initial* flavour at time t_1 of the B meson which decays into the final state f at time t_2 is determined from the semileptonic or K decay of the other B meson. The time dependence of the asymmetry is

$$\boxed{\mathcal{A}(\Delta t) = \frac{\bar{I}(\Delta t) - I(\Delta t)}{\bar{I}(\Delta t) + I(\Delta t)}}, \quad (20.44)$$

where $\bar{I}$ (I) is the intensity into the final state f for an initial $\overline{B}^0$ (B^0) meson. Hence, after introducing (20.43) and (20.37) the asymmetry becomes

$$\mathcal{A}(\Delta t) = \frac{-\left(1 - \left|\lambda_f\right|^2\right)\cos\left(\Delta m_d \Delta t\right) + 2\, \mathrm{Im}\, \lambda_f \cdot \sin\left(\Delta m_d \Delta t\right)}{1 + \left|\lambda_f\right|^2}. \quad (20.45)$$

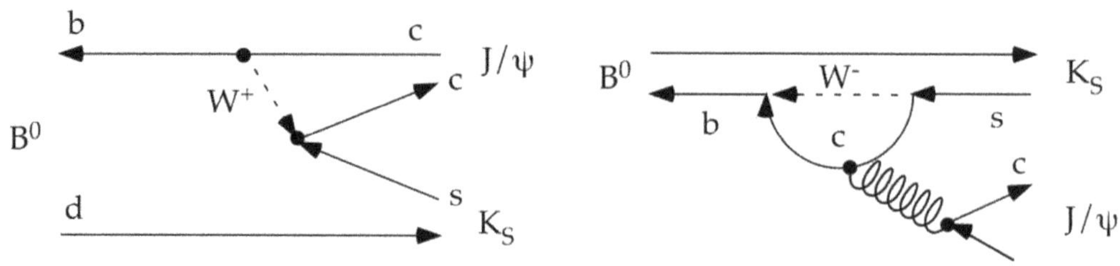

Figure 20.8. Feynman graphs of $B^0 \to J/\psi(1S)K_S$ decay: tree (left) and penguin (right).

Let us now assume that f is also a CP-eigenstate with eigenvalue $\eta_f = \pm 1$. The decay amplitude (20.30) can be expressed as

$$\begin{aligned}\overline{A}_f &= \langle f|H_w|\overline{B}^0\rangle = \langle f|(CPT)^{-1}H_w(CPT)|\overline{B}^0\rangle \\ &= \eta_f\langle f|T^{-1}H_wT|B^0\rangle = \eta_f\langle B^0|H_w|f\rangle = \eta_f\langle f|H_w|B^0\rangle^* = \eta_f A_f^*. \quad (20.46)\end{aligned}$$

The final state amplitude is given by the superposition of amplitudes from the different Feynman graphs that contribute to the decay. If one graph dominates then

$$\lambda_f \simeq \frac{q}{p}\frac{\overline{A}_f}{A_f} = \eta_f \frac{q}{p}\frac{A_f^*}{A_f} \Rightarrow \left|\lambda_f\right|^2 \simeq 1, \quad (20.47)$$

in which case the expression (20.45) for the asymmetry reduces to

$$\mathcal{A}(\Delta t) \simeq \operatorname{Im}\lambda_f \cdot \sin\left(\Delta m_d \Delta t\right). \quad (20.48)$$

The asymmetry is finite when λ_f is *complex*. As shown below this leads to CP violation.

Let us describe the measurement of $B^0 \to J/\psi(1S)K_S$ for which CP violation was discovered in the $B^0 - \overline{B}^0$ system [8]. The final state $J/\psi(1S)K_S$ is a CP-eigenstate with eigenvalue $\eta_f = -1$ since the angular momentum between $J/\psi(1S)(1^{--})$ and $K_S(0^{-+})$ is equal to one (the B^0 being spinless). The amplitudes for the tree and penguin diagrams are depicted in figure 20.8. For the $J/\psi(1S)K_S$ final state the tree diagram dominates, so (20.48) is applicable. This is also true for other B decays into $c\bar{c}$ states (charmonium).

Let us first calculate the contribution from indirect CP violation q/p to λ_f. The mixing matrix for the $B^0 - \overline{B}^0$ oscillation is given in analogy to (19.58):

$$H_w = \begin{pmatrix} M_1 & M_{12} \\ M_{12}^* & M_2 \end{pmatrix} = \begin{pmatrix} \alpha & \frac{\beta}{r} \\ \beta r & \alpha \end{pmatrix}. \quad (20.49)$$

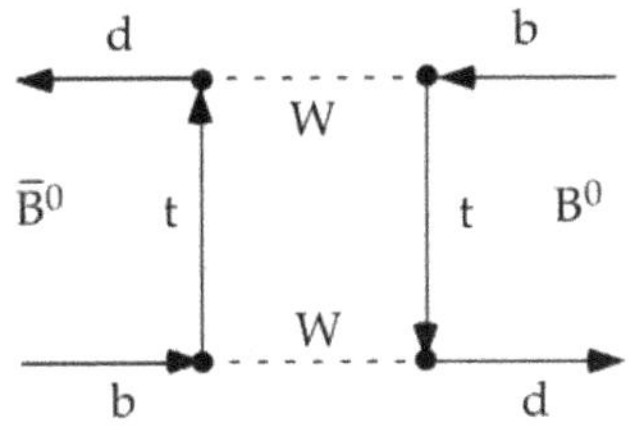

Figure 20.9. Box diagram contributing to $B^0 - \overline{B}^0$ oscillation.

The relations between the parameters p, q and r are, according to (19.81),

$$p = \frac{1+\epsilon}{\sqrt{2\left(1+|\epsilon|^2\right)}}, \qquad q = \frac{1-\epsilon}{\sqrt{2\left(1+|\epsilon|^2\right)}}, \qquad r = \frac{1-\epsilon}{1+\epsilon} \Rightarrow r = \frac{q}{p}, \tag{20.50}$$

therefore

$$\frac{M_{12}}{M_{12}^*} = \frac{1}{r^2} = \frac{p^2}{q^2} \Rightarrow \frac{q}{p} = \sqrt{\frac{M_{12}^*}{M_{12}}}. \tag{20.51}$$

The matrix element M_{12} describes the oscillation from $\overline{B}^0$ to B^0 (figure 20.9):

$$M_{12} \propto \left(V_{tb} V_{td}^*\right)^2 \Rightarrow \sqrt{\frac{M_{12}^*}{M_{12}}} = \frac{V_{tb}^* V_{td}}{V_{tb} V_{td}^*}. \tag{20.52}$$

The ratio A_f^*/A_f (direct CP violation) depends on the couplings V_{cb} and V_{cs} (figure 20.8):

$$\frac{A_f^*}{A_f} = \frac{V_{cb} V_{cs}^*}{V_{cb}^* V_{cs}}, \tag{20.53}$$

so that with (20.47),

$$\operatorname{Im} \lambda_f = \eta_f \operatorname{Im}\left(\frac{V_{tb}^* V_{td}}{V_{tb} V_{td}^*}\right)\left(\frac{V_{cb} V_{cs}^*}{V_{cb}^* V_{cs}}\right)\left(\frac{V_{cs} V_{cd}^*}{V_{cs}^* V_{cd}}\right). \tag{20.54}$$

The third parenthesis arises from $K^0 - \overline{K}^0$ mixing, as B^0 decays into $J/\psi(1S)K^0$ and $\overline{B}^0$ into $J/\psi(1S)\overline{K}^0$. In terms of the Wolfenstein representation (16.69) in which only V_{td} and V_{td}^* are complex (up to order λ^3),

$$\operatorname{Im} \lambda_f = \eta_f \operatorname{Im}\left(\frac{1-\rho-\mathrm{i}\eta}{1-\rho+\mathrm{i}\eta}\right) = -\eta_f \frac{2\eta(1-\rho)}{(1-\rho)^2+\eta^2}. \tag{20.55}$$

The relation between the angle β and the CKM elements is, according to (20.9):

$$\beta = \arg\left(-\frac{V_{cd} V_{cb}^*}{V_{td} V_{tb}^*}\right) = \arg\left[\frac{\lambda A \lambda^2}{A\lambda^3\left(1-\rho-\mathrm{i}\eta\right)}\right] = \arg\left[\frac{1-\rho+\mathrm{i}\eta}{(1-\rho)^2+\eta^2}\right]. \tag{20.56}$$

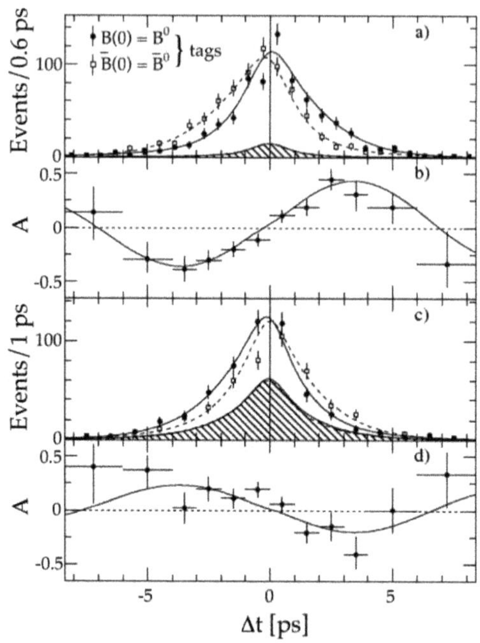

Figure 20.10. *CP*-violating intensities and asymmetries $A \equiv \mathcal{A}(\Delta t)$ in $J/\psi(1S)K_S$ and $\psi(2S)K_S$, $\chi_{c1}K_S$, $\eta_c K_S$ with $\eta_f = -1$ (*a*), (*b*), and in $J/\psi(1S)K_L$ with $\eta_f = +1$ (*c*), (*d*) (BABAR experiment). The B^0 ($\overline{B}^0$) tag corresponds to the production of the $\overline{B}^0$ (B^0) meson which decays into the final state *f*. The hatched areas show the background. After [9].

The angle β (see figure 20.1) is given by

$$\sin\beta \simeq \frac{\eta}{\sqrt{(1-\rho)^2+\eta^2}} \quad \text{and} \quad \cos\beta \simeq \frac{1-\rho}{\sqrt{(1-\rho)^2+\eta^2}}. \tag{20.57}$$

Thus with (20.55):

$$\sin 2\beta = 2\sin\beta\cos\beta = \frac{2\eta(1-\rho)}{\eta^2+(1-\rho)^2} = -\frac{\text{Im}\,\lambda_f}{\eta_f}. \tag{20.58}$$

The imaginary part of λ_f leads to $\beta \neq 0$ and therefore induces *CP* violation. The asymmetry (20.48) finally reads

$$\boxed{\mathcal{A}(\Delta t) \simeq -\eta_f \sin 2\beta \sin(\Delta m_d \Delta t)}. \tag{20.59}$$

Figure 20.10 shows the asymmetry $\mathcal{A}(\Delta t)$ measured by BABAR [9]. The flavour of the tagged *B* meson is determined from its hadronic decay into kaonic final states, or from semileptonic decays. The decay products $J/\psi(1S)$ and K_S from the other *B* meson are detected through their e^+e^- or $\mu^+\mu^-$ and $\pi\pi$ decays, respectively. Other *B* decays into charmonium are also studied, namely into $\psi(2S)$ (1^{--}) K_S, χ_{c1} (1^{++}) K_S,

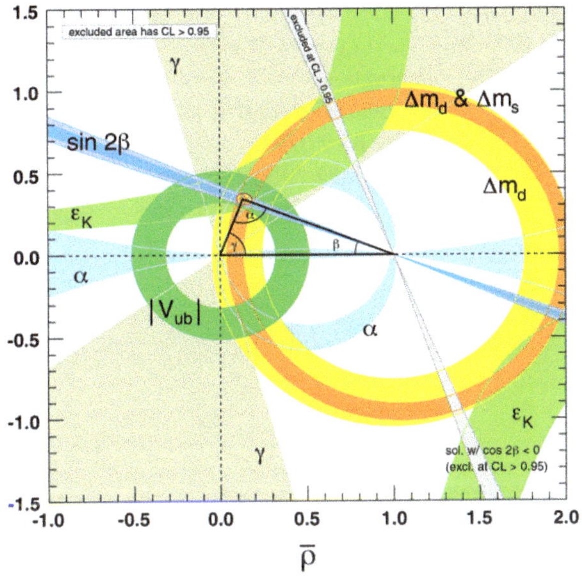

Figure 20.11. The unitarity triangle in the complex plane. From [1].

η_c (0^{-+}) K_S with $\eta_f = -1$, and into $J/\psi(1S)K_L$ with $\eta_f = +1$. As in section 20.2 the oscillation amplitude is reduced by the imperfect flavour identification.

Figure 20.10 shows that the *CP*-violating asymmetry is rather large, in contrast to that in the neutral kaon system. The asymmetry is due to the *interference* of the amplitudes for the two decay paths $B^0 \to \overline{B}^0 \to f$ and $B^0 \to f$. The fit leads to $\sin 2\beta = 0.741 \pm 0.075$. The more precise value from recent data is [1]

$$\sin 2\beta = 0.682 \pm 0.019 \tag{20.60}$$

or $\beta \simeq 21°$. Note that there is a fourfold ambiguity on β ($\beta \to \frac{\pi}{2} - \beta$ and $\beta \to \pi + \beta$) which has been solved with the help of other decay modes [1]. From the sign of the asymmetry (figure 20.10) one finds with (20.59) that

$$\Delta m_d > 0 \Rightarrow m_H > m_L. \tag{20.61}$$

Figure 20.11 shows the allowed physical area in the complex plane from all available data. From ϵ_K (ϵ in the $K^0 - \overline{K}^0$ system) follows that $\beta > 0$. The excellent agreement between all data shows that the CKM matrix is unitary with three quark generations, within measurement errors.

References

[1] Olive K *et al* (Particle Data Group) 2014 *Chin. Phys.* C **38** 090001
[2] Aaij R *et al* 2013 LHCb Collaboration *Phys. Lett.* B **726** 151
[3] Albrecht H *et al* 1991 *Phys. Lett.* B **255** 297
[4] Aubert B *et al* 2002 *Nucl. Instrum. Methods Phys. Res.* A **479** 1

[5] Aubert B *et al* 2002 *Phys. Rev. Lett.* **88** 221802
[6] Abulencia A *et al* 2006 *Phys. Rev. Lett.* **97** 242003
[7] Aaij R *et al* 2013 LHCb Collaboration *New J. Phys.* **15** 053021
[8] Aubert B *et al* 2001 *Phys. Rev. Lett.* **87** 091801
Abe K *et al* 2001 *Phys. Rev. Lett.* **87** 091802
[9] Aubert B *et al* 2002 *Phys. Rev. Lett.* **89** 201802

IOP Publishing

Nuclear and Particle Physics

Claude Amsler

Appendix A

SU(3) wave functions of the light ground state mesons

Table A.1 lists the wave functions of the ground state 1^1S_0 and 1^3S_1 mesons made of the lightest u, d and s quarks. The 1^1S_0 wave functions ϕ_S are symmetric with respect to quark–antiquark exchange while the 1^3S_1 ones (ϕ_A) are antisymmetric. The singlet $|1\rangle$ and octet $|8\rangle$ mix to give the observed states (see (10.14)).

Table A.1. SU(3) wave functions of the light-quark ground state $^1S_0$ and $^3S_1$ mesons.

1^1S_0 (0^{-+})	ϕ_S	1^3S_1 (1^{--})	ϕ_A
K^+	$\frac{1}{\sqrt{2}}\|u\bar{s} + \bar{s}u\rangle$	K^{+*}	$\frac{1}{\sqrt{2}}\|u\bar{s} - \bar{s}u\rangle$
K^0	$\frac{1}{\sqrt{2}}\|d\bar{s} + \bar{s}d\rangle$	K^{0*}	$\frac{1}{\sqrt{2}}\|d\bar{s} - \bar{s}d\rangle$
K^-	$-\frac{1}{\sqrt{2}}\|s\bar{u} + \bar{u}s\rangle$	K^{-*}	$-\frac{1}{\sqrt{2}}\|s\bar{u} - \bar{u}s\rangle$
$\overline{K}^0$	$-\frac{1}{\sqrt{2}}\|s\bar{d} + \bar{d}s\rangle$	$\overline{K}^{0*}$	$-\frac{1}{\sqrt{2}}\|s\bar{d} - \bar{d}s\rangle$
π^+	$\frac{1}{\sqrt{2}}\|u\bar{d} + \bar{d}u\rangle$	ρ^+	$\frac{1}{\sqrt{2}}\|u\bar{d} - \bar{d}u\rangle$
π^-	$-\frac{1}{\sqrt{2}}\|d\bar{u} + \bar{u}d\rangle$	ρ^-	$-\frac{1}{\sqrt{2}}\|d\bar{u} - \bar{u}d\rangle$
π^0	$\frac{1}{2}(\|d\bar{d} - u\bar{u}\rangle + \|\bar{d}d - \bar{u}u\rangle)$	ρ^0	$\frac{1}{2}(\|d\bar{d} - u\bar{u}\rangle - \|\bar{d}d - \bar{u}u\rangle)$
$\|8\rangle$	$\frac{1}{2\sqrt{3}}(\|u\bar{u} + d\bar{d} - 2s\bar{s}\rangle$ $+ \|\bar{u}u + \bar{d}d - 2\bar{s}s\rangle)$	$\|8\rangle$	$\frac{1}{2\sqrt{3}}(\|u\bar{u} + d\bar{d} - 2s\bar{s}\rangle$ $- \|\bar{u}u + \bar{d}d - 2\bar{s}s\rangle)$
$\|1\rangle$	$\frac{1}{\sqrt{6}}(\|u\bar{u} + d\bar{d} + s\bar{s}\rangle$ $+ \|\bar{u}u + \bar{d}d + \bar{s}s\rangle)$	$\|1\rangle$	$\frac{1}{\sqrt{6}}(\|u\bar{u} + d\bar{d} + s\bar{s}\rangle$ $- \|\bar{u}u + \bar{d}d + \bar{s}s\rangle)$

doi:10.1088/978-0-7503-1140-3ch21

© IOP Publishing Ltd 2015

IOP Publishing

Nuclear and Particle Physics

Claude Amsler

Appendix B

Mass formula

In this appendix we derive a sum rule and an alternative to the mass formula (10.25) involving the masses of all mesons in the nonet. We also provide an intuitive explanation for the deviation from ideal mixing in the 1^1S_0 nonet.

The eigenvalues λ_1 and λ_2 of the mass matrix (10.18)

$$M = \langle 8,\, 1|\, H\, |8,\, 1\rangle = \begin{pmatrix} m_8 & m_{81} \\ m_{18} & m_1 \end{pmatrix} \tag{B.1}$$

are found by solving the secular equation (10.19):

$$\left(m_8 - \lambda\right)\left(m_1 - \lambda\right) - m_{18}^2 = 0, \tag{B.2}$$

where $m_{18} = m_{81}$. The solutions are

$$\lambda_{1,2} = \frac{m_1 + m_8 \pm \sqrt{(m_1 + m_8)^2 + 4\left(m_{81}^2 - m_1 m_8\right)}}{2}, \tag{B.3}$$

hence

$$\begin{aligned} \lambda_1 + \lambda_2 &= m_1 + m_8, \\ \lambda_1\lambda_2 &= m_1 m_8 - m_{81}^2. \end{aligned} \tag{B.4}$$

In section 10.2.1 we have expressed m_8 and m_{18} in terms of the constituent quarks. However, oscillations mediated by the exchange of two or more gluons (figure B.1) introduces a perturbation leading to an additional term A in the mass matrix.

doi:10.1088/978-0-7503-1140-3ch22 © IOP Publishing Ltd 2015

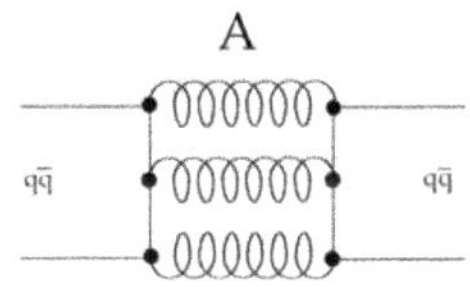

Figure B.1. Perturbation of the wave function by an intermediate state of several gluons.

For example, for the 1^3S_1 nonet:

$$m_8 = \frac{1}{6}(1,\ 1,\ -2)\begin{pmatrix} 2m_u + A & A & A \\ A & 2m_d + A & A \\ A & A & 2m_s + A \end{pmatrix}\begin{pmatrix} 1 \\ 1 \\ -2 \end{pmatrix} = \frac{(2m_u + 4m_s)}{3}$$

$$= \frac{1}{3}\left(m_\rho + 4m_{K^*} - 2m_\rho\right) = \frac{1}{3}\left(4m_{K^*} - m_\rho\right) \tag{B.5}$$

and

$$m_{81} = \frac{1}{\sqrt{18}}(1,\ 1,\ -2)\begin{pmatrix} 2m_u + A & A & A \\ A & 2m_d + A & A \\ A & A & 2m_s + A \end{pmatrix}\begin{pmatrix} 1 \\ 1 \\ 1 \end{pmatrix} = \frac{2\sqrt{2}\,(m_u - m_s)}{3}$$

$$= m_{18} = \frac{\sqrt{2}}{3}\left(m_\rho - 2m_{K^*} + m_\rho\right) = \frac{2\sqrt{2}}{3}(m_\rho - m_{K^*}) \tag{B.6}$$

are independent of A while

$$m_1 = \frac{1}{3}(1,\ 1,\ 1)\begin{pmatrix} 2m_u + A & A & A \\ A & 2m_d + A & A \\ A & A & 2m_s + A \end{pmatrix}\begin{pmatrix} 1 \\ 1 \\ 1 \end{pmatrix} = \frac{(4m_u + 2m_s)}{3} + 3A$$

$$= \frac{1}{3}\left(m_\rho + 2m_{K^*}\right) + 3A. \tag{B.7}$$

We have assumed that $m_u = m_d$. Substituting m_8, m_1 and m_{81} into (B.4) gives with $\lambda_1 = m_\phi$ and $\lambda_2 = m_\omega$:

$$\begin{aligned} m_\phi + m_\omega &= 3A + 2m_{K^*}, \\ m_\phi m_\omega &= 2m_{K^*}m_\rho - m_\rho^2 + A\left(4m_{K^*} - m_\rho\right). \end{aligned} \tag{B.8}$$

Eliminating A leads to the Schwinger sum rule

$$(m_\omega + m_\phi)\left(4m_{K^*} - m_\rho\right) = 3m_\omega m_\phi - 8m_{K^*}m_\rho + 3m_\rho^2 + 8m_{K^*}^2, \tag{B.9}$$

which should be fulfilled by all meson nonets.

In the special case $A = 0$ we recover ideal mixing, see (10.27). This suggests a reason for the strong deviation from ideal mixing in the 1^1S_0 nonet: the interaction shown in figure B.1 proceeds through the exchange of two gluons, while for the ideally mixed 1^3S_1 nonet at least three gluons are required since a spin-1 state does not decay into two spin-1 gluons (section 11.6). This reduces the strength of the perturbation A. For nonets with angular momentum $\ell > 0$ the centrifugal barrier suppresses the $q\bar{q}$ annihilation shown in figure B.1, also leading to $A \to 0$ and ideal mixing. The sum rule (B.9) is satisfied only approximately in the 1^1S_0 nonet.

Finally, we seek an alternative to the mass formula (10.25) involving all nonet masses. The matrix M can be assembled from a matrix U containing in its columns the eigenfunctions of M and from the diagonal matrix D (eigen-decomposition):

$$\begin{aligned} M &= UDU^{-1} \\ &= \begin{pmatrix} \cos\theta & \sin\theta \\ -\sin\theta & \cos\theta \end{pmatrix}\begin{pmatrix} \lambda_1 & 0 \\ 0 & \lambda_2 \end{pmatrix}\begin{pmatrix} \cos\theta & -\sin\theta \\ \sin\theta & \cos\theta \end{pmatrix}. \end{aligned} \tag{B.10}$$

For instance, the octet mass m_8 is given by

$$m_8 = \lambda_1 \cos^2\theta + \lambda_2 \sin^2\theta = \frac{\lambda_1 + \lambda_2 \tan^2\theta}{1 + \tan^2\theta}. \tag{B.11}$$

Solving for θ gives

$$\boxed{\tan^2\theta = \frac{\lambda_1 - m_8}{-\lambda_2 + m_8}}. \tag{B.12}$$

As an illustration one obtains by introducing the 1^3S_1 masses

$$\tan^2\theta = \frac{4m_{K^*} - m_\rho - 3m_\phi}{-4m_{K^*} + m_\rho + 3m_\omega} \Rightarrow \theta \simeq 37°, \tag{B.13}$$

as expected for ideal mixing. This mass formula involves the masses of all the mesons in the nonet, but does not provide the sign of θ.

Appendix C

SU(3) wave functions of light ground state baryons

Table C.1 shows the SU(3) wave functions ϕ_{MS} (mixed symmetric) and ϕ_{MA} (mixed antisymmetric) of the ground state spin-$\frac{1}{2}$ baryons made of u, d and s quarks. According to the symmetric quark model the baryon total wave functions ψ_f are constructed by combining with the spin-$\frac{1}{2}$ wave functions χ_{MS} and χ_{MA}:

$$\begin{aligned}
\chi_{MS}\left(\frac{1}{2}, \frac{1}{2}\right) &= \frac{1}{\sqrt{3}} \left|\left(\frac{\uparrow\downarrow + \downarrow\uparrow}{\sqrt{2}}\right)\uparrow - \uparrow\uparrow\downarrow \sqrt{2}\right\rangle, \\
\chi_{MS}\left(\frac{1}{2}, -\frac{1}{2}\right) &= -\frac{1}{\sqrt{3}} \left|\left(\frac{\uparrow\downarrow + \downarrow\uparrow}{\sqrt{2}}\right)\downarrow - \downarrow\downarrow\uparrow \sqrt{2}\right\rangle, \\
\chi_{MA}\left(\frac{1}{2}, \frac{1}{2}\right) &= \frac{1}{\sqrt{2}} |(\uparrow\downarrow - \downarrow\uparrow)\uparrow\rangle, \\
\chi_{MA}\left(\frac{1}{2}, -\frac{1}{2}\right) &= \frac{1}{\sqrt{2}} |(\uparrow\downarrow - \downarrow\uparrow)\downarrow\rangle.
\end{aligned} \tag{C.1}$$

The functions that are symmetric under the permutation of two quarks are

$$\boxed{\psi_f\left(J = \frac{1}{2}\right) = \frac{1}{\sqrt{2}}\left[\phi_{MS}\chi_{MS} + \phi_{MA}\chi_{MA}\right]}. \tag{C.2}$$

With the addition of colour the wave function becomes antisymmetric, as required for fermions. The antisymmetric state $\phi_A = \frac{1}{\sqrt{6}}|s(du - ud) + (usd - dsu) + (du - ud)s\rangle$ is not realized in the ground state.

doi:10.1088/978-0-7503-1140-3ch23

© IOP Publishing Ltd 2015

Table C.1. SU(3) wave functions of the ground state spin-$\frac{1}{2}$ baryons made of the lightest u, d and s quarks.

	ϕ_{MS}	ϕ_{MA}
p	$\frac{1}{\sqrt{6}}\|(ud+du)u-2uud\rangle$	$\frac{1}{\sqrt{2}}\|(ud-du)u\rangle$
n	$-\frac{1}{\sqrt{6}}\|(ud+du)d-2ddu\rangle$	$\frac{1}{\sqrt{2}}\|(ud-du)d\rangle$
Σ^+	$\frac{1}{\sqrt{6}}\|(us+su)u-2uus\rangle$	$\frac{1}{\sqrt{2}}\|(us-su)u\rangle$
Σ^0	$\frac{1}{\sqrt{6}}\left\|s\left(\frac{du+ud}{\sqrt{2}}\right)+\left(\frac{dsu+usd}{\sqrt{2}}\right)-2\left(\frac{du+ud}{\sqrt{2}}\right)s\right\rangle$	$\frac{1}{\sqrt{2}}\left\|\left(\frac{dsu+usd}{\sqrt{2}}\right)-s\left(\frac{ud+du}{\sqrt{2}}\right)\right\rangle$
Σ^-	$\frac{1}{\sqrt{6}}\|(ds+sd)u-2dds\rangle$	$\frac{1}{\sqrt{2}}\|(ds-sd)d\rangle$
Λ	$\frac{1}{\sqrt{2}}\left\|\left(\frac{dsu-usd}{\sqrt{2}}\right)+s\left(\frac{du-ud}{\sqrt{2}}\right)\right\rangle$	$\frac{1}{\sqrt{6}}\left\|s\left(\frac{du-ud}{\sqrt{2}}\right)+\left(\frac{usd-dsu}{\sqrt{2}}\right)-2\left(\frac{du-ud}{\sqrt{2}}\right)s\right\rangle$
Ξ^-	$-\frac{1}{\sqrt{6}}\|(ds+sd)s-2ssd\rangle$	$\frac{1}{\sqrt{2}}\|(ds-sd)s\rangle$
Ξ^0	$-\frac{1}{\sqrt{6}}\|(us+su)s-2ssu\rangle$	$\frac{1}{\sqrt{2}}\|(us-su)s\rangle$

Table C.2. SU(3) wave functions of the ground state spin-$\frac{3}{2}$ baryons.

	ϕ_S
$\Delta^{++}(1232)$	$\|uuu\rangle$
$\Delta^{+}(1232)$	$\frac{1}{\sqrt{3}}\|uud+udu+duu\rangle$
$\Delta^{0}(1232)$	$\frac{1}{\sqrt{3}}\|udd+dud+ddu\rangle$
$\Delta^{-}(1232)$	$\|ddd\rangle$
$\Sigma^{+}(1383)$	$\frac{1}{\sqrt{3}}\|uus+usu+suu\rangle$
$\Sigma^{0}(1384)$	$\frac{1}{\sqrt{3}}\|uds+usd+sdu\rangle$
$\Sigma^{-}(1387)$	$\frac{1}{\sqrt{3}}\|dds+dsd+sdd\rangle$
$\Xi^{0}(1532)$	$\frac{1}{\sqrt{3}}\|uss+sus+ssu\rangle$
$\Xi^{-}(1535)$	$\frac{1}{\sqrt{3}}\|dss+sds+ssd\rangle$
Ω^-	$\|sss\rangle$

Table C.2 shows the symmetric SU(3) wave functions ϕ_S of the ground state spin-$\frac{3}{2}$ baryons made of u, d and s quarks. The total wave functions are constructed by combining with the spin-$\frac{3}{2}$ wave functions χ_S

$$\begin{aligned}
\chi_S\left(\frac{3}{2}, \frac{3}{2}\right) &= |\uparrow\uparrow\uparrow\rangle, \\
\chi_S\left(\frac{3}{2}, \frac{1}{2}\right) &= \frac{1}{\sqrt{3}}\,|\uparrow\uparrow\downarrow + \uparrow\downarrow\uparrow + \downarrow\uparrow\uparrow\rangle, \\
\chi_S\left(\frac{3}{2}, -\frac{1}{2}\right) &= \frac{1}{\sqrt{3}}\,|\uparrow\downarrow\downarrow + \downarrow\downarrow\uparrow + \downarrow\uparrow\downarrow\rangle, \\
\chi_S\left(\frac{3}{2}, -\frac{3}{2}\right) &= |\downarrow\downarrow\downarrow\rangle.
\end{aligned} \tag{C.3}$$

The symmetric functions under the permutation of two quarks are

$$\boxed{\psi_f\left(J = \frac{3}{2}\right) = \phi_S \chi_S}. \tag{C.4}$$

IOP Publishing

Nuclear and Particle Physics

Claude Amsler

Appendix D

The CKM matrix and *CP* violation

We prove that *CP* conservation requires the CKM matrix (16.67) to be real ($\delta = 0$ or π). Let us consider for example the quark scattering process $ud \rightarrow dc$ with transition matrix $\mathcal{M}$:

$$\mathcal{M} = \frac{4G_F}{\sqrt{2}}\left(\bar{c}\,\gamma^\mu\left[\frac{1-\gamma^5}{2}\right]V_{cd}\,d\right)\left(\bar{d}\,V_{ud}^*\,\gamma_\mu\left[\frac{1-\gamma^5}{2}\right]u\right). \tag{D.1}$$

The transition amplitude for the time reversed reaction $dc \rightarrow ud$

$$\mathcal{M}^\dagger = \frac{4G_F}{\sqrt{2}}\left(\bar{d}\,V_{cd}^*\,\gamma_\mu\left[\frac{1-\gamma^5}{2}\right]c\right)\left(\bar{u}\gamma^\mu\left[\frac{1-\gamma^5}{2}\right]V_{ud}\,d\right) \tag{D.2}$$

is equal to the transition amplitude $\overline{\mathcal{M}}$ for the *CP*-transformed reaction $\bar{u}\bar{d} \rightarrow \bar{d}\bar{c}$, assuming *CP* (and *CPT*) conservation:

$$\overline{\mathcal{M}} = \mathcal{M}^\dagger. \tag{D.3}$$

Let us calculate $\overline{\mathcal{M}}$. According to (15.82), the *CP*-transformed spinor u^{CP} is

$$u^{CP} = P\overbrace{\mathrm{i}\gamma^2 u^*}^{u^c} = \mathrm{i}P\gamma^2\overbrace{\gamma^0(\gamma^0}^{I}u^{\dagger T}) = \mathrm{i}P\gamma^2\gamma^0\left(u^\dagger\gamma^0\right)^T = \mathrm{i}P\gamma^2\gamma^0\bar{u}^T. \tag{D.4}$$

The corresponding adjoint spinor is

$$\bar{u}^{CP} = \left(u^{CP}\right)^\dagger\gamma^0 = \left(\mathrm{i}P\gamma^2u^*\right)^\dagger\gamma^0 = \mathrm{i}u^T\gamma^2P\gamma^0. \tag{D.5}$$

doi:10.1088/978-0-7503-1140-3ch24 © IOP Publishing Ltd 2015

The transition amplitude for the CP-transformed reaction $\bar{u}\bar{d} \to \bar{d}\bar{c}$ is

$$\begin{aligned}
\overline{\mathcal{M}} &= \frac{4G_F}{\sqrt{2}}\left(\bar{c}^{CP}\gamma^\mu\left[\frac{1-\gamma^5}{2}\right] V_{cd}\, d^{CP}\right)\left(\bar{d}^{CP}\, V_{ud}^*\, \gamma^\mu\left[\frac{1-\gamma^5}{2}\right]u^{CP}\right) \\
&= \frac{4G_F}{\sqrt{2}}\left(c^T\gamma^2P\gamma^0\gamma^\mu\left[\frac{1-\gamma^5}{2}\right] V_{cd}\, P\gamma^2\gamma^0\bar{d}^T\right) \\
&\quad\times\left(d^T\, V_{ud}^*\, \gamma^2P\gamma^0\gamma^\mu\left[\frac{1-\gamma^5}{2}\right]\gamma^2\gamma^0\bar{u}^T\right) \\
&= \frac{4G_F}{\sqrt{2}}\left(c^T\gamma^\mu\left[\frac{1+\gamma^5}{2}\right] V_{cd}\, \bar{d}^T\right)\left(d^T\, \gamma^\mu\, V_{ud}^*\left[\frac{1+\gamma^5}{2}\right]\bar{u}^T\right) \\
&= \frac{4G_F}{\sqrt{2}}\left(\bar{d}\, V_{cd}\, \gamma^\mu\left[\frac{1-\gamma^5}{2}\right]c\right)\left(\bar{u}\gamma^\mu\left[\frac{1-\gamma^5}{2}\right] V_{ud}^*\, d\right).
\end{aligned} \qquad \text{(D.6)}$$

Parity P was applied twice: $\vec{p} \to -\vec{p} \to \vec{p}$ and $P^2 = (\gamma^0)^2 = 1$. Also, $(\gamma^5)^T = \gamma^5$ and $\gamma^5(\gamma^\mu)^T = -(\gamma^\mu)^T\gamma^5$. By comparing with (D.2) one finds that

$$\overline{\mathcal{M}} = \mathcal{M}^\dagger \Rightarrow V_{cd} = V_{cd}^* \qquad \text{and} \qquad V_{ud} = V_{ud}^*. \qquad \text{(D.7)}$$

CP conservation requires the CKM matrix elements to be real.

IOP Publishing

Nuclear and Particle Physics

Claude Amsler

Appendix E

Infinitesimal gauge field transformation

Let us prove the lemma (16.121)

$$\boxed{D'^{\mu}\psi' = \left(1 + \mathrm{i}g\vec{\chi}\cdot\frac{\vec{\tau}}{2}\right)D^{\mu}\psi} \tag{E.1}$$

in section 16.6. The field transformations (16.119) can be written as

$$\vec{W}'^{\mu} = \vec{W}^{\mu} + \delta\vec{W}^{\mu} \qquad \text{with} \qquad \delta W_i^{\mu} \equiv -\partial^{\mu}\chi_i - g\epsilon_{ijk}\chi_j W_k^{\mu}. \tag{E.2}$$

The left-hand side of (E.1) gives, when neglecting terms in $\delta\vec{W}^{\mu}\cdot\vec{\chi}$ (infinitesimal transformation with $|\vec{\chi}| \ll 1$) and inserting the covariant derivatives:

$$\begin{aligned}
&\left(\partial^{\mu} + \mathrm{i}g\frac{\vec{\tau}}{2}\cdot\vec{W}'^{\mu}\right)\left(1 + \mathrm{i}g\vec{\chi}\cdot\frac{\vec{\tau}}{2}\right)\psi \\
&= \left(\partial^{\mu} + \mathrm{i}g\frac{\vec{\tau}}{2}\cdot\vec{W}^{\mu} + \mathrm{i}g\frac{\vec{\tau}}{2}\cdot\delta\vec{W}^{\mu}\right)\left(1 + \mathrm{i}g\vec{\chi}\cdot\frac{\vec{\tau}}{2}\right)\psi \\
&= \left[\partial^{\mu} + \mathrm{i}g\frac{\vec{\tau}}{2}\cdot\vec{W}^{\mu} + \mathrm{i}g\frac{\vec{\tau}}{2}\cdot\delta\vec{W}^{\mu} + \mathrm{i}g\frac{\vec{\tau}}{2}\cdot\partial^{\mu}\vec{\chi} + \mathrm{i}g\vec{\chi}\cdot\frac{\vec{\tau}}{2}\partial^{\mu} - g^2\left(\frac{\vec{\tau}}{2}\cdot\vec{W}^{\mu}\right)\left(\frac{\vec{\tau}}{2}\cdot\vec{\chi}\right)\right]\psi \\
&= \left[\overbrace{\partial^{\mu} + \mathrm{i}g\frac{\vec{\tau}}{2}\cdot\vec{W}^{\mu}}^{1} - \overbrace{\mathrm{i}g\frac{\vec{\tau}}{2}\cdot\partial^{\mu}\vec{\chi}}^{2} + \overbrace{\mathrm{i}g\frac{\vec{\tau}}{2}\cdot\partial^{\mu}\vec{\chi}}^{2} + \overbrace{\mathrm{i}g\vec{\chi}\cdot\frac{\vec{\tau}}{2}\partial^{\mu}}^{3}\right]\psi \\
&\quad + \left[-\mathrm{i}\frac{g^2}{2}\tau_i\epsilon_{ijk}\chi_j W_k^{\mu} - g^2\left(\frac{\vec{\tau}}{2}\cdot\vec{W}^{\mu}\right)\left(\frac{\vec{\tau}}{2}\cdot\vec{\chi}\right)\right]\psi.
\end{aligned} \tag{E.3}$$

doi:10.1088/978-0-7503-1140-3ch25

© IOP Publishing Ltd 2015

The right-hand side of (E.1) gives

$$\left(1 + \mathrm{i}g\vec{\chi} \cdot \frac{\vec{\tau}}{2}\right)\left(\partial^\mu + \mathrm{i}g\frac{\vec{\tau}}{2} \cdot \vec{W}^\mu\right)\psi$$

$$= \left[\overbrace{\partial^\mu + \mathrm{i}g\frac{\vec{\tau}}{2} \cdot \vec{W}^\mu}^{1} + \overbrace{\mathrm{i}g\vec{\chi} \cdot \frac{\vec{\tau}}{2}\partial^\mu}^{3} - g^2\left(\vec{\chi} \cdot \frac{\vec{\tau}}{2}\right)\left(\frac{\vec{\tau}}{2} \cdot \vec{W}^\mu\right)\right]\psi. \tag{E.4}$$

The labelled expressions in (E.3) and (E.4) cancel out. One still needs to prove that

$$-\mathrm{i}\frac{g^2}{2}\tau_i\epsilon_{ijk}\chi_j W_k^\mu = \frac{g^2}{4}\Big[\underbrace{(\vec{\tau} \cdot \vec{W}^\mu)(\vec{\tau} \cdot \vec{\chi})}_{(\tau_i W_i^\mu)(\tau_j\chi_j)} - \underbrace{(\vec{\chi} \cdot \vec{\tau})(\vec{\tau} \cdot \vec{W}^\mu)}_{(\tau_j\chi_j)(\tau_i W_i^\mu)}\Big]. \tag{E.5}$$

By using the property $\tau_i\tau_j = \delta_{ij} + \mathrm{i}\epsilon_{ijk}\tau_k$ of the Pauli matrices the terms on the right-hand side can be rewritten as

$$\frac{g^2}{4}\Big[\delta_{ij} W_i^\mu\chi_j + \mathrm{i}\epsilon_{ijk}\tau_k W_i^\mu\chi_j - \delta_{ji}\chi_j W_i^\mu - \mathrm{i}\epsilon_{jik}\tau_k\chi_j W_i^\mu\Big]$$

$$= \frac{g^2}{2}\mathrm{i}\epsilon_{ijk}\tau_k\chi_j W_i^\mu = \mathrm{i}\frac{g^2}{2}\tau_i\underbrace{\epsilon_{kji}}_{-\epsilon_{ijk}}\chi_j W_k^\mu, \tag{E.6}$$

since ϵ is antisymmetric and where we have swapped the indices i and k.

IOP Publishing

Nuclear and Particle Physics

Claude Amsler

Appendix F

Glossary

α. The fine structure constant α (1.12) is a measure for the strength of the electromagnetic interaction and is equal to 1/137 at low energies (section 7.3).

α_s. The (energy-dependent) parameter α_s (figure 7.6) is a measure for the strength of the strong interaction and describes the coupling between quarks (section 7.3).

α series. There are four uranium α-decay chains leading to the stable isotopes of lead (^{206}Pb, ^{207}Pb, ^{208}Pb) and the long-lived bismuth isotope ^{209}Bi (figure 5.8).

Activity. The activity (4.7) of a radioactive source is the number of decays s^{-1}. Activites are measured in becquerel (1 Bq = 1 decay s^{-1}).

Allowed decays. Allowed β-decays are transitions with low log ft-value (figure 13.8).

Am–Be. The americium–beryllium radioactive source is used to produce low energy (<10 MeV) neutrons (figure 5.14).

Analysing power. The analysing power (11.31) is a measure for the left–right asymmetry observed when scattering a beam off a polarized target (figure 11.4).

Anomalous magnetic moment. The anomalous magnetic (dipole) moment of an elementary particle is the ratio $\frac{g-2}{2}$, where g is its g-factor. The value $g = 2$ is predicted in first approximation for spin-$\frac{1}{2}$ fermions by the Dirac equation (section 13.3.3).

Antihydrogen. Antihydrogen is the simplest form of antimatter consisting of a positron (or antielectron) orbiting an antiproton (sections 9.1, 11.9). Antihydrogen was first produced at LEAR (CERN) (see figure 9.8). The CPT theorem (section 11.9) predicts that matter and antimatter should have identical physical properties.

doi:10.1088/978-0-7503-1140-3ch26

© IOP Publishing Ltd 2015

Asymptotic freedom. Quarks are confined in hadrons but the interaction becomes weaker as the distance between them decreases. They become free to move around (section 7.3).

Atomic number and weight. The atomic number Z of a nucleus is its number of protons and the atomic mass number A its number of nucleons. The atomic weight A in the periodic table is the mass of the atom averaged over the isotopic abundance, relative to the mass of the ^{12}C atom (chapter 3).

Auger electron. Auger electrons are electrons expelled during the de-excitation of the atomic shell (section 5.1).

Barkas effect. The energy lost by positively and negatively charged particles in a medium are not exactly equal at low energies (figure 9.10).

Baryon number. Baryons such as the proton or the neutron are made of three quarks, each with baryon number $B = \frac{1}{3}$ which add up to $B = 1$ (section 7.3.1). A violation of the baryon number is possible in GUTs (grand unified theories) which would lead to a finite (but extremely long) mean lifetime of the proton (section 7.4).

BEH mechanism. The Brout–Englert–Higgs mechanism explains how the Higgs boson, discovered in 2012 (section 17.3), endows particles with mass and spontaneously breaks the symmetry in weak interactions (figure 16.19).

Betatron. The betatron is a circular low energy electron accelerator based on induction (see (9.8)), which is mostly used in hospitals for radiotherapy (figure 9.5).

Bethe–Bloch formula. The Bethe–Bloch formula (9.19) describes the energy loss of charged particles (with the exception of electrons and positrons) traversing an absorber (section 9.2).

Bhabha scattering. Bhabha scattering refers to elastic electron–positron scattering (figure 16.2).

Branching ratio. The decay branching ratio (or branching fraction) is the probability for a particle to decay into a specific final state (section 4.1). It is expressed as the ratio of partial to total widths (see (4.34)).

Breeder. The breeder nuclear reactor is based on the fission of ^{239}Pu which is produced from the capture of fast neutrons by ^{238}U. The breeder reactor generates energy while producing simultaneously fissile material (section 5.4.2).

Breit–Wigner distribution. The Breit–Wigner distribution (4.28) describes the shape of resonance excitations, see (10.55, 12.90 and 17.40).

Bremsstrahlung. Bremsstrahlung is the emission of x-rays or γ-rays by charged particles accelerated in electric fields (figure 7.4(c)). Bremsstrahlung is the dominant energy loss mechanism for electrons and positrons at high energies (section 9.4).

***C* parity.** The operation of C parity (or charge conjugation) transforms a particle into its antiparticle (section 11.7).

***CP* violation.** The conservation of the combined operations of parity (P) and charge conjugation (C) is violated in the neutral kaon system (section 19.2 and figure 19.9) and in hadronic decays of B mesons (section 20.4 and figure 20.10).

Cabibbo angle. The Cabibbo mechanism (figure 7.12(b)) states that a u quark interacts mostly with a d quark with a coupling proportional to the cosine of the Cabibbo angle θ_C (sections 13.2.1 and 16.2). The coupling to the s quark is much smaller and is proportional to $\sin \theta_c \simeq 0.23$ (13.60). For heavier quarks the Cabbibo mechanism is generalized to the CKM mechanism (section 16.3).

Callan–Gross relation. The Callan–Gross relation (18.28) relates the two structure functions F_1 and F_2 which describe the energy distributions of the quarks in the nucleon, see chapter 18.

Calorimeter. A calorimeter measures the energy deposited by charged particles (hadronic calorimeter, section 9.13) or by high energy photons, electrons and positrons (electromagnetic calorimeter, section 9.12).

Čerenkov radiation. Čerenkov radiation (figure 9.34) is emitted by a fast charged particle moving through a medium when its velocity is larger than the speed of light in the medium. This radiation is used to identify the type of particle and to measure its velocity (figures 9.37 and 14.18).

Charge independence. Charge independence (or charge symmetry) states that the strong interaction does not distinguish between particles of different charges belonging to the same isopin multiplet (section 12.4 and figure 13.5). This rule holds for the strong and electromagnetic interactions (table 12.1).

Chew–Frautschi. The Chew–Frautschi diagram (figure 10.24) is a plot of the Regge trajectory (10.66) which relates the spin of a hadron to the square of its mass.

Chiral couplings. The chiral couplings (16.169) determine the couplings of left-handed and right-handed quarks to the Z^0 boson (section 16.7, see also table 17.1).

CKM matrix. The Cabibbo–Kobayashi–Maskawa matrix (16.66) describes the couplings of the u, c and t quarks to the d, s and b quarks (sections 16.3 and 20.1). The CKM matrix is complex, leading to CP violation in the quark sector (appendix D).

CNO cycle. The CNO cycle (14.33) describes the genesis of the organic elements in the interior of stars.

Colour. Quarks exist in three colours, red, green and blue and antiquarks in corresponding anticolours. Gluons contain both colour and anticolour (section 10.4).

Confinement. Quarks and gluons are not observed as free entities but are confined in mesons and baryons (section 10.3.3).

Critical energy. The critical energy of a medium (table 9.1) is the energy of the electron at which the energy losses by bremsstrahlung and ionization become equal (see (9.40)).

Cross section. The cross section, measured in barn (1 b = 10^{-24} cm^2), is the effective surface leading to an interaction between a projectile and a target particle (figure 1.5 and sections 12.1, 12.2). Typical cross sections are given in section 7.3.4.

Current. A weak charged current is the current associated with a particle emitting a $W^\pm$ boson. A weak neutral current is the current associated with a particle emitting a Z^0 boson (sections 7.3.3 and 16.2.1).

Cyclotron frequency. The cyclotron frequency (9.3) and (11.128) is the angular frequency of a charged particle rotating in a homogeneous magnetic field.

$\Delta S = \Delta Q$ rule. The rule (13.5) holds for semileptonic decays of hadrons with strangeness S (containing s quarks), where Q is the charge of the hadrons (see also sections 19.1 and 19.3). The similar rules $\Delta C = \Delta Q$ and $\Delta B' = \Delta Q$ hold for semi-leptonic decays involving c and b quarks.

Δi = 1/2 rule. The rule (13.10) states that in non-leptonic weak hadron decays the isospin changes dominantly by $\pm\frac{1}{2}$ (chapter 13 and section 19.2.2).

Dalitz plot. A Dalitz plot (figure 12.3) is a 2D representation of the energy sharing in three body decays (section 12.3).

Dating. The ^{14}C isotope with a half-life of 5730 years is used for dating in archaeology (see (13.41)).

Decay constant. The decay constant λ is the slope of the decay law (4.2) which describes the rate at which a radioactive substance decays. It is equal to the full width at half maximum (4.29) of the decaying state.

Detectors. A description of some of the detectors mentioned in this book can be found in the following sections: ATHENA (11.9), ATLAS (17.3), BABAR (20.2), CMS (17.3), CPLEAR (19.3.1), Crystal Ball (10.3.1), Crystal Barrel (12.3.3), Daya Bay (14.6.5), Gargamelle (16.2.1), SNO (14.6), Super-Kamiokande (14.6), UA1 (17.2).

Dipole moment. The finite electric dipole moment of the neutron which violates T invariance is predicted to be very small (section 11.8.1). The magnetic dipole moment of the neutrino is also expected to be very small (sections 9.6 and 14.3). The magnetic dipole moment of the muon (section 13.3.3) is measured in $g - 2$ experiments (figures 13.17 and 13.18).

Discoveries. Some of the discoveries described in this book can be found here: c quark (figures 10.14(a) and (10.15) left), b quark (figure 10.20), π^0 (figure 8.6), $\pi^\pm$ (figure 7.10), K^0 (figure 10.1), Ω^- (figure 10.5), τ neutrino (figure 14.4), t quark (figure 10.29), $W^\pm$ and Z^0 bosons (figure 17.12), antineutrino (section 14.1), antiproton (section 15.2 and figure 15.2), Higgs (figure 17.21), neutral currents, (figures 16.9 and 16.11), neutron (section 13.1), positron (figure 15.1).

Dripline. The proton and neutron driplines delimit the region of nuclear stability (figure 6.6).

Fermi coupling. The Fermi coupling constant G_F (13.59) determines the strength of the weak interaction (sections 13.2.3 and 16.1). The related g_V (vector) coupling (13.32) applies to Fermi transitions and the g_A (axial vector) coupling (13.35) to Gamow–Teller transitions (section 13.2.1).

Fermi function. The Fermi function (figure 13.6) is an energy and atomic number dependent correction to the energy distribution of the electron or positron in β-decay, which takes into account its Coulomb interaction with the nucleus.

Feynman graph. A Feynman diagram (or graph) is the pictorial representation of an interaction or decay between elementary particles in reactions or decays (section 7.3).

***ft*-value.** The *ft*-value (13.52) estimates the decay probabilty of a radioactive nucleus from its measured mean lifetime and the Fermi function.

Fifth force. The conservation of baryon number (section 7.3.1) suggests the existence of a fifth force which couples to the mass number A (section 11.5). This force has been sought with torsion balances, so far without success.

Fission. The fission of a heavy nucleus such as uranium (section 5.4.2) is induced by thermal neutrons and is used in nuclear reactors (figure 5.18) to produce energy. Spontaneous fission also occurs and is the main decay mechanism of transuranic elements (sections 5.4.1 and figure 6.5).

Flavour. Quarks exist in three generations, each with two flavours (u, d), (c, s), (t, b). Leptons also exist in three families or generations (ν_e, e^-), (ν_μ, μ^-), (ν_τ, τ^-), see section 7.1.

Forbidden decays. Forbidden decays are β-decays in which the leptons carry away angular momentum. They have large log *ft*-values (13.52) and are suppressed (section 13.2.1).

Form factor. The form factor (2.18) of a nucleus describes its charge distribution. It is obtained with a spectrometer (figure 2.8) measuring the angular distribution in high energy electron–nucleus scattering (section 2.2). The electric and magnetic dipole moments of the nucleon have been measured in electron–nucleon scattering (figure 2.9 and section 2.4).

Formation length. The formation length (figure 9.40) determines the optimum radiator thickness needed for a transition radiation detector.

Fragmentation. Fragmentation (or hadronization) is the process in which struck quarks combine with sea quarks to form hadrons (section 10.3.3).

Furry theorem. The Furry theorem states that the number of pions in $\pi\pi$ interactions is even (figure 12.17).

Fusion. Fusion is the process in which light nuclei combine to build heavier ones. The energy production in the Sun (section 14.3) is initiated by a weak interaction, the fusion of two protons (14.28). In contrast, controlled fusion uses strong interaction processes such as deuterium–tritium fusion (14.31).

***g*-factor.** See anomalous magnetic moment.

Gamow factor. The Gamow factor (5.27) is a measure for the height of the potential barrier in α-decay.

Geiger–Nuttal rule. The Geiger–Nuttal rule (5.32) states that the lifetime of an α-emitter increases exponentially with decreasing $\sqrt{T}$, where T is the kinetic energy of the α particle.

Gell–Mann–Nishijima. The Gell–Mann–Nishijima formula (10.44) relates the charge of a quark (or hadron) to its isospin and flavour quantum numbers (section 10.2.3).

Glueball. A glueball is a hadron which contains only gluons (figure 10.25(*b*)). Glueballs are predicted to exist as a consequence of the non-abelian structure of QCD (section 16.6). Candidates for such states have been reported (see for example figure 12.10 (left)).

Golden rule. Fermi's golden rule (13.25) is the general formula to compute the mean lifetime of β-decays.

Goldstone boson. Goldstone bosons are massless and spinless particles that appear in the mechanism of spontaneous symmetry breaking (section 16.5).

Gray. The energy deposited by radiation in living tissues is measured in grays (Gy). 1 Gy corresponds to an energy deposit of 1 J kg^{-1} (section 5.5).

GUT. Grand unified theories predict that the strengths of the strong, electromagnetic and weak interactions become equal at the energy scale of 10^{16} GeV (section 7.4).

Hadron. Hadrons (from the Greek word 'strong') are strongly interacting particles (section 7.2 and chapter 10). They are classified according to baryons composed of three quarks (such as the nucleon), and mesons composed of quark-antiquark pairs (such as the pion).

Half-life. The half-life $t_{1/2}$ is the average time at which half of the radioactive substance has decayed. The mean lifetime τ and the half-life are related by $t_{1/2} = \tau \ln 2$ (section 4.1).

Handedness. Left-handed particles have a negative helicity, right-handed a positive helicity (section 15.3). The weak interaction between charged currents involves only left-handed fermions and right-handed antifermions (section 13.3.1). See also chiral couplings.

Heaviside–Lorentz units. In Heaviside–Lorentz units the permittivity ϵ_0 and permeability μ_0 of the vacuum are set equal to one (section 1.1.1).

Heavy water. Heavy water (D_2O) is used as neutron moderator in CANDU nuclear reactors (section 5.4.2) and as a neutron target by the SNO detector (figure 14.20) to detect electron neutrinos from the Sun.

Helicity. The helicity (15.76) of a particle is the projection $\vec{\sigma}\vec{p}/p$ of its spin onto its momentum. Section 11.6 deals with the helicity of the photon, section 14.4 with that of the neutrino.

Hierarchy. In neutrino oscillations the *normal* hierarchy is the one shown in figure 14.22 where the two light eigenstates ν_1 and ν_2 lie below ν_3. In the *inverted* hierarchy ν_3 is the lightest state. This open issue could be solved e.g. by double-β decay experiments (section 14.5).

Horn. A horn (figure 14.23) is a magnetic device producing a parallel beam of charged particles. Horns are used for the production of antiprotons (figure 9.8) and neutrinos (figure 16.6).

Hoyle resonance. The Hoyle resonance is an excited nuclear state of carbon, the existence of which is crucial to the production of organic elements in the Sun and in heavy stars (section 14.3).

Hypercharge. The strong hypercharge (10.45) is related to the flavour quantum numbers of quarks. The weak hypercharge (defined in (16.127)) is related to the charge and weak isospin (table 16.1).

Hypernucleus. A hypernucleus contains one or more hyperons replacing some of the nucleons (see (11.54)).

Hyperon. A hyperon is a baryon containing one or more heavy quarks (s, c, b) (section 7.2).

Ideal mixing. Neutral mesons are (quantum mechanical) superpositions of quark–antiquark pairs with different flavours. In ideally mixed mesons the $u\bar{u}$ and $d\bar{d}$ pairs decouple from the heavier ones (e.g. (10.16)).

Interval rule. The interval rule (6.28) describes the spacing of atomic levels split by the hyperfine interaction between nuclear and atomic spins.

Isobar. Isobars (chapter 3 and figure 5.6) are nuclei with the same number of nucleons, **isotones** with the same number of neutrons, and **isotopes** with the same number of protons.

Isomer. An isomer is a high-spin nucleus which decays by γ emission and which has a relatively long lifetime (section 5.1).

Isoscalar target. An isoscalar target (such as ^{12}C and deuterium) contains equal numbers of u and d quarks (sections 17.4.2 and 17.4.3).

Isospin. A hadron with *strong* isospin i exists in $2i + 1$ varieties of electric charge (section 10.2.3). Isospin is a conserved quantum number in strong interactions (table 12.1). Left-handed fermions have the *weak* isospin $\frac{1}{2}$ and right-handed fermions the *weak* isospin 0 (section 16.7).

Kurie plot. The Kurie plot (figure 14.5) is a convenient diagram to represent the energy distribution of the electron in β-decay (section 14.2.1).

Landau–Vavilov distribution. The fluctuation in the energy loss of charged particles traversing an absorber is described by the Landau–Vavilov distribution (figure 9.13).

Landau–Yang theorem. The Landau–Yang theorem (section 11.6) states that a spin-1 particle does not decay into two massless spin-1 particles.

Larmor frequency. The Larmor frequency ((11.119) for the proton and (13.75) for the muon) is the angular frequency of the spin rotating in a homogeneous magnetic field (see also Thomas precession).

Leptons. Leptons (from the Greek word 'slim') exist in three families or generations (e^-, ν_e), (μ^-, ν_μ), (τ^-, ν_τ). There are also six corresponding antileptons. They interact only through the weak or electromagnetic interaction (figure 7.1). The lepton number is conserved within each pair, except for neutrinos which oscillate between generations (section 14.6).

Lorentz angle. In a gaseous or silicon detector the Lorentz angle (9.64) is the angle between the electron drift lines and the direction of the electric field when a magnetic field is applied perpendicularly to the electric field (figure 9.33).

Luminosity. In a particle collider the collision rate is obtained by multiplying the luminosity (9.14) by the interaction cross section.

Magic numbers. Nuclei with the magic numbers (6.1) of protons or neutrons are strongly bound. Magic numbers correspond to closed nuclear shells (figure 6.2).

Majorana neutrino. A Majorana neutrino ν is identical to its charge conjugated partner $\bar{\nu}$. Whether neutrinos are Majorana or Dirac particles ($\nu \neq \bar{\nu}$) is an open issue that could be solved by studying double β-decay (section 14.5).

Malter effect. The Malter effect is the ageing of proportional wire chambers (section 9.6) due to polymer deposits on the wires.

Mass formula. The mixing angles in the light meson nonets can be calculated from the meson masses with the formulae (10.30) and (B.13). See also the nuclear mass formula (3.1).

Michel spectrum. The Michel spectrum is the energy distribution (13.67) of the electron in muon decay (section 13.3.2, see also figure 16.4).

Mirror nuclei. Mirror nuclei have mutually interchanged atomic and neutron numbers. As a consequence of charge independence they should have the same excitation spectra (figure 12.12).

Mixing angle. The nonet mixing angle (see equation (10.14)) determines the fraction of u and d quarks in a neutral meson (section 10.2.1). See also mass formula.

Mixing (complete). Complete mixing occurs in flavour oscillations when the oscillation period is comparable with or smaller than the mean lifetime (section 19.1.2), e.g. in the $B_s^0 - \overline{B}_s^0$ system (figure 20.7).

Molière distribution. The deflection angle of a charged particle traversing an absorber follows the Molière distribution which is approximately Gaussian (figure 9.16).

Molière radius. The Molière radius (9.81) is the radial extension of an electromagnetic shower in which 90% of the shower energy is deposited.

Møller scattering. Møller scattering refers to elastic electron–electron scattering (figure 9.18, see also section 15.6.1).

Mott cross section. Electron scattering off spinless nuclei is described by the Mott cross section (2.8).

MSW effect. The Mikheyev–Smirnov–Wolfenstein effect enhances the flavour oscillation of neutrinos which traverse a medium (section 14.6.6) such as the solar interior (section 14.6.7).

Multipole order. The multipole order (5.3) determines the angular momentum carried by the γ in electromagnetic nuclear decays. High multipole orders correspond to long lifetimes, see (5.5).

Muonic atoms. In muonic atoms a shell electron has been replaced by a negatively charged muon. Muonic atoms are used for nuclear structure studies (section 2.3 and figure 6.8) and to measure the muon mass which enters the determination of the Fermi coupling constant (see (16.14)). For the similar pionic atoms see section 11.4.

Natural units. The scientific notation is greatly simplified when using natural units $\hbar = c = 1$ (section 1.1.1). They can be transformed into SI units by using (1.16) and (1.17).

Neutralino. The neutralino is a hypothetical supersymmetric particles (section 7.4). Dark matter in the Universe could be composed of neutralinos.

Neutrino oscillations. Neutrino flavour is not a conserved quantity. Neutrinos oscillate into one another with frequencies determined by the squared mass differences Δm_{23}^2 (section 14.6.3) and Δm_{12}^2 (section 14.6.4) between the neutrino eigenstates ν_1, ν_2 and ν_3.

Nuclear magneton. The magnetic moment of the nucleon is expressed in nuclear magneton μ_N (2.45).

Optical theorem. The optical theorem relates the total cross section to the imaginary part of the forward scattering amplitude (section 19.1.3).

OZI rule. The Okubo–Zweig–Iizuka rule of strong interactions states that processes depicted by disconnected Feynman diagrams are suppressed (figure 10.7(*b*)).

Parity. Parity P flips the sign of the spatial coordinates (section 11.3). Hadrons have the internal parities ± 1 (section 11.4). Fermions and antifermions have opposite parities (section 15.4 and equation (11.96)). Parity is not conserved in weak interactions (section 13.3). The parity of the photon is -1 (section 11.4). The pion and kaon have also negative internal parities, see (11.51) and (11.55).

Partons. The partons are the building blocks of hadrons, the quarks, antiquarks and gluons (sections 10.2, 17.4.2).

PMNS matrix. The Pontecorvo–Maki–Nagakawa–Sakata matrix is the neutrino mixing matrix (14.66), equivalent to the CKM matrix (16.67) in the quark sector.

Positronium. Positronium is an electron–positron pair bound by the Coulomb force. In orthopositronium the spins are parallel, in parapositronium they are antiparallel (figure 10.18 and sections 10.3.3, 11.6 and 11.7).

Proca equations. The Proca equations (16.91) and (16.92) are generalizations of the inhomogeneous Maxwell equations for massive photons.

Propagator. The propagator of the photon (15.107) or of the gauge boson (15.109) transfers momentum between the interacting currents.

Pseudoscalar mesons. The pseudoscalar mesons (10.28) are spin-0 mesons with negative parity.

***Q*-value.** The Q value is the amount of released energy in a decay which is shared among the decay products in the form of kinetic energy.

Quadrupole moment. The deformation of a nucleus is obtained from its quadrupole moment Q (6.23). Spherical nuclei have $Q = 0$, prolate and oblate have $Q > 0$ and $Q < 0$, respectively (figure 6.7).

Quarkonium. Quarkonium (section 10.3.3) is a generic term for bound neutral quark–antiquark pairs such as charmonium (figure 10.17) or bottomonium (figure 10.21).

***R* parity.** The R parity (7.47) is believed to be a conserved quantum number in the decay of supersymmetric particles.

r-process. The r-process is a fast multiple neutron capture process occurring in stars which produces heavy elements beyond iron (chapters 3 and 6).

Radiation length. The radiation length is the thickness of material needed to absorb about 50% of a high energy γ flux (see equation (9.60)). The radiation length (table 9.1) also enters the formula (9.31) to calculate the multiple scattering angle of charged particles behind an absorber.

Range straggling. Range straggling is the relative fluctuation in the range (9.28) of energetic particles stopping in a medium (figure 9.14(*b*)).

Rapidity. The rapidity (8.73) is related to the velocity of a high energy particle and is independent of the reference frame (see also (8.76) for highly relativistic particles).

Regeneration. The short-lived component (K_S) in a neutral kaon beam can be regenerated from the long-lived one (K_L) by inserting an absorber into the beam (section 19.1.3).

Regge trajectory. See Chew–Frautschi plot.

Rosenbluth cross section. Electron scattering off the nucleon is described by the Rosenbluth cross section (2.42).

Rutherford cross section. The cross section for low energy scattering between spinless nuclei is described by the Rutherford cross section (1.14).

Sargent rule. The Sargent rule (13.58) states that the lifetime of semileptonic decays and allowed nuclear decays is inversely proportional to the fifth power of the Q-value.

Saxon–Woods distribution. The Saxon–Woods distribution (2.28) describes the radial charge distribution in medium size and heavy nuclei.

Scaling. Scaling means that the cross section for deep inelastic neutrino or electron scattering off the proton depends only on the dimensionless parameter $x = Q^2/2m_p\nu$ (where Q and ν are the momentum and energy transfers), see figure 18.2. In other words the size of the partons cannot be measured by increasing the momentum transfer because they are point-like (section 17.4.2 and chapter 18).

Sea quark. Sea quarks are transient virtual quark–antiquark pairs popping up from the vacuum which contribute with **valence** quarks to the mass of the hadron (equation (17.84) and figure 18.5).

Sievert. The exposure of the human body to radiation, weighted by the sensitivity of the various organs, is measured in sievert (Sv, section 5.5).

Solenoid retarding spectrometer. The mass of the electron neutrino is being determined accurately with large acceptance solenoid retarding spectrometers (figures 14.8 and 14.9).

Spontaneous symmetry breaking. See BEH mechanism.

Squarks and sleptons. Squarks and sleptons are hypothetical spin-0 supersymmetric partners of quarks and leptons (section 7.4).

Sterile neutrinos. Sterile neutrinos build a hypothetical fourth family of neutrinos which do not interact and therefore remain invisible (sections 14.3 and 17.2.2).

Stochastic cooling. Stochastic cooling is a corrective process used to reduce the dimensions and the momentum byte of beams in storage rings (figure 9.9).

Stopping power. The stopping power $\mathrm{d}E/\mathrm{d}x$ (9.19) is the energy loss per unit path length of charged particle traversing a medium (figure 9.12).

Strangeness. The number of s quarks building a given hadron is determined by the quantum number of strangeness (section 10.1).

Structure function. The structure functions (18.30) describe the momentum distribution of quarks in the nucleon (section 17.4.2 and chapter 18).

SU(3). Hadrons made of u, d and s quarks are classified within the SU(3) *flavour* symmetry group (section 10.2.3). The flavour wave functions are listed in the appendices A and C. The extension to four quarks (SU(4) group) is discussed in section 10.3. The SU(3) *colour* symmetry group forms the basis of quantum chromodynamics (section 10.4).

Supersymmetry. Supersymmetry is a theorized extension of the standard model of particle physics which introduces partner particles to quarks, leptons and gauge bosons (section 7.4), see also squarks.

Tetraquark. A tetraquark is a hadron made of two quark–antiquark pairs. Several candidates containing c quarks have been reported (section 10.4).

Thomas precession. The Thomas spin precession is a relativistic correction to the rotation angle in a magnetic field due to the Lorentz contraction (figure 13.16).

Thomson cross section. The Thomson cross section (equation (9.44)) is the low energy limit of the Compton cross section (section 9.5) which describes the scattering of a photon off a free electron.

Time reversal. Time reversal invariance (section 11.8) holds for all interactions, except for the neutral kaon system and for hadronic B decays.

Transition radiation. Transition radiation is emitted in the form of x-rays by extremely fast charged particles (such as high energy electrons) crossing the interface between two media (section 9.11).

Transversality. The helicity of a transversal spin-1 boson (such as a free photon, section 11.6) is ± 1, while the helicity of a longitudinal spin-1 boson is zero. Transversality refers to the direction of the field.

Ultra-cold neutrons. The velocity of ultra-cold neutrons (obtained by moderating reactor neutrons) is typically a few m s^{-1} (figures 11.17 and 13.1).

Universality. *Lepton* universality states that the couplings to the gauge bosons are independent of lepton generation (sections 13.2.4 and (17.33)). The *quark–antiquark* binding potential (10.60) is universal, that is, independent of quark flavour.

$V - A$ interaction. The interaction of weak charged currents is of the type $V - A$ (section 16.1) which involves only left-handed fermions or right-handed antifermions. This leads to parity violation in nuclear β-decay (section 13.3.1) and in muon decay (section 13.3.2).

Vacuum polarization. Vacuum polarization (figure 7.5(a)) is produced by virtual positrons from transient electron–positron pairs which are attracted by a passing negatively charged particle, thus partially shielding its charge.

Vector mesons. The vector mesons (10.10) are spin-1 mesons with negative parity.

Weak focusing. The condition (9.5) for the field index (9.4) is fulfilled in weak focusing accelerators (such as the cyclotron).

Weinberg angle. The Weinberg (or weak mixing) angle (16.154) determines the amount of mixing between the electromagnetic and the neutral weak fields in the Glashow–Weinberg–Salam model (section 16.7). This angle determines the couplings (16.169) to left- and right-handed neutral currents, and the ratio (16.161) of $W^{\pm}$ to Z^0 masses.

Weyl equations. The relativistic Weyl equation (15.70) couple the left-handed and right-handed components of the wave function of a spin-$\frac{1}{2}$ particle. They follow from the Dirac equation (chapter 15).

Yang–Mills fields. The Yang–Mills fields are needed to compensate the local phase transformation (16.115) applied on the wavefunction of a particle (section 16.6).

Zemach formalism. In the Zemach formalism the transition matrix is expressed as a function of kinematic variables (section 12.3.2).

www.ingramcontent.com/pod-product-compliance
Ingram Content Group UK Ltd.
Pitfield, Milton Keynes, MK11 3LW, UK
UKHW051909100926
8556IPUK00005B/824

9 780750 311410